About The Author

Tristan Donovan was born in Shepherd's Bush, London, in 1975. His first experience of video games was *Space Invaders* and he liked it, which was just as well because that was one of only three games he had on the TI-99/4a computer that saw him through the 1980s.

He disliked English at school and studied ecology at university, so naturally became a journalist after graduating in 1998.

Since 2001 he has worked for Haymarket Media in a number of roles, the latest of which is as deputy editor of *Third Sector*. Tristan has also written for *The Guardian*, *Edge*, *Stuff*, *The Big Issue*, *Games*™, *Game Developer*, *The Gadget Show* and a whole bunch of trade magazines you probably haven't heard of.

He lives in East Sussex, UK with his partner and two dachshunds.

REPLAY

The History Of Video Games

TRISTAN DONOVAN

YELLOW ANT

First published in Great Britain in 2010 by

Yellow Ant
65 Southover High Street, Lewes,
East Sussex, BN7 1JA

www.yellowantmedia.com

Cover by Jay Priest and Tom Homewood
Cover photo © Corbis
Typeset by Yellow Ant

ISBN 978-0-9565072-0-4

To Jay, Mum, Dad and Jade

CONTENTS

Richard Garriott

FOREWORD

By Richard Garriott

Many consider the video games industry a young one. And, indeed, compared to many industries it is. It has developed from being a home-based hobby of the odd computer nerd to a multi-billion dollar business in just 30 years or so. I am old enough, and consider myself lucky enough, to have worked in the industry for much of its history. Astounding achievements in technology and design have driven this business to the forefront of the entertainment industry, surpassing books and movies long ago as not only the preferred medium for entertainment, but the most lucrative as well. Yet, it still has not been recognized as the important cultural art form that it is.

It is important to look back and remember how quickly we got here. Many who consider video game history focus on certain parts, such as consoles and other hardware that helped propel this business into the artistic medium it is today. However, there are many more aspects that are equally important. I believe that Tristan Donovan's account is the most comprehensive thus far. In this book you will see his account of the inception of the video game's true foundations. He details with great insight the people and events that led to what is the most powerful creative field today, and he takes a holistic view of the genre. Tristan's unique approach demonstrates the strength of this field – he focuses on how video games have become a medium for creativity unlike any other industry, and how those creators, artists, storytellers, and developers have impacted culture in not just the US, but worldwide. That is quite a powerful influence and warrants recognition.

This book credits the greatest artistic creators of our time but doesn't limit what they've accomplished to a particular platform. The video game genre spans coin-operated machines, consoles, personal computers, and more recently, the impetus of mobile, web-based, and handheld markets. There are very few venues in life these days you will not see some sort of influence of a video game – from music to film, to education to the military, games have touched the lives of people all over the world. While some cultures prefer a particular game style over another, the common denominator is that the art of the video game is not simply synonymous with entertainment, but with life.

INTRODUCTION

"Why are you writing another book on the history of video games," asked Michael Katz, the former head of Sega of America, when I interviewed him for this book.

There are many reasons why, but two stand out. The first is that the attempts at writing the history of video games to date have been US rather than global histories. In *Replay: The History of Video Games*, I hope to redress the balance, giving the US its due without neglecting the important influence of games developed in Japan, Europe and elsewhere. The second, and more important, reason is that video game history is usually told as a story of hardware not software: a tale of successive generations of game consoles and their manufacturers' battle for market share. I wanted to write a history of video games as an art form rather than as a business product.

In addition, video games do not just exist on consoles. They appear on mobile phones, in arcades, within web browsers and, of course, on computers - formats that lack the distinct generational divides of consoles. Hardware is merely the vehicle for the creativity and vision of the video game developers who have spent the last 50 or so years moulding a new entertainment medium where, unlike almost all other rival media, the user is an active participant rather than a passive observer.

Hardware sets limits on what can be achieved, but it does not dictate what is created. The design of the ZX Spectrum home computer did not guarantee the creation of British surrealist games such as *Jet Set Willy* or *Deus Ex Machina*. The technology of the Nintendo 64 only made *Super Mario 64* possible, it did not ensure that Shigeru Miyamoto would make it.

The real history of video games is a story of human creativity, aided by technological growth. *Replay* sets out to celebrate the vitality and vision of video game creators, and to shed light on why video games have evolved in the way they have. For that reason not all of the games featured in this book will have been popular and conversely some very popular games are not mentioned. The focus is on the innovative, not the commercially successful.

Finally, a note on terminology. I've used the term 'video game' throughout this book with the occasional use of 'game' when there is no risk of confusion with other forms of game such as board games. I chose video game in preference to other terms for several reasons: it remains in every day use, unlike TV game or electronic game; it is broad enough to encompass the entire medium unlike 'computer game', which would exclude games, such as Atari's *Pong*, that did not use microprocessors; and terms such as 'interactive entertainment', while more accurate, have failed to catch on despite repeated attempts over the years.

REPLAY

The History Of Video Games

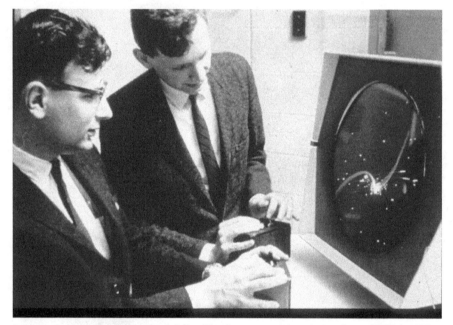

Space race: Spacewar! co-creators Dan Edwards (left) and Peter Samson engage in intergalactic warfare on a PDP-1 circa 1962. Courtesy of the Computer History Museum

CHAPTER 1

Hey! Let's Play Games!

The world changed forever on the morning of the 17th July 1945. At 5.29am the first atomic bomb exploded at the Alamogordo Bombing Range in the Jornada del Muerto desert, New Mexico. The blast swelled into an intimidating mushroom cloud that rose 7.5 kilometres into the sky and ripped out a 3-metre-deep crater lined with irradiated glass formed from melted sand. The explosion marked the consummation of the top-secret Manhattan Project that had tasked some of the Allies' best scientists and engineers with building the ultimate weapon - a weapon that would end the Second World War.

Within weeks of the Alamogordo test, atomic bombs had levelled the Japanese cities of Hiroshima and Nagasaki. The bombs killed thousands instantly and left many more to die slowly from radiation poisoning. Five days after the destruction of Nagasaki on the 9th August 1945, the Japanese government surrendered. The Second World War was over. The world the war left behind was polarised between the communist east, led by the USSR, and the US-led free-market democracies of the west. The relationship between the wartime allies of the USA and USSR soon unravelled resulting in the Cold War, a 40-year standoff that would repeatedly take the world to the brink of nuclear war.

But the Cold War was more than just a military conflict. It was a struggle between two incompatible visions of the future and would be fought not just in diplomacy and warfare but also in economic development, propaganda, espionage and technological progress. And it was in the technological arms race of the Cold War that the video game would be conceived.

* * *

On the 14th February 1946, exactly six months after Japan's surrender, the University of Pennsylvania switched on the first programmable computer: the Electronic Numeric

Integrator and Calculator, or ENIAC for short. The state-of-the-art computer took three years to build, cost $500,000 of US military funding and was created to calculate artillery-firing tables for the army. It was a colossus of a machine, weighing 30 tonnes and requiring 63 square metres of floor space. Its innards contained more than 1,500 mechanical relays and 17,000 vacuum tubes – the automated switches that allowed the ENIAC to carry out instructions and make calculations. Since it had no screen or keyboard, instructions were fed in using punch cards. The ENIAC would reply by printing punch cards of its own. These then had to be fed into an IBM accounting machine to be translated into anything of meaning. The press heralded the ENIAC as a "giant brain".

It was an apt description given that many computer scientists dreamed of creating an artificial intelligence. Foremost among these computer scientists were the British mathematician Alan Turing and the American computing expert Claude Shannon. The pair had worked together during the war decrypting the secret codes used by German U-boats. The pair's ideas and theories would form the foundations of modern computing. They saw artificial intelligence as the ultimate aim of computer research and both agreed that getting a computer to defeat a human at Chess would be an important step towards realising that dream.

The board game's appeal as a tool for artificial intelligence research was simple. While rules of Chess are straightforward, the variety of possible moves and situations meant that even if a computer could play a million games of Chess every second it would take 10^{108} years for it to play every possible version of the game.[1] As a result any computer that could defeat an expert human player at Chess would need to be able to react to and anticipate the moves of that person in an intelligent way. As Shannon put it in his 1950 paper *Programming a Computer for Playing Chess*: "Although perhaps of no practical importance, the question [of computer Chess] is of theoretical interest, and it is hoped that a satisfactory solution of this problem will act as a wedge in attacking other problems of a similar nature and of greater significance."

In 1947, Turing became the first person to write a computer Chess program. However, Turing's code was so advanced none of the primitive computers that existed at the time could run it. Eventually in 1952, Turing resorted to testing his Chess game by playing a match with a colleague where he pretended to be the computer. After hours of painstakingly mimicking his computer code, Turing lost to his colleague. He would never get the opportunity to implement his ideas for computer Chess on a computer.

---

1. That's 1,000,000,000,000,000,000,000,000,000,000,000,000,000,000,000,000,000,000,000,000,000,000,000,0 00,000,000,000,000,000,000,000,000 years. Far, far longer than the 13.7 billion years estimated to have elapsed since The Big Bang.

The same year that he tested his program with his colleague, he was arrested and convicted of homosexuality. Two years later, having been shunned by the scientific establishment because of his sexuality, he committed suicide by eating an apple laced with cyanide.

While Turing never got to make his chess program, computer scientists such as Shannon and Alex Bernstein would spend much of the 1950s investigating artificial intelligence by making computers play games. While Chess remained the ultimate test, others brought simpler games to life.

In 1951 the UK's Labour government launched the Festival of Britain, a sprawling year-long national event that it hoped would instil a sense of hope in a population reeling from the aftermath of the Second World War. With UK cities, particularly London, still marred by ruins and bomb craters, the government hoped its celebration of art, science and culture would persuade the population that a better future was on the horizon. Herbert Morrison, the deputy prime minister who oversaw the festival's creation, said the celebrations would be "a tonic for the nation". Keen to be involved in the celebrations, the British computer company Ferranti promised the government it would contribute to the festival's Exhibition of Science in South Kensington, London. But by late 1950, with the festival just weeks away, Ferranti still lacked an exhibit. John Bennett, an Australian employee of the firm, came to the rescue.

Bennett proposed creating a computer that could play Nim. In this simple parlour game players are presented with several piles of matches. Each player then takes it in turns to remove one or more of the matches from any one of the piles. The player who removes the last match loses. Bennett got the idea of a Nim-playing computer from the Nimatron, an electro-mechanical machine exhibited at the 1940 World's Fair in New York. Despite suggesting Ferranti create a game-playing computer, Bennett's aim was not to entertain but to show off the ability of computers to do maths. And since Nim is based on mathematical principles it seemed a good example. Indeed, the guide book produced to accompany the Nimrod, as the computer exhibit was named, was at pains to explain that it was maths, not fun, that was the machine's purpose: "It may appear that, in trying to make machines play games, we are wasting our time. This is not true as the theory of games is extremely complex and a machine that can play a complex game can also be programmed to carry out very complex practical problems."

Work to create the Nimrod began on the 1st December 1950 with Ferranti engineer Raymond Stuart-Williams turning Bennett's designs into reality. By the 12th April 1951 the Nimrod was ready. It was a huge machine – 12 feet wide, five feet tall and nine feet deep – but the actual computer running the game accounted for no more than two per cent of its size. Instead the bulk of the machine was due to the multitude of vacuum tubes used to display lights, the electronic equivalent of the matches used in Nim. The resulting exhibit, which made its public debut on the 5th May 1951, boasted that the Nimrod was "faster than thought" and challenged the public to pit their wits against Ferranti's "electronic brain". The public was won over, but few showed any interest in the maths and science behind it. They just wanted to play. "Most of the public were quite happy to gawk at the flashing lights and be impressed," said Bennett.

BBC radio journalist Paul Jennings described the Nimrod as a daunting machine in his report on the exhibition: "Like everyone else I came to a standstill before the electric brain or, as they prefer to call it, the Nimrod Digital Computer. This looks like a tremendous grey refrigerator…it's absolutely frightening…I suppose at the next exhibition they'll even have real heaps of matches and awful steel arms will come out of the machine to pick them up."

After the Festival of Britain wound down in October, the Nimrod went on display at the Berlin Industrial Show and generated a similar response. Even West Germany's economics minister Ludwig Erhard tried unsuccessfully to beat the machine. But, having impressed the public, Ferranti dismantled the Nimrod and got back to work on more serious projects.

Another traditional game to make an early transition to computers was Noughts and Crosses, which was recreated on the Electronic Delay Storage Automatic Calculator (EDSAC) at the University of Cambridge in England. Built in 1949 by Professor Maurice Wilkes, the head of the university's mathematical laboratory, the EDSAC was as much a landmark in computing as the ENIAC. It was the first computer with memory that users could read, add or remove information to or from; memory now known as random access memory or RAM. For this Wilkes, who incidentally also tutored Bennett, is rightly regarded as a major figure in the evolution of computers but it would be one of his students who would recreate Noughts and Crosses on the EDSAC. Alexander Douglas wrote his version of the game for his 1952 PhD thesis on the interaction between humans and computers. Once he finished his studies, however, his *Noughts and Crosses* game was quickly forgotten, cast aside as a simple program designed to illustrate a more serious point.

Others tried their hand at Checkers with IBM employee Arthur 'Art' Samuel leading the way. As with all the other games created on computers at this time, Samuel's computer versions of Checkers were not about entertainment but research. Like the Chess programmers, Samuel wanted to create a Checkers game that could defeat a human player. He completed his first Checkers game in 1952 on an IBM 701; the first commercial computer created by the company, and would spend the next two decades refining it. By 1955 he had developed a version that could learn from its mistakes that caused IBM's share price to leap 15 points when it was shown on US television and by 1961 Samuel's programme was defeating US Checkers champions.

* * *

At the same time as the scientists of the 1940s and 1950s were teaching computers to play board games, television sets were rapidly making their way into people's homes. Although the television existed before the Second World War, the conflict saw factories cease production of TV sets to support the war effort by producing radar displays and other equipment for the military. The end of the war, however, produced the perfect conditions for television to take the world by storm. The technological breakthroughs made during the Second World War had brought down the cost of manufacturing TV sets and US consumers now had money to burn after years of austerity. In 1946 just 0.5 per cent of households owned a television. By 1950 this proportion had soared to 9 per cent and by the end of the decade there was a television in almost 90 per cent of US homes. While the shows on offer from the TV networks springing up across the US seemed enough to get sets flying off the shelves, several people involved in the world of TV began to wonder if the sets could be used for anything else beyond receiving programs.

In 1947, the pioneering TV network Dumont became first to try and explore the idea of allowing people to play games on their TV sets. Two of the company's employees – Thomas Goldsmith and Estle Mann – came up with the Cathode-Ray Tube Amusement Device. Based on a simple electronic circuit, the device would allow people to fire missiles at a target, such as an aeroplane, stuck onto the screen by the player. The device would use the cathode-ray tube within the TV set to draw lines representing the trajectory of the missile and to create a virtual explosion if the target was hit.[2] Goldsmith and Mann applied for a patent for the idea in January 1947, which was approved the following year, but Dumont never turned the device into a commercial product.

A few years later another TV engineer had a similar thought. Born in Germany in 1922, Ralph Baer had spent most of his teenage years watching the rise of the Nazi Party in his home country and the subsequent oppression of his fellow Jews. Eventually, in September 1938, his family fled to the US just weeks before Kristallnacht, the moment when the Nazis' oppression turned violent and Germany's Jews began to be

---

2. Cathode-ray tubes are devices that fire electron beams at TV screens to create a picture and were the basis of every TV set right up to the end of the 20th century, when the arrival of plasma and LCD flat screens made them obsolete.

rounded up and sent to die in concentration camps. "My father saw what was coming and got all the paperwork together for us to go to New York," he said. "We went to the American consulate and sat in his office. I spoke pretty good English. I guess being able to have that conversation with the consulate might have made all the difference because the quota for being let into the US was very small. If we hadn't got into the quota then it would have been…[motions slicing of the neck]."

In the US, Baer studied television and radio technology and eventually ended up working at military contractors Loral Electronics, where in 1951 he and some colleagues were asked to build a TV set from scratch. "We used test equipment to check our progress and one of the pieces of equipment we used put horizontal lines, vertical lines, cross-hatch patterns, and colour lines on the screen," he said. "You could move them around to some extent and use them to adjust the television set. Moving these patterns around was kind of neat and the idea came to me that maybe we wanted to build something into a television set. I don't know that I thought about it as a game, more something to fool with and to give you something to do with a television set other than watch stupid network programmes." Baer's idea proved fleeting and he quickly cast it aside. But a seed had been sown.

* * *

By the start of 1958, the video game was still an elusive concept. Computer scientists still saw games as tools for their research and the engineers who saw the potential for TV to be a two-way experience between screen and viewer had failed to develop their ideas further. Bennett's reporter-scaring Nimrod was still the nearest thing to a video game anyone outside engineering workshops or university computer labs had seen. But 1958 would see the concept of the video game come one step closer thanks to William Higinbotham.

Higinbotham had worked on the Manhattan Project, building the timing switches that made the bomb explode at the correct moment. Like many of the scientists who created the bomb, he harboured mixed feelings about what he had done and would spend much of his post-war life campaigning against nuclear proliferation. After the war, he became head of the instrumentation division at the Brookhaven National Laboratory – a US government research facility based on Long Island, New York. Every year Brookhaven would open its doors to the public to show off its work. These visitor days tended to contain static exhibits that did little to excite the public and so, with the 1958 open day looming, Higinbotham decided to make a more engaging attraction.

He came up with the idea for a fun, interactive exhibit: a tennis game played on the screen of an oscilloscope that he built using transistor circuitry with the help of Brookhaven engineer Robert Dvorak. The game, *Tennis for Two*, recreated a side-on view of a tennis court with a net in the middle and thin ghostly lines that represented the players' racquets. The large box-shaped controllers created for the game allowed players to move their racquets using a dial and whack the ball by pressing a button. Brookhaven's visitors loved it. "The high schoolers liked it best, you couldn't pull them away from it," recalled Higinbotham more than 20 years later. In fact *Tennis for Two* was so popular that it returned for a second appearance at Brookhaven's 1959 open day. But neither Higinbotham nor anybody else at Brookhaven thought much of the game and after its 1959 encore it was dismantled so its parts could be used in other projects. With that Higinbotham went back to his efforts to stop nuclear proliferation, eventually forming a division at Brookhaven to advise the US Atomic Energy Agency on how to handle radioactive material.

The 1950s had been a decade of false starts for the video game. Almost as soon as anybody started exploring the idea they walked away, convinced it was a waste of time. Computer Chess had proved a fruitful line of inquiry for artificial intelligence research – indeed many of the principles pioneered by Shannon and others would later be used by video game designers to create challenging computer-controlled opponents for game players – but remained steadfastly about research rather than entertainment.

But as the 1960s dawned, the idea that computers should only be used for serious applications was about to be challenged head on by a group of computing students who rejected the po-faced formality of their professors and saw programming as fun and creative rather than staid and serious.

* * *

The Tech Model Railroad Club lived up to its name. Based in Building 20 of the Massachusetts Institute of Technology (MIT), the students in the club were united by an interest in building elaborate model railroads using complex combinations of relays and switches. Many of the club members also shared a love of computing and trashy sci-fi books such as *Buck Rogers* and, in particular, the work of E.E. Smith. Smith wrote unashamedly trashy novellas telling stories of romance, war and adventure in outer space that were packed with melodramatic dialogue and clichéd plot twists. His *Lensman* and *Skylark* series of books, written in the 1920s and 1930s, helped define the space opera genre of science fiction and fans such as Tech Model Railroad Club member Steve Russell lapped up his trashy tales.

The club members' attitude to computing was in stark contrast to that of their professors and the computer scientists of the previous two decades. They saw merit in creating anything that seemed like a fun idea regardless of its practical value. Club member Robert Wagner's *Expensive Desk Calculator* was typical. Written on MIT's $3 million TX-0 computer, it did what a desktop calculator of the day did but on a machine worth thousands more. Wagner's professors were unimpressed by what they saw as a contemptible misuse of advanced computer technology and gave him a zero grade as a punishment. Such disapproval, however, did little to quash the playful programming spirit of the club's members and, in late 1961, their unorthodox attitude really got a chance to shine when the Digital Equipment Corporation (DEC) gave MIT its latest computer, the PDP-1.

The $120,000 PDP-1 may have been the size of a large car, but with its keyboard and screen it was in many ways the forerunner of the modern desktop computer. The imminent arrival of the cutting-edge machine caught the imagination of the Tech Model Railroad Club. "Long before the PDP-1 was up and running Wayne Witaenem, Steve Russell and I had formed a sort of ad-hoc committee on what to do with it," club member Martin Graetz told *Edge* magazine in 2003. After some debate the students hit on the idea of making a game. "Wayne said: 'Look, you need action and you need some kind of skill level. It should be a game where you have to control things moving around on the screen like, oh, spaceships'," recalled Graetz.

And with that comment *Spacewar!*, a two-player spaceship duel set in outer space, was born. Russell took on the job of programming the game, but his progress was slow. He would repeatedly make excuses about why the game was still not finished when questioned by other club members. Eventually Russell's excuses ran out when he told club member Alan Kotok that he could not start work on the game until he had some routines that could carry out sine-cosine calculations.[3] Kotok went straight to the Digital Equipment Corporation, got the routines and handed them to Russell. "Alan Kotok came to me and said: 'Alright, here are the sine-cosine routines. Now what's your excuse?'," said Russell.

Out of excuses, Russell finally got to work and completed the first version of *Spacewar!* in late 1961, complete with a curvy rocket ship inspired by the stories of Smith and another based on the US military's Redstone Rocket.[4] But the club's members felt *Spacewar!* needed improvement and quickly started adding enhancements. Russell's use

---

3. As part of his studies Kotok created a computer Chess program of his own that in 1962 would become the first one capable of defeating amateur players of the board game.

4. The Redstone Rocket was a direct descendent of Nazi Germany's V-2 rockets and created by many of the same German scientists, who the US government secretly employed after the end of Second World War.

of real-life space physics meant there was no inertia in the game, making it hard to play. so Dan Edwards inserted a star into the play area that had a gravitational pull that players could use to swing their rockets around. The lack of any background in the game made it hard for players to judge how fast the rocket ships were travelling, so Peter Sampson added the star map from another of the club's professor-annoying programs: *Expensive Planetarium*. Kotok and Bob Saunders then created a dedicated controller to replace the PDP-1's in-built bank of 18 switches that made *Spacewar!* uncomfortable to play. By spring 1962 *Spacewar!* was finally finished.

Word of the club's groundbreaking game quickly spread among PDP-1 users at MIT and soon students were staying at the lab well into the night for a fix of *Spacewar!*. For a brief moment Russell and the others thought about trying to sell the game but concluded that since you needed a $120,000 computer to play it there wouldn't be much interest. So they gave it away, handing copies of the game to any PDP-1 user who wanted one. Soon word spread beyond the confines of MIT. In computer labs without a PDP-1, programmers recreated the Tech Model Railroad Club's game for their systems, spreading its reach even further. DEC began using the game to demonstrate the PDP-1 to potential customers and eventually included a copy of the game with every PDP-1 it sold. And despite attempts by computer administrators to delete the time-wasting program that they saw as an affront to the seriousness of computing, *Spacewar!* continued to thrive, growing in influence and popularity all the way.

But while computer students got to sample the delights of *Spacewar!*, few expected it to go any further. After all, computers were simply too big and too expensive for anyone who didn't have some serious application in mind. Few expected the situation to change. When film director Stanley Kubrick consulted more than 100 experts about what the technology of 2001 would look like for his 1968 movie *2001: A Space Odyssey*, he came back with tales of intelligent machines that would play Chess to grandmaster standard and would be capable of voice recognition. But they would still be huge. *Spacewar!*, it seemed, was destined to remain a treat for the computing elite.

* * *

While *Spacewar!* was imprisoned by the technology needed to run it, the idea Ralph Baer had as an engineer at Loral back in 1951 was about to come of age. In August 1966 Baer, now head of instrument design at New Hampshire-based military contractors Sanders Associates, went on a business trip to New York City. After finishing his work, he headed to the East Side Bus Terminal to wait for his ride back home. And while he waited, Baer had a brainwave. "I remember sitting on a stoop somewhere at the bus station in New York waiting for my bus to come in. The idea came full-blown: 'Hey! Let's play games',," he recalled. The next morning he set about writing a four-page proposal setting out his ideas for a $19.95 game-playing device that would plug into a TV set. "I was a bit conflicted when writing the proposal," he recalled. "I am the chief engineer and a division manager at a big military company, so how the hell do I write this stuff? I start off calling it by some terminology that sounds like military terminology, by the time I get halfway through it changes and by the end I'm calling it Channel LP – for let's play."

Unsure how his bosses would react, Baer used his position as the head of a large division in Sanders to start work on Channel LP in secret. He acquired a room and brought in one of his technicians, Bill Harrison, to help out with the project. "My division was on the fifth floor of a large building. On the sixth floor, right opposite the elevator, there was an empty room that I commandeered and I gave Bill Harrison keys. Later Bill Rusch joined us as chief engineer. Rusch was constructive, creative and a pain in the ass. He'd come in late and break off for an hour before he got started, no discipline. I hated that, but he was very creative and very smart. There were just the three of us and nobody knew what we were doing in that room."

By March 1967 the trio had a working machine and bunch of game ideas. There was a chase game where players controlled dots trying to dodge or catch each other. Another game was a remake of Ping-Pong where players controlled bats at either side of the screen to deflect a ball that bounced around the screen. Baer and his team also devised a game where players used a plastic rifle to shoot on-screen targets and another where the player had to furiously pump a plunger-type controller to make the screen change colour. With a working prototype complete and a selection of games on offer, Baer decided to face the music and show his bosses what he had been doing. He showed his games machine to Herbert Campman, the corporate director of research and development at Sanders, in the hope of getting funding. Interested, but unsure where Baer's work would lead, Campman agreed a small amount of investment. "He gave me $2,000 and five months of labour on it," said Baer. "It wasn't very generous, but it made it official." As the project progressed Campman kept a close eye on the developments made by the team, becoming a fan of their shooting game in particular. "He would shoot from the hip and was pretty good at it," said Baer.

Other bosses were less supportive: "I had to tell my boss, who was the executive vice-president at the time, about the project. At regular intervals he would ask me: 'Are you still screwing around with this stuff?'. Of course a few years later when the licence money started rolling in, everybody was telling me how supportive they'd been." Baer also had to demonstrate his creation, which was now being called the Brown Box, to the company's executive board, including founder Royden Sanders. "Everybody was stone-faced during the demonstration, especially Royden Sanders," said Baer. "But there were two guys among the directors who got very enthusiastic and said 'that's great'. Everybody else thought I was nuts."

By the end of 1967 the Brown Box was nearing completion and had attracted the interest of TelePrompter Corporation, a cable TV company that saw it during a visit to Sanders. Sanders' position as a military contractor meant it couldn't just start making Baer's toy, so the hope was that TelePrompter would buy the rights to produce it. But after two months of talks, cash-flow problems at TelePrompter resulted in the talks being abandoned. And since neither Baer nor Sanders had any idea who else might want to buy the rights, the Brown Box was left to gather dust.

Handmade: Bill Pitts (left) and Hugh Tuck constructing the first coin-op video game, Galaxy Game.
Courtesy of Bill Pitts

CHAPTER 2

Avoid Missing Ball For High Score

As a student Bill Pitts lived for life underground. Instead of attending lectures, Pitts spent his time at Stanford University, California, combing the sprawling network of steam tunnels beneath the 8,000-acre campus for access points into off-limits buildings. "I went to Stanford in the fall of '64 and for the first two years my hobby was breaking into buildings," he recalled.

While Pitts was not the only student exploring the ill-lit and noisy tunnels, his expeditions were mainly a solitary affair. "There were others, but we didn't really know each other," he said. "Sometimes there would be a brick wall and the tunnel would go through and others before me had knocked the bricks out so you could crawl through." Exploring the tunnels was a risky business: "It was pretty dangerous. I had a very heavy leather jacket; it was all raggedy, the lining on the inside was falling out. I would wear it in the steam tunnels even though it was hotter than 120° Fahrenheit down there. If any of the steam pipes broke I thought it would protect me, but actually I would have just cooked a little bit more slowly."

Pitts' interest in exploring Stanford's campus would prove fateful. One evening in 1966, while driving to meet some friends at a bar, he spotted a driveway going up into the hills about five miles from the centre of Stanford. "I could tell by the sign right at the front that this was a Stanford facility," he said. "It was also a building I hadn't broken into yet, so I figured I needed to come back later that night and break into this building." Armed with the toolkit he used for picking locks and unscrewing grates on his adventures, Pitts returned to the mystery site at 11pm that night to break into the laboratory. His initial reaction was disappointment. "It's all lit up and there's lots of doors and they are all unlocked, but I go inside and what's inside is the Stanford Artificial Intelligence Project. They had a big huge time-sharing computer system called the PDP-6 – one big computer and probably 20 Teletypes connected to it so lots of people could each be developing code simultaneously and each one thought they had the computer to themselves. Back then it was magical. It was amazing that this single computer could be servicing 20 people at the same time. I was enthralled by it."[1]

Pitts had done some introductory computing courses and wanted to get to grips with the space-age computer he had discovered. He persuaded Lester Earnest, the head of the artificial intelligence project, to let him use the machine when no one else was waiting. "Les said: 'You can use it as long as no-one else is using it'," said Pitts. "So I ended up going up there every night probably at eight or nine o'clock and working through 'til six or seven in the morning when other people showed up. I didn't go to classes anymore. I couldn't care less about classes; I wanted to play with computers. My dad was going crazy, my parents were well aware of the fact that I wasn't going to classes. My dad would tell me you're just going to be a computer bum."

At the facility Pitts saw first-hand the cutting edge of computer science. He worked with Arthur Samuel, who had quit IBM for academia at the start of the 1960s, on the latest incarnation of his Checkers game. He heard the first music created by the software that would form the basis of Yamaha's keyboards. He watched postgraduate students connect robotic arms and cameras to the PDP-6 and teach it to recognise, pick up and stack blocks. And he got to play *Spacewar!*.

"*Spacewar!* was one of the cool things at the A.I. lab," he recalled. "I had a friend from high school, Hugh Tuck, and when he was in town I'd take him to the A.I. lab and we'd play *Spacewar!*" And it was during one of these *Spacewar!* sessions in 1966 that Tuck remarked that if only they could make a coin-operated version of the game they would get rich. With computers still hugely expensive and large, the idea was little more than a daydream. But then, in 1969, the Digital Equipment Corporation unveiled the $20,000 PDP-11. At that price, Pitts thought, a coin-op version of *Spacewar!* might be possible: "I called Hugh up and said we could now build one of these things."

While $20,000 was still prohibitive for arcades that were used to buying slot machines for around $1,000, the pair figured they could make one and work out how cheap they would need to make the machine for it to be commercially viable. With money from Tuck's wealthy parents, the pair started adapting a PDP-11 to create their coin-operated version of *Spacewar!*, which they named *Galaxy Game*. They decided to charge players 10 cents a game or a quarter for three games. The winner

---

1. Teletypes were a brand of teleprinter. Teleprinters were electric typewriters that connected to early computers and were used in place of screens. Users would type out their commands on a roll of paper in the teleprinter, which would then print out responses to their commands. Teleprinters also formed the basis of newswires, allowing news agencies such as Reuters to send news reports over the wire to teleprinters in newspaper offices.

of each game would get a free game. The idea was to ensure the machine was in constant use and therefore always taking money.

By August 1971 everything was almost in place: The Tresidder student union on the Stanford University campus had agreed to be the test site for *Galaxy Game* and the final touches were being made. Then the pair got a call from a man named Nolan Bushnell, who worked for a company called Nutting Associates. "He had heard of us through mutual contacts," said Pitts. "He called me up and said 'Hey, come on over and see what I'm doing. I know you're building a version of *Spacewar!* using a whole PDP-11 and that's gotta cost a lot of money and I just want to show you the one I'm doing because I think you're going to lose a lot of money.'"

* * *

Bushnell, like Pitts, discovered *Spacewar!* during his student days at the University of Utah in the mid-1960s and had fallen in love with the game. But, unlike Pitts, Bushnell had long-standing interest in the amusements business. At school he wanted to design rides for Disney's amusement parks and, after gambling away his tuition fees at university, he had started working for the Lagoon Amusement Park in Farmington, a town just north of Salt Lake City where the University of Utah was based. Bushnell's love of *Spacewar!*, interest in electrical engineering and involvement with the amusement business, coupled with his entrepreneurial spirit, caused him to immediately think about turning the Tech Model Railroad Club's game into a coin-operated machine. "When I first saw *Spacewar!* on the PDP-1, I was working summers at Lagoon so I was intimately aware of arcade economics," he said. "It occurred to me that if I could put that game on a computer screen and into the arcades, it would make a lot of money. But with the million-dollar computers of the time it wouldn't work."

But the idea refused to go away. After graduating in 1968, Bushnell became an engineer for Ampex Corporation, a company best known for its breakthroughs in audio and video recording technology. While working there he read about the Data General Nova, a computer that cost $3,995, and immediately thought again of *Spacewar!*. "I thought if I could get that computer to run four monitors and have four coin slots, it would make enough money to pay for itself," said Bushnell. Bushnell teamed up with Ted Dabney, another Ampex engineer, to try and design his *Spacewar!* coin-op machine on paper. "We were good friends and Ted had a lot of analogue computer skills that I didn't have," said Bushnell. "I was a digital guy. I knew how to deal with bits and bytes and logic and things like that and Ted really understood a lot more about how to interface with a consumer television set and power supplies and things like that."

Using the Nova proved to be a dead end. For a start the computer was so slow it couldn't update the television screen quickly enough to keep the game moving at the necessary speed. Bushnell and Dabney sought to ease the demands on the computer by creating separate pieces of hardware to handle jobs such as displaying the stars that formed the backdrop of the game. It still didn't work. Even reducing the number of screens supported by the computer failed to get the game working. By Thanksgiving 1970, Bushnell concluded the project was doomed to failure. "I got frustrated and decided to abandon it," said Bushnell. "But I kept worrying about the problem and thinking about it and then I had that 'a-ha' moment where I thought I'm going to get rid of the computer and do it all in hardware. From that point, it just flew together."

Bit by bit Dabney and Bushnell created dedicated circuits to perform each of the functions they originally hoped Data General's computer would handle. The approach not only overcame the technological difficulties but also made the machine a lot cheaper to build. So much cheaper that it no longer needed to support multiple screens to justify its price tag to arcade owners. But the new approach did force a rethink of the game itself. Out went the two-player duelling and the gravitational field of *Spacewar!*. Instead players controlled one spaceship that had to shoot down two flying saucers controlled by the hardware. In short it was no longer *Spacewar!*.

By the summer of 1971 the game was nearing completion and Bushnell was starting to wonder who they could sell the game to. A trip to the dentist solved that problem. "I was at my dentist and, with a mouthful of cotton, I told him what I was doing and he said 'you should talk to this guy'," said Bushnell. "One of his other patients was the sales guy at Nutting Associates, so he gave me the telephone number and I called him up, told him what I was doing and we went in and negotiated a deal."

* * *

Nutting Associates started after Bill Nutting, a resident of the Californian city of Palo Alto, invested some money in a local company that made teaching equipment for the US Navy. Among the company's products was a multiple-choice quiz machine that projected film with the questions on a screen and then prompted naval trainees to press a button to give their answer. He figured that if a coin slot was added to the machine it could be popular bar game and turned to his brother Dave Nutting, a former first lieutenant in the Army Corps of Engineers, to adapt the technology.

"It appeared to me as a fun challenge. I re-engineered and repackaged the concept and we then called it *Computer Quiz*," said Dave. "In the meantime Bill contacted various coin-op distributors who liked the idea."

With interest high, Dave moved to Milwaukee to start a manufacturing operation closer to Chicago, the hub of the amusements business. "I rented space and began to build up inventory when Bill announced his wife Claire did not go along with the plan," said Dave. "Claire was a complete control freak and I was a threat to her." The clash led the brothers to part ways. Dave formed his own company Nutting Industries to make the same machine under the name *I.Q. Computer* while Bill went ahead with *Computer Quiz*. Both games became a success with around 4,400 *Computer Quiz* and 3,600 *I.Q. Computer* machines being built at a time when a popular pinball table would have a production run of 2,000 to 3,000.

Computer Quiz got Nutting Associates off to a good start, but by 1971 it needed a new hit and Bushnell and Dabney's radical video game machine looked just the ticket. So in August 1971 Bushnell left Ampex for Nutting Associates to complete work on the game he believed would transform the amusements business. And in a nod to *Computer Quiz*, the game was named *Computer Space*. It was then that Bushnell got word of the video game being made by Pitts and Tuck.[2] He decided to call them up: "I was curious. I didn't know what was inside their game and I expected it to be a PDP-8 or PDP-10 at the time. I was curious about what their economics were."

Pitts and Tuck accepted Bushnell's invite and headed to Nutting's building in Mountain View, California. "We went in there and Nolan was literally an engineer with an oscilloscope in his hand working on *Computer Space*," said Pitts. "It was at a point where he could demonstrate it to us, although it was still in development." Bushnell's hopes of learning from the pair came to nothing. "I thought they were clever guys but I was hoping they had cut costs down somehow and they hadn't. I left a little disappointed that they hadn't and yet at the same time relived because I felt they weren't going to be competition for me." Pitts thought Bushnell's technology was great but believed he and Tuck had a better game: "I was very impressed by his engineering skills but our game was absolutely true to *Spacewar!*. It was a real version of *Spacewar!*. Nolan's thing was a totally bastardised version."

A few weeks later, in September 1971, *Galaxy Game*, the first coin-operated video game, made its debut at the Tresidder Union. From the moment it was switched on

---

2. It should be noted that at this time, and throughout most of the 1970s, 'TV games' was the more common term. The term 'video game' eventually came to the fore later in the 1970s and the term 'TV game' faded away in the early 1980s. 'Computer games' were also sometimes talked about but, since most video games did not use microprocessors before the late 1970s, it's a misleading term. As Ralph Baer put it: "People began calling them computer games. They weren't. There were no computers!"

the machine attracted a crowd. "We had people 10-deep, packed around the machine trying to look over each other to watch the guys play the game," said Pitts. The generous approach to charging meant *Galaxy Game* earned nowhere near enough to justify its cost, but the game's popularity encouraged Pitts and Tuck to persevere.

"Everybody was really excited about it, so Hugh and I decided to build version number two," said Pitts. The pair went to town on version two, constructing proper fibreglass casing and reprogramming the computer so it could support two games at once just like Bushnell originally planned to do with *Computer Space* to cut costs.

By the time version two was complete, Tuck's family had spent $65,000 on the project – a huge sum in 1971 – but the machine still couldn't justify its cost and soon the pair had to give up. "The truth is Hugh and I were both engineers and we didn't pay attention to business issues at all, my driving goal was to recreate *Spacewar!* with coin receptors on it," said Pitts. "Nolan was much more of a businessman than I was. His emphasis was to take *Spacewar!* and try to drive it down a business path, whereas I was trying to drive it down a geek path by being honest to the game."

* * *

In November 1971, two months after the launch of *Galaxy Game*, the first *Computer Space* machine was installed at the Dutch Goose bar near the Stanford University campus. Its black and white TV screen sat encased in colourful and curvy fibreglass that could have come straight from the set of the 1968 sci-fi film *Barbarella*. *Computer Space* screamed the future and to Bushnell's delight the drinkers at the Dutch Goose seemed to like it. "The Dutch Goose was the first location where we tested *Computer Space* and it did fantastically well. What we didn't realise is that it had a very high percentage of college students," said Bushnell.

With the initial test having gone well, Nutting Associates pushed ahead with the production of *Computer Space* hoping to woo arcade operators with its revolutionary technology and lack of moving parts.[3] Nutting Associates produced more than 1,500 *Computer Space* units expecting a smash hit, but the reaction away from student bars proved less favourable. "When we put it in a few working man's beer bars it did no money," said Bushnell. "It didn't do anything because it was too complex."

People in the arcade business were equally confused by the game. "In 1971, my brother Bill came out with *Computer Space*," recalled Dave Nutting. "Empire Distributing

---

3. The electro-mechanical arcade games that were popular at the time were notoriously prone to breaking down due to the various moving parts they were built out of.

was handling my electro-mechanical game *Flying Ace* and was also distributor for Nutting Associates. I was at Empire meeting the principals Gil Kitt and Joe Robbins when a call came through from Bill and Nolan Bushnell asking for their response on receiving their first *Computer Space*. Gil and Joe had the speakerphone on so I could hear. Joe responded that the game play was very confusing and his people were having trouble understanding the controls. Nolan came on to say that *Computer Space* was just the beginning of a new era and the future of the coin amusement would be video games and pinball would no longer be the industry staple. Gil stood up and loudly stated: 'There is no future in video games and if the day comes that video games take over, I will eat my hat'. Several years later at a convention I ran into Gil and asked him if he remembered his comment. He blushed and laughed and said: 'Boy was I wrong, it is a good thing I retired'."

Computer Space did have fans though. Owen Rubin, who would later work at Atari, was one: "It was the first video game I ever saw. I was always hooked on pinball and other coin amusements in arcades near me, so when I saw this, I was immediately hooked." Another future Atari employee Dave Shepperd also fell in love with the game: "I remember thinking it was the coolest thing I had ever seen. I loved that space-age, shaped-metal, flaked-fibreglass cabinet too." Inspired, Shepperd built a video game himself: "Being basically a cheapskate and not wanting to drop any more quarters into such a thing, I went home and proceeded to design and build my own video game using parts scrounged from junk bins."

For Bushnell, *Computer Space* had done well enough: "Compared to the games that came after it looks like a flop. But I had never created a million-dollar product before. It represented a reasonable royalty stream for me." His experience at Nutting Associates also inspired him to form his own business: "I got to see Nutting operating and they gave me a huge amount of confidence to go out on my own because I knew I couldn't screw it up more than they did." And with that Bushnell and Dabney decided to form Syzygy Engineering with the goal of delivering on Bushnell's claim that video games would replace pinball as the mainstay of the arcades.[4]

* * *

Meanwhile, Ralph Baer's Brown Box was about to finally make it into the shops. Efforts by his employer Sanders Associates to find a licensee for the games console

---

4. Syzygy is the term for a straight-line alignment of three celestial bodies, such as when the Earth, Moon and Sun line up during a solar eclipse.

had hit the buffers in early 1968 when the potential buyer TelePrompter went bust. "Nothing happened for a year and a half because we didn't know what the hell to do with it," said Baer. "It finally dawned on me that television manufacturers were the companies most likely to manufacture, advertise, distribute, and sell something that's made with exactly the components and manufacturing techniques as the television sets themselves." Sanders demoed the Brown Box to the television manufacturers who dominated the US market at the time: General Electric; Magnavox; Motorola; Philco; RCA; and Sylvania. "When we demonstrated to these companies in '69 everyone of them went 'that's great', but nobody would offer a dime except RCA and when we worked out the agreement we said we couldn't live with that and walked away," said Baer.

Once again it looked like the Brown Box was destined for the scrapheap. Then Bill Enders, one of the RCA executives who had been involved in the talks with Sanders, left to join Magnavox and convinced his new employer to look again. The Brown Box's creators – Baer, Bill Harrison and Bill Rusch – headed to Magnavox's headquarters in Fort Wayne, Indiana, to demonstrate their work once again. This time Magnavox said yes. In January 1971, Magnavox signed a preliminary deal with Sanders and began work on turning the Brown Box into a marketable product. Magnavox redesigned the casing for the machine and briefly renamed it the Skill-O-Vision before settling on the Odyssey.

The Brown Box's collection of seven games was built up to 12 titles including the maze-chase game *Cat & Mouse*, an educational title called *States!* and the *Ping-Pong* game developed back in 1967. The rifle game that convinced Sanders to keep the project alive became the sold-separately *Shooting Gallery* add-on for the Odyssey. Magnavox then decided to add paper money, playing cards and poker chips to enhance the games and plastic overlays that attached to the TV screen to make up for the Odyssey's primitive visuals. And with so much packed in with the game console, the $19.95 price tag Baer originally hoped for became $99.95. Baer was appalled: "I saw the box and out comes 10,000 playing cards, paper money and all this crap. I just knew nobody's ever going to use this stuff."

With the enhancements in place Magnavox set a launch date of August 1972 for the world's first games console, which the company decided would only be available through Magnavox dealerships. In the build up to the launch, Magnavox demonstrated the Odyssey to Magnavox dealerships and the media. On the 24th May 1972 it put the Odyssey on display at the Airport Marina in Burlingame, California, near San Francisco. One of the people who decided to take a look was Nolan Bushnell.

At the time Syzygy, the company Bushnell founded with Dabney, had struck a deal to create video games for the Chicago-based pinball giant Bally Midway. Bushnell wanted Syzygy to make a driving video game for Bally Midway, convinced this would win over the punters alienated by *Computer Space*. Seeing the Odyssey and its *Ping-Pong* game in Burlingame did little to change his mind and so the following month Syzygy, which had been getting by repairing broken arcade machines and running *Computer Space* machines in arcades near its rented offices in Santa Clara, started preparing to create Bushnell's driving game. Dabney and Bushnell agreed to invest $250 each in the company to incorporate it only to find that another company already had the Syzygy name. Bushnell turned to his favourite game – the Japanese board game Go – for inspiration and suggested the company's new name should be Atari, a term from Go similar to check in Chess. Dabney agreed and on 27th June 1972 Atari Incorporated was born.

That same day Atari hired Al Alcorn, a young engineer who had worked for Dabney and Bushnell at Ampex as a trainee. Bushnell wanted to give Alcorn a very simple game to get him used to the basics of video game technology and thought of *Ping-Pong*, the Odyssey game he had played the month before. He described the game to Alcorn and told him it was a part of a deal he had done with General Electric. "I thought it would be a good way of getting him through the whole process because the circuits I'd designed were pretty complex," said Bushnell. There was no deal, however, and Bushnell had no intention of doing anything with the game. He thought the bat-and-ball action was too simplistic to be popular and saw it as no more than on-the-job training for the young employee. Alcorn, however, threw himself into the project. He improved on Bushnell's brief by making the ball bounce off the player's bats at different angles depending on which part of the bat it hit. He also added scores and crude sound effects. The result had just one instruction: "Avoid missing ball for high score". These minor improvements did not drastically change the game, but were enough to make Bushnell and Dabney change their plans. "My mind changed the minute it got really fun, when we found ourselves playing it for an hour or two after work every night," said Bushnell, who named Alcorn's game *Pong*.

That September Atari decided to test *Pong* on the customers of Andy Capp's Tavern in Sunnyvale, California. At the same time Bushnell headed to Chicago to show Bally Midway the game, hoping it would fulfil Atari's contract with the pinball manufacturer. Bally Midway, however, was unimpressed. "They didn't want it," said Bushnell. "First of all, it was only two-player and no coin-op game at the time was only a two-player game, some had two-players but there had to be a one-player option. That was the big veto in their minds."

Back in California, Alcorn also got some bad news from the owner of Andy Capp's – *Pong* had stopped working. Alcorn drove over to the bar to investigate. On arrival he opened the coin box so he could give himself free games while trying to diagnose the problem and out gushed a flood of coins; spilling, spinning and sliding all over the barroom floor. The sheer amount of coins put into the *Pong* machine had caused it to seize up. The customers at Andy Capp's had gone crazy for *Pong*, people had even begun queuing outside the bar waiting for it to open just so they could play the game.

At a time when the average coin-op machine would make $50 a week, *Pong* was raking in more than $200 a week. Atari now knew it had a hit on its hands; the only problem was how to get it into the arcades. Hoping the game's takings would persuade Bally Midway to change their mind, Bushnell went back to the pinball firm. Worried the company wouldn't believe the real figure, Bushnell told them it was making a third of what it actually was. Bally Midway once again rejected the game. Atari then offered *Pong* to Nutting Associates in exchange for a 10 per cent royalty, only to be rejected again on the grounds that the royalty demand was too high.

With options drying up, Atari decided to make the game itself. It was a big leap for the young firm: it had next-to-no money, no production line and no links with arcade machine distributors. Bushnell was nervous about the move but figured the game's simple design meant it would be easy to build. Atari gambled everything on its first run of *Pong* machines. "Our first run was 11 units, which was 100 per cent of the money that we had," said Bushnell. Each machine cost $280 to make but sold for $900.

"We sold the 11 units immediately for cash, so all of a sudden we had our cash back. The next release was 50 units and we completely ran out of space," said Bushnell. Luckily for Atari, the company in the business unit adjacent to their offices went bust just as space got tight. "We went from 2,000 square feet to 4,000 square feet and knocked a hole in the wall to link the two," said Bushnell.

By now word about *Pong* had spread through the arcade business. "We had distributors all over the country who were just screaming for the units," said Bushnell. Atari needed a proper production line fast if it was going to meet the soaring demand for *Pong*, but lacked the cash needed to set up a proper manufacturing facility. So Bushnell headed to the banks to ask for a credit facility. The banks were, however, disinterested – put off by Bushnell's long hair and the dubious image of the amusements business, which had become linked in the public mind with gangsters and gambling. Back in the 1930s gangsters had close ties to the amusements business, none more famously than Frank Costello – a notorious mobster nicknamed 'the prime minister of the underworld'. Costello owned a network of 25,000 slot machines located in cafés, gas stations, bars, restaurants and drug stores across New York City that earned him millions of dollars every year and helped bankroll his less legitimate activities.

The authorities had long been worried about the connection between the Mafia and the slot machine industry, so when several manufacturers started producing pinball machines that offered cash prizes they decided to act. New York City's Republican mayor Fiorella La Guardia led the charge. A year after becoming mayor in 1934, La Guardia began petitioning courts for a ban on pinball, arguing it was an extension of gambling. After years of legal battles, La Guardia got his way in 1942 when a Bronx court sided with him and banned pinball – a ban that would stay in force until 1976. To celebrate his victory La Guardia held a press conference by the city's waterside where he smashed up a confiscated pinball machine with a sledgehammer before throwing it into the East River. Over the next three weeks police impounded more than 3,000 pinball machines, dealing a severe blow to Costello's slot machine empire. Other US cities and towns began to follow New York's lead, fixing the idea that pinball and arcades were inextricably linked with gangsters, gambling and moral decline.

So when Bushnell asked banks for a loan to help build his amusements machine business, they showed him the door. Eventually Bushnell persuaded the bank Wells Fargo to lend Atari $50,000 on the back of an order for 150 *Pong* machines. It was less than Atari had hoped for, but enough to get a production line going.

With funding in place, the company turned a disused roller-skating rink into its new manufacturing arm and headed to the local unemployment office to recruit an instantly available workforce. "They were horrible," said Bushnell of the staff Atari hired to man the *Pong* production line. "We had a bunch of heroin addicts and things like that. They were stealing our TVs. We were young and dumb is what I like to say. But we learned quickly. They didn't last very long."

Soon *Pong* had taken the nation by storm, introducing millions to the idea of the video game. Other amusement machine manufacturers quickly started producing their own versions of the game, hoping to cash in on the new craze. Pinball firms such as Chicago Coin and Williams released thinly veiled remakes of Atari's hit. Bally Midway went back to Atari and signed a licensing deal that gave the Californian start-up a 5 per cent cut from sales of its *Pong* clone.

Nutting Associates, doubtless regretting its decision to turn down Bushnell's offer of *Pong*, released *Computer Space Ball*. Some of these clones achieved sales comparable to the 8,000-plus *Pong* machines sold by Atari. *Paddle Battle* and *Tennis Tourney*, for example, transformed the fortunes of Florida-based Allied Leisure, increasing its annual sales of $1.5 million in 1972 to $11.4 million in 1973.

Pong soon went global. In Japan, Taito, an amusements manufacturer built off the back of jukeboxes, peanut vending machines and crane games, looked at *Pong* and produced *Elepong* – the first Japanese arcade game. French billiards table makers René Pierre jumped on the *Pong* bandwagon with *Smatch* and in Italy, Bologna-based pinball company Zaccaria entered the digital age with *TV Joker*, a *Pong* copy produced under licence from Atari. "In 1972, *Pong* arrived in Italy and it was a great success," recalled Natale Zaccaria, co-founder of Zaccaria. "Zaccaria produced pinballs and sold them all over the world, so we had a wide net of contacts. When the video games started, we were ready to start selling and producing them under licence. Zaccaria assembled a cabinet for Italy and called it *TV Joker*. At the start we were buying the motherboards from the US and building just the cabinets."

Pong also helped Magnavox sell its Odyssey console and by 1974 some 200,000 had been sold, largely on the back of its *Ping-Pong* game. "Everybody played *Ping-Pong* and that's it," said Baer. "It was a good game but what made it really popular was *Pong*. That's when we realised 'hell, all we had to do was stop after game number six'." Magnavox eventually threatened to sue Atari for infringing Baer's patents but, feeling the young company didn't have much money, it agreed to give the firm the rights to make the game for a one-off payment of $700,000. Magnavox's lawyers were less forgiving of Atari rivals such as Allied Leisure, Bally Midway, Nutting Associates and Williams.

By September 1974 an estimated 100,000 coin-operated video games were in operation across the US, raking in around $250 million a year. For the amusements business, long shamed by being connected to gambling and gangsters, the video game offered a new start, attracting a new demographic to the arcades. "For years, our games – pinballs, shuffle alley, pool – appealed mainly to, you know, the labouring class. Now with the video games you have a broader patronage," Howard Robinson, the manager of an Atlanta coin-op distributor, told *The Ledger* newspaper in September 1974. "A lot of lounges will take a video game that never would have let a pinball machine in the door."

As Frank Ballouz, sales manager for Atari, remarked a couple of years later: "Many arcades used to be in rat-hole locations. Now they have turned into family amusement centres where you can take your wife and six-year-old daughter and 14-year-old son."

The idea that video games were somehow separate from the seedy arcade machines of old was something Atari deliberately pushed. "We fostered the idea that it was a more sophisticated thing to do because we thought it was better marketing," said Bushnell.

Bushnell had delivered on his promise that *Computer Space* was just the start of a new era for the amusements business. The only question now was how to follow up such a megahit.

Fun Inc.: Nolan Bushnell watches Gran Trak 10 games roll off the Atari production line, July 1974
Tony Korody / Sygma / Corbis

CHAPTER 3

A Good Home Recreation Thing

Pong's popularity sent shockwaves through the amusements business. In less than six months Atari had gone from an unknown start-up to the leaders of a revolution in the arcades. For the game-playing public, video games embodied the technological dreams of the Cold War in a way pinball tables and electro-mechanical games never could. No longer was TV just for watching, now the viewer could take control. As Florida's *Ocala Star-Banner* newspaper put it: "What better evidence is there that Americans are living in the space age than the growing application of electronics in games that are played?"

The success of *Pong* restructured the amusements business. Arcade owners turned their backs on the cranky and unreliable electro-mechanical games that once filled their game rooms and embraced the video game. "Video games offered a wider assortment of entertainment and, since video games had fewer moving pieces, they were more reliable," said Bob Lawton, who founded the Funspot Arcade in Wiers Beach, New Hampshire, in 1952. "Ask anyone who ran electro-mechanical games back in the day and they will tell you the same thing. You can do so much more with a video game than you can with a plastic car, electric motors and relays."

Within a year of *Pong*'s debut in Andy Capp's Tavern, more than 15 companies had piled into the coin-operated video game business that once was Atari's alone. Not that these companies strayed far from the bat-and-ball formula of *Pong*. Instead they produced barely disguised copies and various new twists on Atari's game such as Chicago Coin's *TV Pingame*, a fusion of *Pong* and pinball where players used the bat to hit the virtual ball into digital pins to score points, and Ramtek's *Clean Sweep*, where the goal was to clear dots from the screen by hitting the ball over them. But with competition intensifying, Atari knew it needed to expand its range of games beyond *Pong* remakes.[1] "We

---

1. Atari did produce several *Pong* variants of its own, including the four-player *Quadrapong* and the volleyball-inspired *Rebound*, where players had to hit the ball over a virtual net.

knew that we understood the technology and everybody else pretty much just xeroxed our technology," said Atari boss Nolan Bushnell.[2] "I felt we could out-innovate them."

To encourage this innovation, Bushnell sought to mould Atari into a business based on egalitarian values and fostered a working culture based on fun and creativity. He spelled out his thinking in a two-page company manifesto that drew on the ideas of the hippy movement of the late 1960s. The manifesto declared "an unethical corporation has no right to existence in any social framework" and promised that Atari would "maintain a social atmosphere where we can be friends and comrades apart from the organizational hierarchy". It also stated that Atari would not tolerate discrimination of any kind including "the short hairs against the long hairs or the long hairs against the short hairs". "This is slightly after the days of Aquarius and the hippy revolution and we all wanted to create this wonderful, idealistic meritocracy," explained Bushnell.

In practice, these values translated into a lack of fixed working hours, an anything-goes dress code and parties with free beer that the company threw if targets were met. "We were all very young," said Bushnell. "The management team were all in their late 20s to early 30s and most of the employees were in their early 20s. With that kind of demographic, a corporate culture of fun naturally evolves. Then we found out our employees would respond to having a party for hitting quotas as much as having a bonus. We became known as a party company because we'd have beer kegs on the back lot all the time because we were hitting quotas all the time."

Steve Bristow, who joined Atari as an engineer in June 1973, felt the company's attitude was a world away from the big technology firms of the day. "At Atari it didn't matter if you had tattoos or rode in on a motorcycle," he said. "At that time in IBM you had to wear a white shirt, dark pants and a black tie with your badge stapled to your shoulder or something. At Atari the work people did counted more than how they looked."

The company also turned a blind eye to the use of illegal drugs by employees. "There was absolutely no drug use in the factory, but we did have parties and, along with beer, some people preferred marijuana and we closed our eyes to it. It was pretty wild," said Bushnell. Bristow felt it reflected the times: "This was California in the 1970s. It wasn't company policy or anything, but at company parties one could detect certain odours and some people had sniffles. It was more of the times than of Atari."

---

2. Bushnell gained full control of Atari shortly after *Pong* became a success when the company's other founder Ted Dabney quit in 1973 because he disliked running a large business. Dabney sold his share of Atari to Bushnell for $250,000.

Despite Atari's laid-back management style its staff worked hard, putting in long hours because they enjoyed their jobs. "It was quite common to have people working through the night. Sometimes we'd work 24 hours just because we were excited about what we were doing," said Dave Shepperd, who became an Atari game designer in 1976. Noah Anglin, who quit IBM to become manager at Atari in 1976, remembered being impressed by the commitment of Atari's employees: "What I saw was these absolutely brilliant hard-working guys. They redefined hard working and the ability to work hard."

This blurring of work and life, coupled with Bushnell's non-conformist management, helped Atari stay one step ahead of the bigger manufacturers now seeking to conquer the video game business. While other companies rehashed *Pong*, Atari began releasing new types of video game. It challenged arcade goers to steer through meteor storms against the clock with *Space Race*. *Pong* creator Al Alcorn's *Gotcha* got people playing virtual kiss chase in a maze, using joysticks encased in pink rubber domes designed to look like breasts. In *Qwak!*, Atari handed players a rifle-shaped light gun for a virtual duck hunt. All three sold thousands.

Only Nutting Associates, Bushnell's former employer, tried to explore what more could be done with video games in the immediate wake of *Pong*. It produced *Missile Radar*, a game where players had to shoot down incoming missiles. Atari later reworked the idea to create *Missile Command*. In March 1974 Atari's experiments with video games resulted in the release of *Gran Trak 10* – the first driving video game. *Gran Trak 10* showed a bird's eye view of a racecourse and asked players to drive their virtual racing car round the track using a steering wheel, gear stick and the game's accelerate and brake pedals. It became Atari's biggest-selling game since *Pong* but, thanks to an accounting error, the company underpriced the machine and lost money on every one sold. The resulting losses pushed Atari to the brink of collapse.

Matters were not helped by Atari's decision to go global in 1973 by opening Atari Japan in Tokyo. "The Atari Japan excursion was an unmitigated, unbridled disaster," said Bushnell. "We were young and thinking that everything was possible. We probably violated every international trade law with Japan. We actually funded the thing with cash and bought a factory without worrying about permits and things like that, which are so difficult to get in Japan." Like many foreign companies that tried to enter the Japanese market, Atari found itself hampered by a legal system and business culture that openly conspired against overseas firms. In the early 1960s, Ikeda Hayato, the Japanese prime minister who played a crucial role in the nation's post-Second World War economic success, had introduced laws that restricted the activities of foreign companies in a bid to protect Japanese businesses. On top of this, Japanese coin-op distributors refused to work with the cocky American business. "The distribution over there was really closed to us," said Bushnell. "Sega didn't like us. Taito didn't like us. They were doing everything they could to throw obstacles in our way. They were entrenched and they were Japanese. We were American and stupid."

Taito in particular was working hard to turn itself into the Japanese answer to Atari. After the success of its 1973 *Pong* clones *Elepong* and *Soccer*, the company started to explore new video game concepts. In 1974 Tomohiro Nishikado, the designer of *Soccer*, created the company's first truly original game: the racing game *Speed Race*. As with *Gran Trak 10*, the action was viewed from above but instead of squashing a whole track into one screen, *Speed Race* created the impression of a larger course by having rival cars move down the screen as the player accelerated. The player, whose car could only be moved left to right and spent the whole game at the bottom of the straight-line race track, had to weave in and out of the traffic as he or she overtook the other racers. "Until then there had not been any games that differed greatly from *Pong* in Japan," said Nishikado, who started his career at Taito making electro-mechanical games. *Speed Race* proved popular both in Japan and in the US, where Bally Midway released it as *Wheels*, providing the earliest indication that Japan was destined to become a major force in video games. "Until then we only imported games from the US and with this game we managed to start exporting games to the US," said Nishikado.

Atari Japan, meanwhile, ate through $500,000 before Bushnell admitted defeat in 1974, just as the company's failure to go through the right channels began to catch up with it. "Basically, to keep from going to jail we had to sell it," said Bushnell. Atari Japan was sold to Nakamura Manufacturing, a Japanese coin-op manufacturer and distributor formed in 1955 by Masaya Nakamura that would rename itself Namco in 1977. The buy out made Nakamura Manufacturing the exclusive distributor of Atari games in Japan for 10 years. The Atari Japan disaster, the under-pricing of *Gran Trak 10* and the slowing sales of *Pong* games left Atari teetering on the edge of closure. Then, just as it looked like Atari was doomed, one of Bushnell's more wily business moves came to the rescue.

* * *

Back in late 1972 when *Pong* became a runaway hit, Atari discovered that the coin-op distribution system in the US limited its ability to profit from the game. The coin-op business was based around distributors who bought the machines and then sold the machines to, or installed them in, the various bars, arcades and other outlets they supplied. To attract these locations to their network, distributors demanded exclusive deals from manufacturers for the geographical areas they covered so that only they, and not competing distributors, had access to certain machines whether that was Atari video games, Bally pinball tables or Rock-Ola jukeboxes. So in a town with two distributors, a coin-op manufacturer could only hope to get its machines in the locations in one of those distributors' networks.

Bushnell worried this system not only meant Atari sold fewer games but would also encourage the formation of a serious competitor. He hit on a novel solution – he would form a bogus rival that would repackage Atari's games and sell them to the distributors that Atari could not work with because of pre-existing deals. "It was a defensive strategy as much as an offensive strategy," said Bushnell. "I was always looking to put anybody who copied us out of business if I possibly could. That was kind of my ethic. I found that the distributors that we did not have in each of these cities were desperate to find somebody to knock our products off to be able to compete with the guy across the town that had our stuff. I said this is a gigantic demand that is going to create a competitor that might be somebody that's actually good, so let me make it so much harder for them by satisfying that demand. That's what Kee Games was all about. I wanted to cut off distribution to would-be competitors."

Kee Games was named after Joe Keenan, the friend of Bushnell's who agreed to head up the pretend competitor. Bushnell also appointed Bristow as the new company's vice-president of engineering. To convince the coin-op industry that Kee Games was a real competitor to Atari, Bushnell concocted a cover story about how Bristow and other Atari employees had jumped ship to form their own company. "The original thing we leaked was that some of our best people had left and started a competitor. That seems very logical to a lot of people," said Bushnell. "Then we floated the rumour that we were suing them for theft of trade secrets. That also sounded very logical to everybody. A couple of months later we said we had settled the lawsuits and the settlement was we owned a piece of Kee Games."

To maintain the pretence Kee Games had its own offices, salespeople and a small game development team, but its main activity was re-releasing Atari games under different names such as *Spike* – the Kee Games' version of *Rebound*. With so many companies copying Atari at the time, few questioned the similarities between the games. "All the Kee Games' circuit boards were manufactured in the Atari facility. We had our own cabinets and developed our own games, but it was part of the same thing," said Bristow. The deception worked and soon Kee Games was striking deals with the distributors Atari couldn't reach. "The distributors swallowed it and Atari was able to bust open the distribution model so that now we were selling to everybody," said Bristow. Only one person – Joe Robbins of Empire Distributing – saw through the spin, according to Bushnell: "I remember him coming up to me in a trade show and he says: 'You think you're so smart, but I know what you did'. He did it in such a way that you knew he had a lot of respect for what we were able to accomplish."

Having bypassed the restrictions of the distribution system, Kee Games then provided the big hit Atari needed to repair its finances with *Tank*, a two-player game where players steered tanks around a mine-infested maze trying to shoot each other. The idea grew out of Bristow's desire to update Bushnell and Dabney's first video game, *Computer Space*. "*Computer Space* was a really good fighting game, but many people found it hard to play. The idea of a free-floating spaceship that you had to counter velocity with rotation and counter thrust wasn't easy," said Bristow. "As a youth my uncle had put me to work clearing his orchard using a Caterpillar tractor, which drove like a tank. I thought that could be turned into *Computer Space* done right."

Bristow got Lyle Rains, one of Kee Games' engineers, to turn the idea into a working game. Rains enhanced Bristow's basic idea by adding a maze littered with deadly mines. Released in November 1974, *Tank* became the most popular video game since *Pong* with more than 15,000 sold. With Kee Games now awash with cash, Bushnell used the opportunity to officially merge it with Atari. As part of the deal Keenan became Atari's new president.

The profits from *Tank* repaired Atari's battered balance sheet and the combination of the two company's distribution networks gave the reinvigorated Atari unparalleled reach within the coin-op market. It also erased the costs involved in pretending that the two were separate businesses. The timing was fortuitous as Atari was about to launch itself into the consumer electronics business with a version of *Pong* for the home.

* * *

The idea to take *Pong* into people's living rooms was suggested by Atari engineer Harold Lee. Given the game's original inspiration – the Magnavox Odyssey games console – the idea to make a home version of *Pong* was an obvious one, but Lee believed Atari could improve on the Odyssey by using integrated circuits.[3] The Odyssey had

---

3. Invented in the late 1950s, integrated circuits – also called microchips – allowed the discrete components that used to form electronic circuits to be shrunk and flattened onto a silicon chip. The result was a massive breakthrough in electronics. Integrated circuits were not only much smaller but were easy to mass produce (the chips could essentially be printed en masse), used less electricity and were more reliable.

been developed in the late 1960s when integrated circuits were far too expensive to use in consumer products. By the start of the 1970s they still remained prohibitively expensive, but had become cheap enough to use in arcade video games such as *Pong*. Lee, however, believed the cost of integrated circuits would soon fall enough to make a *Pong* console that could be plugged into home TVs. *Pong*'s creator Al Alcorn agreed with Lee's assessment and the pair asked Bushnell to fund the project.

Bushnell was sceptical: "The technology was expensive. The integrated circuit boards by themselves cost almost $200, so that was clearly never a consumer product." Despite Bushnell's doubts the pair remained convinced the plan would work and set about making a prototype to prove it could be done, with help from Atari engineer Bob Brown. "It was really a skunk works project," said Bushnell. "We put very little money into it until we were pretty sure we could do it."

With next to no funding the trio spent most of 1974 building a prototype home *Pong* console that could be sold at an acceptable price point. By late 1974 it was clear Lee's idea really would work and, most impressively of all, the whole game could be fitted on a single integrated circuit – a breakthrough that drastically reduced the production costs. Atari wanted to manufacture the *Pong* consoles itself but needed to invest in a bigger and more advanced production line to produce the machines in the quantities needed for the consumer market.

Getting the necessary funding for this was proving difficult until Atari sorted out its finances by merging with Kee Games. This in turn helped the company secure $20 million of funding from technology investor Don Valentine, the founder of venture capitalists Sequoia Capital. By early 1975 Atari was ready to start touting its new games machine to retailers.

But retailers didn't want Atari's $99.95 mini-*Pong*. "We took the first *Pong* to the toy trade fair and we sold none," said Bushnell. "The toy stores at the time, their most expensive product was $29 and so the toy channel was closed to us." Rejected by toy stores, Atari hawked it to the television and hi-fi stores only to find they were also disinterested. Increasingly desperate for retailer support, Atari pitched its home video game to the department store chain Sears Roebuck, the largest retailer in the US at the time.

"We called Sears really as a last resort," said Bushnell. Atari ended up being pointed to the buyer for the company's sporting goods departments. "The sporting goods department of Sears turns into a ping-pong, pool table type of department around Christmas and it turned out that the year before they had successfully sold out of a home pinball," said Bushnell. "The buyer said pinballs are in bars, *Pong* is in bars, this will be a good home recreation thing. "

It was the break Atari needed. Sears struck an exclusive deal with Atari for the console. Sears would stock the game in its 900 stores and promote it heavily in the run-up to Christmas 1975. In exchange Atari rebranded the game as the Sears Tele-Games Pong and agreed not to release its own Atari-branded version until the new year.

That Christmas 150,000 Sears Tele-Games Pong consoles flew off the shelves as customers went crazy for the chance to play *Pong* in their own homes. While the Odyssey had already offered consumers the chance to play video games at home, the arrival of Atari's console was the moment where millions suddenly realised that video games could be played on their own TV sets as well as in bars and arcades. "It's the first time people have been able to talk back to their television set, and make it do what they want it to do," Bushnell told the *Wilmington Morning Star.* "It gives you a sense of control, whereas before all you could do was sit and watch channels."

The console ushered in a second wave of *Pong*-mania that turned Atari, a near bankrupt business just over a year before, into a household name. And just as with the coin-operated version of *Pong*, Atari was quickly joined by a stampede of imitators hoping to cash in on the TV games craze. Atari's competitors were aided by the arrival of General Instruments' AY-3-8500 microchip. "The AY-3-8500 chip did much the same thing as in the Atari machine, but General Instruments independently developed it," said Ralph Baer, the creator of the original Odyssey console. "Two guys did it in Glenrothes, Scotland, against the better judgment of management. This General Instruments guy in Long Island, New York, the general manager there, heard about what was going on and told those guys to come over and bring their demo with them. He moved it."

The AY-3-8500, and the rival chips that followed, allowed any company to produce a home *Pong* without having to design an integrated circuit from scratch. Provided they could get hold of the chips that is. The home *Pong* boom caught the chip manufacturers by surprise and they simply could not produce enough to satisfy demand.

Companies such as toy manufacturers Coleco and Magnavox which ordered their chips early received them on time, while the late comers were left in the lurch unable to get their consoles on the shop shelves in time for Christmas 1976, when the excitement about home *Pong* peaked. Despite the microchip supply problems, millions of people brought home *Pong* games. By Christmas 1977 there were more than 60 *Pong*-style consoles on sale around the world and nearly 13 million had been sold in the US alone.

But the implications of the microchip for video games did not end there. As the mid-1970s turned in the late 1970s, the arrival of a new type of microchip – the microprocessor – would reshape not just the video game business but also the very nature of what and how people played.[4]

4. Microprocessors are a type of integrated circuit that effectively put the functions of a computer on a single silicon chip. Unlike normal integrated circuits, they could be programmed to perform different functions without any need to redesign the circuit design.

Computer on a chip: Manufacturing Intel's 8080 microprocessor
Courtesy of Intel Corporation

CHAPTER 4

Chewing Gum, Bailing Wire And Spit

Victor Gruen was angry. The 1950s were changing America and the Austrian-born socialist architect believed it was changing for the worst. He felt the growth of car ownership and suburban living was ripping out the heart of society, isolating people from their communities. But he had an idea that he believed would challenge these economic forces: the shopping mall.

Drawing inspiration from the covered shopping arcades of European cities, Gruen envisaged a new kind of retail environment – a city centre for the modern world. The shopping mall would, he imagined, bring communities together to shop, socialise and be entertained within an enclosed and climate-controlled building. And in 1954 he got his chance to put his ideas to the test in the Minneapolis suburb of Edina, Minnesota, as the guinea pig. His Southdale Center mall, which opened in 1956, proved to be the starting gun for a transformation of American cities and towns. Over the next two decades, malls sprung up across the nation ushering in a social and retail revolution as they went. The spread of malls became an unstoppable juggernaut. By the end of 1964 around 7,600 malls had opened across the US. By 1972 that number had almost doubled to 13,174.

But for Dave Nutting, the proliferation of the mall meant it was time to start over again. After the breakdown of his business partnership with his brother – Bill Nutting of Nutting Associates – he had formed MCI Milwaukee Coin, a manufacturer of electro-mechanical games. But when the company's investors got wind of how arcade operator Aladdin's Castle was building an amusements empire on the back of the expansion of shopping malls, they decided MCI should change direction. "MCI was selling direct to Aladdin's Castle, who were establishing coin-operated arcades in the new shopping malls emerging throughout the country," said Nutting. "My sales manager said we should do that. We created Red Baron game rooms in 20 locations from Ohio in the east to Phoenix in the west. My investors then decided we should shut down the MCI game manufacturing arm and concentrate on the game rooms."

Nutting, an engineer by trade, decided it was time to go: "Over the years I had become acquainted with the people at Bally Midway and they suggested we work out a consulting relationship. I took my young electronics engineer Jeff Fredriksen and two techs and created Dave Nutting Associates." Shortly after parting ways from MCI, a representative from an up-and-coming technology firm called Intel invited Nutting and Fredriksen to attend a talk about its latest product. "The rep representing Intel began to tell us about a revolutionary new technology called a microprocessor," said Nutting. "Intel engineers were travelling the country giving lectures on this new technology. Jeff and I drove down to Chicago and attended one of these lectures."

Intel's product, the 4004, was the first functioning microprocessor and although it could do little more than add and subtract, the potential fired Nutting's imagination: "I immediately became convinced it was the future of all coin amusement devices. Design one microprocessor hardware system and all games would be created in software." Nutting quickly set about building a relationship with Intel: "I convinced the Intel marketing person that the microprocessor would revolutionise the coin amusement industry from pinballs to slot machines to video games and that my group was an advanced R&D group for Bally. Our local Intel rep then convinced Intel to send us one of the first 50 development units." The development unit arrived at Dave Nutting Associates in early 1974. The company used it to build a microprocessor-based pinball machine to persuade Bally to invest further in his company's exploration of the technology.

"My overall game plan for my grand presentation to Bally's management was to obtain two in-production Bally pinballs and strip one of all electro-mechanical components and leave the other for comparative play," said Nutting. The company bought two of Bally's movie-themed *Flicker* pinball machines, gutted one and rebuilt it around Intel's microprocessor. By September 1974 the enhanced *Flicker* table was ready and Bally's management were invited in to see the results. "I had the two *Flickers* side-by-side," said Nutting. "Both played exactly the same. The only visual difference was the back panel had LED read outs versus the mechanical drum scoring of a conventional pinball. The inside of the cabinet was empty except for a transformer." Bally's executives couldn't believe what they were seeing. "I found John Britz, Bally's executive vice-president, wandering around opening closet doors looking for the main computer running the pinball," said Nutting.

But Bally worried that arcade owners would not understand microprocessor pinballs and decided to phase in their introduction slowly. It also decided to get its own engineers to build their own microprocessor-based hardware rather than using the system developed by Dave Nutting Associates. In response, Dave Nutting Associates teamed up with a small pinball company from Phoenix called Micro Games to create

Spirit of 76 – the first pinball game designed for a microprocessor. *Spirit of 76* made its debut at the 1975 Amusement & Music Operators Association trade show, where its low cost design quickly attracted the industry's interest. "The units were lighter and easier to service and were 30 per cent cheaper to manufacture," said Nutting. Soon arcade owners stopped buying electro-mechanical pinball tables, preferring to wait for microprocessor-based tables to reach the market. Soon every significant pinball manufacturer was following Dave Nutting Associates' lead.

By then, however, Nutting was preparing to do what he did for pinball to video games. The video games of the time were made using transistor-transistor logic (TTL) circuits that had to be made from scratch for each game. While this was adequate for simple *Pong* games, by the mid-1970s the limits of these simple circuits were holding video games back. "Game designers tried to create more sophisticated game play but they found themselves pushing the limits of TTL," said Nutting. "The dedicated circuits could not be manufactured. The electric noise generated by the circuits would confuse the logic and the game play would go off and do its own thing."

While the 4004 microprocessor lacked the power to display images on a TV, by 1975 Intel had come up with the 8080, a microprocessor capable of controlling the on-screen action of a video game. All Dave Nutting Associates needed now was a video game it could use to prove its plan would work. As luck would have it Bally Midway had just the machine. As part of its relationship with Japanese video game manufacturers Taito, Bally had obtained the North American rights to *Western Gun*, the latest game devised by *Speed Race* creator Tomohiro Nishikado. *Western Gun* pitted two players as Wild West gunmen trying to shoot the other in a showdown and was popular in Japan.

But the game was afflicted with many of the problems that plagued TTL video games and Bally couldn't put the game into production as a result. Bally asked Dave Nutting Associates to redesign the game using Intel's 8080 microprocessor. Using a microprocessor turned the video game development process on its head. No longer would engineers armed with soldering irons build games out of hardware. Instead computer programmers would write the game in software that told the flexible hardware of microprocessors how the hardware should work.

"TTL logic was a hard-wired system, to make a change in game play meant redoing the circuit. Once we established the microprocessor hardware system all game logic was done in software," said Nutting. To help with the programming, Nutting enlisted the help of two student volunteers from the University of Wisconsin's computer science course: Jay Fenton and Tom McHugh. Fenton, a transsexual who became Jamie in early 1990s, was suspicious of getting involved with the amusements business. "I was worried about working for the Mafia. The amusements device industry had a much shoddier reputation back then. It didn't take long for me to realise how silly that stereotype was." McHugh became the main programmer of *Gun Fight*, Dave Nutting Associates' remake of *Western Gun*, with Fenton concentrating on programming the company's pinballs. For Nutting himself, working with programmers was liberating: "I, as the game designer and director, could literally sit with a software programmer like Jay Fenton and mould the game flow. It was like giving me play dough."

By the middle of 1975 *Gun Fight* was ready to go into production. Bally, however, was getting nervous. "RAM was, at that time, expensive," said Nutting. "Marcine 'Iggy' Wolverton, the president of the Midway, asked Jeff Fredriksen and I out to lunch and he appeared nervous. Iggy looked at us and stated 'I hope you guys know what you are doing because I am about to commit to purchasing $3 million of RAM in order to get a good price'. Of course we nodded yes." Bally's RAM order was a major purchase. Nutting estimated it swallowed up around 60 per cent of the memory chips available in the world at the time. Wolverton needn't have worried though. *Gun Fight* became a popular arcade game and soon every video game manufacturer was looking at how they could use microprocessors in their products, Nishikado included: "Quite frankly I thought the play of *Gun Fight* was not really good and in Japan my version of *Western Gun* was better received. But I was very impressed with the use of the microprocessor technology and couldn't wait to learn this skill. I started analysing the game as soon as I could."

The days of TTL video games were finished. One by one the world's video game manufacturers embraced the new world of the microprocessor. 1976 saw the release of the last two significant TTL games: Atari's *Breakout* and *Death Race*, the creation of Exidy – a small coin-op business in Mountain View, California.

Exidy came up with the idea for *Death Race* after licensing its game *Destruction Derby* to the far bigger Chicago Coin, who released it as *Demolition Derby*. Chicago Coin's version destroyed sales of Exidy's original. "We had to do something," said Howell Ivy, one of Exidy's game developers at the time. "Someone jokingly said 'why don't we make a people-chase game?' We had a steering wheel on the game, so let's drive to chase the people." The idea was simple enough that Exidy could easily adapt the design of *Destruction Derby*, saving it the trouble and cost of building a brand new game. The reworked game would, they decided, give players points every time they ran over one of the people and leave a headstone-like cross marking the spot where the person was hit. They named it *Death Race*. "We had no clue that it would cause any controversy," said Ivy. "The game was fun and challenging. There was no underlying motivation or thoughts in creating the first controversial video game. It was created out of necessity and defence of our own product licensing." The media and public, however, didn't agree and *Death Race* provoked the first major moral panic over the content of a video game. "The controversy began with a reporter in Seattle," said Ivy. "The reporter interviewed a mother in an arcade and she said the game was teaching kids to run over and kill people. The story was placed on the Associated Press news wire and then escalated nationwide. The first indications were requests for interviews with us at Exidy."

Exidy's media handling did little to quell the outrage. "If people get a kick out of running down pedestrians, you have to let them do it," Paul Jacobs, the company's director of marketing, told one reporter. Psychologists, journalists and politicians lined up to condemn the game. Dr Gerals Driessen, manager of the National Safety Council's research department, described *Death Race* as part of an "insidious" shift that was seeing people move from watching violence on TV to participating in violence in video games. It was a charge still being levelled at video games more than 30 years later. As the criticism mounted, Exidy hastily concocted a story that it wasn't people being run over, but gremlins and ghouls. The lie fooled no one and soon the controversy began making its way onto national US TV news programmes such as *60 Minutes*. Exidy received dozens of letters about the game. Nearly all condemned *Death Race*. One neatly handwritten letter threatened to bomb Exidy and its facilities. "We did not take this threat lightly, we asked ourselves 'what have we done?'," said Ivy. "The police were called and for several weeks we did have security guards at our facility both day and night. The letter was not signed and the person was never caught or heard of again."

The rest of the video game industry watched the controversy carefully. While several distributors and arcade owners refused to touch *Death Race*, video game manufacturers kept quiet – preferring to see what could be learned from the controversy. The main lesson was that controversy sells. "The height of the controversy lasted for about two months then slowly died as other news stories became more important," said Ivy. "During this time the demand for the game actually increased. We did have customers cancel their order while others increased their orders. The controversy increased the public awareness and demand for the game. Negative as it was, we felt the press coverage did increase the demand for the game and established Exidy as a major provider of video game products at that time."

Around the same time as *Death Race* arrived in a blaze of controversy, Atari was enjoying major success with *Breakout*. *Breakout* came out of another of Nolan Bushnell's attempts to instil a creative working culture at Atari: away days where staff would debate new ideas. "We'll take the engineering team out to resorts on the ocean for a weekend or three days and do what we called brainstorming," said Noah Anglin, a manager in Atari's coin-op division. "Everything went up on the board no matter how crazy the idea was and some of them were really far out." There was only one rule, according to Atari engineer Howard Delman: "Nothing could be criticised, but anyone could elaborate or enhance someone else's idea."

At one of these away days someone proposed *Breakout*, a game that took the bat-and-ball format of *Pong* but challenged players to use the ball to smash bricks. "The idea didn't meet our first clutch of games we were going to work on, but Nolan really liked it," said Steve Bristow, Atari's vice-president of engineering. With Bushnell keen to see *Breakout* put into production, Bristow handed the job of developing the game to Steve Jobs, a young hippy who had taken a technician's job at Atari so he could earn enough money to go backpacking in India in search of spiritual enlightenment. "Jobs always had a sense of his own self-worth that people found a little put off-ish," said Bristow. "He was not allowed to go onto the factory floor because he wouldn't wear shoes. He had these open-toe sandals that workplace inspectors would not allow in an area where there are forklift trucks around and heavy lifting."

Atari expected the game to require dozens of microchips, so to keep costs low Jobs was offered a bonus for every integrated circuit he culled from the game. Jobs asked his friend Steve Wozniak for help, offering to give him half of the bonus payment. Wozniak, a technical genius who worked for the business technology firm Hewlett Packard, agreed. "Wozniak spent his evenings working on a prototype for *Breakout* and he delivered a very compact design," said Bristow. Wozniak slashed the number of integrated circuits in half and netted Jobs a bonus worth several thousand dollars. Jobs, however, told Wozniak he got $700 and gave his friend $350 for his effort. Wozniak would only learn of his friend's deceit after the pair formed Apple Computer.

Atari never used Wozniak's prototype of *Breakout*. The design was too complex to manufacture and the company decided to make some changes to the game after he had worked on it as well. On release *Breakout* became the biggest arcade game of 1976 and the following year was included on the Video Pinball home games console.[1] But the rise of microprocessor-based video games, however, meant *Breakout* would be Atari's last TTL game. Bushnell saw the microprocessor as the natural technology for the video game. "I made the games business happen eight years sooner than it would have happened," said Bushnell. "I think my patents were unique and bizarre enough that it's not for sure that someone else would have come up with something like it, but I'm sure that as soon as microprocessors were ubiquitous somebody would have done a video game system."

The move to microprocessors also required a different set of skills from video

---

1. The Video Pinball console was a home *Pong*-type console featuring seven games, including *Breakout*, *Rebound* and four pinball games. The console was also released as the Sears Pinball Breakaway. In Japan, Epoch released it as the Epoch TV-Block in 1979.

game developers, shifting the focus away from electrical engineers towards computer programmers. "Initially many of the programmers, including me, were also hardware engineers," said Delman. "But after a few years, the two disciplines became distinct." The need for programming skills prompted Atari to embark on a recruitment drive in 1976 to find the people who could make the new generation of video games. One of these recruits was Dave Shepperd, the electrical engineer who had started making video games at home after playing *Computer Space* in the early 1970s.

"By late '75 and early '76, it was clear to Atari the future was in microprocessors. They put an ad in the paper and I happened to see it," said Shepperd. Prior to seeing the advert, Shepperd had begun experimenting with the Altair 8800, one of the very first microprocessor-based home computers. Available in the kit form via mail order, the Altair 8800 was nothing if not basic. Released in 1975 by MITS, it had no video output beyond a number of LED lights and just 256 bytes of memory.[2] It had no keyboard so users had to program it using a bank of switches on the front of the computer.

Despite its user-unfriendliness, thousands of computer hobbyists bought an Altair and set about building hardware and writing software for the system, which was powered by the same microprocessor used in *Gun Fight*. Among them were Paul Allen and Bill Gates who wrote a version of the programming language BASIC for the Altair and formed Microsoft to sell it. Shepperd, meanwhile, was making games for the system. "I designed and built a new video subsystem integrated into the Altair," he said. "I got it working and coded up a few very simple games. Many of my neighbours would come over and we'd play games on it until the early hours in the morning. We'd make up new rules as we went along and I'd just patch them into the code and put them into the computer using only the toggle switches on the front panel and, later, from an adapted old electric typewriter keyboard I found in a dumpster."

With his experience of writing games on a computer, Shepperd landed the Atari job and on Monday 2nd February 1976 he turned up for his first day of work bursting with excitement at the prospect of using the advanced computer equipment he imagined was lurking within Atari. "Atari's cabinets looked real cool. The games were loads of fun. It seemed like the neatest, newest, most interesting thing I could be doing," said Shepperd. "I had been working for a company that made products for IBM and Sperry Univac. The test equipment we had in our labs was pretty high end. The computers I was using were multi-million dollar IBM mainframes housed in very large climate-controlled, raised-floor computer rooms. For reasons I cannot explain,

2. Less than the memory used by an email with no text or subject line.

perhaps because the product was so new, I imagined Atari had engineering labs with even more state-of-the-art development tools and test equipment."

The reality brought Shepperd crashing back to earth: "The development labs were just tiny rooms in an old office building. The computer systems we had to use were third-hand PDP-11s. All of the test equipment we had was old and pretty beat up. It was tough to find an oscilloscope with working probes. The office building had no air conditioning and with all the people and equipment jammed into the tiny office spaces it made the rooms almost unbearable, especially in the summer months. They were operating under a very limited budget and it seemed they were just keeping things together with chewing gum, bailing wire and spit."

Despite the conditions and ropey equipment Shepperd, like the other program-mers who joined Atari at that time, was excited to be making games rather business software. His first project was to make *Flyball*, a simple baseball game that did little to demonstrate the potential of the new era of microprocessor video games. But the second project he was assigned to, *Night Driver*, would ram home the potential of the new technology. Unlike earlier driving games such as *Gran Trak 10* and *Speed Race*, which were viewed from above, *Night Driver* would be viewed from the driver's seat. The idea came from a photocopy of a flyer for another arcade game that Shepperd was briefly shown. "I have no recollection of what words were printed on the paper, so I cannot say what game it was and it could easily have been in a foreign language, German perhaps," said Shepperd. "The game's screen was only partially visible in the picture, but I could see little white boxes which were enough for me to imagine them as roadside reflectors."[3]

Shepperd never got to play the game that inspired *Night Driver*. "The flyer had nothing in the way of describing a game play. At no time did anybody suggest, either inside or outside Atari, how I was to make an actual game out of moving little white boxes around the screen. That I had to dream up on my own." To work out how the game should look, Shepperd opted for learning from first-hand experience. "I remem-ber driving around at various times and various speeds – research, you know – watch-ing what the things on the side of the road appeared to be doing as they passed my peripheral vision," said Shepperd. His solution was to have the white boxes emerging from a flat virtual horizon and growing bigger and further apart as the player's car

3. It remains unclear what the game on the flyer was, but the most likely candidate is *Nürburgring/1*, an arcade video game released in West Germany earlier in 1976 by the company Dr Ing Reiner Forest. Created by the company's founder Reiner Forest and named after the famous German racetrack, *Nürburgring/1* pioneered the driver's perspective viewpoint used in *Night Driver*. Forest later created *Nürburgring/2*, a motorcycle-based driving game, and *Nürburgring/3*, a more advanced ver-sion of the original game. But by the early 1980s, Forest's company had quit the video game business to focus on making driving simulations.

moved towards them. Once the boxes reached the edges of the screen they disappeared. The first time Shepperd got the movement effect working, he and his boss were stunned.

"The little white boxes spilled out from a point I had chosen for a horizon at ever increasing speed. We both sat there mesmerized by the sight," he said. "It was quite cool even though there was no steering or accelerator then. The project leader probably thought we had a winner right then and there but I wasn't sure because at the time I still had no idea of how to score the game. All I knew was it looked really cool."

Once Shepperd got the steering and acceleration working, *Night Driver* seemed destined to be a successful game. "One thing that nearly always was true at Atari, especially in the early days, was if the game was popular amongst the people in the labs, it was probably going to do quite well," he said. "I often had to kick visitors off the prototype in order that I could continue with development. The visitors were not only other engineering folks, but word had spread and I had visitors from marketing and sales and all over all the time."

Night Driver introduced the idea of first-person perspective driving games, which are still widespread today, to a wider audience. The illusion of fast movement the game conjured up also showed just how microprocessors had released video games from the constraints of hardware-based design. But as well as ushering in changes in the arcades, microprocessors were about to alter the nature of home video games as well.

Dungeon master: Richard Garriott, aka Lord British
Courtesy of Richard Garriott

CHAPTER 5

The Biggest Eureka Moment Ever

While the video game set about conquering the arcades of early 1970s, the birthplace of the medium – the computer – remained the preserve of the elite: sealed behind the closed doors of academia, government and business. Yet, unknown to players lining up to spend their loose change on *Pong*, video games were also thriving on the computer. The post-*Spacewar!* generation of computer programmers had picked up the baton of the Tech Model Railroad Club and begun to hone their coding skills by creating games.

Unlike their counterparts in Atari and Bally, these game makers faced none of the commercial pressures of the arcades, where the demand was for simple, attention-grabbing games designed to extract cash from punters' pockets as fast as possible. Their only limitation was the capabilities of the computers they used. While the *Spacewar!* team enjoyed the luxury of a screen, most users were still interacting with computers via teleprinters even as late as the mid-1970s.

The reliance of teleprinters meant the only visuals their games could offer came in the form of text printed out on rolls of paper. "Whatever the machine had to say or display was printed on a narrow roll of newsprint paper that would click up the teleprinter painfully slowly," said Don Daglow, who started making games while studying playwriting at Pomona College in Claremont, California. "We had a terminal that printed at 30 characters per second on paper 80 characters wide. It would print a new line every two seconds. It was so fast it took our breath away. When you've never seen it before it's like magic – speed doesn't enter into it."

This lack of speed, however, ruled out the creation of action games similar to those in the arcades. Instead, computer programmers had little choice but to make turn-based games. The vast majority of these games were incredibly crude. There were countless versions of tic-tac-toe, hangman and roulette, dozens of copies of board games such as *Battleships* and swarms of games that challenged players to guess numbers or words selected at random by the computer. But as the culture of game making spread amongst computer users, programmers began to explore more innovative ideas. Soon players could take part in Wild West shoot outs in *Highnoon*, take command of the USS Enterprise in *Star Trek*, manage virtual cities in *The Sumer Game*, search for monsters that lurked within digital caves in *Hunt the Wumpus* and try to land an Apollo Lunar Module on the moon in *Lunar*. The action in all these games took place turn by turn, with the text describing the outcomes of each player decision pecked out slowly on teleprinters.

Even sport got the text-and-turns treatment, thanks to Daglow's 1971 game *Baseball*. "Simulating things on computers was one of the things people did – if you read about something being done on a computer in a newspaper, very often you'd read they did a simulation of this or that," said Daglow. "Once I understood what the computer could do, the idea of *Baseball* came from there – because baseball is such a mathematical game."[1] *Baseball*'s simulation approach to sport was worlds away from the sports video games of the early arcades, which were, by-and-large, variations on *Pong*.

Other programmers took the concept of simulations even further, pushing at the very limits of what could be regarded as a game. *Eliza* was one such experiment. Written in 1966 by Joseph Weizenbaum, a computer science professor at the Massachusetts Institute of Technology, *Eliza* turned the computer into a virtual psychotherapist that would ask users about their feelings and then use their typed replies to try and create a meaningful conversation.[2] Although it was often unconvincing, *Eliza*'s attempt at letting people to interact with a computer using everyday language fired the imaginations of programmers across the world.

One programmer *Eliza* influenced was Will Crowther. Crowther was a programmer at defence contractor Bolt Beranek and Newman where he helped lay the foundations for the internet by creating data transfer routines for the US military computer network ARPAnet. In 1975 Crowther and his wife Pat divorced and his two daughters went to live with their mother. Crowther worried he was drifting away from his daughters and began searching for a way to connect with them. He homed in on the idea of writing a game for them on his workplace computer.

He based the game on the Bed Quilt Cave, part of the Mammoth-Flint Ridge caves of Kentucky that he and his wife used to explore together. He divided the caves into separate locations and gave each a text description before adding treasure to find, puzzles to solve and roaming monsters to fight. To make the game easy for his children to play he decided that, like *Eliza*, it should let players use everyday English and got the game to recognise a small number of two-word verb-noun commands such as 'go north' or 'get

---

1. So much so that baseball has its own field of statistics called sabermetrics, which Daglow plundered to create *Baseball*.

2. *Eliza* was actually the name of the software Weizenbaum's that ran the code that created his virtual psychotherapist, which was actually called *Doctor*. However most people called it *Eliza* and the real name became near forgotten.

treasure'. Crowther hoped this 'natural language' approach would make the game less intimidating to non-computer users.

The result, *Adventure*, was a giant leap forward for text games. While *Hunt the Wumpus* let people explore a virtual cave and *Highnoon* had described in-game events in text, none had used writing to try and create a world in the mind of players or let them interact with it using plain English. Yet while his daughters loved the game, Crowther thought it was nothing special. After completing *Adventure* in 1976, he left it on the computer system at work and headed to Alaska for a holiday. It could have ended there. Disapproving system administrators regularly deleted games they found to save precious memory space on the computers they managed. Indeed many of the computer games created during the 1960s and 1970s were lost forever thanks to these purges. "Only a small minority actually survived," said Daglow. "The ones that got spread around by Decus – the Digital Equipment Corporation User Society – are the most likely to have survived, in part because a lot of those were reformatted or republished in the earliest computer hobbyist magazines." But while Crowther took in the icy sights of Alaska, his colleagues discovered *Adventure* and began sharing it with other computer users. Soon it began turning up on computer networks in universities and workplaces throughout the world.

In early 1977 *Adventure* arrived at Stanford University where it caught the attention of computer science student Don Woods. "A fellow student who had an account on the medical school's time-sharing computer had discovered a copy in the 'games' folder there and described it briefly," said Woods. "I was intrigued and got him to transfer a copy to the artificial intelligence lab's computer where I had an account. It was definitely different from other computer games of the time. Some computer games included the element of exploration, but they were generally abstract and limited 'worlds' such as the 20 randomly connected rooms of *Hunt the Wumpus*. The descriptions in Crowther's game really drew me into it and the various puzzles hooked me." At the time it was common for the programmers to enhance or make alterations to games made by other people, after all no one had any expectation of making money from their creations. Woods believed he could improve *Adventure* and, after getting Crowther's blessing, began to reprogram it.

He changed the layout of the caves, added new puzzles to solve and made the dwarves of the original roam the caves at random rather than follow pre-defined routes. Woods' roommate Robert Pariseau also contributed ideas for enhancements, one of which was to make the maze of indistinguishable caverns created by Crowther even harder to solve. By April 1977 Woods had completed his alterations. He made the new version of *Adventure* available for others to play and copy, and went away for the university's spring break. Woods was in for a surprise when he returned. "I was told the lab computer had been overloaded due to people connecting from all over to play *Adventure*," said Woods. The new version of *Adventure* generated even more interest than Crowther's original and inspired others to write their own 'text adventures'.

Among them were a four members of the Dynamic Modelling group at the Massachusetts Institute of Technology's computer science lab: Tim Anderson; Marc Blank; Bruce Daniels; and Dave Lebling. "We decided to write a follow on to *Adventure* because we were simultaneously entranced and captivated by *Adventure* and annoyed at how hard it was to guess the right word to use, how few objects that were mentioned in the text could actually be referenced and how many things we wanted to say to the game that it couldn't understand," said Lebling. "We also wanted to see if we could do one; this is a typical reaction to a chunk of new code or a new idea if you are a software person."

Lebling had been making games for some time when the four decided to create their own take on *Adventure*, which they gave the work-in-progress title of *Zork!*. Unlike some of his computer game making peers he, as a student at MIT, had access to the one of the more cutting-edge systems of the era: the Imlac PDS-1. The Imlac could not only do graphics on their built-in displays, but was one of the first computers that offered a Windows-style interface although it used a light pen instead of a mouse and required users to press a foot pedal to click. After making an enhanced version of *Spacewar!* and creating a graphical version of the previously text-only *Hunt the Wumpus*, Lebling started helping fellow MIT programmer Greg Thompson update a game called *Maze*.

Steve Colley and Howard Palmer, two programmers at NASA Ames Research Centre in California, created *Maze* in 1973 on an Imlac. It took full advantage of the Imlac's visual capabilities to create a 3D maze viewed from a first-person perspective that players had to escape from. Later Palmer and Thompson, who worked at NASA Ames at the time, changed the game so that two Imlacs could be linked together and players – represented by floating eyeballs – could move around the maze trying to shoot each other. When Thompson ended up leaving NASA Ames to join MIT's Dynamic Modelling team in early 1974, he brought *Maze* with him. "*Maze* was based on a graphical maze-running game, Greg had brought from NASA Ames. We decided it would be much more fun if multiple people could play it and shoot each other," Lebling said. The pair reworked *Maze* again so that up to eight people could play it at once. They created computer-controlled 'robot' players to make up the numbers when there weren't enough real players and let players send each other text messages during while playing. Lebling and Thompson's 1974 update of *Maze* pre-dated the online player versus player 'death matches' of first-person shooters, which would come to dominate the video games after the success of *Doom*, by nearly 20 years. "We actually played it a few times with colleagues on the West Coast, though ARPAnet was rather slow and the lag was horrible. *Maze* became so popular that the management of our group tried to suppress it," said Lebling.

In stark contrast to *Maze*, however, *Zork!* – Lebling, Blank, Daniels and Anderson's attempt to outdo *Adventure* – was a text only. To outshine *Adventure*, the quartet invented a new fantasy world to explore and improved on the writing of *Adventure* seeking to give *Zork!* a more literary feel. The four also reworked the way the computer read players' instructions so that people could use more complex sentences, such as 'pick up the axe and chop down the tree' rather than two-word commands. After completing the game they renamed it *Dungeon*. It wasn't long before the lawyers of TSR, the makers of the pen-and-paper role-playing game *Dungeons & Dragons*, came knocking.

* * *

Dungeons & Dragons – a fusion of tabletop war games, J.R.R. Tolkien's *The Lord of the Rings* books and amateur dramatics – had become a phenomenon since its 1974 launch, recruiting millions of fans who spent hours acting out adventures based on complex statistical rules and rolls of polyhedral dice. Being a *Dungeons & Dragons* player required serious commitment, even if you were not the dungeon master – the player who had the job of designing the quest, running the game and handling the numerous probability equations that decide the outcomes of player decisions. Games of *Dungeons & Dragons* could take weeks with each play session lasting hours. Much of this would be taken up with debates about the calculations that accompanied the actions of players, said Richard Garriott, who joined the legions of *Dungeons & Dragons* fans in 1977 aged 17. "When you watch most people play paper *Dungeons & Dragons* they would sit down and go I've got a +3 sword, I'm standing behind you and I surprised you so I have initiative – that gives me +2," said Garriott. "They go through this amazingly detailed argument about what the probability of a hit or miss should be. Finally, when they resolve that after five to 10 minutes, they roll a die and go 'look I hit' or 'oops I missed' and then they would start the argument all over. So the frequency of a turn of play is stunningly low."

The amount of number crunching and frustration involved made *Dungeons & Dragons* perfect for computerisation. "It was so well suited to simulate on a computer," said Daglow, who in 1975 created *Dungeon*, one of the earliest computer role-playing games, after getting fed up with the difficulty of getting players together for a game of *Dungeons & Dragons*.[3] "With *Dungeons & Dragons* a lot of the things that were most frustrating on paper and time consuming, the computer does all that for you." *Dungeon* gave Daglow

---

3. Daglow was not the first person to think of transferring TSR's game to a computer. Rusty Rutherford's *Pedit5*, the earliest known computer role-playing game, appeared on the PLATO computer network in 1974, shortly after *Dungeons & Dragons* was launched. *Pedit5's* unremarkable name was an attempt to avoid deletion by fooling system administrators into thinking it was a serious program rather than a game. The disguise failed and the pioneering game was erased forever without a second thought. It did, however, survive long enough to inspire other PLATO users to develop more role-playing games.

the chance to make a game for the new computer monitor terminals that were arriving at Stanford at the time rather than for a teleprinter. These terminals, however, could only display monochrome text and it could take up to 20 to 30 seconds for the screen to change. But the screen allowed Daglow to give his game some visuals in the form of a map composed of punctuation marks and mathematical symbols. It was an approach many subsequent video games, particularly role-playing games, would revisit time and time again in the late 1970s and early 1980s.

TSR had paid no attention to Daglow's game or many of the other computer role-playing games that copied *Dungeons & Dragons*, but when it got wind of MIT's *Dungeon* it decided to send in the lawyers. "TSR had a trademark on the word 'dungeon', which they decided to defend," said Lebling. "MIT's lawyers told them at great length that they were being silly, but we decided to change the name back to *Zork!* anyway as it was more distinctive and unusual." By this time the idea of having an unusual name had grown in appeal since the game's creators were thinking about forming a software publishing business to cash in on the latest by-product of the microprocessor: the home computer.

Kit computers such as the Altair 8800 and KIM-1 had already brought the idea of home computers closer to reality, but as more advanced microprocessors came onto the market in the second half of the 1970s the vision really began to gather momentum. Soon pioneering companies and technologically minded entrepreneurs were investigating the idea of creating computers small and cheap enough that everyone could own one.

One of the first people to really push the idea forward was Steve Wozniak. After completing his prototype of Atari's coin-op game *Breakout*, he decided to make his own computer. He spent his evenings and weekends building the Apple I, a microprocessor-based computer that could connect to a keyboard and a home TV. He showed the proto-type to his friend Steve Jobs, who had just returned from his trip to India. Jobs suggested they form a company to sell it to other computer enthusiasts and on 1st April 1976 they formed Apple Computer. The company produced more than 150 hand-made Apple Is but, by the time it went on sale in the summer of 1976, Wozniak was already close to completing work on a better computer that could appeal to a wider audience: the Apple II. Wozniak set himself the goal of designing a computer powerful enough to allow people to create state-of-the-art video games.

It would, he decided, have colour graphics, proper sound and connections for game controllers and plug into home TVs. In particular he wanted it to be good enough to run a version of *Breakout* created in BASIC – a slow but relatively easy programming language. It was a wildly ambitious goal. Home computers were still a new concept and the idea that he could make one that could run arcade video games and still have a price tag acceptable to the general public seemed crazy. But by August 1976 he, almost to his own amazement, had created just that. In his biography *iWoz*, Wozniak described getting *Breakout* running on his computer as "the biggest, earth-shaking, Eureka moment ever". Being the canny businessman he was, Jobs saw that the Apple II was a machine that would appeal to more than just technically minded computer geeks and started searching for an investor who could help put it on the shop shelves throughout the US.

Apple's first port of call was Chuck Peddle, an engineer at Commodore Business Machines. Jack Tramiel, a Polish immigrant who had survived the Nazis' Auschwitz concentration camp, formed Commodore in 1955 as a typewriter repair shop in the Bronx, New York City, and built it into a leading manufacturer of office equipment. For Peddle, the call from Jobs was well timed. Commodore had recently bought microprocessor manufacturer MOS Technologies, the maker of the KIM-1, and Peddle was trying to persuade Tramiel to forget about pocket calculators and get into home computers. Peddle arranged for Jobs and Wozniak to show the Apple II to Commodore's board. Impressed, the board asked how much they wanted for it. Jobs demanded several hundred thousand dollars and the pair were promptly shown the door. Commodore decided it would make its first home computer itself instead. Undeterred, Jobs and Wozniak decided to see if Atari would back them. "The decision Nolan Bushnell and Joe Keenan came up with was that this was outside our area but we have this investor on our board – Don Valentine – and we'll put you in touch with Don," said Steve Bristow, Atari's vice-president of engineering. Valentine also declined to invest, but arranged a meeting between Apple and Mike Markkula, a 30-year-old who had just left Intel having made his fortune working for the firm. Markkula was convinced the Apple II would be a success and provided the funds Apple needed to start manufacturing the computer and his business expertise.

By the time the Apple II finally started rolling off the production line, however, Commodore had already got its home computer on the market. The $599 Commodore PET was an all-in-one system that fused keyboard, monitor, tape cassette player and computer together in curvy beige plastic. Despite its monochrome visuals, the PET attracted $3 million of pre-orders - enough to make it an instant success. Apple also faced competition from Tandy, the owners of electronics retailer Radio Shack, which had released another monochrome home computer: the TRS-80. As the smallest of the three companies, Apple could easily have struggled, but Wozniak's video game-inspired inclusion of colour graphics and the company's clever marketing gave it the edge. By 1981 the Apple II had claimed 23 per cent of the US home computer market compared to Tandy's 16 per cent and Commodore's 10 per cent.

The arrival of the Apple II, TRS-80 and PET brought a swift end to days when computers were only found in large institutions. Now anyone could potentially have a computer in their home. But while most people agreed computers were the future, few had any idea what households would do with them. Would they calculate their tax returns or catalogue record collections? Would they teach their children to program the machines in the hope that they would have the skills that would be needed in the workplace of the future? Or would they store family recipes or address books on a cassette tape?

It turned out that early home computers would be used almost exclusively for one purpose alone: playing video games. And many of the games they played were versions of those once locked away on the computers of academia, government and business. These games first started to migrate into the home through magazines and books that contained listings of computer programs for people to type in line by line. Then these games began to be sold in stores. Computer Chess, arguably the original video game, was among the first to go on sale thanks to a Canadian company called Micro-Ware, which released *Microchess* on the KIM-1 in 1976. Other forms of computer game quickly followed, among them educational titles such as *The Oregon Trail* – a 1971 game developed by three student teachers to teach elementary school children in Minnesota about the life and trials faced by the settlers who led the US's western expansion in the mid-1800s. It became a staple of classrooms across the US in the 1980s and early 1990s. But one of the most popular forms of computer game to reach the home was the text adventure.

Scott Adams, a computer programmer from Florida, brought the text adventure to the home after hearing work colleagues discussing *Adventure* while working at telecommunications firm Stromberg-Carlson. "I came in early and stayed late for a week and played it. I was hooked on the concept, it was great fun," he said. Adams had already made a game on his TRS-80 computer that he was selling through a local Radio Shack store. "It was a dog racing game, with a random number generator and some text, that had you betting on which dog would finish first," he said. "The game was a real dog itself. I sold maybe 10 copies. It was junk."

Unsurprisingly, Adams felt an *Adventure*-type game might be more popular and set about making a similar game. His programmer pals thought he was wasting his time. "I was told it would be impossible to make anything like *Adventure* fit into a computer with 16k of memory space," he said. His sceptical programmer friends had a point; *Adventure* took up 256k of memory, far more than the TRS-80 could cope with. But Adams figured out a number of memory saving tricks that allowed him to squash his game, *Adventure-land*, onto the TRS-80, such as getting the computer to recognise the players' commands from the first three letters alone. *Adventureland* played much like *Adventure* although the story was set outdoors rather than within underground caves. Adams did, however, drop the idea of fighting monsters and concentrated on the puzzle solving after objections from some of his friends. "In the very first version of *Adventureland* you ended up killing the bear after it fell off the ledge," he said. "One of my friends said that was too harsh and could I change it? I did and thereafter all my games were more orientated towards full family fun."

For a game-playing public used to action-based arcade games, 1978's *Adventureland* was an unusual and exciting concept. But while it eventually became a popular game, it took Adams some time to get it into shops. "There were very few companies making home computer software and even fewer selling games," he said. "I started small with an ad in a computer magazine. I remember my first large order. It was from Manny Garcia who ran a Radio Shack in Chicago and he ordered 50 tapes. At the time I had no idea about wholesale-retail and he had to explain the concepts. It took a week to make all the tapes and send them to him. When he got them he called back and asked where was the packaging?"

Adams was not alone. Across the US, business-naive computer enthusiasts were beginning to write games they hoped to sell to the growing ranks of home computer owners. Few had any idea they were building an industry. They copied their games onto cassette tapes or 5.25-inch floppy disks on their own computers. They drove or posted their games to shops, photocopied instructions and packaged their work in Ziploc bags that were more commonly used to keep sandwiches fresh. The shortage of games, however, meant many of these game makers started earning significant sums from their work. Bill Budge, a student at Berkeley University in California, was one. He started out by writing a bunch of simple games, including a copy of *Pong*, on his Apple II. After selling Apple the rights to three of his games, which got released in 1979 as *Penny Arcade*, in return for a $700 printer, he started selling his work to Stoneware, a small game publisher run by Barney Stone. "Barney said I think I can sell these games in computer stores, which were springing up all over the place," said Budge. "I remember my family went on vacation to Hawaii and I was so interested in writing these games that I decided not to go. I just stayed with my Apple and programmed for two weeks solid with nobody to bother me. Then he turned up one day with a cheque for $7,000 – my monthly royalties."

On the other side of the US, the creators of *Zork!* had also joined the fledgling game business by forming Infocom. "There was no plan to make games the focus, but after casting about for product ideas, Marc Blank and co-founder Joel Berez suggested *Zork!* might be a good choice to get us going," said Lebling. Like Adams' IT friends, Infocom worried that getting a huge game like *Zork!* onto a home computer was impossible. "There were lots of objections," said Lebling. "Microcomputer memories were really, really small and *Zork!* was huge. We weren't sure it was possible." Despite the reservations, Infocom gave it a shot and, after chopping up the original into three separate games, managed to squash the game on the primitive home computers of the day. While *Adventureland* had introduced computer owners to the concept of the text adventure, *Zork!*'s recognition of proper sentences and detailed descriptions were a marked improvement. The first game in the *Zork!* trilogy sold hundreds of thousands of copies across various computer formats and turned Infocom into one of the biggest names in computer gaming.

Around the same time as the home computer versions of *Zork!* were released in 1980, Ken and Roberta Williams – a husband and wife from Los Angeles, California – took the idea of the text adventure in a new direction with their debut game *Mystery House*. The Williams' leap into the nascent video game business began when Ken, a freelance computer programmer, introduced Roberta to *Adventure*. "I showed it to Roberta and she grabbed the keyboard and played it all night. She was addicted. When she finished the game, she wanted me to program a similar game that she would design," said Ken. Roberta saw the text adventure as an exciting new way of storytelling and set about designing a murder mystery inspired by the board game *Cluedo* and Agatha Christie's 1939 best-selling novel *And Then There Were None*.

She drew out the game's locations and plot twists on the back of large sheets of wrapping paper while Ken set about turning her ideas into a working game on their Apple II. Unlike Adams and Infocom, Roberta decided that text alone would not do her game justice and insisted Ken allowed her to include black and white line drawings that illustrated each location alongside the text, despite the memory limitations of the Apple II. This refusal to bend to the technology at a time when most game makers built their creations around their programming skills would come to define Roberta's approach to game design. "I always thought of the story, characters and game world," she said. "I needed to understand those before I could even think about any game framework, engine or interface. The game engine was built around my ideas, not the other way around."

The pair released the game through Ken's company On-Line Systems and turned their kitchen table into a makeshift factory floor where their Apple II produced copy after copy of the game. Each copy was packaged in a Ziploc bag with a photocopied set of instructions. They then called every computer store they could find to ask them to stock the game. "There were literally only about eight places that sold software. It was easy to call them and there was no software available, so they were thrilled to hear from us," said Ken. They sold more than 3,000 copies of the $24.95 game in just six months and soon had enough money to turn their game making into a full-time business and move out of the Los Angeles sprawl to the outskirts of the Sierra Nevada Mountains. They later renamed their company Sierra Online in honour of their new home. Their second game, 1981's fairytale-themed *The Wizard and the Princess*, took the idea of illustrated text adventures a step further by including full colour visuals. Sierra's use of graphics provoked very different reactions from Infocom and Adams' company Adventure International. Adams eventually followed Sierra's example and start adding visuals to try and compete. Infocom, however, went to the other extreme and sought to make its reliance on text a virtue, running adverts that declared, "we unleash the world's most powerful graphics technology" next to an illustration of a big glowing brain.

Text adventures, however, were not the only games making a splash with home computer users. Flight simulators also made the transition. Flight simulations had always lived a double life somewhere between training and entertainment. Edwin Link Jr, a pipe organ maker from Binghamton in New York state, created the first flight sim, the Link Trainer, in 1929. The Link Trainer consisted of a cockpit perched on a moveable platform and used motors, organ bellows and recorded sound effects to mimic the experience and sensation of flying a plane. Link originally envisaged it as a coin-operated carnival ride that might also be used to teach would-be pilots the basics of flying before they took to the skies. His 1931 patent for the machine described it as a "combination training device for student aviators and entertainment apparatus".

The outbreak of the Second World War, however, saw its use as a training tool come to the fore after the US Air Force ordered more than 10,000 Link Trainers. Over the course of the war it would be used to deliver basic training to more than 500,000 pilots. The flight sim came on in leaps and bounds after the war as the growth of commercial aviation and the arms race of the Cold War fuelled investment in more advanced simulators. By the start of the 1960s, flight simulators had moveable cameras that scanned over model landscapes in line with the users' controls to replicate the visual experience of flying. Despite these improvements, the increasing complexity of aircraft meant that these mechanical simulators were struggling to replicate the experience in a way that was useful for training. So when computers with visual displays started becoming a realistic option in the late 1960s, the flight sim transferred into the digital realm. This transition not only improved the effectiveness of flight simulators but also allowed amateur and would-be pilots with computer access to use them. One of these users was Bruce Artwick, a physics student and pilot. When the first home computers arrived Artwick believed other amateur pilots would jump at the chance to have a flight sim in their own home. He formed his own software company SubLogic and wrote *Flight Simulator*, the first home computer flight sim, which debuted on the Apple II in early 1978. *Flight Simulator* sought to replicate reality as closely as the Apple II could, using real-life physics and offering a wide range of planes, from crop dusters through to fighter jets, to fly. The popularity of Artwick's creation would inspire others to produce more home computer flight sims, which quickly divided into military and civilian aviation, and simulations of other vehicles including submarines, space shuttles, helicopters and tanks.

Recreations of tabletop war games were another regular sight in the early days of home computers. As with *Dungeons & Dragons*, the motivation behind transferring these to computers was mathematical. Tabletop war games had evolved out of Kriegsspiel, a game created for the Prussian army in the 18th century as a military training aid for its officers. Kriegsspiel became a national obsession. Sets with detailed figurines of soldiers were sent to every military division, the Kaiser attended tournaments and the original 60-page rulebook was later enhanced with data from real conflicts. When Prussia won the Six Weeks War against Austria in 1866 and defeated France in 1870's Franco-Prussian War, the country thanked Kriegsspiel for its victories. Impressed, rival nations quickly adopted the game including Japan, which credited its success in the Russo-Japanese War of 1905 to Kriegsspiel. Families across Europe also began playing the game using toy soldiers.

Such was the craze that in 1913 science fiction writer H.G. Wells wrote *Little Wars*, a rulebook for toy soldiers that is sometimes credited as the basis of the modern tabletop war game. The craze faded from military prominence after Germany's defeat in the First World War, but for a dedicated core of fans it never went away and there was still a loyal hobbyist following in the 1970s. But like the players of *Dungeons & Dragons*, itself the creation of three war game designers, war gamers had to grapple with lengthy games where huge amounts of time were spent calculating complex equations that decided the outcomes of the battles they replicated. It didn't take long for war game fans to realise that computers could make their lives a lot easier and by the early 1970s simple war games such as *Civil War*, a recreation of the American Civil War, were appearing.

Home computers encouraged further growth in computerised war games, but few sought to do anything more than recreate the tabletop experience. "They were pretty grim," said Chris Crawford, a war gamer who started making video games in the late 1970s on university computers. "Most commercial war games were written in BASIC and relied on a conventional board for the placement of the pieces." Crawford's answer to the lack of vision exercised by the early war game creators was *Tanktics*, a tank versus tank war game he created in 1977 on a IBM 1130 computer at his workplace - the University of California. "I was playing board war games and I was acutely aware of the absence of the fog of war, which I consider to be crucial to simulation of warfare," he said.[4] "I considered that computers could solve the problem. I don't think people fully appreciated just how big a leap this was. Most had become accustomed to the absence of fog of war and took full knowledge for granted. They didn't like the idea of fog of war." Crawford took his ideas further with *Eastern Front 1941*, which he wrote in 1981 after joining Atari's

---

4. Fog of war is a military term that refers to the level of uncertainty in a battle. For example not knowing the location of your enemy or the amount of supplies their capabilities. Tabletop war games tended not to simulate fog of war due to the difficulty of keeping the other players' moves secret.

home computer division.[5] *Eastern Front 1941* introduced the idea of real-time conflict into the war game. Tabletop war games were turn based and most computer war games had blindly followed suit. Crawford realised that on a computer players could make their decisions, but the actions themselves did not need to happen immediately. Instead the game could wait until all the decisions were made and then carry out each player's move at the same time, replicating the real-time nature of war.

The last major genre to make the leap to the home was the role-playing game and leading the way was Richard Garriott, a teenager from Houston, Texas. 1977 was a pivotal year for Garriott. Those 12 months saw all the ingredients that would make Garriott one of the world's most recognised game designers come together. "It happened in fairly quick succession," he said. "First my sister-in-law gave me a copy of *The Lord of the Rings* and right after I read the book, in the summer of 1977, I took a seven-week summer course for high-school students at the University of Oklahoma in things like computer programming and mathematics and statistics. When I arrived there, all the other students had not only read *The Lord of the Rings* but they were all playing this game, *Dungeons & Dragons*, which became our evening activity. We also had access to some of the early computers that were around at universities before they were available in high school."

On arrival at the university, he was greeted by a group of students who mistook his lack of a southern US accent for an English one and nicknamed him British. Garriott embraced the nickname and eventually named his *Dungeons & Dragons* alter ego Lord British. Inspired by his trio of discoveries – Tolkien, *Dungeons & Dragons* and computers – Garriott returned to Houston and began writing games on the primitive teleprinter computer at his school. "I began to write games I used to call *D&D1*, *D&D2*, *D&D3*, etc, in homage, of course, to *Dungeons & Dragons*," he said. "Because it was very hard to create this software on a Teletype, you generally wrote out every line of the program on paper first." Garriott's father Owen, a NASA astronaut, noticed the program his son was working on and warned him what he was trying to do might be too ambitious. "It was a huge program compared to what anybody else bothered to write back in those days. It must have been a whopping 1,000 lines of code or something – it pales today but at the time seemed huge," he said. "My dad said 'Richard, there's a pretty low probability that you'll get that to work because it's going to be so complicated'. I said well I'm pretty motivated to pull this off, so he said I'll make you a bet. The bet was if I could get *D&D1* working pretty much straight away he would cover half the price of a personal computer, right as the Apple II was coming out."

---

5. Having turned down the Apple II, Atari entered the home computer business in 1979 with the Atari 400 and Atari 800 computers.

The bet spurred Garriott on and he managed to get the game working on the PDP-11 computer that his school's Teletype was connected to. Even though *D&D1* was the first game Garriott had written, the beginnings of *Ultima*, the video game series that would propel Garriott to fame and fortune, were already evident. "Even though it was printed on paper it still looked a lot like *Ultima*," said Garriott. "It would print out a little asterisk for walls, spaces for corridors, power sign for treasure on a 10x10 character map. You would say if you wanted to move north, south, east or west and you would wait about 10 seconds for it to print out the new 10x10 map and off you would go fighting monsters and finding treasure."

Garriott's dad stayed true to his word and stumped up the cash to help his son buy an Apple II. By the time the computer arrived, Garriott had already produced numerous new versions of his game and was up to *D&D28*. On getting his Apple II, Garriott started work on *D&D28b*, which he soon renamed *Alakabeth: World of Doom*, with the goal of adding graphics to his previously text-only game. He came up with the idea of giving players a first-person view of the dungeons and the monsters within after playing another Apple II game called *Escape*: "In *Escape* you saw a top-down viewpoint screen, watched a maze being generated and then it just dropped you in the middle of this 3D maze and you had to walk out of it."

Garriott, however, had no intention of selling his game: "It was really written for myself and for my friends to play. We were playing *Dungeons & Dragons* in the evenings and I'd set up the computer nearby so people could play the game." In the summer of 1979, having completed *Alakabeth*, Garriott got a summer job as an assistant at the Computer-Land store in Clearlake City, Texas. One evening after work he decided to play his game and loaded it onto one of the store's Apple IIs. The store's manager John Mayer noticed it immediately. "He said: 'Richard, this game is far better than any game we sell here, you should seriously think about distributing it'," said Garriott. Mayer agreed to stock the game, so Garriott spent $200 on Ziploc bags, floppy disks and photocopied instructions so he could produce copies for the shop to sell. "I thought it was a huge amount of money," he said. One of the copies of *Alakabeth* that Garriott produced for the Compu-terLand store ended up in the hands of California Pacific Computer, one of the largest software distributors in the US at the time. "They called me on the phone, sent me tickets to fly to California so I could sign contracts and they agreed to pay me $5 per unit that was sold," said Garriott. California Pacific also hit on the idea of using Garriott's Lord British character, which he had included in the credits of *Alakabeth* alongside his real name, to help market the game. "They said: 'You know Richard Garriott is a perfectly fine name, but not nearly as memorable as Lord British would be. So why don't we just drop Richard Garriott from the credits'," said Garriott, who gave the marketing ploy the go-ahead.

The appeal of a game based on *Dungeons & Dragons*, a new concept to home computer owners, was huge. The game sold 30,000 copies in total, earning Garriott $150,000 – considerably more than his space-travelling father did in a year. "I was still in high school, so I didn't really conceive how much money that was," said Garriott. "I was just kinda doing my thing. But it was enough money for friends and family to notice and it became obvious that I should do this again with an eye to making a game I intended to be seen by the consumer." His 1980 follow-up to *Alakabeth*, *Ultima: The First Age of Darkness* became an even bigger success, selling around 50,000 copies, but Garriott soon had competition.

In 1981 a company called Sir Tech released a rival role-playing game called *Wizardry: Proving Grounds of the Mad Overlord* that offered better graphics and had players leading a party of adventurers rather than the lone hero of Garriott's games. It outsold Garriott's game by more than two to one, and soon the competition between role-playing game makers became intense as they tried to outdo each other with new features. In 1982's *Ultima II: The Revenge of the Enchantress*, Garriott introduced the idea of letting players talk to, as well as fight, computer-controlled characters. That same year Texas Instruments replaced the black and white line drawings that *Wizardry* and *Ultima* used for their dungeons with solid colour tunnels in *Tunnels of Doom*. *Wizardry II: The Knight of Diamonds*, also released in 1982, gave players the option to import their characters from the first game. By the time *Ultima III: Exodus* arrived in 1983, home computer game publishing was starting to look like a proper business. The number of computer owners had grown massively and so had the number of games being released.

The home computer had freed the games of computer researchers from the networks of academia and allowed them to enrich the range of video games as an entertainment form. Yet as the 1980s dawned, no-one was paying much attention to the games packaged in Ziploc bags on computer store shelves as they were too busy looking at the arcades and the new generation of game consoles that were about to send the US video game crazy.

Invasion of the coin snatchers: A Space Invaders contest
Courtesy of Funspot

CHAPTER 6

High-Strung Prima Donnas

Stella was a bicycle. The bicycle owned by Atari engineer Joe Decuir to be precise. It was also the codename of what was Atari's most important project at the start of 1976: a new form of home game console.

Unlike the home *Pong* console that took the world by storm in Christmas 1975, this new console would harness the flexibility of the microprocessor to allow it to play a whole range of games. And instead of these games being built into the machine, they would be sold separately on memory chips housed in plastic cartridges that could be plugged in whenever people wanted to play a different game. Atari figured that such a system would be a great money spinner even if they sold it for little or no profit, because once people bought the console, they would be more likely to buy game cartridges rather than a new machine when they tired of the games they owned. And since these cartridges cost a few dollars to make but could be sold for $30, Atari stood to make enormous profits if it could get enough of its consoles, which it later named the VCS 2600, into people's homes. But progress on Stella was slow, hampered by Atari's tight finances. Home *Pong* may have made Atari millions, but these profits were rapidly eaten up by expansion and the development of new products. "One of Atari's weaknesses was it was primarily self-financed," said Steve Bristow, the company's vice-president of engineering. "Most of the money it made came from selling products, which it then spent making the next product. There wasn't a pool of capital. There were times when it was practically cheque-to-cheque and there would be a race out of the parking lot to make sure your pay cheque would get cashed."

Atari founder Nolan Bushnell was well aware of the problem: "We were getting ready to do the 2600, which required a lot of cash and we just didn't have a lot of cash."

The company toyed with the idea of floating on the stock market, but backtracked after deciding that the depressed market wouldn't bring in the kind of money needed. Then, as it started examining alternatives, Atari received the news it feared most: another company had come up with exactly the same idea and was close to launching its system.

That company was Fairchild Semiconductor, a Silicon Valley electronics component manufacturer had intended to use its new F8 microprocessor as the basis of its console: the Fairchild Channel F. Fairchild had given the task of creating the Channel F to Jerry Lawson, an African-American field applications engineer for the firm who had already made a microprocessor-based coin-op video game.

"Fairchild wanted to start a new concept – field application engineers, people who were able to help its customers with design in the field. I was hired as the first field application engineer," said Lawson. "At that time the microprocessor came and reared its ugly head. I said the microprocessor is a great tool to be used for displays, but everyone said: 'Oh no, it's too slow'. To prove my case I came up with a design concept to build a video game."

Using Fairchild's F8 microprocessor, which launched in 1975, Lawson created *Demolition Derby*, an overhead view driving game that foreshadowed the action of the non-microprocessor coin-op *Death Race*. "You drove cars around the track and the cars crashed into each other. It was later changed to be called *Death Race*. Instead of hitting cars you hit people and a tombstone popped up. That's how grotesque people get to be when they want to."

Lawson sold his game to a company called Major Manufacturing who tested it out on the customers of a pizza parlour in Campbell, California.[1] "Fairchild heard I was doing it and said 'look we have a concept: we'd like to go into games ourselves'," said Lawson. Like many other microchip manufacturers in the 1970s, Fairchild had decided to move into the consumer electronics market. "The semiconductor industry would put more and more into an integrated circuit and, when you do that, you get to a point where the only thing that's left to do is put power into it," said Lawson. "They finally went 'to heck with this'. Why should they do all this engineering, all this development, so someone else can turn around and put it in a case?"

Lawson became the director of Fairchild's new video game division and set about designing a console based on the F8 microprocessor. The idea of a home games console that could play different games wasn't new. The Magnavox Odyssey, the first home video game machine, let people play different games by inserting different circuit boards. But the technology in Odyssey was nearly five years old when it was released in 1972 and required players to attach overlays to their TV screens to enhance the limited graphics. Lawson's colour video game console needed no such add-ons.

Fairchild launched the $169.95 Channel F in August 1976 with two *Pong*-type games

---

1. Lawson's *Demolition Derby* game didn't get much further than Campbell as Major Manufacturing closed down shortly after it was installed in the pizzeria.

built in and a selection of $19.95 game cartridges to buy separately, including a *Tank* clone called *Desert Fox* and the self-explanatory *Video Blackjack*. The idea of interchangeable game cartridges, or videocarts as Fairchild called them, took some explaining to a public still getting used to the concept of home video games.

"There's an event that takes place the day after Christmas that the retailers call 'Hell Day'," said Lawson. "That's when people bring back all these wonderful gifts that they can't do what they want to do with or don't understand. I made the mistake of coming in the day after Christmas 1976 and our marketing department were on vacation. The phones started ringing. The security guard, he's getting calls left and right."

Soon Lawson's phone was jammed with people struggling to operate their Channel F. "People thought it was an eight-track tape player but an eight-track tape wouldn't fit, so they tried to jam it in there," said Lawson.[2] "One woman called up about grandpa's teeth being stuck in the cartridge chute. Another said: 'Urine hurt the game'. Rover had lifted his leg and peed on it. One guy called me and said: 'I can't find the batteries'. I went: 'It plugs into the wall. What do you mean batteries?' He'd taken it apart because he couldn't find the batteries. By the end of the day, I was so frustrated that a woman called up and said 'My game hums. Do you know why?' and I said: 'Because it don't know the words, lady'."[3]

Fairchild did not have the market to itself for long, however. Shortly after the Channel F reached the shops, TV set manufacturer RCA announced it was going to release its own cartridge-based console in early 1977: the RCA Studio II. Atari knew it was only a matter of time before other rivals appeared. Suddenly the company that had always led the way on video games faced being left behind. With the clock ticking, Bushnell and Atari president Joe Keenan put Atari up for sale, hoping to find a large parent company that could bankroll the 2600.

The news that Atari was up for sale soon reached the New York City headquarters of Warner Communications, a large entertainment conglomerate that had a strong presence in the music, film and comics business. "I got a phone call from one of our large institutional investors," said Manny Gerard, the Warner executive in charge of new acquisitions at the time. "He said something like 'would you be interested in a technology-based entertainment company?'. The words sounded right so I said yes. I didn't know what I was saying yes to, but the description fitted us. Next thing I know I'm going to Los Gatos,

2. Eight-track tape cartridges were a popular music format in North America during the 1960s and 1970s, particularly in car stereos.

3. Customers were just as bewildered a more than a year later when Atari launched its console. "Atari had a very good attract mode to attract you to the game," said Lawson. "People used to play the attract mode and not realise they weren't playing the machine."

California, and I'm taken to this company called Atari." During his visit to Atari, Gerard got a glimpse of Stella, the 2600 prototype. "I wrote an internal memo saying 'I have seen the future and it is called Stella'," he said. "When I saw it – a programmable video game – I said 'oh yeah, this is a big deal'."

Warner decided to buy Atari and in October 1976 paid $28 million for the company, turning Bushnell into a multi-millionaire in the process. Bushnell was excited by the deal. Warner were prepared to spend the millions needed to get the 2600 into people's homes and the daily cash flow battles that he had been fighting ever since forming Atari were finally over. On top of that his initial worries about the competition were easing.

"When I first heard Fairchild was doing this thing it scared me. Then when I saw what they had done I thought these guys are clueless," said Bushnell. Trip Hawkins, a young Harvard graduate hired by Fairchild to carry out market research on the Channel F, had come to the same conclusion. "Fairchild was a semiconductor company," he said. "In the '70s those companies, including Hewlett-Packard and National Semiconductor, all got into manufacturing consumer products that would use their chips – calculators, watches, *Pong* games, etc. Fairchild didn't really understand consumer product development or marketing and they certainly did not understand games. They had no concrete commitment to being in the games business, whereas Atari was founded to be a games company."

Lawson agreed: "The problem was Fairchild still had a component brain, not a systems brain. They always thought themselves more important because they developed the circuit, but to a systems guy like I was, it doesn't make a difference to me who develops that little chip. I couldn't care less. But they didn't understand that because they were coming from a different end of the spectrum. As a result, they didn't understand marketing when it came to dealing with end customers. Their customers were manufacturers – the Honeywells, the IBMs – they'd never had to deal with an end customer."

And when RCA revealed its console would only do black and white visuals, but would cost only $20 less than the Channel F, Atari's confidence grew even more. Atari had Warner's money and a collection of hit arcade games that its rivals could only dream of. By the time Atari launched the $199 VCS 2600 in October 1977 with *Combat*, an in-built version of its arcade hit *Tank*, its competitors were floundering. RCA's black and white console had bombed and Fairchild's Channel F had failed to win over a large audience. But even though the opposition crumbled, sales of the 2600 in Christmas 1977 were a disappointment. Manufacturing delays meant few of the consoles reached the shelves in time for Christmas and only a few hundred thousand 2600s had been sold by early 1978 despite the millions Warner lavished on the system. "We missed the Christmas season," said Bushnell. "We were late in producing units and then there were some huge snowstorms that stopped us getting past the Sierra mountains. As a result we ended up not having enough at Christmas and too many in January and February."

Matters were not helped by the excitement surrounding a new type of video game: the handheld electronic game. In 1977 it was these, not the new generation of video game consoles, that were topping Christmas lists across the US. Toy company Mattel kickstarted the handheld games craze in 1976 when one of its marketing directors – Michael Katz – came up with the idea for a portable electronic game.

"It was the mid-'70s – a time when pocket calculators were a new product and were getting smaller and smaller and less expensive," said Katz. "Everyone had to have a little handheld calculator. I said to Richard Channing, Mattel's director of preliminary design: 'Can you design a new type of game that uses LED technology similar to that in a calculator but that could be portable, battery powered and the size of a handheld calculator?' He went away and came back with the prototype of what was the first handheld game – an obstacle avoidance game where LEDs were coming down at you. You were at the bottom of the screen and had to try and avoid them and make your way to the top."

Excited by their invention, Katz and Channing developed two prototypes games – one based on American football, the other on motor racing. After gauging reaction from consumers, Mattel decided to launch the racing game, *Auto Race*, first and began touting the product to stores.

The retailers loved it. "It was incredible because a lot of them would have managers of all levels come to our presentations, which wasn't normal when you were presenting a toy product," said Katz. "They wanted to see what the first portable electronic game looked like."

Children were just as excited at the opportunity to play video games wherever they went and hundreds of thousands of *Auto Race* games were sold during Christmas 1976. The following June, Mattel released *Football* to even greater success. Millions of *Football* games flew off the shop shelves and the company's new Mattel Electronics division quickly became a significant part of the toy giant's business. The excitement about portable games showed little sign of stopping for the next two years as dozens of companies sought to grab a slice of the action.

Texas Instruments came up with *Speak & Spell*, an educational toy that used a speech synthesizer to challenge kids to spell words on a touchpad keyboard in a robotic monotone. It sold in huge numbers inspiring Texas Instruments to release *Speak & Read* and *Speak & Math*.[4] Just as successful was Milton Bradley's *Simon*, a disc-shaped electronic

4. *Speak & Spell* also inspired numerous musicians to use its robotic tones in their music. Among them were the Pet Shop Boys, Kraftwerk, Limp Bizkit and Beck. British synthpop act Depeche Mode even named their 1981 debut album after the toy.

toy invented by Howard Morrison and Magnavox Odyssey inventor Ralph Baer, who got the idea from *Touch-Me*, an Atari coin-op game released in 1974 that later became the company's sole, unsuccessful venture into handheld games. *Simon* consisted of four large primary coloured buttons that would light up the buttons in a random sequence and play a musical note to accompany each one. The player then had to repeat the sequence. It sold millions and would become a pop cultural icon thanks to its distinctive looks. Another smash hit was Parker Brothers' *Merlin*, a multi-purpose LED game that could play blackjack, tic-tac-toe and *Simon*-esque memory games. It even doubled as a musical instrument, allowing users to play tunes by pressing its buttons. Demand for these games was so huge that stores quickly sold out, leaving parents desperately combing shops and retail parks in search of the elusive toys.

George Ditomassi, the general manager of Milton Bradley's game division, told the press the clamour for *Simon* had taken the company by surprise. "It was just impossible for us to foresee this kind of demand. We knew we had a good item from the day we saw it, but we had no thinking of anything like this," he told the *St Petersburg Independent* as retailers and parents alike bemoaned the lack of *Simon*s on the shelves. The demand for portable games peaked in Christmas 1979 with estimated sales of $400 million in the US alone – up from $35-$40 million in 1977. The craze left Atari struggling to get the public to buy the 2600, which in turn caused the relationship between Warner and Bushnell to break down. Bushnell's initial optimism about the Warner deal had faded fast.

"Warner had this way of saying nothing's going to change, you're going to manage it just the way you've always managed it but you can use our cash and all our properties. It felt like you got to play in a bigger sandbox without the stress," he said. "At the time I believed it was just a financial transaction and that beyond having enough liquidity to buy some stuff, not a lot would change. I was naive, but that's what I believed." Warner, however, had other ideas. For a start Gerard believed Atari's internal operations needed massive changes: "Bushnell and Keenan were not managers, they were hot-shot entrepreneurs. They were really engineers and understood the engineering side, but they weren't managers and this place needed a manager." Warner started pushing for a bigger marketing operation, more managers and formal financial controls. Part of the conflict between Bushnell and Warner stemmed from the differences in East Coast and West Coast management styles.

"The biggest difference was a marketing-centric versus an engineering-centric company. The egalitarian 'we're all in this together' approach versus hierarchy," said Bushnell. "East Coast guys, if they came into the engineering department at 8.30am and there was no-one there, they would say: 'Boy, what a lax place'. But they were out having Martinis at nine o'clock when those guys were still working at night. East Coast is much more about form over substance and that was one of the things that Atari really tried to move away from. We wanted to be all substance and no form."

But not every pre-Warner Atari employee was thinking along the same lines as Bushnell. "You know the old story about are you saluting the person or are you saluting the hat? I always thought the company had consensus. In my naïve way I thought the people who were working for me would agree with me because I was the boss," he said. "I first realised that power was shifting to Warner was at one of our early planning sessions right after the purchase when we were having a planning session and Manny was there. He came up with a couple of things I thought were absolutely ridiculous and my people kind of saluted it. I thought 'Woah! What's going on here?'. It was really surprising to me. Everyone's willing to salute the hat and I thought it was the substance of my arguments and ideas. Naive me."

Unhappy with the way Atari was going, Warner decided to bring in a consultant who it felt could help knock the company into shape. It chose Ray Kassar, a vice-president of textiles manufacturer Burlington Industries. Kassar didn't want the job: "The Atari job was offered to me from a recommendation of a friend who was working for Warner. My reaction was that I had no interest at all in what he was offering. My friend insisted I meet with Manny Gerard and, after four hours of talk, I said I'll take it under certain conditions and they agreed. I said I'll only go for a couple of weeks."

On his arrival in California, Kassar was shocked at Atari's business practices: "The company had no infrastructure. No chief financial officer, no manufacturing person, no human resources, there was nothing. I had no idea how bad it was." Just like Bushnell, Kassar felt there was a clash between Atari's West Coast culture and the East Coast approach of Warner and himself. "We're all more serious in the east, you have a job and you do it the best you can – it's not a playground," he said. "In California at that time things were very casual. They still are. That's ok, that wasn't a problem for me, but someone had to be a grown-up. They were a bunch of kids playing games." It was a divide that Kassar noticed from day one: "When I arrived on the first day, I was dressed in a business suit and a tie, and I met Nolan Bushnell. He had a t-shirt on. The t-shirt said 'I love to fuck'. That was my introduction to Atari."

The divisions were not just at board level. Staff in the company's coin-op games division, once the heart of Atari's activities, felt spurned by Warner's focus on the consumer division and 2600. "Warner, it seemed to us, fell in love with the consumer side and the computer side and thought the coin-op group was old time," said Noah Anglin, a manager in the coin-op division when Warner bought Atari.

The rift between Bushnell and Warner came to a head in November 1978 in a budget meeting in the Atari owners' New York offices. By then Bushnell had started to lose interest in Atari, partly because of his newfound wealth but also because of his increasing frustration with the direction Warner was taking the business. "There was a kind of disengagement," said Bushnell. "I found it very difficult to support activities I thought were stupid. I took a lot of trips, usually to a trade show, but then a week's holiday one side or the other."

But when Bushnell arrived in New York that November, he was in a fighting mood. "There were two or three things that were really bugging me," he said. "First of all I felt that we needed to replace the VCS as quickly as possible – the VCS was already obsolete. I was afraid someone else was going to come and totally outclass it and it took two to three years to get a new product like that through the engineering cycle. The second issue was that we were about ready to start marketing the Atari 800 home computer and Warner was adamant that it was going to have it as a closed computer system and if anyone wanted to buy software they had to buy it from Atari. They would prosecute and sue third-party software developers. I just thought that was mad. The third big issue was the Atari pinball division. We had created these wide-body pinballs and Manny wanted to get into what I called the standard pinball business. The reality was that our cost to manufacture in California was $150 higher than in Chicago, and when you added in the extra freight to the east coast, it was an almost $200 disadvantage."[5]

Annoyed and angry at what he saw as stupidity on Warner's part, Bushnell went into the meeting with fire in his belly. "The Warner board, and Manny particularly, just didn't want to hear the fact that the 2600 was obsolete and I didn't choose my words very well. I said: 'The 2600 is obsolete. It's a piece of shit'."

The Warner board was shocked. "Nolan sits in the meeting and looks up at one point and says sell all your remaining 2600s, the market is saturated. It's all over," said Gerard. "Everybody in the room, including me, kinda stares at him and doesn't know what to say. The guys from Atari didn't know what to say. It was stunning. So raw."

Bushnell's dismissal of the 2600's prospects unnerved Steve Ross, Warner's chairman. "Steve was in a panic," said Gerard. "Steve was a very good guy and very smart man. He said: 'The guy I bought the company from says it's all saturated'. I said 'Steve, you don't know what the fuck you're talking about'. He looked at me. I said: 'Steve, listen to me, it's now the 8th December. On December 26th there are two possibilities. Either there will not be a 2600 on the shelf of any retailer in America, in which case you have the biggest business you ever saw, or there were going to be plenty of them and we're fucked. So let's all relax because in 18 days we're going to know the answer."

---

5. Chicago was home to most of the pinball business including the three largest manufacturers: Bally Midway, Gottlieb and Williams.

Regardless of how sales went that Christmas, Bushnell's days at Atari were numbered. "I had had it up to my ears," said Bushnell. "It's not clear whether I was fired or whether I quit. But Manny and I had a talk after the meeting and he says 'we've got to do something, Nolan' and I said 'yeah, I should get out of here'." In January 1979 Bushnell turned up for work at Atari for the last time to finalise his severance package. As part of the deal Bushnell was barred from working in the video games industry until 1983 – effectively exiled from the business he created. Keenan quit shortly after. The pair teamed up to set up Chuck E Cheese, a children's pizza restaurant chain that used arcade video games and robots to lure in customers. Warner offered Kassar the post of chief executive – the New Yorker's few weeks of consultancy work in California was about to turn into a three-year stint.

For Kassar, Bushnell's exit was a relief. "I couldn't have accomplished what I did with Nolan in the picture. Atari couldn't have two bosses. Two people can't run a company, one person has to have the final responsibility. Nolan would say one thing and I would say another. How do you resolve that? You have to have either him or me. I have nothing against Bushnell. He is a charming, bright guy, very capable and, after all, he started the whole thing, I didn't." By the time Kassar took over, it was clear the 2600 had sold in large enough quantities during Christmas 1978 to allow Warner to dismiss Bushnell's warning that the market was oversaturated, but problems still remained. "The 2600 was selling, but not in any great volume. The problem was the quality of the hardware, which was terrible. The return rate was excessively high," said Kassar, who made improving the reliability of the console his first priority after taking control of Atari. With the potential profits from the 2600 so huge, Kassar focused most of his attention on making the console a success. The coin-op division that Atari built itself on and the company's new home computer operation, found themselves sidelined. The coin-op division, used to being at the heart of the company, took the shift in focus badly. Its employees saw the division as the hit factory of Atari. After all, it was in their division that the great games that had made Atari's reputation were born. Anglin felt the coin-op division harboured the soul of Atari: "If you talk to the coin-op guys you get the same love, passion. That passion, to me, didn't exist in the other groups." The change from Bushnell to Kassar was dramatic, said Atari coin-op engineer Howard Delman: "Nolan understood the value of his engineers. He knew that we were the engine propelling the company. We were like kids in a candy store playing with fabulous technologies and doing things that no-one had ever done before. Under Warner and Kassar, the attitude changed significantly. The new engine propelling Atari was the marketing department and profit became the most important goal." Not that this stopped the coin-op division from carrying the torch for Bushnell's vision of Atari as a company of fun. "Apple were next door to us and one night some of our guys went over and painted worms on the big Apple sign. The next day Steve Jobs and all them were all upset about it," said Anglin. "It was like 'hey come on guys, have a sense of humor'."

The split between Kassar and Atari's game makers would grow even wider when the new Atari boss agreed to be interviewed by *Fortune* magazine. In the article Kassar described Atari's game designers as a bunch of "high-strung prima donnas". "It was a mistake," said Kassar. "When I said that, it was an off-the-record comment and unfortunately it got on the record. I had great respect for the designers and the idea that I didn't is a totally blown-up image of me by engineers who really hated the fact that I wasn't an engineer and came from New York. I really did all I could to encourage the programmers, to cheer them up, to inspire them. Once this particularly bright programmer came to see me and I spent five hours with him because he was so critical and crucial and he was reading his poetry to me. It was a little off the wall but he was a great programmer and that's all I cared about. Without the games we wouldn't have had a business. The programmers had a lot of respect, they were left alone, they did what they pleased. As long as they produced, that was fine with me."

The coin-op division responded to Kassar's public criticism with its trademark playfulness. "Kassar was not an engineer and he made that perfectly clear when he called us 'high-strung prima donnas'," said coin-op game designer Ed Logg. "When that came out we all had t-shirts made saying 'I'm just another high-strung prima donna'. Everybody in coin-op had one. That was one of our cheap shots back." Not that every game designer rejected the high-strung prima donna tag. "We totally were," said Rob Fulop, a game designer in Atari's 2600 games division at the time. "Isn't every actor or actress? So were The Beatles, so was Michael Jackson. People that create things are whacked out, high-strung prima donnas – that's kind of how it works. I remember not feeling insulted at all."

Despite its dislike of the new regime, the coin-op division continued to produce hit after hit during the late 1970s and one of the biggest was 1978's *Atari Football*. Atari's American football game began life in 1974 as *Xs and Os*, but the project had stalled. "I started *Xs and Os* using discrete circuits like in the early games such as *Pong*, but as we got further with the prototype we wanted more objects on the screen," said Bristow, who led the work on that early version. The arrival of microprocessors gave Atari the chance to dig out Bristow's abandoned game and try again. The task of completing the game was given to Michael Albaugh, an engineer who had joined Atari from the telecoms industry. While putting together the game, Albaugh came up with the idea of using a trackball as the controller instead of a joystick. Unlike joysticks, trackballs could measure the speed at which players spun the ball as well as the direction they wanted to move. Albaugh thought it was perfect for *Atari Football*. "It allowed a more direct control of the player objects and added physicality to the game." Atari's senior managers were less convinced that adding this relatively expensive control mechanism was worth it.

"Nolan Bushnell was opposed to it, thinking a joystick would be adequate. I won by threatening to quit," said Albaugh. Atari engineer Jerry Lichac got the job of designing a custom trackball for Atari that would be robust and cheap enough for the company to include in *Atari Football*. "In those days the only ones available were the military things and our engineers actually designed a very low-cost trackball using a cue ball from pool," said Anglin. Almost as soon as the prototype was tested on the public, Atari knew it had another game destined for success. "We got this thing out on test and me, Bristow and coin-op executive Lyle Rains watched this game, as we only wanted products people loved going out," said Anglin. "There were these guys playing *Atari Football*. One guy was slamming the trackball so much his watch flew off of his arm and across the room. There were crowds of people watching people play. We kinda thought this might be a hit."

And it was. As 1978 drew to a close *Atari Football* looked set to be the biggest arcade game of the year by a long margin. But then *Space Invaders* arrived.

* * *

After seeing *Gun Fight*, Dave Nutting Associates' microprocessor reworking of his game *Western Gun*, Tomohiro Nishikado knew he wanted to use the same technology in his next creation. He diligently researched the capabilities of microprocessors and built a computer that would allow him to program games for this new technology. After getting to grips with the technology, he turned his thoughts to what kind of game he wanted to make and homed in on the advantages microprocessors offered in terms of animation. "With microprocessors, the animation is smoother and there are so many more complex physical movements that can be reproduced, so the category of games that we could now create was so much more," he said. Nishikado decided to make a shooting game: "The targets that came to mind were military tanks, ships and airplanes. I decided on airplanes but I just couldn't get the movement of the airplane in flight to look smooth, so I tried many different targets and found that the human form was the smoothest movement."

Taito's president was far from impressed by Nishikado's plan to create a game where players shot people. "I was prevented from using the human form, so I thought of aliens so I could use the similar form, and therefore a smooth movement, while getting around the problem of shooting humans," said Nishikado. He took inspiration from the 1953 film of H.G. Wells' novel *The War of the Worlds* that he had seen as a child: "The bug-like aliens made a great impression on me, so I created my aliens based on that image." The invertebrate alien forms Nishikado eventually created also resembled sea creatures such as crabs, octopuses and squid.

Another big influence was Atari's *Breakout*. Nishikado decided that, like the bat from *Breakout*, the player's missile launcher would be stuck at the bottom of the screen and only capable of moving left or right. In place of *Breakout*'s static bricks, he arranged a phalanx of space invaders – 11 aliens wide and five aliens deep – and got them to march ominously from one edge of the screen to the other while raining laser fire on the player below. And when this extraterrestrial army reached the screen's edge, it would drop down in menacing unity one step closer to the player and its ultimate goal of reaching earth. To help the player, Nishikado added four shields that could provide some cover from the alien barrage, although these would be slowly ripped and torn apart by the onslaught from above. The player's task was straightforward: defeat the aliens before they reached earth, but it was a hopeless battle for survival, as the aliens would never stop. Even if the player killed the whole alien army another would simply take its place. The only reward for the ultimately doomed player was the chance to take down as many aliens as possible before defeat in order to add their name to the game's roll of honour – its high score table.

For players used to the tame, innocent fun of *Pong* and the ponderous battles of *Tank*, *Space Invaders* was a powerful experience. This wasn't just a bit of fun, this was ferocious human-versus-machine action. Exhilarating, stressful, adrenaline-pumping and intimidating in equal measure.

Like the invaders within its virtual world, *Space Invaders* conquered Japan within weeks of its launch in July 1978. Children, teenagers and adults alike flocked to the arcades to join the battle against the alien threat. Pachinko parlours, bowling alleys and even grocery stores reinvented themselves as dedicated *Space Invaders* arcades. Cafés swapped their tables for *Space Invaders* cocktail cabinets. Novelty pop act Funny Stuff took the invasion onto the airwaves with *Disco Space Invaders*, a hit single backed with dance moves inspired by the jerky movements of Nishikado's aliens. Within three months of its launch, *Space Invaders* had gobbled up so many ¥100 coins it brought Japan to a standstill, preventing people from buying subway tickets or using public telephone boxes. A panicked Bank of Japan responded by ordering an investigation of Taito, which would sell more than 100,000 *Space Invaders* machines in Japan alone. Nishikado, however, paid little attention to the fuss his game was causing: "I don't remember being particularly happy or pleased at the time. I was more concerned with the low quality of the hardware for this game and was concentrating my efforts on creating better hardware."

Space Invaders' formula would prove no less potent in North America and Europe.

Bally Midway, Taito's US distributor, sold around 60,000 *Space Invaders* machines and watched its profits soar. Eugene Jarvis, who was a pinball designer for Atari when *Space Invaders* reached the US in late 1978, responded by abandoning the world of flippers and pins. "I was a real pinball fanatic, but when *Space Invaders* came out I knew the future was in video games," he said. "I was instantly addicted by the possibilities of computer intelligence applied to video games. This was a huge advance from the first generation of 'dumb' games like *Pong*, which relied solely on the intelligence of human players."

The impact of *Space Invaders* could also be seen in the US sales figures for coin-op games. In 1978 the business generated revenues of $472 million, slightly down on the previous year's $551 million. In 1979 the figure had more than tripled to $1,333 million – with *Space Invaders* accounting for a large proportion of that total. And having conquered the world's arcades, *Space Invaders* then helped Atari conquer the home.

By late 1979 competition in the home video game business was hotting up. Fairchild and RCA's consoles had bitten the dust but new machines had taken their place. Atari's biggest rival in the arcades, Bally, had released the Professional Arcade. Designed by Dave Nutting Associates, the Bally Professional Arcade was more powerful than Atari's machine – a fact the company hoped would give it the edge. "We knew we were miles ahead of Atari technically," said Jay Fenton, an engineer at Dave Nutting Associates who helped create Bally's console. "Nothing else came close to our console until the Nintendo."

But that technology came at a higher price and, unlike Atari, which sold the 2600 at cost price, Bally was determined to make a profit on every console sold. "What really killed us was being more expensive – like double what the VCS went for," said Fenton. The citizens of New Jersey also delivered the console an unintentional blow, said Dave Nutting. In 1978 the state's voters backed a law allowing casino gambling in Atlantic City. The vote turned the East Coast city into a new Las Vegas and for Bally, which also made fruit machines, it was a major business opportunity. "Bill O'Donnell was the president of Bally and his dream was for Bally to get into owning and operating casinos," said Nutting. "He now had the financial resources, from Bally's incredible success in the commercial video game market, and now had the place. Bally lost interest in pursuing the consumer market and decided to abandon the project." The Professional Arcade was sold off to a group of small businessmen who relaunched it as the Astrocade only to watch it fade into oblivion.

Internal politics also crippled the Magnavox Odyssey², the TV set manufacturer's answer to the 2600 and Fairchild's Channel F. To give its console the edge, Magnavox decided to base the system on the 8244 graphics chip that Intel was developing. This chip, one of the first graphics chips to be created, would handle much of the work involved in generating on-screen images and audio, leaving the Odyssey^2's main microprocessor to concentrate on running games. "It was by far the most advanced graphics chip of its day and gave a huge advantage to the Odyssey^2," said Ed Averett, one of the Intel team that created the chip. But the chip's development was plagued by delays that kept pushing back the launch of the Odyssey^2 until 1978. Despite the delays, Averett was upbeat about the console's chances and quit Intel to make games for the console. "With incredible hardware for that time and distribution in place, the only thing missing was software," he explained. "So I left Intel to design games for the Odyssey^2. Everyone thought I was crazy. Intel, Magnavox, even my family, except my wife and our one-year-old daughter Ashley."

But by the time the Odyssey^2 launched, Magnavox was already trying to extract itself from the video game business. The project had nearly been cancelled before launch, until Ralph Baer intervened and persuaded the company, which was now part of the Dutch electronics giant Philips, to stick with it. And even though it went ahead with the launch, Magnavox was still looking for a way out. "By the time the chip arrived, Magnavox was seriously thinking about getting out of the video game business as soon as its obligation to Intel was fulfilled," said Averett. "All of their engineers had been told to stop designing games and most were reassigned. The lights were going out for the Odyssey^2 before it was even born." The console's only internal support at senior level came from Mike Staup, one of Magnavox's vice-presidents, and he faced an uphill battle trying to keep the rest of the company's upper management from pulling the plug on the system. Averett, however, did quite well out of Magnavox's decision to stop making games internally: "When the Odyssey^2 finally hit the market it sold out immediately, so Magnavox said 'ok, design one more game, but this is the last game we want, ever'. This philosophy of just one more game prevailed for three years."

Averett ended up the sole game creator for the Odyssey^2 – a one-man freelancer working for royalties taking on Atari's dedicated pool of VCS game designers. "It was incredibly frustrating since the Odyssey^2 was vastly superior to the VCS," he said. "Atari deserves huge credit for taking on Magnavox and then Philips with an inferior product and beating them soundly in the marketplace. While frustrated, I did get a lot of satisfaction about being part of one of the best-kept secrets at the time in the industry – being one guy going toe-to-toe against Atari design teams. We went to some lengths to keep that secret for obvious reasons: one of the biggest being I had no time for anything but designing games." The only support Averett got was technical help from his wife and criticism from the kids in his neighbourhood, who he used as play testers. "It was as brutal as you might imagine – kids don't mince words," he said. The Odyssey^2 would eventually crawl past the million sales mark and did well in Europe where Philips released it as the Videopac G7000, but the lack of corporate support ensured the console never came close to matching the sales of the VCS.

But while Bally and Magnavox had been doing their best to help Atari finish off their own consoles, by late 1979 Atari finally found itself facing a serious challenger: Mattel. Flushed with its success in the handheld games business, Mattel decided it wanted a slice of the video game console business too. "Handhelds had established Mattel in the electronic game area, which made it a sensible add-on to go and compete on the console side against Atari," said Katz. In late 1979 Mattel launched its Intellivision console in Fresno, California, to test the market ahead of the full US-wide launch in 1980. Mattel had no intention of letting Atari have an easy ride. It developed an advertising campaign that highlighted how superior the Intellivision's graphics were to the VCS. It paid sporting bodies to endorse the sports games that would be central to its bid for sales. And it formed an internal development team headed by Don Daglow, the pioneering computer programmer who had written *Baseball* and *Dungeon*. "We absolutely felt we could catch up with Atari because the Intellivision was next generation compared to the Atari 2600 – it was that much better," said Daglow.

But just as Mattel was gearing up for its assault on Atari, Manny Gerard had a brainwave. "The single best thing I ever did at Atari was go over to the coin-op building one day in 1979," said Gerard. "They had a coin-op version of *Space Invaders* and they're all playing it. I walked back across the street to Kassar's office and I said 'I'll tell you what I want Ray – take the fucking *Space Invaders*, send it up to consumer engineering, engineer it for the 2600 and licence the name, and if you can't licence the name steal the game play'. He looked at me and said 'oh my god, why didn't I think of that?'. I said 'Because you're too busy running the company'." Atari moved quickly, bought the rights off Taito and, in January 1980, released *Space Invaders* on the 2600. Any question marks about Atari's hold on the console market melted away. "It was the *Space Invaders* cart that blew the 2600 to the Moon," said Gerard. The fuss over the electronic handheld games that had stolen the thunder of video game consoles in the late 1970s evaporated and every kid in every town in America wanted an Atari 2600. And over the next couple of years millions of them would get their wish.

Pop idols: Buckner & Garcia meet Pac-Man
Courtesy of Buckner & Garcia

CHAPTER 7

Pac-Man Fever

It's the summer of 1982 and North America is in the grip of video game mania. In the four years since *Space Invaders* made its Japanese debut, video games had exploded in popularity. Back in 1978 the US sales of home and coin-operated games stood at $454 million; 48 months later in 1982 that figure had soared to $5,313 million. To put it another way, the video game business was expanding by a massive 5 per cent a month.

Excitement about video games pervaded every corner of American life. The public's seemingly insatiable appetite for electronic play had transformed the retail landscape. Arcades had sprung up in every mall and high street. Coin-op games could be found in launderettes, movie theatres, cocktail lounges, hotels and restaurants. Even supermarkets were installing video games for their customers to play. "Arcade locations were like Starbucks back then – literally everywhere," said Scott Miller, who wrote columns for the *Dallas Morning News* about video games at the time. There was no respite at home either as millions upon millions of Atari VCS 2600 consoles had embedded themselves under the nation's TV sets.

Journalists marvelled at the dazzling success of the video game. They pored over analyst reports suggesting that video games would soon be bigger than film and music combined. They interviewed fresh-faced game designers who boasted about how they had spent royalty cheques and bonus payments worth tens of thousands of dollars on a celebrity lifestyle of fast cars and flash pads. And they wrote about the new 'pinball wizards' – the hot-shot players who were the masters of the arcades. "The public and the media were fascinated by the video game," said Walter Day, founder of Twin Galaxies, which started life as a small arcade in Ottumwa, Iowa, before turning itself into the official keeper of video game high score records. "The media, in particular, was amazed by players who could actually beat the games. It was this perception of 'man versus machine' that made many news stories so intriguing to the public."

Everyone wanted a piece of video games, from the movers and shakers of Washington D.C. to the studio bosses of Hollywood. *Star Wars* director George Lucas set about forming a games division at his company Lucasfilm. Walt Disney Pictures sought to cash in with *Tron*, a film about a man trapped inside a video game that was touted as a summer blockbuster. Guides explaining how to beat arcade machines clogged up the bestseller lists. Quaker Oats, Parker Brothers, 20th Century Fox and Thorn EMI formed video game divisions. McDonald's started serving Atari-themed burger meals where "thanks to McDonald's and Atari, the old-fashioned TV dinner is being replaced by an exciting video-dinner that could make you a winner". And if a burger, fries and shake were too much, you could snack on a packet of Universal Foods' Pretzel Invaders. In Washington D.C., a group of young Democrats – including future presidential candidate Al Gore – became known as the Atari Democrats for their support for giving tax breaks to high-tech industries rather than older manufacturing industries such as steel and cars. As *Time* magazine's cover declared in late 1981: "Gronk! Flash! Zap! Video Games Are Blitzing The World'.

The blitz began with *Space Invaders*. Its success reignited interest in video games just as a trinity of technological and cost breakthroughs allowed for a major leap forward in the quality and vision of games being released in the arcades. The first development was the microprocessor and the design freedom it granted game developers, the second and third were improvements in video game visuals: high-resolution vector graphics and colour games. Both came to fruition in 1979.

Vector graphics had existed for years, but had always been too expensive for use in the arcades.[1] Standard TVs, also known as raster scan monitors, build images out of a series of horizontal lines that are drawn in turn left to right starting from the top. Using this method a TV can create a full-screen image once every 50th or 60th of a second. Vector monitors take a different approach.

Instead of building complete pictures, they draw pencil-thin white lines between two co-ordinates on the screen. While poor at drawing complete images, vector graphics were perfect for drawing crisp, smooth outlines that were also brighter than the images created by standard TVs. "The resolution of raster games was not so great in those days," said Owen Rubin, an Atari engineer who started out making vector graphics games on his university's computers. "The graphics of a vector monitor were extremely sharp and, for the time, very high resolution. They just looked very good."

Vector graphics first came to the arcade thanks to Larry Rosenthal, an engineer who, like Atari founder Nolan Bushnell, wanted to bring *Spacewar!* to the arcades. He built the Vectorbeam system that made vector graphics cheap enough to use in arcade games and used it to make *Space Wars*, an arcade version of the Tech Model Railroad Club's game.

---

1. *Spacewar!*, for example, was created on a computer that had a vector graphics monitor.

Rosenthal hoped arcade manufacturers would buy the rights to the game and most were interested. But when Rosenthal insisted on getting half of the profits, potential buyers such as Atari walked away. Having alienated the big players, Rosenthal found himself pitching the system to small-fry video game companies. One of these was Cinematronics of El Cajon, California. Cinematronics was in bad shape when Rosenthal got in touch. The company had released two unsuccessful games and was on its last legs, so figuring it had nothing to lose, it accepted Rosenthal's high price. In October 1977 *Space Wars* went on sale, introducing vector graphics to the arcades for the first time. The game's distinctive ghostly outline visuals helped Cinematronics shift 10,000 machines, saving it from the brink of closure. Cinematronics' relationship with Rosenthal would be short lived. Rosenthal felt he wasn't earning enough from the game and walked out taking his Vectorbeam system with him. After a legal tussle, Cinematronics paid Rosenthal for the rights to use the Vectorbeam technology and set about trying to become the premier creator of vector graphics arcade games. To help it develop more vector games the company hired Tim Skelly, a programmer whose journey into video games began with a night out at The Sub's Pub in Kansas City. "A guy walked into the bar room with a computer under his arm. Seriously," he said. "Of course you talk when someone walks into a bar with odd company or artefacts."

The man with the computer was Douglas Pratt and he planned to open a video game arcade. Skelly decided to go into business with him: "I had my doubts, but almost anything was better than just making sandwiches." The venture failed but gave Skelly enough experience to land a job at Cinematronics designing their new vector games. Skelly loved the visuals: "It was different from what other games were using. The best part was that we could do smooth rotations at high speed. Vector games were much more fluid and fine-grained. Raster, chunky. Vector, smooth. I liked smooth." Skelly's first vector games started rolling off the production line in early 1979. They ranged from the 3D dogfights in space of *Tailgunner* to *Warrior*, an overhead view sword-fighting game where players controlled two smoothly animated warriors carrying long swords.

By then, however, Atari had caught up. In the wake of *Space Wars*, Atari's research and development team in Grass Valley, California, had got to work on vector graphics technology of its own and by early 1978 had a working prototype to show the company's coin-op team. "It wasn't much more than a demonstration test bed, but it clearly demonstrated that cool vector images could be displayed," said Atari engineer Howard Delman, who teamed up with fellow coin-op engineer, Rick Moncrief, to turn the prototype into a useable device. Having refined the prototype, Delman decided Atari's first venture into vectors should be a remake of the moon landing game *Lunar Lander*, the 1973 remake of the 1969 text-only computer game *Lunar* that used the vector graphics abilities of the

DEC GT40 terminal. "I had previously seen the game and thought it would be a good choice to demonstrate the look and feel of our new technology," he said. Released in early 1979, Atari's *Lunar Lander* was a delicate real-time battle against gravity that challenged players to land their craft on the moon's mountainous landscape before their limited supply of fuel ran dry. It was an impressive demonstration of what vectors could do but it would be *Asteroids*, Atari's second venture into vector graphics, that really caught the public imagination.

Asteroids began with a meeting between programmer Ed Logg, who had done some of the work on *Lunar Lander*, and Lyle Rains, vice-president of the coin-op games division. "I get called into Lyle's office and he goes: 'I've got an idea for a game'," said Logg. Rains suggested a game where players controlled a spaceship that had to blow up asteroids, splitting them into smaller and smaller chunks of cosmic debris until they vanished altogether. The challenge would be to avoid colliding into the asteroid fragments. Logg decided it should use vectors: "Vector monitors are high resolution. They are 1064 by 728 pixels whereas standard rasters are 320 by 240 – a big difference in resolution so when you turn your ship you can tell which direction it's facing, which is really important."

Logg developed the rock smashing idea by turning it into a balancing act. Trigger happy players risked being overwhelmed by the volume of asteroids floating around the screen while those who did too little would find themselves under attack from the flying saucers that Logg created to force players to act. The tension between action and inaction was enhanced by the sound effects created by Delman, which echoed the ominous thumping beat that the aliens of *Space Invaders* marched to. "I tried to create the sound of a heartbeat," said Delman. "My sense was that the player's heart rate would be increasing as the game got more frenetic, and I wanted the player, subconsciously, to be hearing his own heart racing." *Asteroids* became the most popular game ever made by Atari and the second biggest arcade game of 1979 – outdone only by *Space Invaders*.

Atari followed it up with a spate of popular vector games, most notably Ed Rotberg's 1980 game *Battlezone*, a futuristic tank battle viewed from within the player's tank. "Given that we now had the vector generator technology, it seemed like a natural follow on to the successful *Tank* and *Tank-8* arcade games for Atari," said Rotberg. The game's 3D visuals inspired a group of retired US Army generals to ask Atari to remake it as a training simulation to help soldiers learn to drive the Bradley Infantry Fighting Vehicle.[2] Atari's management readily agreed to the idea and then told Rotberg.

"I was told about it after the prototype had been promised – and on a very aggressive

2. The combat vehicle built by the US military in response to the Soviet Union's Boyevaya Mashina Pekhoty vehicles, which combined the features of light tanks with armoured personnel carriers.

schedule," he said. "I was not pleased. I felt that Atari should not be doing government/military products. Back at the time, most of us could have gotten jobs in the military-industrial complex if we had wanted to. Many of us were still very much affected by what had happened during the Vietnam War. Most of us had pacifistic leanings at that time, myself included. I simply did not want to work on a product that would help people learn how to kill other people." As the only person capable to meeting the generals' deadline, Rotberg agreed to do the prototype but on the condition that he would never work on a similar product. After three months of toil he completed the *Bradley Trainer* prototype, but it never went beyond the prototype stage.

The crisp outlines of vector games were an exciting departure from the blocky monochrome of old. But by the end of 1979 the arrival of colour graphics was proving even more exciting. Prior to 1979, almost every arcade video game was black and white. The closest they got was the use of transparent coloured plastic to create an illusion of colour in particular areas of the screen. *Breakout* used this approach to make its bricks different colours, while *Space Invaders* had a strip of green plastic glued to the bottom of the screen to colour in the player's missile launcher and shields. "Colour was not added for some time because of cost, both for the monitor and the additional hardware needed to support colour," said Rubin. "At the time, it was not a trivial change. A few games, like an eight-player *Tank* game were tested in colour – it was the only way to have eight players look different – but for most of the games we were doing, colour did not add a lot."

But on the other side of the Pacific from Atari, Namco – the company that bought the wreckage of Atari Japan in 1974 – had come up with one of the first full-colour video games: *Galaxian*. It was a *Space Invaders* clone that removed the shields and added aliens that dive-bombed players. Colour proved a powerful selling point and made *Galaxian* a huge success. Other game developers were inspired to follow suit. Dona Bailey, a car sensors programmer at General Motors, was inspired to leave the car industry for Atari after seeing the colours of *Galaxian*: "I adored *Galaxian*, I thought it was intensely beautiful. Its repetition of patterns, its colours and its swooping and swerving motions. I wanted to make something that seemed as beautiful to me." *Galaxian*'s rougher, tougher remake of *Space Invaders* also proved influential, marking the start of a rapid evolution in shoot 'em ups that saw them crank up the intensity of their man versus machine challenge. Atari's Dave Theurer served up energetic shooting games based on nightmares. His 1980 game *Missile Command*, a trackball-enhanced scramble to protect cities from never-ending barrages of nuclear missiles, came out of Cold War nightmares of nuclear war.[3] Theurer's

---

3. The all-too-real threat of nuclear war between the US and USSR inspired Theurer, but Atari played down the atomic armageddon theme. Officially the game was about defending space bases on planet Zardon.

next creation, 1981's *Tempest*, was a colour vector graphics game based on a nightmare he had about monsters coming out of a hole. It challenged players to zap strange abstract shapes that crawled out of a cylindrical 3D pit. Not to be outdone, Namco continued to hone the aggression of *Galaxian* with 1981's *Galaga* – a sequel that handed the aliens new tricks such as tractor beams used to try and capture the player's craft.

But none of these were as angry as *Defender*, the ferocious shoot 'em up created by Eugene Jarvis that marked leading pinball manufacturer Williams' return to the video game business. Jarvis joined Williams as a pinball designer after a stint at Atari's ill-fated pinball division. Williams had dabbled in video games in the wake of *Pong*, but quickly reverted back to pinball tables. By the end of the 1970s it was clear the decision to walk away was a mistake. "We all could see a revolution happening in video games. It was a no brainer to bullshit management into blowing a few hundred grand on a video game," said Jarvis, who had decided he wanted to make video games rather than pinball tables after playing *Space Invaders*. Jarvis soon found himself charged with developing Williams' comeback game. The game designer had clear ideas about what he wanted to do. He wanted to make what he called "sperm games" – video games that bristled with testosterone, stimulated adrenal glands, and would terrify and thrill in equal measure. He sought to make *Defender* the embodiment of his vision. "The inspiration for *Defender* was to somehow capture the physical rush and freedom of flying in a 2D game and throw in a believable world with cool enemies," said Jarvis. "And then, most importantly, give the player a real purpose – something to defend. The idea of defence as opposed to offence is so much more emotional. Protecting something precious from attack is much more visceral than randomly raping and pillaging aliens."

Defender was a high-speed race to destroy waves of alien attackers who were determined to capture the humans, spread across the game's horizontally scrolling game world. Captured humans would be lifted into the skies and used to turn weak aliens into fast, angry, laser-spitting mutants that would seek out the player. As a result, it was in players' self-interest to stop the aliens from capturing humans. Jarvis completed *Defender* just hours before its debut at the October 1980 Amusement and Music Operators Association trade show in Chicago – the highlight of the US arcade industry's calendar. Williams' return to the video game business was a big deal and the industry was keen to see what the company had come up with. Jarvis and the team were nervous: "None of us really had a clue whether the game was any good or not. Everything was so new at that time."

Defender's macho swagger proved too much for the trade show delegates. The sight of the game's controls – a joystick and five buttons at a time when one or two buttons were standard – scared off numerous delegates. Those who dared to step up to the daunting control panel found themselves beaten to a pulp within seconds of pressing the start button. "The show goers were old shiny-suit guys and blonde spokes-models," said Jarvis. "They didn't know a video game from a TV set. They played for 10 seconds and died." The delegates dismissed *Defender* as a failure. It was too hard and too complex to be a hit, they agreed. So they consigned it to their lists of no-hope games, the titles the industry expected to flop. Another game on that year's list was Namco's *Pac-Man*, the feminine yin to *Defender*'s masculine yang.[4]

Toru Iwatani, *Pac-Man*'s designer, had set out to challenge the status quo of the arcades with his maze game. "Most arcade video games of the time were violent and focused on the male player, so the game centres became places frequented mainly by men," he said. "We decided to change that demographic by designing games that could appeal to women and thus to couples, therefore making game centres desirable places to go on a date." After giving it some thought, Iwatani decided his game should be about eating. "When I imagined what women enjoy, the image of them eating cakes and desserts came to mind so I used 'eating' as a keyword," he said. "When I was doing research with this keyword I came across the image of a pizza with a slice taken out of it and had that eureka moment. So I based the *Pac-Man* character design on that shape."

For the look of the characters in his maze game Iwatani drew on the Japanese kawaii[5] art style he had already used in his previous game, the *Pong*-influenced *Cutie Q*. The cute, kitsch characters of kawaii originated in the art of early manga comics and anime films, but really took off in 1974 when the fashion accessories company Sanrio launched its Hello Kitty range of merchandise aimed primarily, but not exclusively, at teenage girls.

Kawaii characters resonated culturally with the Japanese so much that, by the dawn of the 1980s, the interest was growing rather than fading. Kawaii became so integrated into Japanese culture that kawaii characters can be found on everything from government posters and bank literature to computers and cooking pans. For Iwatani kawaii visuals had two advantages: "The hardware specifications at the time, compared to the present time, were very limited, so we could only have artwork in a very simplistic style and it was very difficult to create a sense of empathy for the player with this limited artistic style. But we wanted as many people as possible to enjoy the game, so by creating kawaii characters we thought we could appeal to women as well."

Iwatani's ideas resulted in a maze chase game where the player, as Pac-Man, has to eat all the dots in the maze while dodging four cute ghosts. Pac-Man's only defence was four power pills located in the far corners of the maze. If eaten, these pills allowed Pac-

---

4. *Pac-Man* was originally called *Puck-Man* and was released under that name in Japan. The game's US distributor, Bally Midway, worried people might vandalise the cabinet and change the P to an F. So they renamed it *Pac-Man* – the name used for the game ever since.

5. The literal translation of kawaii is 'cuteness'.

Man to eat the ghosts for a limited period of time, turning the player from the pursued to the pursuer. It was a simple but elegant game lifted by its charming kawaii looks. But few thought it would be popular. Namco doubted its potential. Namco's US distributor Bally Midway believed no one wanted to play maze games. The delegates at the Chicago trade show agreed. Instead they reckoned the hit in waiting was Namco's other offering *Rally-X*, a colour game where players had to drive a car around a maze spread over several screens to collect flags while being chased by other cars.

"Unlike the other exciting games that were around at the time, *Pac-Man* was designed for people to play with ease and when relaxed without 'excitement'," said Iwatani. "So when it was launched we didn't get the kind of review that other games got. I guess *Pac-Man* didn't have the 'sensational' image. I myself could not imagine that it would be loved by so many people and be such an international hit."

The industry veterans at the trade show were, however, wrong. Very wrong. *Defender* became a huge success as players sought to master the game in the hope of gaining kudos from conquering the most vicious game in the arcade. "Kids used to steal rare silver collectible quarters from their parents' coin collections, which were worth 10 to 100 times a regular quarter, to stick into *Defender*," said Jarvis. "The average *Defender* cabinet in the US would take in about 2,500 quarters a week. Since there were 60,000 *Defender* games out there, you would have up to 150 million quarters in the games every week. That is a lot of quarters." But even *Defender*'s success paled before the commercial juggernaut that was *Pac-Man*. For Twin Galaxies' Day, *Pac-Man* was the moment when the already rapid growth of arcades went into overdrive: "When *Pac-Man* came on the scene, it brought the female audience into the arcade and made the amount of income so great that businessmen started opening up arcades and, thereby, making games available in more places."

Pac-Man's cute kawaii characters were also ideal for merchandising and soon the pizza-inspired hero and the ghosts of Iwatani's mega-hit game started appearing everywhere. ABC-TV started showing a *Pac-Man* cartoon series that attracted 20 million viewers on its first broadcast. *Pac-Man* turned up on lunchboxes, Frisbees, stickers, yo-yos, sleeping bags and 'I brake for Pac-Man' bumper stickers. *Pac-Man* even scaled the heights of the pop charts thanks to Jerry Buckner and Gary Garcia, a song-writing duo from Arkon, Ohio, who worked under the name Buckner & Garcia. The pair discovered *Pac-Man* at their local bar. "We were drawn to the video game craze like everyone at that time and played most of the games," recalled Buckner. "There was a bar near a recording studio we worked at with a *Pac-Man* machine that we played every chance we got. At some point the idea for the song sprang up." Big record labels initially rejected their *Pac-Man Fever* song, but after Buckner & Garcia released it locally and sold 12,000 copies in a week, CBS offered them a deal. CBS re-released *Pac-Man Fever* in December 1981, the following March it hit number 9 in the Billboard Hot 100 chart selling more than a million copies in the process. CBS pushed Buckner & Garcia to make a whole album of songs about video games as quickly as possible to capitalise the success of their novelty single. "With only three weeks to complete the album we would go to a game room and look for a game that was hot and have the good players explain how to play it," said Garcia. "We would then go home and write the music for it and by the next day be laying the basic tracks for the song." The result of those rushed sessions was the *Pac-Man Fever* album: a saccharine pop snapshot of arcade life in early 1982.

Its eight tracks of sugary melodies name-checked some of the biggest games of the time from Sega's traffic dodging *Frogger* (*Froggy's Lament*) to *Centipede*, a shoot 'em up set amid the mushroom-strewn detritus of a forest that was created by former General Motors employee Bailey and *Asteroids* designer Logg (*Ode to Centipede*). Buckner & Garcia's lyrics captured a world of pockets brimming with quarters, intergalactic battles and calloused fingers. Sound effects taken from the games punctuated the tracks with blasts of white noise, eldritch beeps and the robotic monotone of synthesized speech. The album sold nearly a million copies and made Buckner & Garcia stars of the video game boom. They appeared on TV shows such as the Dick Clark-presented chart show *American Bandstand* and a special *Pac-Man Fever* day on MTV, an exciting new TV channel dedicated to music videos that had started broadcasting in August 1981.

Buckner & Garcia weren't the only people sharing in the success of *Pac-Man*. Atari, more by fluke than design, had found itself the holder of the exclusive rights to make *Pac-Man* on home consoles and computers thanks to a $1 million deal signed in 1978 when Namco had no hit games to its name. Atari couldn't believe its luck. For a relative pittance the company had gained control of the biggest game of the past decade. In April 1982, *Pac-Man* arrived on the VCS 2600 sending sales of the console through the roof. More than 12 million *Pac-Man* cartridges were sold worldwide. "*Pac-Man* was our all-time best seller. It was a phenomenon," said Ray Kassar, Atari's president. And with Namco owed no more than 50 cents from each of the $25 cartridges, most of the profit ended up in Atari's coffers. The *Pac-Man* cartridge confirmed the 2600's utter dominance of the home games market. The 2600's lead over its nearest rival, the Mattel Intellivision, was now approaching 20 million units. Atari had pretty much stopped worrying about rival consoles, it was now more concerned about the video game companies that had started releasing 2600 games to cash in on the captive audience Atari had built up with its console.

The challenge to Atari's control of the games released on its console started with a memo innocently sent by the company's product marketing group to the game developers in the home console division. The memo detailed the sales figures for 2600 game cartridges and was meant to help the team understand what types of video game were most popular. But instead of inspiring more successful products, it sparked a rebellion. The hackles of the division's game developers were already up when the memo landed on their desks. "The frustration began when Atari refused to pay a bonus program that was believed to be in place," said David Crane, the programmer who had converted Atari's bomb-dropping arcade game *Canyon Bomber* to the 2600. "Our department manager had negotiated a small royalty based on unit sales and when he later asked about that, he was asked 'what royalty?'. To stop the grumbling, managers went through and gave raises to key employees, but a line had been crossed." The product marketing group's memo reopened the royalties issue. "The memo was a one-page list of the top 20 selling cartridges from the previous year, with their per cent of sales. The purpose of the memo was the hint: 'These type of games are selling best…do more like these'. But this memo also showed us whose games did well, not just the game type. We noticed that four of the designers in a department of 30 were responsible for over 60 per cent of sales. And since we knew that Atari's cartridge sales for the prior year was $100 million, it was a shock to know that four guys making $30,000 per year made the company $60 million."

The four guys in question – Crane, Larry Kaplan, Alan Miller and Bob Whitehead – decided enough was enough and took the matter straight to Kassar. Miller put together a revised employment contract to present to Kassar, based on the kind of deals record labels gave their artists. "The four of us took this little sales statistic up to Kassar," said Crane. "Our point was that the statistics showed we must be doing something better than others. Since a game is a creative product, it is possible that one person is more creative than another and, therefore should be compensated accordingly. We were told that 'you are no more important to Atari than the guy on the assembly line who puts them together – without him we have no sales either'."

Furious at Kassar's dismissal of their arguments, the four quit Atari a few days later. With help from former music industry executive Jim Levy and $750,000 of venture capital investment the four rebels formed Activision, a company that would create and publish games for the 2600. It was a bold step. Until that moment only the manufacturers of video game consoles released the games. Indeed, Atari never even thought anyone else would make games for the 2600 and so had created nothing within the console that could prevent it. Activision's founders had declared war on their former employer and set out to smash Atari's monopoly on 2600 games. When Activision went public with its plans, Atari sued, hoping to crucify the fledging company and maintain its iron grip on the lucrative pool of 2600 owners it had spent millions cultivating. Atari's legal challenge backfired. The court backed Activision and ruled that Atari had no right to stop others developing games for the 2600. In July 1980 Activision's first three games – Crane's

Fishing Derby and *Dragster* plus Whitehead's *Boxing* – reached the shelves packaged in distinctive boxes that prominently displayed the names of their creators.

Activision's public promotion of each game's creator addressed one of the main complaints of Atari's programmers about their employer: the policy of keeping their names out of the public eye. "The fear was either that another company would try to steal them away or that the engineers would get an inflated sense of their worth and start making outrageous demands," said Howard Delman, co-creator of *Lunar Lander*. The reasoning may have made sense to Atari's management, but it angered its game developers who were starting to see themselves as the artistic pioneers of a new form of entertainment. The policy would prompt another of the company's leading VCS 2600 developers to resign in late 1979. Warren Robinett joined Atari in 1977 after completing a masters degree in computer science at Berkeley University, California.

After completing *Slot Racers*, a car-themed remake of *Combat*, Robinett was searching around for an idea for his next game when he encountered Don Woods and Will Crowther's text game *Adventure*. "I played *Adventure* at the Stanford Artificial Intelligence Lab in early 1978. My housemate Julius Smith was a grad student at Stanford and he took me up there," said Robinett. "Crowther and Woods' game took the nerd world by storm in 1978. I was just finishing *Slot Racers* then and needed to come up with an idea for my next game. The idea of exploration through a network of rooms, with useful objects you could find and bring with you and obstacles to get past, and monsters to fight – I thought this could work as a console game."

The 2600's limited capabilities and lack of a keyboard ruled out a direct remake of the text game, so Robinett reworked the ideas into visual form. The turns-and-text original was transformed into an action game where players ran around the screen dodging and fighting monsters and finding objects to allow them to access new areas as they searched for an enchanted chalice.[6] Officially there were 29 rooms in Robinett's *Adventure*, but, unknown to his colleagues, there were actually 30. The secret room was Robinett's protest against Atari's attempts to hide away its game creators. "Atari was keeping us game designers anonymous, which I found irritating," he said.

To access the 30th room players had to discover a hidden dot and use it in the right place to open an invisible doorway. Inside awaited the flashing words: 'Created by Warren Robinett'. "Atari had the power to keep my name off the box, but I had the power to put it on screen," he explained. *Adventure*'s concealed message was one of the earliest 'easter eggs' – a hidden secret within a video game for players who search carefully enough to

---

6. Robinett originally made it about finding the Holy Grail, but Atari's marketing department changed it to an enchanted chalice.

discover them.[7] Such secrets have since become a standard part of video games. Robinett was proud of his game. During its development, Atari's management felt he was being too ambitious and tried to stop him working on it. Halfway through its development, his boss told him to turn it into a game to tie in with the Warner's 1978 *Superman* film. His colleague John Dunn stepped in and used a copy of the half-finished game to create the *Superman* game, so that Robinett could finish his game. When *Adventure* eventually came out in late 1979, it became a big success selling more than a million copies worldwide.

Robinett, however, had already quit by the time it came out: "I thought I had done a pretty good job in creating the *Adventure* cartridge and did not get the slightest bit of positive feedback when I completed it. My boss initially thought it was impossible to do and told me not to do it; when I went and did it anyway, he did not see this as a good thing. He told me I was 'hard to direct'. When I told him I was quitting, he smiled. I guess I forgot to tell him that I had my name hidden in the final game code for *Adventure* that I had handed over to him." Robinett went on to join educational software publisher The Learning Company, where in 1982 he would create *Rocky's Boots*, one of the first successful educational games that taught Boolean logic using a puzzle game format.

* * *

Activision's decision to muscle in on Atari's console audience was well timed. Atari had released its *Space Invaders* cartridge a few months before the first Activision games arrived, causing 2600 sales to rocket. Activision's clever marketing coupled high-quality games such as the bomb-catching action of *Kaboom!* and the jungle adventure *Pitfall!* soon gouged out a sizeable share of the multi-million dollar 2600 game cartridge market. In 1981 Activision had achieved sales of $6.3 million, in 1982 this soared to $66 million.

The public profile of their developers soared in tandem with sales, leaving the company snowed under by thousands of fan mail letters every week. "Publicising our names provided all of the positives of celebrity and none of the negatives," said Crane. "I was never chased by the paparazzi but, in certain circles, there was pretty good name recognition. But the real thrill is hearing directly from a game player that your work touched them in some way. Because there was a name and a face behind the game, players were able to let me know directly how much they enjoyed playing my games."

Other Atari employees took note of Activision's success. Coin-op developers Howard Delman, Ed Rotberg and Roger Hector quit to form Videa in 1981 to make games for

---

7. Other games had contained easter eggs before *Adventure*. *Video Whizball*, a 1978 game for the Fairchild Channel F, also had an easter egg that displayed the name of its creator Brad Reid-Selth. The 1973 DEC GT40 version of *Lunar Lander* that inspired Atari's arcade remake featured a McDonald's restaurant that appeared if the player landed in the right spot.

Atari and other arcade companies. "There was a lot of money being made in the industry, but the fraction coming to the engineers was small relative to the profits," explained Delman. "It occurred to some of us that being a contractor to Atari, or any game company for that matter, could be far more lucrative than being an employee."

That same year another group of employees from the home console division decided to follow Activision's example. Backed with $2 million of venture capital, they founded Imagic on 17th July 1981 with the goal of publishing games for the 2600. Among the Imagic team was Rob Fulop, the author of the 2600 version of *Space Invaders*. "We were authors and we didn't feel like authors at all. We weren't compensated based on how good our work was perceived; our name wasn't on the game. So we left. I wasn't involved in getting the funding for Imagic; someone else did that and invited me to the party. It took me about two seconds to say yeah."

Imagic's debut game, Fulop's *Galaxian*-inspired *Demon Attack*, became one of the best-selling 2600 games of 1982. Manny Gerard, the Warner executive responsible for overseeing Atari, felt the exodus of talent at that time was inevitable. "Entrepreneurial guys go off and that's exactly what happened," he said. "Guys see a way to make money and they run off and they build companies. Atari was getting bigger and it was not as entrepreneurial as it was. It happens. It's the natural evolution of things."

But Activision didn't just inspire Atari employees to walk. It also encouraged companies unconnected to Atari to start releasing 2600 games, creating new rivals such as Quaker Oats' U.S. Games division, Xonox and Fox Video Games.

Atari may have resented the companies seeking to grab a slice of what it regarded as its market, but their existence did little to damage the video game giant's income. By 1982 Atari had become the single biggest business in the Warner Communications conglomerate. It had spent $75 million promoting its products in 1982, more than Coca-Cola and McDonald's. Its sales were more than five times that of Warner's film and music businesses and 70 per cent of Warner's profits came from Atari. As a consequence Warner's share price ballooned from just under $5 a share in 1976 to $63 in 1982. "We made more money than god," said Noah Anglin, a manager in Atari's coin-op division. "We made more money than Warner's movie division. We went from being a mention in their corporate magazine to where we were their corporate magazine."

And with cinema ticket and record sales being hit as teenagers swapped vinyl and the silver screen for the electronic thrills of the arcade, the video game looked unstoppable. In the 48 months since *Space Invaders*' release, the video game had conquered North America. Its relentless ascent marked the biggest revolution in entertainment since the arrival of the TV set. And then, suddenly, everything fell apart.

City to Atari: 'E.T.' trash go home

By M.E. McQUIDDY
Daily News Staff Writer

Alamogordo Daily News

Vol 89—No. 221   Alamogordo, New Mexico, Tuesday, September 27, 1983   25° (Less by Carrier)

Planet forum set October 1

Rail merger not takeover

Alamogordo reacts to influx of Atari junk
Alamogordo Daily News, 27 September 1983

CHAPTER 8

Devilish Contraptions

On the 9th November 1982, the US Surgeon General Dr Everett Koop took to the stage at the Western Psychiatric Institute and Clinic in Pittsburgh. Dr Koop's distinctive bushy beard gave him the air of an Old Testament prophet, which was apt given the impassioned plea he was about to make. On stage he railed against society's failure to challenge domestic violence and child abuse. "If we truly care about human life, if we truly care about the future of our society, then we have to move to confront the terrible implications of family violence," he declared before urging the medical professionals who had gathered to hear him to look out for the signs of such abuse.

After finishing his speech, Dr Koop took questions from the audience. One questioner asked what he thought about the effect of video games on young people. There may be mental and physical harm because teenagers were becoming addicted "body and soul" to these games, Dr Koop replied. "Everything is 'zap the enemy', there's nothing constructive," he added, before conceding there was no scientific evidence to support his view. The next day the newspapers reported how the surgeon general had let rip on video games. His call for action against domestic violence and child abuse went ignored. "Surgeon General sees danger in video games," reported the Associated Press news agency while *The News & Observer* in North Carolina ran a cartoon called *Koop-Man*, showing Dr Koop's bearded and open-mouthed head chasing a worried-looking Pac-Man. Dr Koop immediately released a statement emphasising that his comments were not government policy: "The comments represented my purely personal judgment and was not based on any accumulated scientific evidence. Nothing in my remarks should be interpreted as implying that video games are, per se, violent in nature or harmful to children."

The surgeon general's views did, however, echo widespread concerns about video games. Parents, teachers and officials worried that video game arcades were hubs of delinquency, places where children would be led into a life of crime or drug addiction. Reports in medical journals of new aliments connected to video games, such as *'Space*

Invaders wrist', fuelled the distrust as did rumours of teenagers dying from heart attacks after playing games for hours on end.

By 1981 these fears were resulting in action as communities across the US attempted to suppress video game arcades. From New York and Texas to Florida and Milwaukee, arcades were being hit with new restrictions and, in a few places, outright bans. These concerns were by no means limited to the US. In the UK, Labour MP George Foulkes tabled a motion in Parliament calling for a law that would give local authorities the power to ban arcades. He accused video games of extracting "blood money" from "the weakness of thousands of children". His call prompted a furious retort from Conservative MP and *Space Invaders* fan Michael Brown who labelled the motion a "petty-minded, socialist measure". Foulkes lost the vote. Some countries, however, did introduce bans. In late 1981 both Indonesia and the Philippines outlawed video games on the grounds of protecting the morals of the young. The Philippines government called video games "devilish contraptions" and threatened those who flouted the ban with up to 12 years in prison.

In response to the rising tide of restrictions, arcades began calling themselves 'family entertainment centres', sought to brighten up their poorly lit facilities, and imposed strict rules on behaviour to reassure parents. Some started requiring players to become members in order to play. Atari responded with its Community Awareness Program, a service that supplied its customers with information they could use to combat local attempts to restrict arcades. Only a few brave politicians swam against the tide of moral panic about the arcades. One was Jerry Parker, the mayor of Ottumwa in Ohio. After being lobbied by the city's Twin Galaxies arcade, Parker became an outspoken defender of video games. "He was a very bold man," said Walter Day, the owner of Twin Galaxies. "Hundreds – if not thousands – of other communities and governmental bodies were legislating against video games. Jerry Parker bucked the international trend and proved himself a world-class leader who was willing to take a chance with his career."

But those clamouring for a clampdown need not have worried, because just 28 days after Dr Koop's speech, the video game bubble burst. And it was Atari that brought the boom to a swift end. On the afternoon of the 7th December 1982, Atari announced its expected growth figures for the fourth quarter of the year. Up until then, investors had been led to expect growth of around 50 per cent thanks to the new Atari 5200 console and the release of the *E.T. The Extra-Terrestrial* game on the VCS 2600. Instead, Atari slashed its growth prediction to between 10 and 15 per cent. Investors were shocked. The share price of Atari's parent company Warner Communications collapsed by more than 30 per cent. Atari's announcement crushed investor confidence in the prospects of video games. The investors who had bankrolled the rapid expansion of the business pulled their money out and North America's video game industry imploded. During the next two years, many of the companies that built the business would be destroyed or left as shrivelled wrecks. Atari received much of the blame for the crash, but the causes were far more complex and multi-faceted than the failings of one company.

Some of the seeds of destruction were sown when *Pac-Man* took the US by storm. The game's enormous popularity prompted companies all over the country to start buying arcade machines on credit in the hope of earning fortunes by installing them in locations as unlikely as golf clubhouses and dentist receptions. In the hubris of the boom years it seemed like a great idea, but soon it became clear these locations lacked the volume of passing trade needed to make the machines profitable.

With no money coming in, these companies soon began to default on their debts leaving coin-operated game manufacturers with bad debts worth millions of dollars. "Everything's based on a pyramid scheme to a degree, everyone coming in and financing the next expansion on credit," said Ed Logg, the Atari coin-op designer who made *Asteroids*.

The sheer volume of arcades that opened added to the problems, spreading the finite audience for video games too thin for any arcade operator to make a living. Desperate for customers, some arcades began offering eight rather than four goes per dollar on their video games – reducing income even further. "Too many arcades had opened," said Day. "They were taking customers away from each other at the same time that more and more people were investing in home game systems. Eventually there were about four arcades in Ottumwa by 1984 and the city could not support so many arcades. We all went out of business at the same time."

Arcade goers were also tiring of the increasingly difficulty of the games on offer. "Games were becoming too hard," said Scott Miller, the co-author of *Shootout: Zap the Video Games*, a 1982 guide to beating arcade games. "Arcade makers figured out that too many players could master their games and play for hours on one quarter, so they defeated these players – including me – by greatly increasing the difficulty." The evolution of shoot 'em ups during the boom was typical. By 1982, *Space Invaders* looked sluggish and tame next to the fearsome *Defender* and dizzying fury of *Tempest*. "You had a player base that lived for the challenge and were becoming more and more highly skilled. So you had to up the ante with each game to continue the challenge and thrill the players," said Eugene Jarvis, the designer of *Defender*.

Noah Falstein, the co-designer of intimidating 1982 shoot 'em up *Sinistar*, agreed: "As players got better at them, coin-op games got more challenging in order to keep the coin drop high. In the case of *Sinistar*, the development team actually had an easier version ready to release, but our management insisted on making it tougher to keep it more profitable. I don't actually disagree with this, you have to be careful about profitability, but I do think it contributed to the collapse of the arcade market."

Dedicated video game players thrived on the ever-greater challenges thrown at them, but the mainstream audience, upon whom the boom was built, found them too demanding, poor value for money and not much fun. The final blow that felled the arcades was the growth of home console ownership, which sucked players out of the arcades. In 1981 the coin-op video game business in the US peaked with annual sales of $4,862 million, in 1984 sales had nearly halved to $2,500 million.

The home console market's fall was not far behind, however. The success of Activision, the company formed by four ex-Atari employees to make VCS 2600 games, had encouraged dozens of other businesses to follow suit. These companies churned out poor-quality games in the hope of making a fast buck from the excitement surrounding video games.[1]

"Activision was the main cause of the crash – although indirectly," said Activision co-founder David Crane. "We showed that you didn't have to spend $100 million to produce a game console to make money in video games. In one six-month period 30 new companies sprang up trying to duplicate our success." The volume of games and the dubious quality of many of them started to put customers off. "There was way too much product, some of it inappropriate," said Manny Gerard, the Warner Communications' vice-president who oversaw Atari. "The single greatest failing was built into the 2600 from the very beginning, although nobody understood it at that point, which was we couldn't control the software for our system. People were putting out cartridges for the 2600 – one was called *Custer's Revenge*."

Custer's Revenge was one of three sex-themed games released in the autumn of 1982 by American Multiple Industries, a video game publisher formed by porn filmmakers Caballero Control Corporation. It was both terrible and downright offensive: the aim was to rape a Native American woman tied to a post. The game's launch in New York City attracted 100 protestors armed with placards declaring: "*Custer's Revenge* says rape is fun" and "Pornographers are pimps". Atari was furious, but could do nothing to stop the release of *Custer's Revenge*. "There wasn't anything we could do about it – it was terrible. We had no control over that because we couldn't control the software," said Gerard. *Custer's Revenge* and dozens of other dismal games, including adverts disguised as games such as the toothpaste promotion *Johnson & Johnson Presents Tooth Protectors*

---

1. Profiteering is only one reason why these games were poor quality. Knowledge of what made a great video game was limited and game developers usually worked alone, building games how they saw fit without knowing if there was a better way of working. Atari game designer Chris Crawford's 1982 book *The Art of Computer Game Design*, the first book ever published on the subject, highlights just how haphazard game development was at this time. At one point Crawford pleads with game designers to use professional play testers rather than just asking friends what they thought of their latest creation.

and the pet food plugging *Chase the Chuck Wagon*, delivered death by a thousand cuts to Atari's console.

With so much dross clogging up the shelves, sales stalled and retailers found themselves lumbered with piles of unsold games. Shops did what shops do with unsold goods – they discounted them in the hope of getting rid of the excess stock. Soon games that once sold for $30 could be bought for less than $10. Retailers also stopped ordering new games, causing cartridges to pile up in video game companies' warehouses. These warehouses full of unsellable games were a ticking time bomb for the video game business. "We predicted the crash. I remember saying that 'none of these new companies will be in business in a year'," said Crane. "What we didn't realise is that each company already had a million game cartridges in their warehouse when they went under. It was the sale of these games by liquidators that flooded the market. The liquidators bought them out of bankruptcy for $3, sold them to retailers for $4 and the retailers put them in barrels at the front of the store for $5. When dad went in to buy junior the latest Activision game for $40, he saw that he could be a hero and get eight games for the same money. Sales of new games went to near zero."

Companies such as Activision and Atari had no choice but to slash their prices to shift the cartridges now building up in their own warehouses. A vicious cycle from which no company could escape had begun.

The ageing technology of the 2600 did little to help. The five-year-old system was looking its age and people were growing bored by its limitations. Yet by the start of 1982 no convincing alternative had emerged. Mattel's Intellivision – the nearest challenger – offered too small a technological leap to pry gamers away from their Ataris and, instead of seeking to build its own successor to the 2600, as suggested by the company's founder Nolan Bushnell in 1978, Atari had stifled research projects that might undermine its flagship product.

"There were a number of projects that were started and brought to the point of being ready for production and then stopped," said Steve Bristow, vice-president of engineering at Atari. "I heard words to the effect of 'why should we take risks on introducing a new video game that is possibly going to cannibalise our sales?'" Only by 1982 did it become clear that Atari needed a replacement for the 2600 and fast. Its answer was the Atari 5200, a repackaging of its 1979 home computer the Atari 400. It was too little, too late. An Atari focus group held just before its launch confirmed the worse when the Atari 5200 was put up against the newest console on the market: the Colecovision. "Overall, consumer reactions after game play was that Colecovision performed somewhat better than expected," reported an internal Atari memo about the focus group. "The 5200 did not come out as definitely superior to Colecovision despite some initial expectations that it would be a better system." The Colecovision, created by toy company Coleco, arrived in August 1982 in a blaze of publicity. It was more advanced than the Atari 5200 and, most importantly, came with a copy of Nintendo's hugely popular *Donkey Kong*.

Donkey Kong was the first game designed by Shigeru Miyamoto, who would go on to be regarded as one of the world's very best game designers. The Japanese designer's debut game was commissioned to pull Nintendo's US operation out of a hole. Nintendo of America had bet everything on *Radar Scope*, a *Space Invaders*-style shoot 'em up that had been a hit in Japanese arcades, but sold only 1,000 of the 2,000 machines it built for the US market. Nintendo decided to create a new game to run on the technology used by *Radar Scope* in the hope of shifting the unsold machines.

Miyamoto was originally told to make a game based on *Popeye*, but when Nintendo failed to get the rights to the comic strip, he devised an entirely new game inspired by the 1933 film *King Kong* and the fairy tale *Beauty and the Beast*. It revolved around three characters: Jumpman, a moustached and stumpy carpenter who the player controlled; Donkey Kong, an escaped giant gorilla owned by Jumpman; and Pauline, the object of both Donkey Kong and Jumpman's affections.[2] Players had to help Jumpman climb scaffolding and ladders to reach the top of the screen, where Donkey Kong was holding Pauline hostage, while dodging barrels thrown by the angry ape and other dangers. Miyamoto's distinctive characters and bizarre love triangle plot – told in short animated sequences reminiscent of a silent movie – were revolutionary. The game's jumping action and platform-based levels were equally influential, establishing a new genre of game: the platform game.[3] Following the game's success, Nintendo changed Jumpman's name to Mario in honour of its US landlord Mario Segale, who had agreed to give the company's struggling US arm more time to pay its rent prior to Donkey Kong's release.

Michael Katz, Coleco's vice-president of marketing, felt the *Donkey Kong* deal was vital to the Colecovision: "I don't think the Colecovision would have been launched as successfully as it had if we didn't have the exclusive console rights to *Donkey Kong*. We made it so it was the only way you could get *Donkey Kong* for the home." By Easter 1983, more than a million Colecovisions had been sold off the back of *Donkey Kong*. The release of an adaptor that allowed VCS 2600 games to be played on the Colecovision

2. Miyamoto picked the name after using a Japanese-to-English dictionary to get a translation of the word 'stubborn' and getting the word 'donkey'.

3. *Donkey Kong* was not the first platform game, but it did popularise the genre. The first platform game is probably Universal's 1980 arcade game *Space Panic*, although it lacks the jumping action that would become a hallmark of platform games. Earlier games also had a significant influence on the genre.

spurred sales on even further. Atari had been offered the rights to *Donkey Kong* by Nintendo but turned it down on the grounds that the Japanese company wanted too much money. The decision left Atari facing a powerful new rival that had wiped out its Atari 5200 system just as the 2600 market began to unravel.

Not all the problems affecting the video game industry were of its own making. The US had been in a deep recession and by December 1982 one in 10 American adults were out of work. Petrol prices were also rising, eroding households' disposable income even further. "The gasoline shortage just sapped money away from kids," said Gerard. "If you're an average kid and the way you get around America is in your car and suddenly gasoline prices go nuts, which they did, that hurt."

On top of that, the video game console had lost its position as the most exciting thing in home entertainment to the video cassette recorder, or the VCR for short. The VCR reinvented television, giving people control over what they watched and when for the first time. "It was a major thing," said Rob Fulop, a programmer at game publisher Imagic. "All of a sudden you could see a movie at home whenever. It was amazing. Kids were watching and taping movies, computer games weren't what they did anymore."

As the games business plunged, the VCR went from strength to strength. In the first quarter of 1982, Americans bought 491,000 VCRs. The first quarter of 1983 saw 958,000 sold, an increase of 95 per cent.

The final blow came from the home computer manufacturers who became embroiled in a bitter price war just as the console market hit the skids. The price war began in April 1980 when Jack Tramiel, the founder of Commodore, paid a visit to London, England. Together with Apple and Tandy, Commodore had started the home computer business but while its PET computer did well in Europe it was lagging behind its two key rivals in the US. To add to its American woes, two other big players had entered the home computer business.

One was Atari. The other was Texas Instruments. For Tramiel, who was fond of saying 'business is war', Texas Instruments' entry into the computer business offered an opportunity for revenge. In the mid-1970s the two companies had fought for dominance of the pocket calculator market and the resulting price war almost destroyed Commodore. Tramiel was determined to make sure that this time round it would be Texas Instruments that would be left in ruins.

During his business trip to the UK, Tramiel saw the idea that would form the basis of his assault on his corporate nemesis: the Sinclair ZX80. The British-made computer had outdated technology and was sold as a kit so buyers had to assemble it themselves at home, yet it was hugely successful for one reason: its unbelievably low

£99.95 price tag. Excited by the idea of a computer anyone could afford, Tramiel tore up Commodore's plans for a new business machine and ordered his engineers to make the computing equivalent of the Ford Model T, the 1908 car that introduced the idea of mass car ownership.

Commodore was in a good position to deliver on such a vision. It owned microprocessor manufacturer MOS Technologies and so could get prices that Tandy, Atari and Apple could only dream of. Only Texas Instruments had such an advantage, but its $1,150 TI-99/4 computer was far from mass market. Commodore's engineers built the VIC-20, a colour computer that cost just $299.95, and launched it in 1980. The VIC-20 was an attack on two fronts. It undercut Commodore's computer rivals by hundreds of dollars, forcing them to slash their prices.[4] It was also cheap enough to compete with video game consoles on price, a fact Commodore emphasised with adverts asking: "Why buy just a video game?"

Texas Instruments responded in 1981 by replacing the overpriced TI-99/4 with the $525 TI-99/4a, a home computer designed for the mass market. The war was on. The fight, however, was not just a contest between rival manufacturers, but a struggle between two different and incompatible visions of home computing. Commodore embraced a philosophy of openness, allowing anyone to create software for its computer. Texas Instruments, meanwhile, believed in control. It wanted computing to follow the video game console model, where it and it alone would make and profit from TI-99/4a software. To enforce its beliefs, the Texan giant publicly threatened to sue any company that released TI-99/4a software without its approval.

Video games would prove crucial in Commodore's battle with Texas Instruments. While Texas Instruments pushed the educational benefits of its computer, Commodore embraced fun as a way of attracting buyers. "In 1982 no one was buying computers for the software, other than for games or something like that," said Bob Yannes, who would help design the Commodore 64, the computer Tramiel used to finish off Texas Instruments.

Launched in August 1982, the Commodore 64 was a $595 powerhouse for its era. Armed with a large amount of memory, strong graphics capabilities and an advanced sound chip, it seemed as if it was designed for games. Its release prompted Texas Instruments to offer a $100 rebate on sales of its TI-99/4a, sparking a frenzy of cost cutting as rival computer firms repeatedly tried to undercut each other in a deadly game of corporate Russian roulette.

---

4. Only Apple decided not to get involved in the battle for the mass market, preferring to carve out a more expensive and aspirational niche for its computers.

By early 1983 the benefits of Commodore's openness were starting to shine through. While Commodore owners enjoyed a wide choice of software and games, TI-99/4a owners were being starved of choice. Texas Instruments, which outsold Commodore in Christmas 1982, started to fall behind. By the summer the Commodore 64 was on sale for just $200 and Texas Instruments had been pushed into selling its computer for just $99. In November 1983, having lost $100 million in the second quarter of the year alone trying to keep the TI-99/4a alive, Texas Instruments threw in the towel and shut its home computer operation.[5] Tramiel had won and Commodore now had a 38 per cent share of the fast-growing market for home computers costing less than $1,000.

The computer wars delivered another nail in the coffin of home consoles. "Home computers replaced home video games," said Chris Crawford, a member of Atari's corporate research team at the time. "The price of a home computer system was only about twice that of a home video game system and the software was cheaper and much more plentiful. Combine that with the loss of confidence in the Atari VCS engendered by disasters such as *E.T.* and you can see why sales of the VCS simply collapsed."

Home computers did not, however, offer the big profits game makers were used to. While computer games were cheaper to produce, the market was smaller, the sale price lower and games stored on floppy disks were easier to copy illegally than cartridges. The move to home computers may have offered refuge from the chaos elsewhere, but it came at the cost of massively reduced profits, which in turn forced massive lay offs of developers. Sierra Online was one of the companies forced into such a situation. At the height of the boom it joined the console game bandwagon, only to lose huge sums when the bottom fell out of the market. It retreated to its computer game origins. "We were essentially bankrupt," said Ken Williams, Sierra's co-founder. "Luckily we had never had a bank line and weren't really in a hole, we just had no money. By hunkering down and laying off almost everyone, we were able to start over." Al Lowe was one of the game designers laid off by Sierra in the spring of 1984. "One Friday that spring they went from 120 employees to 40 in an afternoon. It was a black Friday," said Lowe.

The crash also spelled the end for the vector graphics game. Arcade operators were already fed up with the unreliability of these machines by the time the bubble burst. "A vector monitor has to control and dissipate much more energy than a raster monitor

5. Texas Instruments' exit from the computer business also ended board game firm Milton Bradley's involvement in the video game business. The company had already tried to crack the market with 1979's MicroVision, the first handheld games console, and the vector graphics console, the Vectrex. Just prior to Texas Instruments' decision to quit the home computer business, the Texan firm and Milton Bradley had launched the MBX, a video game console with voice recognition and speech synthesis features that came in the form of an add on to the TI-99/4a. Texas Instruments' abandonment of the TI-99/4a killed the MBX as well.

of the same size. More energy means more heat and more expensive parts. Bad for reliability," said Atari coin-op game engineer Michael Albaugh. The difficulty of repairing vector games added to arcade operators' frustration, said Logg: "Vector monitors are hard to replace. If your regular colour monitor goes down fine, no problem, every TV in the world is a monitor. Just fix the appropriate controls to make it work. But a vector monitor, that's a different matter."

As revenue from video games fell, arcades stopped ordering vector games, which were also under pressure from rapid improvements in the visuals of games that used standard raster TVs. "Raster graphics were getting much better: more colours, better graphics resolution," said Tim Skelly, a designer of vector coin-ops for Cinematronics. "Vector graphics had nowhere to go except colour and that didn't add much. The screen was still black or, at best, a static background like *Warrior*."

By 1982 games such as Sega's shoot 'em up *Zaxxon* and Gottlieb's *Q*bert* were ramming home the advantages of standard TVs. Both games pioneered the use of axonometric projection; a drawing technique that let game designers create 3D worlds using 2D images.[6] Until then, the technique had mainly been used for technical drawings where there was a need to show depth as well as height and width. Or, most famously, in the perspective trickery of Dutch artist M.C. Escher's work, which inspired Jeff Lee, the artist on *Q*bert*, to use the approach: "Being a fan of the great Dutch artist M.C. Escher, the master of optical illusions, I constructed a stack of triad-based cubes. Admiring my derivative handiwork, it struck me there's a game in here somewhere. The pseudo-3D look was quite compelling." The game challenged players to help a fuzzy orange creature with two spindly legs and an elephant-like trunk hop around a pyramid built from Lee's Escher-inspired cubes until all had been stepped on, while dodging other strange but deadly beasts. *Q*bert*'s cute and cuddly looks made the game popular enough to spawn a spate of merchandise adorned by its bizarre hero.

Another important innovator in standard TV visuals was Namco's *Xevious*. The shoot 'em up was a labour of love for its Japanese creator Masanobu Endo – he even wrote an entire novel just to flesh out the back story to his game of aerial combat on alien planets. *Xevious* was a visual feast. Its action took place above green grasslands cut up by alien highways and dusty deserts where huge geoglyphs similar to the Nazca Lines of Peru had been etched into the dirt. The metallic alien craft and defensive bases the player fought with were equally impressive, particularly the spinning, shimmering flying saucers that marked the player's first encounter with the extra-terrestrial forces.

---

6. Games that use axonometric projection are sometimes referred to as having an isometric viewpoint. Isometric projection is just one of type of axonometric projection and many of the games described as isometric use different types of axonometric visuals.

The game was, however, not just about looks. *Xevious* set the template for post-crash shoot 'em ups. The player's craft was set on an unstoppable, pre-defined journey – travelling up the screen at a steady pace. With the decision about where to go removed, the player could concentrate on weaving around the screen to avoid enemy fire and picking off enemies who attacked in predictable patterns. The only time the movement stopped was when the player came face-to-face with a boss – a big, super-powered opponent – that took large amounts of firepower and agility to defeat. Together with Konami's *Scramble*, Endo's game became the dominant blueprint for shoot 'em ups, especially those made in Japan, for the best part of a decade.[7] The fixed-screen action of *Space Invaders* and its clones, and the player-directed travel and openness of *Defender* became the hallmarks of an earlier era.

The stunning visuals of games such as *Xevious* ensured a swift end to the vector game. Atari's grandiose *Major Havoc* marked the last gasp for the graphical approach. Designed by Owen Rubin with help from Mark Cerny, *Major Havoc* sprawled across game genres. It opened with a 3D space fight against 'robot fish' then became a *Lunar Lander*-type game before changing into a platform-maze game hybrid where the player had to guide the character Major Havoc through low-gravity mazes filled with traps to set a nuclear reactor to blow up and then escape before it exploded. There was even a version of *Breakout* hidden within the game. "It was rather ambitious," said Rubin. "A normal game took six to eight months, this one took almost 18 months. The game kept evolving and was put on test several times and tested well, but was incomplete. Why did Atari let me just keep working on it? I have no idea. In hindsight, I am glad they did. I only wish we had done the game in non-vector graphics though because it would have sold so many more. Vector games had a bad reputation by then because they broke all the time."

By the time *Major Havoc* finally made it to the arcades in November 1983, the arcade and console game industry was in ruins. Few had seen the crash coming. Bill Grubb, the president of Imagic, started 1983 boasting to the press about his plans to spend $10 million advertising the company's games that year. By the end of the year Imagic was mortally wounded. "We thought the boom would go on forever," said Fulop. "Like any hot thing, the people who are there assume it's going to go on forever. And especially when you're young, you can't imagine anything would change. It was a total shock. I still haven't got over it."

Imagic tried to survive by making home computer games but the damage was too

---

7. *Scramble* also had a fixed direction and speed of travel for the player's spaceship, but the game's world moved horizontally rather than vertically as in *Xevious*.

deep and the company eventually closed down. Quaker Oats killed off its U.S. Games division in April 1983. Gottlieb watched its sequel to *Q*bert*, *Q*bert Qubes*, sink. Nintendo's *Donkey Kong 3* was met with apathy. Unable to pay its debts, Cinematronics filed for Chapter 11 bankruptcy protection to try and stave off its demise.

Soon some of the industry's leading developers started looking elsewhere for employment. Eugene Jarvis, the designer of the incredibly successful *Defender*, had scored one of the last big hit coin-ops with 1983's *Robotron: 2084* – a claustrophobic shoot 'em up where players battled swarms of robotic attackers. "Just as *Defender* was about freedom and speed, *Robotron* was about confinement with slower, more precise motion," said Jarvis. "It is amazing how often in *Robotron* you think you're dead and then realise somehow you escaped an incredibly tight spot. The adrenalin is intense." George Orwell's dystopian novel *1984* inspired the game's story of defending the last human family from killer robots. "It was working off the Orwellian theme. It was clear in '82 that nothing was going to happen in 1984 so it had to be 2084 and instead of humans pulling the strings it would be robots," said Jarvis. "The theme was based upon an extrapolation of Moore's law, the inexorable doubling of computer power every 18 months. Sooner or later they will be smarter than us because instead of trying to double our intelligence we are trying to halve it with meth and medical marijuana."

But after his next game, the gaudy *Blaster*, flopped Jarvis decided it was time to abandon the sinking ship. "*Blaster* was an early attempt at a 3D space flight genre," he said. "It was pretty fun, but was released after the event horizon was crossed by the industry into the black hole of the mid-'80s. It really seemed that the industry was done. Every possible game had been invented and all creativity was exhausted. Just like the Hula-Hoop, Pet Rock and disco crazes of earlier eras, it was over. I thought I would recycle myself with an MBA and get a regular job."

For those who stayed behind, hopes that the situation would improve faded fast. "People weren't aware of the speed or magnitude of the crash. Once a company gets big, there's a feeling that 'we can do no wrong'. I certainly was taken by surprise at the velocity of events," said Atari's Crawford. "We kept scaling back and thinking 'this time we've gotten on top of the problem' and things just kept getting worse."

The only glimmer of hope amid the darkness came in the summer of 1983 in the form of a knight in shining armour. His name was Dirk the Daring and he was the animated star of *Dragon's Lair*, an arcade game released by Cinematronics that used a new data storage format called laserdisc. Just like CDs or DVDs, laserdiscs were a type of optical disc that had a diameter of 30 centimetres and were designed for playing movies.

Dragon's Lair looked like an interactive cartoon and was created with the help of

Don Bluth, a former Disney animator who had formed his own studio, Sullivan Bluth Studios, which had recently made the animated feature film *The Secret of NIMH*. *Dragon's Lair* was the idea of Rick Dyer, owner of Rick Dyer Industries, who was looking for ways to use laserdisc's ability to store pre-recorded video in a game. "Dyer felt that using the laserdisc player would allow the creation of an 'interactive movie game' and that our animation would be the perfect format for the game," said Bluth. "In October 1982, Rick brought Cinematronics' co-owner Jim Pierce to our facility in Studio City, California, to discuss the possibilities. When they left, we kind of stared at each other and wondered what they were talking about. We were filmmakers and knew very little about video games, let alone video game production. But, once Rick explained what he was trying to achieve, we felt that we could figure it out."

The answer was to create a cartoon and chop it into pieces. Players would choose how to react at the appropriate moment, triggering the next slice of animation. But since the story was fixed, the game really consisted of players working out which action they had to do to see the next section of the game. "Laser games didn't really provide true interactivity," said Bluth. "It was more of a 'memory' game, learning when and which way to move the joystick or when to hit the action button."

Despite only having a veneer of interactivity, *Dragon's Lair*'s cartoon visuals made it hugely successful.[8] Cinematronics sold 10,000 *Dragon's Lair* machines in a little over three months. For a moment it looked like laserdisc could save the video game business. Other laserdisc games followed, such as Stern Electronics' *Goal to Go*, an American football game that used footage from real-life matches, and Sega's *Astron Belt*, which cannibalised special effects-laden scenes from the movie *Star Trek II: The Wrath of Khan*. Dyer started work on the Halcyon, a computer, console and laserdisc player hybrid designed to bring laserdisc games into the home. "Laser games were very popular," said Bob Lawton, owner of the Funspot arcade in Weirs Beach, New Hampshire. "We had one set up by our front entrance and even had a television on top of it so other people could see what was being played. Huge crowds would gather whenever a good player got on the machine and they would cheer when he did well."

But the laserdisc fad died almost as quickly as it arrived. "After a while, the problems of a home laserdisc player started to surface and the constant breakdowns of the laserdisc player spelled the end of laser games for us," said Lawton. "The laserdisc players being used were never intended for commercial use or constantly searching for scenes to play. This constant scene searching eventually wore out the disc player."

---

8. *Dragon's Lair* was not the first laserdisc game. A Californian coin-op manufacturer called Electro-Sport got there first in 1982 with *Quarter Horse*, where players had to guess which horse would win the races stored on its laserdisc.

The lack of interactivity also meant players soon lost interest. It was the final straw for Cinematronics, who went bust when the excitement about laserdisc games evaporated. Dyer's company Rick Dyer Industries closed down shortly after the release of the $2,500 Halcyon in January 1985.

The crash devastated the home console market: it peaked in 1983 with US sales of $3,200 million before withering away to a $100 million-a-year industry in 1986. As the money disappeared, many of the companies that had built the video game business during the 1970s and early 1980s disappeared. Magnavox, the company that released the first game console, cancelled the release of its Odyssey 3 system and left the business. Mattel, the birthplace of handheld gaming, gave up on the Intellivision after losing tens of millions in 1983. Adventure International, the company that brought text adventures to home computers, vanished after risking everything betting on Texas Instruments' bid to conquer the personal computer market. "To me the crash came when the TI-99/4a was discontinued," said founder Scott Adams. "That was the real big dip that did us in. We didn't have the deep pockets to ride out that period." Coleco gave up on the Colecovision after getting burned in the home computer market and decided to concentrate on its Cabbage Patch Kids line of dolls. Arcade giant Bally Midway's video game revenues plunged by 60 per cent and it responded by shutting down Dave Nutting Associates, the company that pioneered the use of microprocessors in video games.

Atari was transformed from one of the biggest business success stories ever seen into one of the biggest disasters in corporate history – losing so much money so fast that it threatened to bring down the whole of Warner Communications. Warner desperately tried to save its ailing cash cow. It cancelled the long-held plan to move Atari into a purpose-built campus in Silicon Valley. It moved the manufacturing arm to Hong Kong and fired thousands of employees. It fired Atari president Ray Kassar, slashed marketing budgets and cancelled research and development projects. And when all of this failed to stem the losses, Warner broke Atari in two. The profit-making coin-op division became Atari Games. The computer and console divisions became Atari Corporation and were sold off in July 1984 for $240 million to Tramiel, who had resigned from Commodore that January after clashes with the company's principal shareholder Irving Gould. "Tramiel bought it and he basically abandoned the games business. It's one of the great mistakes in history because there was still a business and his walking out left it wide open for Nintendo," said Gerard. The company that had built and dominated the video game industry would never fully recover.

If any single game summed up both the excesses of the boom years and the pain of the fall, it was *E.T. The Extra-Terrestrial* – Atari's big VCS 2600 game for Christmas

1982. Steven Spielberg's 1982 summer blockbuster, a tale of a friendly alien stranded on Earth, had become one of the biggest grossing films of all time. In a bid to ingratiate himself with the hottest director in Hollywood, Warner chairman Steve Ross struck a $25 million deal with Spielberg for the rights to make a game based on the movie and then informed Atari of what he had done. Kassar was shocked: "Ross forced me to make *E.T.*. He called me and said I've guaranteed Spielberg $25 million to work on this project. I said: 'Steve, we've never guaranteed anybody any money. Why would you want to guarantee $25 million?'"

Kassar argued that the film's lack of action didn't lend itself to a video game, but Ross had already made his mind up. Atari was told to get the game out before Christmas, leaving the company with barely any time to make the game. "We didn't have enough lead time," said Kassar. "This was in August, he wanted it for Christmas. Normally we had a six-month lead time."

Kassar persuaded VCS game programmer Howard Scott Warshaw to knock the game together in six weeks in return for a hefty bonus. The result was terrible but on time. "Maybe better engineers working 100 hours a day, nine days a week could have hit the window and done it better," said Gerard. "The real answer, probably, was don't put the product out until it's good."

As per the deal with Spielberg, Atari flooded the shops with five million *E.T.* cartridges that Christmas. "Most of them came back from the retailers," said Kassar. It was a financial disaster and Kassar took the flak. "I was fired," he said. "They tried to blame me for the *E.T.* fiasco. Somebody had to be the fall guy and it wasn't going to be Steve, he was chairman of Warner. Somebody had to be the fall guy and it was me."

In September 1983 the returned *E.T.* cartridges, along with mountains of unsold and defective Atari game cartridges, consoles, computers and accessories, were loaded on to more than 20 semi-trailer trucks at Atari's plant in El Paso, Texas.[9] From there the trucks headed to Alamogordo in New Mexico where the detritus of Atari's glory days was dumped into a landfill. "What else are you going to do with them? You had to get rid of them. You reach a point where you couldn't even sell them into the second-hand market. There were way too many of them," said Gerard. On 29th September 1983, concrete was poured over the crushed remains of Atari's golden age that filled the landfill site. The video game was dead and quite literally buried.

9. The El Paso facility used to be a manufacturing plant but became a 'recycling' centre for the company after the industry crashed.

Uncle Clive: The British inventor demonstrates his ZX80 computer
© Phillip Jackson / Daily Mail / Rex Features

CHAPTER 9

Uncle Clive

In July 1978 Bruce Everiss took on the lease to 25 Brunswick Street, Liverpool, England. A qualified accountant, Everiss had been running a computer-based book-keeping company for several years when he decided to enter the retail business by opening Microdigital, one of the first computer stores in Europe. "I had started reading the UK trade magazines about computing – *Computing* and *Computer Weekly*," he said. "In there I started reading the very first articles about these microcomputers that were causing a bit of a stir in America and the first computer stores that were setting up." Galvanised by the potential of computing for all, Everiss decided he should open a computer store of his own. "I thought this is obviously going to be a coming thing. I didn't know how big it was going to be, no one did. I begged, borrowed and stole as much money as I could get my hands on and set up a computer store in Liverpool," he said.

At the time the whole of Europe was lagging behind the US in the rise of home computers and video games. While the US and Japan forged ahead building a new entertainment industry on the back of the digital revolution, Europeans had largely settled into the role of consumers rather than producers of video games.

"By the late '70s, the European market had become pretty big," said Noah Anglin, the Atari executive who set up its European factory in Ireland in 1978. "We were shipping a lot of games to Europe and it was just too expensive, so we needed a factory to build products for the European market and that's where Tipperary came in." The Irish government pulled out all the stops to lure Atari to the rural town. It tracked down a suitable building and offered Atari Ireland tax-free status for several years. Atari's decision to make Tipperary its home was a welcome boost for a country blighted by poverty and unemployment. "In those days Ireland was really bad – 40 per cent unemployment. Tipperary didn't even have a stop light in it," said Anglin. "We hired local people. They were the most loyal, most hardworking guys you have ever seen in your life. I guess when you've got 14 other guys waiting for your job it's a pretty good incentive." Atari Ireland, however, was no more than a manufacturing base. Game development remained the preserve of the company's teams on the US West Coast. Of the few home grown arcade game makers in Europe, most took their cues from America and Japan. "Almost all of the video games in Europe were produced under license of the original producers," said Natale Zaccaria, co-founder of Zaccaria; the Italian pinball manufacturer that started producing video games after *Pong* took the world by storm.

Europe had been more active when it came to home consoles. After the Magnavox Odyssey reached Europe in 1973, several companies started making *Pong*-based home game machines using analogue technology similar to that used in the American-made machine. The first of these appeared in early 1974, almost two years before Atari released its home *Pong* console. The UK's Videomaster Home T.V. Game was the earliest but others quickly followed including Italian kitchen appliance manufacturer Zanussi's Ping-O-Tronic and the VideoSport MK2 from British hi-fi and television retailer Henry's, which was housed in wood grain casing. Europe's involvement in the home games business continued throughout the 1970s. Even as late as 1978 Germany's Interton VC-4000 console, which was released throughout Europe under different names by various companies, was offering a European-made alternative to the Atari VCS 2600.[1]

Europe, however, lagged behind when it came to home computers. There was no European system to match the Apple II, Commodore PET or TRS-80. Instead there were the MK14 and Nascom 1, primitive UK-made computers that consisted of bare circuit boards with calculator-style keypads. Everiss's shop stocked the flashy and expensive new American machines, primitive British kit computers and hard-to-find books that explained the mysterious inner workings of computers. Microdigital became a Mecca for aspiring computer owners. "People came from all over the country to the store," recalled Everiss. "You'll get people who had driven the length of the country to be there and see it. People would fly in from all over Europe as well. People would just come along and chat really."

By 1979 Microdigital was producing its own newsletter, the *Liverpool Software Gazette*, to keep the cabal of geeks and technophiles who formed its customer base informed about the latest developments. In one issue Everiss used his editorial to praise the democratisation of computing that the home computer had ushered in. Yet few had any idea where this was all leading. "We didn't know there was going to

---

1. Europe's console manufacturers, however, rapidly lost ground to US-designed consoles after the arrival of the Philips Videopac G7000 the following year. The G7000, the European name for the Magnavox Odyssey2, proved almost as popular as the Atari VCS 2600 in Europe.

be a games industry. Games were just touted as being a possible use for these micro-computer things, so was cataloguing stamp collections," said Everiss. "The displays, processing power and memory were very, very weak. What you could do was very limited." For Everiss the first inklings that games might be more than just another application came during a trip to California to check out the latest developments in computer retail in US: "I went to a computer store called Computer Components of Orange County round 1979-ish and at the back of the shop they had some polythene bags and in them there were cassettes with Apple II games that had been duplicated in someone's home or bedroom. These were the first commercial games I'd ever come across."

But it would be Everiss's employees and customers who first saw the way things were going. In 1980 Tony Milner and Tony Badin, a pair of chemistry graduates who were regular visitors to Everiss's store, formed one of the UK's first video game companies, Bug-Byte. To help them they hired two of Microdigital's employees, Mark Butler and Eugene Evans, to help make and sell games for the company.

Badin and Milner's inspiration was the Sinclair ZX80 home computer, the latest creation of British inventor Clive Sinclair. Affectionately nicknamed Uncle Clive by his fans, Sinclair was the living embodiment of a British boffin with his thin spectacles, balding scalp and ginger beard. He built his reputation during the 1960s and 1970s by making super-cheap versions of the latest cutting-edge consumer electronics from cut-price hi-fis and portable TVs to digital wristwatches and pocket calculators. The low price was sometimes reflected in unreliability, most notably the digital Black Watch that didn't work and almost bankrupted his company due to the volume of returns, but items such as his pocket calculators brought expensive electronics to the mass market.

"Clive Sinclair was really a bit like a mad scientist," said Alfred Milgrom, the Australian co-founder of London-based book publishers Melbourne House. "It seemed to me that his main interest was not so much in marketing his products but more in the development of his inventions. It was almost as if the main purpose of each product was to fund the research for the next one." The ZX80 was no exception. Sinclair saw it as a way to generate funds to bankroll research into one of his pet projects: a flat-screen TV. In keeping with Sinclair's belief in low-cost electronics, the ZX80 cost just £99.95 fully assembled, or £79.95 if bought as a kit to assemble at home with a soldering iron, but it offered features comparable to rival systems costing hundreds of pounds more. It quickly became the UK's biggest-selling home computer. The success of the ZX80 marked the moment when affordable home computing became a reality in Britain. "Before the ZX80 there was no computer industry in the UK," said Milgrom, who would write and publish the book *30 Programs for the ZX80* shortly after the ZX80 became a success. "The ZX80 was tremendously important. It struck a note with the UK public. It was a simple machine with only 1Kb of memory and was released with no software or books for people who bought it."

The ZX80 was the machine that the UK's legions of would-be computer enthusiasts had been waiting for. "Machines like the Commodore PET and Apple II were a bit too far out of reach for the average interested school kid to buy," said Jeff Minter, a Basingstoke teenager who had fed his interest in home computers by making games on his school's Commodore PET until the ZX80 arrived. "Uncle Clive gave us affordable computing for the first time in the shape of the ZX80."

By the time Sinclair released the cheaper, more powerful and even more successful ZX81 in March 1981 many of those who bought the ZX80 had reached the same conclusion: they should make and sell games. And since few shops sold games, they copied their games onto blank cassette tapes and sold them via mail order.[2]

One of the first people to make an impact with a mail order game was Kevin Toms, a programmer from the seaside town of Bournemouth, who in 1981 released *Football Manager* on Sinclair's follow-up to the ZX80, the ZX81. The game evolved out of a board game Toms had designed about running a soccer club that was part inspired by *Soccerama*, a 1968 board game about football management. "It started when I was about 11 years old and I did several iterations right through into my twenties," said Toms. "I used to cut up cereal packets to try out ideas on and I remember buying blank card decks from [stationers] WH Smith."

After getting a ZX81, Toms realised his board game would make a better computer game: "It gave me a much better tool to run the game on, especially for automating things like league table calculations and fixtures. It also helped me to make the simulation of what was happening more realistic and interesting." *Football Manager* was text-only, but it captured the drama of football in a way that the era's basic action-based soccer games could not, swinging from the highs of steering your club to the top of the league to the lows of seeing your star player injured in a match. And, just as was the case for everyone else, selling the game via mail order was really the only option open to Toms. "When I started there were no retailers at all," said Toms. "The best way to sell a game was by mail order direct to the public. So that is what I did." Toms' three editions of *Football Manager* would go on to sell

---

2. The ZX80 used cassettes to store and load programs. While the US was already starting to move away from cassette storage by the early 1980s, computers with disk drives were rare in the UK until the latter half of the decade, mainly due to cost.

close to two million copies across a number of computer formats and create a video game genre that is still a top-seller today.

The ZX80 also attracted the curiosity of Mel Croucher, a former architect from Portsmouth. "When Uncle Clive came up with the ZX80, I already had an entertainment cassette business," said Croucher. "An advert appeared for some computer software on a cassette and I think that cassette cost four or five quid. I had already produced audiocassettes for about 30p a throw including the labels and packaging, so I thought I'd give games a go and switched to computer software that day. The reason was a mixture of avarice and ignorance."

Croucher's debut games *The Bible*, *Can of Worms* and *Love and Death* were 1Kb exercises in the surreal. In *Can of Worms* players used whoopee-cushions to give a wheelchair-bound Hitler a heart attack, performed vasectomies and tried to guess how much water would empty the King's blocked loo. "They were piss takes, at least as good as what else was on the market in the early days but turned inside out. The themes were overtly stupid with a bit of propaganda chucked in," said Croucher. "*The Sunday People* accused me of peddling pornography to kids. Great publicity." He followed these experiments with *Pimania*, a bizarre text adventure based around the character of PiMan, a pink naked cartoon man with a bulbous nose. The game offered players the chance to win a £6,000 golden sundial if they could solve the riddle within the game and work out where and when the prize would appear. "I was trying to blur fantasy and reality, but my method was to take those dreary traditional game plays and get the player laughing as they went on those idiotic quests," he said. "Pimaniacs turned up all over the place, convinced they had cracked the quest for the golden sundial with its diamond bauble. Stonehenge was a favourite at solstice, so was Jerusalem on Christmas Eve."

Eventually a teacher and music shop proprietor from Ilkely, West Yorkshire, solved the riddle. They arrived at the white horse cut in the chalk hill of the Sussex Downs near the village of Alfriston on 22nd July 1985, three years after the game's release, to claim their prize. "They stood in the horse's mouth. I didn't have the heart to tell them the exact location was in the horse's arse," said Croucher.

Croucher's strange games foreshadowed a taste for the bizarre and surreal among British developers and players that would really come to the fore after Sinclair launched his ZX Spectrum computer in April 1982. Costing £125 to £175 depending on the amount of built-in memory, the Spectrum was Sinclair's response to the BBC Micro. Developed by Acorn Computers, the computer firm founded by former Sinclair employee Chris Curry, the BBC Micro was part of a bid by the state-owned British Broadcasting Corporation to create a standard computer format as part of a government push to increase children's computer literacy. Although he had sought to win the contract to create the BBC's computer, Sinclair later accused the corporation of using taxpayers' money to undermine the nation's computer manufacturers. "They should not be making computers, any more than they should be making BBC cars or BBC toothpaste," he raged.

Despite his fears, the Spectrum's low price made it the UK's home computer of choice, outselling both the Commodore 64 and BBC Micro. For a brief moment it was thought to be the world's best-selling computer. Such was its success that Britain's Prime Minister Margaret Thatcher even showed the Spectrum to the visiting Japanese premier as an example of the UK's technological superiority. The Spectrum's sales encouraged an explosion in the number of games being made in the UK. Game companies sprung up in every corner of the country from St Austell in Cornwall (Microdeal) to the Isle of Harris in the Western Isles of Scotland (Bamby Software). A total of 226 British-made Spectrum games were released in 1982 alone. The following year the number of games released soared to 1,188 and the number of companies making them rocketed from 95 to 458. "The games industry was being dragged along on the back of the Sinclair Spectrum, which was a thousand times more successful than Sinclair expected it to be," said Everiss, who sold Microdigital to hi-fi chain Lasky's in 1981 off the back of the rising interest in home computing. "He thought people would be cataloguing their stamp collections on the back of it. The fact that the Spectrum became 99 per cent used for game playing took him by surprise."

The Spectrum gave Bug-Byte its first major success, *Manic Miner*. Created by Matthew Smith, a teenage programmer from Wallasey, Merseyside, *Manic Miner* typified the 'anything goes' approach of the fledgling UK games industry featuring a world of mutant telephones and deadly toilets. At heart it was a remake of a popular US-made platform game called *Miner 2049'er*, but Smith's version was enlivened by a taste for the surreal and the bizarre that would become common among early British games.

Manic Miner became a best seller and Smith responded by forming his own company, Software Projects, through which he released *Jet Set Willy*, an even weirder sequel that pitted players against wobbling jellies, rolling eggs, angry Greek housekeepers and feet lifted straight out of the anarchic TV comedy show *Monty Python's Flying Circus*. Smith's bizarre game was only the start of a wider embrace of the surreal among British game designers. In 1984 Peter Harrap came up with *Wanted: Monty Mole*, a strange take on one of the most divisive events in British history: The Miner's Strike. The strike was a long and often-violent showdown between the

UK government and the National Union of Mineworkers led by socialist firebrand Arthur Scargill. It was an all-out battle for supremacy between government and the union movement, which had twice brought down the British government in the 1970s. The defeat of the striking miners broke the union movement's grip on the levers of power in the UK. Harrap's game, released at the height of the strike, cast players as a mole who breaks the picket lines to get coal direct from a fictional secret mine owned by Scargill. The game's theme attracted widespread media interest but it was more absurd than political – Scargill's mine was packed with bizarre enemies such as hairspray cans, leaping sharks and bathroom taps.

Minter, who had started making games after getting his ZX80, also embraced the strange. After making straightforward versions of popular arcade games such as *Centipede*, he formed Llamasoft and started releasing games that fused his obsession with Pink Floyd lightshows, furry ruminants and adrenaline-pumping shoot 'em ups such as *Defender* and *Tempest*. "I liked the simplicity of these games and how, in the best games, complex behaviours and strategies could emerge from the interaction of a small rule set," he said. "Older shooters, although arguably more primitive, were often more creative in terms of controls and enemy behaviours than before everything became a series of reworkings of *Xevious*. It's almost an attempt to imagine how such games might have evolved if their evolution hadn't been stunted by endless versions of *Xevious* and bosses."

Minter built up a cult following with games such as *Attack of the Mutant Camels*, a psychedelic shoot 'em up where players battle giant camels, and *Metagalactic Llamas Battle at the Edge of Time*, where a laser-spitting llama has to kill spiders before they turn into killer weevils. Sheep, llamas, giraffes and camels became hallmarks of his work. "I just liked the animals really and I'd already called the company Llamasoft, so it made sense to start bringing the animals into the game," he explained. "It certainly did distinguish us – we were typically the only ones bringing life-sized sheep models to computer shows. A Llamasoft game with no sheepies would just be kind of odd."

The taste for strangeness became so widespread that 'British surrealism' became a loose stylistic movement that decorated familiar game concepts in the outlandish imaginations of their creators. Yet despite the psychedelic trappings, the movement was more influenced by *Monty Python* than hallucinogenic drugs. "A lot of us in the nascent games biz grew up watching *Monty Python* on telly and I think that probably inspired a lot of the 'British surrealism' you saw in a lot of games," said Minter. "Certainly I'd cop to the Pythons being the major influence on stuff like *Revenge of the Mutant Camels* and the same is probably true of *Manic Miner* too. Drug use that

I was aware of back then was pretty low-key stuff, a couple of spliffs with the lads rather than dropping acid and tripping out, so I genuinely doubt that the surrealism was down to use of psychedelics."

Gary Penn, a journalist who was part of the team that launched Britain's anarchic Commodore 64 games magazine *Zzap! 64*, agreed that drugs were not a major feature: "There was mainly a lot of drinking. There were circles of drugs, but it wasn't as prevalent as in the music industry." For Croucher the surrealism was inherent within the British culture: "We are a surreal nation, left to our own devices. We are not at all what we seem to be – politically, linguistically, historically and, above all, in terms of humour. It's not that we distort the truth; it's more puckish than that. We're a bunch of pucks."

While others dabbled in a veneer of surrealism, Croucher's agit-prop games continued to push back the boundaries reaching their zenith with *Deus Ex Machina*, a work so unusual it is debatable whether it really could be called a video game. Inspired by E.M. Forster's 1909 short story *The Machine Stops* and the 'seven ages of man' described in William Shakespeare's *As You Like It*, *Deus Ex Machina* told of a future where an all-powerful computer controls the world and all births are genetically engineered to the machine's ideal. But after a mouse dropping contaminates the computer's fertilisation system, a mutant embryo forms. The player's role is to protect the embryo from the Defect Police, the computer's eugenic enforcers, by playing a series of seven abstract mini-games that represent the seven ages of man. "I thought that by the mid-1980s all cutting-edge computer games would be like interactive movies with proper structures, real characters, half-decent original stories, an acceptable soundtrack, a variety of user-defined narratives and variable outcomes," he explained. "I thought I'd better get in first and produce the computer game equivalent to *Metropolis* and *Citizen Kane* before the bastards started churning out dross."

Deus Ex Machina included an audio cassette that contained the game's soundtrack, which mixed story-setting voiceovers from British TV celebrities such as *Doctor Who* actor Jon Pertwee and comedian Frankie Howerd, as head of the Defect Police, and strange songs about a sperm fertilising an egg while dreaming of fish and chips. "When I was a kid I was very frightened by Frankie Howerd's performances on the radio and it was a cathartic experience to hire him for the day and order him to kill babies," said Croucher. "Originally I wanted TV astronomer Sir Patrick Moore to play the part of the sperm. Now that would have been utterly surreal."

While many dabbled in the surreal, the two most significant games to emerge from the UK at this time were unconnected to the heady experimentation of Minter,

Smith and Croucher. One of these was *Knight Lore*, a game written by Chris and Tim Stamper, the founders of Leicestershire-based Ultimate Play The Game. *Knight Lore* built on the ideas first explored in Atari's VCS 2600 game *Adventure*, which reinterpreted the exploration and puzzle-solving of text adventures within the context of an action game, by combining them with the axonometric visuals pioneered by arcade games *Zaxxon* and *Q*bert*.

The visual approach had already made it onto the Spectrum via *Ant Attack*, where players rescued people trapped in a M.C. Escher-inspired city overrun with giant ants, but the Stampers' cartoon visuals and addition of adventure game elements inspired many British game developers. "As soon as I saw *Knight Lore* and had picked my jaw up from the floor, I knew I had to use a similar system. It looked fabulous," said Jon Ritman, one of the many game designers who followed the Stamper brothers' lead in producing what the British games press called 'arcade adventures'. After creating the *Knight Lore*-inspired Spectrum title *Batman*, Ritman teamed up with artist Bernie Drummond to create *Head Over Heels*, an intersection of British surrealism and the arcade adventure genre. The game revolved around the puzzle-solving adventures of two symbiotic creatures, but came dressed in a world that fused Disney and Dali. There were stairs constructed out of sleeping dogs, toy rabbits that gave special powers and Prince Charles Daleks, which welded the big-eared head of the heir to the British throne to the body of the robotic aliens from *Doctor Who*. Ritman put the game's strangeness down to Drummond: "Mad visions just leak out of his head."

The other important game to emerge from the UK in the first half of the 1980s was *Elite*, a 1984 BBC Micro game written by Cambridge University students Ian Bell and David Braben. *Elite* evolved out of Braben's efforts to create a space combat game that used wireframe 3D visuals similar to those used in vector arcade games. Previous attempts to take the space combat games into the third dimension had rarely lived up to the promise of the idea. Vector games such as *Tailgunner* and *Star Wars* had restricted players to manning the guns rather than piloting their virtual spacecraft. *Star Raiders*, a 1979 game for the Atari 400 computer, offered movement but used flat sprites that changed in size to give an illusion of a 3D world. "There were a few sprite-based shooting games that implied a 3D effect, where the sprites were made to get bigger and smaller, and you centred the sights on them, but these were very different," said Braben. Braben's space combat game, however, was visually more exciting and closer in spirit to Atari's 3D tank warfare game *Battlezone*. While the visuals Braben produced were technically impressive, the rudimentary space game he created with it seemed too limited to keep players excited for long.

So Braben joined forces with Bell, who had already had a couple of his games published, to turn it into a better game. After further tweaks failed to counter the eventual boredom of relentless space battles, they decided they needed to add more for the player to do. "We had a clear idea of what we were trying to do, which is to put a framework around space combat to make it compelling, but it took quite a lot of thought and discussion to work out exactly how we would do it," said Braben.

One of the first additions were space stations where players could relive the spacecraft docking sequence from the film *2001: A Space Odyssey*. But what would players do once they had docked in the space station? Braben and Bell decided to let them upgrade their spacecraft with better guns but this immediately raised the question of how players would get these upgrades. Money, the pair concluded. Another question immediately emerged: How would players earn money?

Ideas spilled forward. Players could trade goods between space stations in different parts of the universe or carry out odd jobs or earn bounties from killing space pirates or mine asteroids for minerals. "The idea of score equals money seemed utterly logical, especially as we had settled on trading as the way the player was grounded in the game," said Braben. "Very quickly we contextualised the additional sources of money – so the reward for shooting a ship became bounty and so on. This was at the time of the Miner's Strike and somehow a dog-eat-dog mentality for the game felt appropriate."

The idea of being a space-age trucker had already been explored by a few trading games, such as 1974's *Star Traders* – a text game for mainframe computers – and 1980's *Galactic Trader* on the Apple II, but neither Bell and Braben knew of their existence. "To my knowledge there were no space trading games. The main way games influenced me, at least, was in terms of what I didn't want in a game," said Braben. "I felt games had got into a bit of a rut, always with three lives, a score that went up in 10s with a free life at 10,000 and a play time aimed at 10 minutes or so. I strongly felt games didn't have to be that way. There were some text adventure games with a much longer play life and a story and this contrast showed other approaches were possible."

Bell and Braben spent two years making *Elite* in-between their university studies, perfecting the game's mix of combat, moneymaking, ship-upgrading and intergalactic trading to create a game where players could make their own way and decide their own priorities. The player began the game with 100 credits and a basic spacecraft but from then on the universe was their oyster. They could be a hero, a space pirate, a miner, an entrepreneur, a gun for hire or simply head out into the void and explore the stars. *Elite*'s world in a box opened up a new avenue for game designers to explore: the concept of open-ended worlds where players decide what to do and where to go, rather than being required to complete pre-decided goals in worlds that restricted their choices.

Elite, *Knight Lore*, *Deus Ex Machina* and *Jet Set Willy* typified the atmosphere of creativity and opportunity that powered the early UK games industry, which for most of the early 1980s was still an unstructured cottage industry. "People could afford to take risks, the barriers to entry were really low," said Penn. "Literally anyone in their bedroom who had half a brain and some passion could make something and get it in the hands of people. There was a lot of inventiveness, not all of it necessarily good. That was part of the joy of being around in that period – the amount of innovation that was going on was quite something."

The energy of the UK market at this time also encouraged the growth of video game industries in Spain and Australia.

Spain had also embraced the ZX Spectrum, exposing the Spanish to many of the games being released in the UK while giving the Iberian nation's ambitious game companies the chance to sell their work to the much larger pool of British players. For companies like Madrid's Dinamic, the UK game business showed the way forward. "Imagine and Ocean were our idols," said co-founder Victor Ruiz, referring to the UK's two biggest game publishers around the time of Dinamic's formation in 1984. Dinamic and its rivals such as Opera Soft and Indescomp would make Spain a leader in European game development in the 1980s, thanks to visually impressive games such as the *Rambo*-inspired *Army Moves* and the bank-robbing action game *Goody*.

Looks were a big focus for many Spanish games. "The visuals and graphic effects are very important in video games and we paid special attention to them," said Pedro Ruiz, the director of Opera Soft. "At the end of the day, a video game is a visual experience and some spectacular graphics can make up for a game that is not particularly good. We wanted to develop games that got people hooked and were hard to play – sometimes too hard – and to use the latest technologies available at the time."

The high point for Spanish games in the 1980s was Paco Menéndez's *Knight Lore*-inspired *La Abadía del Crimen*, which Opera Soft published in 1988. "It was inspired by a novel, Umberto Eco's *The Name of the Rose*. We got in contact with the writer so we could give the game the same name as the novel, but received no reply. This is why we changed the title," said Pedro Ruiz. "It was not an arcade-type game of skill, but a game of intelligence." Set in a medieval monastery, the player took on the role of a Franciscan monk who must solve a series of murders while carrying out religious duties. "The game was special both in terms of the result and how it was made," said fellow Opera Soft employee Gonzalo Suárez, who left the Spanish movie business to make games starting with 1987's *Goody*. "*La Abadía del Crimen* was a graphic adventure but with a freedom of movement unknown until that time with a recreation of the abbey in isometric graphics that far outstripped the production kings of the time like Ultimate." Never released outside Spain, *La Abadía del Crimen* was a commercial disappointment that only achieved the recognition it deserved years later.

The UK market also proved crucial to Australia. Having watched the rise of the ZX80, Milgrom and his wife Naomi Besen returned to their native Australia in December 1980 with a plan to start Beam Software, a business that would create games for the UK market that Melbourne House would then publish. "At the very beginning the idea was not to develop software, but rather to develop content for computer books," said Milgrom. "Then one day I thought more about the concept of publishing and I realised that there was very little difference between developing material and putting that content onto paper or onto a cassette tape."

Beam Software became the focal point for the Australian game industry. It recruited graduates from the University of Melbourne's computing courses and rapidly expanded off the back of hits such as its million-selling text adventure remake of J.R.R. Tolkien's *The Hobbit*.[3]

Soon almost every would-be game developer in Australia was moving to Melbourne to join the swelling ranks of Beam Software. "We were doubling in staff every year for almost eight years. Every three years we had to move offices," said Milgrom. "People from all over Australia would write to us and come over for interviews. You have to understand we were offering people a job. In the UK a lot of people were doing it as a hobby, but you couldn't do that from Australia because there was no means of distribution. You couldn't just expect your games to be sold as there's no major market in Australia. They could move over to the UK, but it made a lot more sense to come and work for us."

Off the back of Beam Software's hit games, Melbourne House ditched book publishing and became one of the UK's largest game publishers of the 1980s. The talent that built Beam would go on to create the bulk of the Australian games industry. "One of the things I am especially proud of is that Beam effectively started the

---

3. *The Hobbit* introduced several new concepts to text adventures, including characters that would carry out actions and make decisions independently of the player, a marked change from the usually static worlds of these games. It also assigned physical properties to the various objects in the game. "The puzzles were based on using those properties," said Milgrom. "But it also meant that some totally unintended things could be done by the players because the physics of the environment allowed it to happen, such as tricking Thorin to get in the chest and locking him in."

games industry in Australia," said Milgrom. "Almost all of the development studios in Australia since then were started by ex-Beam employees or have been substantially staffed by ex-Beam employees."

Powered by Sinclair's cheap computers, the UK's bedroom programmers had turned their country into a hotbed of experimental game design, inspired developers in Spain and laid the foundations for the Australian industry at the same time. But the UK was not the only European country forging new ground in video games.

Froggy Software: (left to right) Clotilde Marion, Jean-Louis Le Breton and Tristan Cazenave
Courtesy of Jean-Louis Le Breton

CHAPTER 10

The French Touch

Paris was a war zone. Egged on by the Vietnam War and the rebellious rhetoric of the Situationist International, thousands marched on the city streets demanding revolution.[1] They spray painted slogans onto the city's walls: 'DEMAND THE IMPOSSIBLE', 'IMAGINATION IS SEIZING POWER', 'MAKE LOVE, NOT WAR' and 'BOREDOM IS COUNTER-REVOLUNTIONARY'.

They constructed makeshift barricades out of parked cars and started fires. They battled with France's quasi-military riot police, the Compagnies Républicaines de Sécurité, which sought to suppress the uprising with tear gas and beatings with batons. The protestors responded by hurling bottles, bricks and paving stones ripped up from the streets. France's trade unions sided with the protestors and encouraged wildcat strikes across the nation in a show of solidarity. The government had lost control and France teetered on the brink of revolution. For a few days in May 1968 it looked as if the motley coalition of students, trade unions, Trotskyites, anti-capitalists, situationists, anarchists and Maoists would win their fight for revolution. Ultimately they did not. In early June, the protests died out thanks to a combination of government capitulation and renewed crackdowns on the protestors.

But the failed revolution inspired many. Among them was Jean-Louis Le Breton, a Parisian teenager whose worldview was shaped by the idealism of the revolutionaries who took to the streets that May. "I was 16 in '68 and part of the protests in Paris," he said. "I spent most of my time in the Latin Quarter with other students. Our teachers were on strike and we had a lot of discussions. We thought we could change the world. It was both a period of political consciousness and of utopia. We used to mix flower power with throwing cobblestones at policemen. Many things changed after '68: women could wear trousers, radios and TV felt more free and able to criticise the

---

1. The situationists were a revolutionary group of French artists and philosophers that began as an artistic movement but evolved into a political movement led by Guy Debord, a French intellectual and war game enthusiast. His manifesto *The Society of Spectacle* summed up the movement's politics with its theory that people had become spectators in their own lives

government." During the late 1970s and early 1980s Le Breton explored his desire to challenge the status quo via music. He experimented with synthesizers in his band Dicotylédon before delving into avant-garde rock 'n' roll with another act, Los Gonococcos. Then in 1982 he found a new outlet. "Los Gonococcos split in 1982 and I exchanged my synthesizers for the first Apple computer delivered in France, the Apple II," he said. "At that time, Steve Jobs and Steve Wozniak were presented as two guys working in their garage – such a pleasant image in opposition with IBM. I found that programming in BASIC was easy and fun and I could imagine a lot of amusements with this fascinating machine. It was possible to take power over computers and bring them into the mad galaxy of my young and open mind."

Le Breton had played video games before but didn't like them: "I've never been interested in playing games. The first game I played was a game that took place in Egypt – I don't remember the title. I was interested by the fact that you could move the character, but it was no fun. Too many fights. Not for me." But after playing Sierra's illustrated text adventure *Mystery House*, he decided to write a game of his own. "The graphics and scenario of *Mystery House* were such bad quality that I thought I could easily produce the same kind of game," he said. The result was 1983's *Le Vampire Fou*, the first text adventure written in French. "You had to enter the castle of Le Vampire to kill him before he killed you," said Breton. "It was the kind of game that made you crazy before you could find the right answer."

Le Breton earned nothing from *Le Vampire Fou*. Its publisher Ciel Bleu – an importer of Canadian educational software – went bust shortly after its release. With Ciel Bleu gone, Le Breton teamed up with his friend Fabrice Gille in 1984 to form his own game publishing company Froggy Software, which summed up the essence of its games as 'aventure, humour, décalage et déconnade'.[2] From their base in Le Breton's home, an old bar in the 20th district of Paris, the pair dreamed big. "We felt both like modern young people and artisans. The ideas of the games came out of our brains and were directly translated into the computer. I was personally happy to use the computer in a literary way. I thought we should not let computers be only in engineers' hands," said Le Breton. The spirit of May 1968 lurked within Froggy's DNA. "May 1968 surely had an influence on the way we started the company, with a completely free and open state of mind and a bit of craziness. We wanted to change the mentalities, the old-fashioned way of thinking. Humour, politics and new technologies seemed to be an interesting way to spread our state of mind," said Le Breton. Almost all of Froggy's games were text adventures, but with their humour and political themes they were a world away

---

2. Adventure, humour, leftfield and 'a willingness to making fun of anything'.

from the fantasy and sci-fi tales that typified the genre in the UK and US. *Même les Pommes de Terre ont des Yeux* offered a comic take on South American revolutionary politics. *La Souris Golote* revelled in puns about cheese. The sordid murder mystery of *Le Crime du Parking* touched on rape, drug addiction and homosexuality while *Paranoïak* had players battling against their character's smorgasbord of mental illnesses. Le Breton's efforts prompted French games magazine *Tilt* to dub him the Alfred Hitchcock of gaming.

Le Breton and Gille were not the only French game designers taking games in a more highbrow and consciously artistic direction. Muriel Tramis, an African-Caribbean woman who grew up on the French-Caribbean island of Martinique, was also exploring the medium's potential. She left Martinique for France in the 1970s to study engineering at university and, after several years working in the aerospace industry, became interested in the potential of video games and joined Parisian game publisher Coktel Vision. She decided her own heritage should be the subject of her debut game *Méwilo*, an 1987 adventure game written with help from another former Martinique resident Patrick Chamoiseau, one of the founding figures of the black literary movement Créolité.

"The game was inspired by the Carib legend of jars of gold," explained Tramis. "At the height of the slave revolts, plantation masters saved their gold in the worst way. They got their most faithful slave to dig a hole and then killed and buried him with the gold in order that the ghost of the unfortunate slave would keep the curious away from the treasure." In the game the player took on the role of Méwilo, a parapsychologist who travels to the Martinique city of Saint-Pierre in 1902 to investigate reports of a haunting, just days before the settlement's destruction at the hands of the Mount Pelée volcano.

"This synopsis is a pre-text for visiting the city and discovering the daily economic, political and religious life of this legendary city," said Tramis. The game's exploration of French-Caribbean culture won Tramis a silver medal from the Parisian department of culture – making it one of the first games to receive official recognition for its artistic merit.

Tramis and Chamoiseau probed the history of slavery further in 1988's *Freedom: Rebels in the Darkness*. Set once again in the French Caribbean, the game cast the player as a black slave on a sugar cane plantation who must lead an uprising against the plantation's owner. *Freedom* mixed action, strategy and role-playing into what Tramis summed up as a "war game".

"Fugitive slaves, my ancestors, were true warriors that I had to pay tribute to as a descendant of slaves," she said. "At the time I made the game, these stories were not known because they were hidden. Today the official recognition of slavery as a crime against humanity has changed the world, people are aware now. I could talk through the game at a time when the subject was still painful. It was my duty to remember. A journalist wrote that this game was as important as *Little Big Man* has been in film for the culture of American Indians. I was flattered."

Tramis and Froggy's attempts to elevate video games beyond the simple thrills of the arcades formed part of a wider search during the 1980s amongst French game developers for a style of their own. Unlike their counterparts in the UK, France's game industry had been slow to develop. In the UK the instant success of Clive Sinclair's computers had acted as a catalyst for the thousands of games spewed out by bedroom programmers, but France lacked a clear market leader. Only in 1983 did systems such as the British-made Oric-1 and French-designed Thomson TO7 finally start to emerge as France's computers of choice.[3]

Until then the French had flirted with a bewildering range of contenders from the ZX81 and Apple II to home-grown systems such as the Exelvision EXL100 and Hector. But once the Oric-1, TO7 and, later and most successfully, the Amstrad CPC, gained a sizeable following, game publishers started to form with Loriciels, Ere Informatique and Infogrames leading the way in 1983. Within months of their formation, however, the country's video game pioneers started asking themselves what defined a French game and how they could set themselves apart from the creations of American and British programmers. A summer 1984 article in *Tilt* reported how French game designers, having cut their teeth on simple arcade games, now wanted to create something more personal, more rooted in reality, more French. Inevitably, opinion was divided about what this meant in practice, but many homed in on strong narratives, real-life settings and visuals inspired by the art of France's vibrant comic book industry. Text adventures provided the natural home for such content. "Back then the adventure game was king," said Tramis. "There were many more scenarios with literary rich universes and characters. There was a ferment of ideas and lots of originality. France loves stories."

The focus on real-world scenarios reflected France's relative disinterest in fantasy compared to the British or Americans. "I have always wanted to base my titles on a historical, geographical or scientific reality," said Bertrand Brocard, the founder of game publisher Cobra Soft and author of 1985's *Meurtre à Grande Vitesse*, a popular murder mystery adventure game set on a high-speed French TGV train. "The TGV

3. Thomson's computers became France's equivalent of the UK's BBC Micro after the French government made them the basis of a national programme to put computers in every school.

was still a novelty then as it had been running for less than two years and it was something ultra-modern. At the time the driver would announce to the passengers: 'We have just reached 260 kilometers per hour'. Nowadays it goes at 300 kilometres per hour with no announcement. The player had two hours of travel between Lyon and Paris to solve the mystery and arrest the culprit, who could not escape from the train during the journey." Other Cobra Soft games reflected current affairs. Among them were *Dossier G.: L'Affaire du Rainbow-Warrior*, a game inspired by French intelligence service's sinking of Greenpeace's Rainbow Warrior ship in 1985, and *Cessna Over Moscow*, a 1987 game inspired by Mathias Rust – the West German pilot who flew a light aircraft into the Soviet Union and landed in Red Square to the USSR's embarrassment that same year. For Brocard, however, the French style was more a reflection of the personal interests and tastes of the small group of companies and individuals who were making games rather a reflection of France itself. "Game production in France was not very extensive," he said. "When video games started in France, production involved such a small number of people that chance, I think, led things in certain directions. To my mind this issue of the 'French touch' is associated with Ere Informatique and the charisma of Philippe Ulrich. He is an artist through and through. He had managed to 'formalize' this difference."

Ulrich was the co-founder of Ere Informatique and, like Le Breton, he was a musician before he was a game designer. "In 1978 I published my first music album with CBS, *Le Roi du Gasoil*, I often slept in the Paris Métro at that time," he said. "I wanted to cut a second album with more electronic music. To that effect I took to soldering together my own rhythm boxes. When Clive Sinclair put his ZX80 on the market, I emptied my piggy bank to buy one. I could barely believe the hallucinating results I got after coding my first lines of BASIC."

Ulrich threw himself into learning all he could about his new computer, digesting 500-page guides to the inner workings of machine code on the ZX80's Z80 microprocessor. His first machine code game was a version of the board game Reversi, which he swapped the rights to in exchange for a 8Kb memory expansion pack for his ZX80. Soon after he met Emmanuel Viau, another aspiring game designer, and the pair decided to form their own publishing house: Ere Informatique. Like others on the French games scene, Ulrich wanted to give his work a distinctive style but, unlike his peers, he had one eye firmly on the larger UK market. "In France, authors would create games related to their culture, while the bulk of our market was the United Kingdom, so I came up with a vague concept of world culture," he said. Ulrich wanted Ere Informatique's games to have international appeal – something of a no-no amongst France's cultural elitists – while retaining a French flavour: "Our games didn't have the excellent game play of original English-language games but graphically their aesthetics were superior, which spawned the term French Touch – later reused by musicians such as Daft Punk and Air."

Ulrich's most notable realisation of the 'French Touch' was 1988's *Captain Blood*, a cinematic space adventure created with artist and programmer Didier Bouchon. The game tells the story of the space-travelling Captain Blood who must hunt down and destroy five clones of himself to stay alive. To hunt down the clones, the player must travel the galaxy and converse with aliens using Bluddian, an alien language created specifically for the game that was based on the use of 150 icons, each of which represented a word. With its H.R. Giger-inspired visuals, fractal-enhanced explosions, accompanying novella and a theme tune composed by French synthesizer musician Jean Michel Jarre, *Captain Blood* was nothing short of an epic, although its bizarreness often confused.

"I wanted to be an example and to invent new stuff that stood out," said Ulrich. "I wanted to impress the player. I wanted the extra-terrestrials to be alive in the computer. When playing *The Hobbit* I hated the stereotyped answers such as 'I don't understand' or 'what is your name?'. The challenge was to make it intelligent. The incredible thing is that the aliens answered all questions, were funny and never repeated the same thing twice." During the game's development Ere Informatique ran into financial problems and was bought by its more commercially minded rival Infogrames, which was less than keen on Ulrich and Bouchon's strange game. "At Infogrames they bought licences and developed more classic games and it was marketing that boosted the sales," said Ulrich. With little funding from Infogrames, the pair holed themselves up in the Landes forest in southwest France to finish the game.

"We worked ourselves to the point of exhaustion to complete *Captain Blood*. It was really tough. I covered several reams of paper with Bluddian dialog; Didier would code the programs and created the graphics. When I showed the game to Infogrames they did not understand. 'Is that a UFO or what'? 'You're crazy,' they told me. After it was released the sales people at Infogrames told me that the game was selling by the hundreds, they had never seen anything like it."

Narrative-based games dominated France's output during the 1980s but the French Touch could be seen in other forms of game as well, such as 1985's *L'Aigle d'Or*, a marriage of action and adventure that had an influence in France comparable to that of *Knight Lore* in the UK. The French Touch could also be seen in Eric Chahi's gory platform game *Infernal Runner*, the French comic book visuals of strip poker game *Teenage Queen* and *North & South*, a simple strategy game based on a Belgian comic about the American Civil War.

Across the border in West Germany, however, game developers were heading in an altogether different direction, partly out of necessity. For West Germany the legacy of the Third Reich would have an important influence on the types of games the country developed. American, French, Spanish and British games regularly dealt in death and destruction with little dissent. But for a country still living in the shadow of the Nazi atrocities of the Second World War, anything that glorified violence or military conflict was frowned upon both culturally and legally. This post-war aversion to violence led to the formation of youth media watchdog the Bundesprüfstelle für Jugendgefährdende Schriften in 1954.

The watchdog's role was to assess any media that could corrupt the nation's young and it had two powers at its disposal. First, it could seek outright bans for extremely offensive content – such as Nazi propaganda and excessive violence – in the courts. Second, it could place media it considered harmful on its list of indexed media, which meant the product could only be sold to adults and could not be advertised, promoted or put on display in shops. Initially the watchdog focused on the media of the 1950s – comics, magazines, vinyl records and books – but as new forms of media, including video games, emerged these too came under its jurisdiction. Eventually the watchdog renamed itself the Bundesprüfstelle für Jugendgefährdende Medien (BPJM) to reflect its widening remit.[4]

On the 19th December 1984, the BPJM named the first three video games to be added to the index: Activision's aerial combat game *River Raid*; Atari's coin-op tank sim *Battlezone*; and *Speed Racer*, a Commodore 64 driving game that let players run over pedestrians. "*Battlezone* was indexed because of the glorification of war propagated by its content and because the board stated that the content propagated aggressive behaviour," said Petra Meier, vice-president of the BPJM. "*River Raid* was also indexed because of content seen as a glorification of war and an enhancement of violent behaviour." Over the years the BPJM has indexed several hundred games, largely because of violent content. "Probably 90 per cent of the games that were indexed have been indexed because of the portrayal of violence," said Meier. "Of course as far as violence is concerned the decision of what will be considered a 'detailed portrayal of violence' might have undergone some change over the years."

The threat of being indexed by the BPJM, together with a wider cultural aversion to violence, was a big disincentive to game developers thinking of producing more traditional action games, particularly in the 1980s when children and teenagers formed

---

4. The word schriften in the watchdog's original name referred to print or printed media although the law that created the regulatory body never limited its role to this.

the bulk of video game players. "The BPJM has influenced the games produced in Germany," said Cerat Yerli, the founder of German game studio Crytek, the makers of *Far Cry* – an action-packed first-person shooter released in 2004. "I think companies have changed the way they develop. The laws definitely have an impact on design and production. Germany is a very social country, the government takes on responsibility for social elements more than governments that are, for example, like the US. The US doesn't really care about what elements are in your car as long as your car can drive. In Germany there is this responsibility about your car because they say it impacts social security and there are more laws. Every area of life in Germany is much more controlled socially or in law and I think Germany therefore thinks it has to take all the responsibility about entertainment or communication channels that could potentially impact culture or young people."

West Germany's game developers were also heavily influenced by the nation's fondness for board games, of which it is the world's biggest consumer per capita. Germans in particular like social board games with simple rules and economic or strategic themes such as *The Settlers of Catan*. And this, coupled with the aversion to violence, encouraged West Germany's video game developers to start creating trading and management games. "This strand of games came out of Germany because Germans tend to have high interest in or affinity with management simulations or strategic games," said Nils-Holger Henning, director of business development at German online games publisher Bigpoint. Yerli added: "Trading games, strategy games and manager games sum up the German style of games. There was a game in Germany called *Hanse* that was very successful."

Based on the Hanseatic League that dominated trade along the coast of the Baltic Sea from the 13th to 17th century, *Hanse* was a trading game that along with *Kaiser*, a historically themed kingdom management game, helped establish trading and management games as the distinctive feature of West Germany's game business. Instead of going to war, players built their business empires or kingdoms through trade, diplomacy and careful management. In many ways the roots of these games could be seen in *The Sumer Game*, the 1969 city management game developed by American programmer Richard Merrill, but their origins had more to do with Germany's board game market.

But while West Germany, Britain, France and Spain developed distinctive styles of game, few other western European countries followed suit. This was especially surprising in Italy's case. During the 1970s Italy had been the continent's leading producer of video games. It had three arcade game manufacturers, although none developed their own games, and had produced some of the earliest home consoles to reach European shelves. Even in 1980 things looked good for Italy: Bologna-based coin-op manufacturers Zaccaria had decided to start making its own games starting with the shoot 'em up *Quasar*. But *Quasar* proved to be a false dawn and Zaccaria never created the video game hit it hoped to make. "Zaccaria made a great investment to develop video games in Italy, but the competition from the US and Japan was too great," said Natale Zaccaria, the company's co-founder. By the end of 1984 Zaccaria had given up on video games altogether. And for some inexplicable reason it would take until the start of the 1990s before another Italian game company of note emerged. "Maybe if Zaccaria was not forced to close, the programming sector of the company would have better developed and other Italian producers could have followed the example," said Zaccaria. "Probably Italy was just missing the leading example."

The only other European country to really have a major influence on video games in the 1980s was the Netherlands, where instead of making games, amateur programmers spent their time creating demos to show off their coding skills. The trend for demos originated in 1985 among a group of Dutch programmers from the city of Alkmaar who got their kicks from hacking into commercial games, disabling the copy protection measures and distributing free copies via the post or bulletin board systems that computer owners with modems could log on to. They called themselves The 1001 Crew and called what they did cracking.

Joost Hoing was one of the crew's members: "We competed with other crackers around the world to crack a game fast, good and small.[5] I enjoyed the fact that if you 'won' by doing the best crack, the whole world copied and played your version of the game and saw your name on the screen." As competition between crackers intensified, The 1001 Crew started adding intros to games to let the world know who cracked them. "We basically 'tagged' the game to show who did it," said Hoing. "Before the game started it showed something like 'Cracked in 1983 by 1001 Crew – Hit the Space Bar'. These intros had to be very small in size since it had to fit with the complete game. In order to show our programming skills, we created more and more impressive intros with bouncing logos, colour bars, music, etc. Again as small as possible in size." The intro demos created by The 1001 Crew and another Dutch group called The Judges spread across Europe as computer users shared illegal copies of games the Dutch teams had cracked. "Everyone in the Commodore 64 world knew our name. We were famous," said Hoing.

The work of the 1001 Crew and The Judges was a direct challenge to other crackers. 'Match this,' their demos effectively said. Within a year dozens of demo crews had

5. Crackers often sought to compress games into smaller amounts of memory, so they took less time to download from bulletin board systems or loaded quicker.

formed to try and do just that, spending their nights cracking games and creating new demos in a programming and hacking arms race, spawning a pan-European subculture of cracking and demo making called the demoscene. While the Netherlands was its spiritual home, the demoscene was a European-wide movement that could be found everywhere from Scandinavia and Italy to Britain and West Germany. By the late 1980s demo crews were travelling to demoparties, weekend-long sessions of non-stop programming – a geeky version of the illegal rave parties that emerged around the same time across Europe on the back of acid house music. The demoparties were marathon contests of programming one-upmanship that culminated in individuals' work being shown on video projectors attached to large speakers. Not that attendees spent all their time bathing in the light of their computer screens. "I went to a few but hardly touched a Commodore 64 there," said Hoing. "We were around 18 at the time so we were more into discos, music, girls, beers, etc."

For Europe's game industry the demoscene was both angel and devil. On the plus side, game designers could enhance their games by plundering the numerous programming breakthroughs of the demoscene. "I know music and sprite routines from demos were used in a lot of games," said Hoing.

Many demo makers later renounced their connections with the cracking scene and became professional game developers. By the mid-1990s the diverging interests of those interested in writing demos and those who enjoyed cracking had caused the movement to split in two. With the demoscene going legit, even more of those who cut their teeth making demos resurfaced in game studios. Finland's Future Crew, which started making demos on the Commodore 64 in 1986, is a case in point. After the crew fizzled out in 1994 some of its members joined Finland's leading game development companies including Remedy Entertainment, the makers of *Max Payne,* a film noir action game released in 2001 and built around impressive 'bullet-time' slow motion effects similar to those seen in *The Matrix* movies.

"We're all over the place," said Alex Evans, a British game developer who started out making demos under the name Statix before joining Peter Molyneux's Bullfrog studio in the late 1990s. "If you're in the games industry or in the demoscene you can see the interconnection is very, very strong. There has been a huge crossover into things like mobile and downloadable games, where you have to fit brilliant experiences into tiny spaces, which is what the demoscene has been doing for many years."

Before the split, however, game companies saw the demoscene and its crackers as the enemy: law-breakers who smashed their expensive attempts to prevent illegal copying and gave away free copies of their games, cutting into sales and profits. "Piracy held the industry back," said Bruce Everiss, who became the operations director of short-lived UK games publisher Imagine after selling off his Liverpool computer store Microdigital. "If no-one's paying for stuff then stuff doesn't get done. It's that simple." Not that the game industry's dislike of the crackers resulted in any direct action. "Did game companies attempt to stop groups like 1001? Never," said Hoing. "Police had other things to do than going after a bunch of kids cracking games." The crackers were only just part of the widespread piracy of games in the 1980s. In schools across Europe children swapped games with abandon, aided by the ease of tape-to-tape copying. "Anyone who was at school at that time will remember the swapping of games," said Everiss. "It came from nowhere. One year people weren't swapping games, the next year they were. You would only sell so many games because once it was out and about everyone swapped it."

For Everiss, the sudden rise of schoolyard pirating of games killed Imagine, a Liverpool firm that dominated the UK games industry during its brief two-year existence. Founded by Mark Butler and Dominic Lawson, both former employees of Liverpool's first game publisher Bug-Byte, Imagine achieved instant success with its debut release: a run-of-the-mill shoot 'em up called *Arcadia*. *Arcadia* became the best-selling Spectrum game of Christmas 1982 and turned Imagine into one of the wealthiest game companies in Europe. But success was followed by excess. "It was really very, very heady," said Everiss. "We were inventing the industry as we went along. Up until Imagine, the industry had been a kitchen-table industry. Imagine was the first UK company to have things like a sales team, marketing people. We were the first to do multi-lingual packaging. We put programmers into offices, which was a new thing, and then started using sound and graphics artists."

Imagine became living proof of the dream of many bedroom programmers: that they could get filthy rich making video games. The company's plush offices boasted a garage filled with fast sports cars. At the age of 23, Butler was a symbol of the 1980s yuppie dream: a young man who had become rich through his entrepreneurialism. They formed their own advertising agency and started expanding across Europe. At one point they tried to rent the disused revolving restaurant on top of Liverpool's St John's Beacon tower, only to be put off by the excessive rent demanded by its landlord, the local council. "That was typical of us," said Everiss. "We thought it would make a good executive office being up in the air going round and round in circles." Most excessive of all was Imagine's decision to pour huge sums of money into developing *Bandersnatch* and *Psyclapse*, which it described as the first 'mega-games'. These games would come with hardware add-ons that, Imagine claimed, would enhance the abilities of home computers such as the Spectrum and usher in a new era in video games. It was not to be. In July 1984 Imagine went bust, its money drained away by over-expansion, the slow progress on developing the mega-games and falling sales due, at least in part, to piracy. The implosion was captured blow-by-blow by a BBC TV documentary crew who had set out to tell the story of Imagine's success, but instead recorded its very public demise.

Imagine weren't the only company to bite the dust around that time. The number of UK companies publishing Spectrum games peaked at 474 in 1984. The following year just 281 remained and by 1988 the number had tumbled to just 101. The industry became increasingly polarised between big publishers such as Ocean Software, who built business empires on the back of games based on blockbuster movies and popular TV shows such as *Robocop*, *Miami Vice* and *Knight Rider*, and budget publishers such as Mastertronic, which sold games for as little as £1.99 compared to the usual £8.99. By 1987 around 60 per cent of games sold in the UK were thought to be budget games. "At £1.99 it was hardly worth copying the game, you could have the real thing," said Everiss. The middle ground of companies that released full-price original games steadily lost ground, unable to compete on price or recognition. "At one stage we tried to launch a mid-price range and were just stuck in the middle. It was difficult, you had to be in one camp or the other," said David Darling, who founded Warwickshire-based budget game publisher Codemasters with his brother Richard in 1985.

The same was starting to happen in France. Infogrames, whose founders were laughed at by French venture capitalists when they asked for investment back in 1983, swallowed up Cobra Soft as well as Ere Informatique. Meanwhile, Guillemot Informatique, a leading distributor of computer equipment based in Montreuil, launched a game publishing business called Ubisoft in 1986 that quickly expanded across Europe. Both Infogrames and Ubisoft would go on to become multinational gaming giants. The wilder elements of Europe's early games industry started to leave the business. Surrealist game maker Mel Croucher sold off his game company Automata UK for 10 pence in 1985, while Jean-Louis Le Breton quit games to become a journalist. The European industry was growing up. Companies merged, expanded, created marketing teams and professionalised. Soon the games business was dominated by companies such as Ocean, Infogrames and US Gold, a UK publisher that rose to prominence converting American games onto home computers that were popular in Europe.

Formed in Birmingham by Geoff Brown, a former teacher and singer in progressive rock band Galliard, US Gold was a triumph of business nous over creativity. Brown bought his first home computer, an Atari 800, just as home computers began to take off in the UK. "There weren't many people in the UK owning an Atari, so those who did were enthusiasts and if you were an enthusiast you were prepared to look for the games," he said. "I got hold of a US magazine called *Compute!* that had all these wonderful games I had never heard of. The screenshots looked brilliant, so I thought I'm going to get myself one of those." The game he chose was *Galactic Chase*, a 1981 game from Stedek Software. It was a straightforward copy of the arcade game *Galaxian*, but its production quality was miles ahead of what was being developed in the UK. "A lot of the UK programmers were still writing in BASIC. These guys were writing totally in machine language," said Brown. "It was light years ahead of anything the UK was doing."

After making some money importing *Galactic Chase* to the UK, Brown bought an airplane ticket and headed to the US to sign up more of the games being made by the North American computer game business that had come to the fore after the spectacular collapse of Atari.

Arcade action: A British teenager tries out Yu Suzuki's 1989 coin-op Turbo Out Run
Paul Brown / Rex Features

CHAPTER 11

Macintoshization

One afternoon in 1975 a Harvard University student decided to write a seven-year plan that would result in the birth of one of the world's biggest game publishers. It may have been the days of *Pong* but Trip Hawkins, the student in question, was already electrified by the new world of video games. "From the moment I saw my first computer in 1972, I knew I wanted to make video games," he said. "I had a strong feeling that people were meant to interact, not to sit passively like plants in front of the TV. I was already designing board games but saw instantly that a computer would allow me to put 'real life in a box'." Just before he wrote his plan, Hawkins had read about the opening of one of the first computer stores and the Intel microprocessor, a computer-on-a-chip. He knew then that video games would become a mainstream form of entertainment. Technology, however, was against him. The computers of the day were still too expensive and too primitive to allow Hawkins to realise his dreams.

So instead Hawkins decided to spend the next seven years preparing for 1982, the year in which he believed computer technology would have caught up with his vision for video games. "By then, I figured, there would be enough hardware in homes to support a game software company," he said. He adhered to his plan with religious devotion. He tailored his degree in strategy and applied game theory so that he could learn how to make video games. He took an MBA course to get the business skills he needed to run his future company and carried out market research into the computer and games console business. In 1978 he joined Apple Computer where he honed his business skills and, thanks to the stock options he got when the company floated on the stock exchange in 1980, the funds he needed to start his game business. "I made enough in my four years at Apple to know I could completely fund the company if I wanted," he said.

And as 1981 came to a close, Hawkins was finally ready, but by then the video game boom was already well under way. "I actually felt late," he said. "Because of the success of Atari's early hardware and a cottage industry of Apple II software companies,

I counted 135 companies already making video games but I had a unique vision and thought I could compete and become one of the leaders. This is what happens to you after you hang around with Steve Jobs for a few years."

Sticking rigidly to his plan, Hawkins quit Apple on New Year's Day 1982 and set about forming Electronic Arts. Hawkins' vision for Electronic Arts echoed the old Hollywood studio system that emerged in the 1920s, with its plan to control game development, publishing and distribution. Electronic Arts would make games on multiple platforms, package them in boxes not plastic bags, and distribute them direct to retailers. It would also promote its game designers as if they were movie directors – artistic visionaries of the new era of interactive entertainment. The company's publicity materials set out its 'games as art' rhetoric: "We are an association of electronic artists united by a common goal. The work we publish will be work that appeals to the imagination as opposed to instincts for gratuitous destruction." Other publicity materials asked "can a computer make you cry?" and promised games that would "blur the traditional distinctions between art and entertainment and education and fantasy".[1]

But by the time Electronic Arts released its first games on 21st March 1983, the North American game business was going down the tubes. "Atari officially crashed in December 1982," said Hawkins. "The media, retailers and consumers vacated the console market in 1983, leaving Electronic Arts in a void. Start-ups like Electronic Arts had to focus on the Apple II, Commodore 64, etc. But those markets never got very big because the computers were more expensive and harder to use. They were really a hobby market more than a consumer market." The post-Atari world of the home computers was an inhospitable landscape for those hoping to make a livelihood out of video games. "It was a brutal time," said Bing Gordon, Electronic Arts' head of marketing and product development at the time. "We entered the dark ages of interactive entertainment. The five years between 1982 and 1987 were hard, hard, hard. Each Christmas, all the experts at leading newspapers reminded potential customers that the video game business had died with Atari and would never return."

What market did exist was splintered; fragmented across myriad home computer systems each with different technology and capabilities. It was also a market riddled with piracy, unlike the cartridge-based consoles of old. "People would steal your game. They wouldn't buy it, they would copy it," said Rob Fulop, a game designer at Imagic, the former console starlet that tried unsuccessfully to survive the crash by making

---

1. Electronic Arts' promotion of its game designers as artists was short lived. "Even though we got some publicity, it didn't really catch on with the public," said Hawkins. "For consumers it was really more about which games were fun and not who made them and why. As a result, the approach was phased out over time and the products became bigger brands than the artists."

computer games. The differences between the hardware of computers and consoles, meanwhile, required game designers to rethink their work. Controls shifted from joysticks to keyboards. Games moved from being stored on microchips in cartridges to floppy disks. "You had long load times, a lot more memory and higher resolution visuals than you did on video game consoles," said Don Daglow, who became a producer for Electronic Arts after Mattel abandoned the Intellivision console. "You had the ability to save a game on disk, so we could do games that could take longer because you could save. Floppy disks allowed us to be more ambitious." But computers were also slower. "Game companies had been concentrating on action games for consoles and computers weren't fast enough at that time to really do a good job with an action game," said Michael Katz, who quit Coleco as the crash set in to become the president of San Francisco-based computer game specialists Epyx.

Home computer users were also a different type of consumer compared to the console owners game companies grew up with. They were older, more educated and more technically minded.[2] "The video games before the crash were all specifically directed at young people, while computer games were directed at an older audience," said Chris Crawford, who became a freelance game designer after Atari's implosion. The differences in hardware and consumer tastes led game designers to move away from action games towards more cerebral, complex and slower forms of game. "Games prior to the crash sought to appeal to the mass market, but post-crash games became increasingly geared towards dedicated game players who wanted complexity and this further alienated the non-hardcore audience," said David Crane, co-founder of game publisher Activision.

Most of Electronic Arts' debut games reflected this new era of complexity. Foremost among these games were *M.U.L.E.* and *Pinball Construction Set*. *M.U.L.E.* was a computerised multiplayer board game based on supply and demand economics that cast players as colonisers of a faraway planet, trying to scratch out a living. Its transgender creator Dan Bunten, who later became Dani Bunten Berry after a sex change, drew inspiration from *Monopoly* and Richard Heinlein's novel *Time Enough for Love*, a sci-fi retelling of the trials of America's old west pioneers. In the game each of the four players commandeered plots of land to produce energy, grow food and mine ore in a bid to become the richest. But while *Monopoly* was about cut-throat competition, *M.U.L.E.* was tempered by the need for players to work together to ensure there was enough food and energy for all of them to survive. *M.U.L.E.* was a commercial failure, but its

---

2. They had to be. Home computers at the time were intimidating. There were no windows or mice, just a > prompt and a flashing cursor impatiently waiting for the user to type in commands in computer language, usually BASIC.

careful balance of player competition and co-operation made it a seminal example of multiplayer gaming. *Pinball Construction Set* on the other hand used the extra memory and save features of computers to let people design and play their own pinball tables. Together with the same year's *Lode Runner*, a platform game with a level-creation tool, it pioneered the idea of allowing players to create game content – a concept that would be taken further by games such as *Quake* and *LittleBigPlanet*. *Pinball Construction Set*'s creator Bill Budge came up with the idea after spending some time working for Apple: "The people at Apple liked to go and play pinball at lunch – it was a big fad at the time. The engineers would spend time perfecting their moves on these pinball machines – typical obsessive-compulsive programmer behaviour. I would go with them and watch. It occurred to me you could make a pinball game on the Apple II." The result was 1981's *Raster Blaster*, a pinball game based on a single table, that Budge released through his own company BudgeCo. He then figured that a pinball game that let people create new tables would be even better and thanks to his time at Apple, he knew exactly how the table-creation element should work. "I was watching the Macintosh develop and I was really familiar with the Lisa. That introduced me to the graphical user interface and how cool all that was," he said. "I thought you could do a lot of the same stuff on the Apple II."

The Lisa, and its still-in-development successor the Macintosh, were Apple's latest computers. Both used a new approach to computer interfaces: the graphical user interface or GUI. The concept of the GUI dated back to 1950 when electric engineer Douglas Engelbart concluded that computers would be easier to use if people interacted with them via television screens rather than keyboards, punch cards or switches. But in an era where computers and television were still so new, his ideas were dismissed as bizarre and unrealistic. Then the Cold War intervened.

In August 1957 the Soviet Union launched the first successful intercontinental ballistic missile and on the 4th October that same year launched the world's first artificial satellite, Sputnik 1, into orbit. The next step was obvious: putting nuclear warheads on intercontinental ballistic missiles. The US government responded by forming the Advanced Research Projects Agency (APRA) to bankroll research to help the US regain its technological superiority over its superpower rival. And in 1964 APRA decided to fund Engelbart's research to the tune of $1 million a year. Using the money Engelbart created the GUI, the basis of almost every computer since the mid-1990s. He invented the mouse, the idea of windows that users could reshape and move around the screen, designed the word processor, came up with the concept of cutting and pasting, and devised icons that could be pointed at and clicked on using the mouse. In short, he produced the template for modern GUIs such as Microsoft Windows and Mac OS. In

1973 the Xerox PARC research institute in Palo Alto used Engelbart's ideas to come up with the Alto, one of the earliest GUI computers. Xerox did little to turn the Alto into a commercial product, but when Apple co-founder Steve Jobs paid a visit to the facility he saw the potential of the GUI. Apple's first attempt at a GUI-based computer, the Apple Lisa, went on sale in 1983. It introduced Engelbart's concepts to a wider audience but its high price – $9,995 – meant it was a commercial failure. The following year, however, Apple tried again with the Apple Macintosh. Unlike the Lisa, the $1,995 Macintosh made an immediate and lasting impact. For those used to the unfriendly and intimidating computers of the late 1970s and early 1980s it was a liberating moment.

"The human interface of a computer as we know it today, with windows and a mouse, was new to the world of personal computers when the Lisa and Mac came out," said Darin Adler, a programmer at Illinois game developers ICOM Simulations. The Macintosh sparked a revolution in computer design as Apple's rivals began to create GUIs for their next generation home computers.[3] The Macintosh was also a big influence on game designers, many of whom saw GUIs as a way to make more complex games easier to understand. Its influence was such that *Computer Gaming World* journalist Charles Ardai argued that video games were undergoing a process of 'Macintoshization'. "GUIs served to regularise the interface and make it a bit more indirect," said Crawford. "Most games had direct interfaces: push the joystick left and your player moved left. GUIs moved us a bit further towards abstraction by putting some of the verbs onscreen as buttons or menus. This in turn greatly expanded the size of the verb list that we could present to the player."

Crawford took advantage of the Macintosh's GUI with his 1985 game *Balance of Power*, a simulation of Cold War geopolitics where players took charge of the US or USSR. "Actually, it wasn't so much the GUI that appealed to me as the raw computational power of the Mac," said Crawford. "I went from an Atari with an 8-bit processor and 48Kb of RAM to a Mac with a 16-bit processor and 128Kb." At the time of *Balance of Power*'s release the Cold War had been under way for 40 years and showed no sign of ending. If anything the aggressive and uncompromising stance of President Ronald Reagan led many to suspect that nuclear war was becoming more, not less, likely. "The militaristic rhetoric of the Reagan administration led me to fear the prospect of a nuclear war with the Soviet Union," said Crawford. "A lot of people in those days shared that fear. It seemed as if the Cold War was heating up and might become a hot war. Ever since my student days I had tried to understand how nations could get

---

3. This post-Mac generation included the Acorn Archimedes, Atari ST and Commodore Amiga. The IBM PC and its compatibles eventually caught up thanks to the success of the third version of Microsoft's Windows operating system, which was released in 1990.

themselves caught up in the idiocy of war. I had studied lots of military and diplomatic history and I was finally coming to understand the basic principles. I wanted to communicate those principles."

The goal of *Balance of Power* was to defeat the rival superpower by increasing your standing among the world's nations. Players could use diplomacy, military muscle, espionage, money or insurgency to try to bend nations to their will, but they had to avoid confrontations with the opposing superpower that could end in nuclear war. The outbreak of nuclear war ended the game with a simple message: "You have ignited a nuclear war. And no, there is no animated display of a mushroom cloud with parts of bodies flying through the air. We do not reward failure." Crawford's simulation sought to model global reality as closely as possible, even including obscure political science concepts such as Finlandisation – the term used to describe how Finland sought to appease the neighbouring Soviet Union during the Cold War by censoring anti-communist media and refusing to grant asylum to political refugees from the USSR. Such was the complexity hidden beneath its simple interface of pull-down menus and icons to click on that Crawford later wrote and published a book explaining the game's inner workings. "My hope was that players would appreciate the complexity of it all, that they would understand that military action is, in fact, occasionally desirable but it had to be used judiciously and in the context of a larger diplomatic strategy," said Crawford. Despite its complexity and political subject matter, *Balance of Power* sold around 250,000 copies – a significant amount for the time.

ICOM Simulations, meanwhile, used the Mac's GUI to rethink the text adventure. "Our idea was to do an adventure game that fitted into the Mac user interface," said Adler. "Programs like *MacPaint* and the *Mac Finder* concentrate on mouse clicks and drags for user interface. We wanted to do the same for an adventure game. One of our ideas was 'when in doubt, make it work the same way the *Finder* does'. Another was to choose a game with a style that fitted well with the black and white display of the Mac. That's why we used a film noir story – we figured those movies were black and white already."

When ICOM began work on their film noir adventure in 1984, little had changed in the way adventure games worked since they first appeared in the late 1970s. Instead of rethinking the method of interaction, adventure game specialists Infocom had concentrated on improving the writing and creating 'feelies' – items packaged with the game that were intended to enhance the experience. Infocom's feelies first appeared in the company's 1982 murder mystery *Deadline*, which came with pieces of evidence from the crime scene such as police interview notes and a photo of the murder scene. "The items in the package became a trademark of our games and were also a small anti-piracy aid as just copying the disks wouldn't get you all you needed to solve the puzzles," said Infocom co-founder Dave Lebling.

But core to Infocom's efforts to stay ahead of the pack was better storytelling. To help it achieve this goal it began to working with professional authors such as Douglas Adams, who helped Infocom turn his comedy sci-fi novel *The Hitchhiker's Guide To The Galaxy* into a video game. "It was quite a close collaboration, not like the typical author-game designer collaboration where the author talks to the designer for an hour and then plays the game months later with PR people snapping photos to show off the 'collaboration'," said Steve Meretzky, the Infocom game designer who worked with Adams on *The Hitchhiker's Guide To The Galaxy*. "In general he was a delight to collaborate with because he understood the medium but didn't feel as bound by its conventions as someone who'd already been working in it for several years like me. He came up with all sorts of crazy and inventive ideas like the game lying to you. On the other hand he was the world's worst procrastinator. He would wait until the last minute and then wait another six months. As he once said: 'I love deadlines. I especially love the whooshing sound they make as they pass by'."

After completing *The Hitchhiker's Guide To The Galaxy* in 1984, Meretzky went on to write *A Mind Forever Voyaging* that, along with 1986's *Trinity*, would mark the pinnacle of Infocom's quest for literary excellence. Like *Balance of Power*, President Reagan inspired *A Mind Forever Voyaging*. "Reagan had just been re-elected in a landslide and I was completely horrified because I despised him and his administration," said Meretzky. "Text adventures were such a compelling medium. While playing one you just thought about the game day and night, mulling over different solutions for puzzles, so I thought that it might be a particularly effective medium for getting a message across. I wanted to show Reagan as the right-wing, war-mongering, fundamentalist-coddling, budget-exploding, wedge-driving, environment-destroying, intellectual lightweight he was."

In the game players took on the role of Perry Simm, a software program with human intelligence created to live in a computer simulation that extrapolated the effects of a government social policy that echoed Reagan's free-market and socially conservative stance. As the players explored various simulations of the future, they watched how the policy would destroy freedom and peace in America while grappling with the ethical dilemma of what would happen to Simm when the simulation was over. While Crawford had struggled to find a publisher for his take on the Cold War, Infocom backed Meretzsky's political critique. "A few people expressed concern about a game that might make some players angry, but Dave Lebling, who's quite conservative, stuck up for the idea and said that perhaps someday he might want to make a game that attacked liberal principles and wanted to be free to do so," said Meretzky.

"I am politically well to Steve's right and very strong on free speech. He designed a good game and it was worth producing," said Lebling. "One could imagine a sequel about the rebellion against the state set up by listening to Perry Simm and his lousy socio-economic models."

Brian Moriarty's *Trinity*, meanwhile, was a fantastical time-travelling adventure about the dangers of nuclear bombs and atomic war. "I first conceived the idea of an adventure game based on the Trinity Test in 1983," said Moriarty.[4] "It was inspired by a history book I'd read years before in my high school library: *Day of Trinity* by Lansing Lamont. The dramatic story of the creation of the first atomic bomb really captured my imagination for some reason. It seemed the perfect setting for an interactive story in which a character from the future finds himself facing the possibility of changing history. It was about the mystery of choice."

But while Infocom concentrated on writing, its foremost rival – Sierra Online – had begun looking for an escape from the restrictions of text as its co-founder and lead game designer Roberta Williams became increasingly frustrated with the genre's limitations. Sierra's change in direction began with *Time Zone*, a $99 sci-fi adventure game spread across six double-sized floppy disks that Williams envisaged as a video game equivalent of the epic movies of Cecil B. DeMille. Having completed her epic she publicly admitted she was burnt out and couldn't bear the thought of looking at another text adventure. So after creating a couple more adventures based on Disney licences, she made her bid for freedom with 1984's fairytale adventure *King's Quest*. For the game Williams ditched the Apple II so she could take advantage of the more powerful features of the PCjr, IBM's low-cost version of its standard PC. The PCjr allowed Williams to fulfil her long-held desire to introduce animation to her games.

She also used it as a chance to reduce the reliance on text input by letting players move the lead character around the screen using the arrow keys on the PCjr's keyboard. But she stopped short of abandoning text altogether, requiring players to type in the usual verb-noun commands to perform actions other than moving. *King's Quest*'s fusion of animation and adventure was a watershed moment for the genre, but it would take ICOM's 1985 film noir game, *Déjà Vu: A Nightmare Comes True*, to finally free players from the tyranny of text commands. *Déjà Vu* dropped text input altogether. Instead of having to type in commands in the hope that the game would understand, players could click on a selection of action words and then click on the object or person they wanted the action to apply to. *Déjà Vu*'s story was no match for the works of Sierra or

4. The Trinity Test was the first nuclear explosion, which was carried out near Alamogordo, New Mexico, on the 17th July 1945.

Infocom, but ICOM had showed the way forward. By the end of the 1980s the text adventure would be on its last legs and Infocom with it, thanks to the company's aversion to animation and GUIs. "We were very text-oriented and were happy to spend more memory space on words than on pictures," said Lebling. "In those days, the games with lots of graphics had very few words and we thought that the personal computer technology of the day was better suited to words. It made us somewhat hostile to graphics in general, which was a bad thing."

Adventure games were not the only genre where designers were seeking to explore the narrative horizons of video games. Richard Garriott was also seeking to make story a central feature of his role-playing game series *Ultima*. After releasing 1983's *Ultima III: Exodus* through his own company, Origin Systems, he got to see his fan mail for the first time. The letters shocked him. "I found it fascinating to read what people were doing in my games," said Garriott. "People would say I bought it and really enjoyed it and after I solved the main plot I had a great deal of fun going back and killing everybody or people would write in about the shortcuts they found to achieve solutions where they did not play a good guy but won by killing all the villagers in town because it was the fastest way to advance." At the same time Garriott also started receiving hate mail from supporters of Bothered About Dungeons & Dragons (BADD), a pressure group that claimed *Dungeons & Dragons* was spreading Satanism and was formed by Patricia Pulling, a grieving mother from Virginia who believed her 19-year-old son Irving killed himself because of the role-playing game. Pulling didn't pick up on *Ultima III: Exodus*, which came in a box with a demon on the front, but others who agreed with her views did. "This was when the Christian right in the United States was coming out very strongly against role-playing games," he said. "I received hate mail from religious groups describing me as the satanic perverter of America's youth." To Garriott it seemed that both the writers of the hate mail, and some of his fans, had misunderstood his work. "I found it so ironic and laughable," he said. "Here I am writing games that I believe are, on the whole, quite positive and yet there's a clearly this segment of the population so divergent from that belief."

Garriott decided that the fourth *Ultima* game would mark a change in direction. "I sat down and thought real hard about what I could do that would reward people in the way the real world reacts," he said. His solution was to make a game about virtue. Garriott pored over every book on philosophy and morality he could lay his hands on in a search for some simple truths that he could put at the heart of the game. He boiled down the ideas he read about into eight virtues based on three broad principles: truth, love and courage. His confidence was also boosted when he noticed that one of his favourite movies, *The Wizard of Oz*, also homed in on the same ideas with its scarecrow, tin man and lion characters. "I independently arrived at truth, love and courage but L. Frank Baum had clearly arrived at a similar conclusion," he said.[5] "I was given confidence that one of my creative heroes had arrived at a very clearly parallel conclusion, so I resolved to stick with truth, love and courage." Garriott also resolved to track players' behaviour in the game to assess how virtuous they were rather than simply encouraging players to build up numerical attributes, as was the norm in role-playing games. "I tried to think of very rational, not necessarily terribly obvious ways where if people played the game with an eye towards being good the game would reward them, and if they deviated from that way the game would not," he said. "For example, I would set a test where, if you ran into a creature that was evil and weaker than you and you ran away, I called you a coward. On the other hand if you faced it, whether or not you lived, I gave you greater valour. But if you walked away from a wolf or something that wasn't evil I did not deduct your valour because I do not feel it is a courageous thing to go kill a wild animal that goes for you because it happens to be looking for food."

Among the tests Garriott set players was a blind shopkeeper where, after buying goods, players could short change the sightless woman. "If players left a multiple of any kind she would accept it and let you walk out. She wouldn't make any comment about it whatsoever," said Garriott. "But even though she didn't say anything, the computer records the fact that you're a lying, cheating, thieving bastard. Much later in the game you need to go back to the same woman to get a major clue. If you had been short changing her, she'd go 'I'd love to help the avatar, this guy on the path of virtue, but you're the most dishonest, thieving scumbag I've ever met so I'm not telling you'. All of those tests I tried to set up in the same way so you were never told immediately whether that was the right or wrong thing to do. It only happened over time as your behaviour accumulated."

The result, 1985's *Ultima IV: Quest of the Avatar*, was a major departure from every other role-playing game created up until that point. Its moral backbone and quest for enlightenment added a new dimension to the *Ultima* series. It also saw Garriott make a concerted effort to move away from the traditional Tolkien-inspired worlds of other fantasy games for one of his own design. Garriott and his colleagues worried the change in direction would anger players. "It was so different to every other game that had come out I was sincerely worried that no one would like it and no one would understand it," said Garriott. "I even had people in my own office who would express very specific doubts." To Garriott's relief, *Ultima IV* became the series' biggest seller by a long way. The fan mail also showed that players had embraced the change. One

5. The author of *The Wonderful Wizard of Oz*, the 1900 children's novel that the film was based on.

letter in particular stuck in Garriott's mind: "This mother wrote me a letter and said: 'My daughter was actually having personal life issues with some of the issues you brought forth in your game. Your game showed her cause and effect that is completely appropriate even in the real world. My daughter has grown through this and I just wanted to write you a letter expressing that I, as a parent, recognise that fully and appreciate it. I'm touched by the work you've done'."

By the end of 1986 it was clear that video games were no longer moral vacuums. From the morality play of *Ultima IV*, to the political commentary of *A Mind Forever Voyaging* and the economic and social allegory of *M.U.L.E.*, game designers were discovering hidden depths to their medium thanks to the move to home computers. The early 1980s industry crash had ushered in a new richness to the video game.

* * *

Change was also happening outside the home as coin-op game designers adjusted to the post-crash world of fewer arcades and fewer players. "1982 and '83 were very difficult years for Exidy and the overall amusement industry as a whole," said Howell Ivy, a game designer at the American coin-op manufacturer. "The industry needed a change. The strategies were the development of new, faster, better game systems."

The coin-op video game companies that had survived the shake-out focused on their big advantage over home systems: hardware. Unlike the makers of home games, coin-op designers could decide what technology suited their games rather than having to work within the limitations of the popular platforms. It was an advantage arcade game makers pushed hard to keep people coming to the arcades. They offered graphics and sound that no home computer or console could match, designed elaborate cabinets and built controllers designed specifically for their games.

Exidy's 1983 game *Crossbow*, for example, handed players a pretend crossbow that they could use to shoot on-screen enemies and revived interest in light gun games.[6] Soon the arcades shook to the sound of gunfire as players got to blast terrorists, zombies and criminals with a variety of plastic guns thanks to games such as *Operation Wolf*, *Beast Busters* and *Virtua Cop*.

Atari Games' Ed Logg, meanwhile, used cabinet design to emphasise the social advantages of the arcade with his 1985 fantasy shoot 'em up *Gauntlet*. "The idea for *Gauntlet* came from two major sources," said Logg. "My son was heavily into *Dungeons*

6. Light gun games pre-date video games. The first was the Seeburg Ray-O-Lite, a duck-shooting electro-mechanical coin-op game released in 1936. The Magnavox Odyssey's *Shooting Gallery* game, which came with a toy rifle, was the first video game to use the idea.

& Dragons at the time and he wanted me to do a *Dungeons & Dragons* game so bad. The other thing that came along was a computer game called *Dandy*. It was a four-player, co-operative-style game. I melded the two ideas together to form *Gauntlet*." The result was a game that had the trappings of *Dungeons & Dragons* with its warriors, wizards and monsters but was actually a shoot 'em up set in a large maze. Logg's stroke of genius, however, was creating a cabinet that let up to four people play at once as a team. *Gauntlet* became the social hub of the arcades, a game that brought players together. People could join in games at any point simply by inserting a coin and so complete strangers ended up playing together. For arcade operators it was the kind of cash cow they hadn't seen since the glory days of the early 1980s, thanks largely to its ability to take money from four people at once. At the height of its popularity, the average *Gauntlet* machine was raking in an estimated $900 a week.

Logg's cabinet design, however, looked tame next to the work of Sega's Yu Suzuki, a Japanese game designer who came to prominence in the wake of the crash. Suzuki enhanced his games with cabinets that resembled fairground rides. His 1985 motorbike racing game *Hang-On* marked the start of a five-year exploration of the intersection between video games and theme park rides. *Hang-On*'s cabinet was a replica motorcycle with a screen mounted into the windshield. Players steered by leaning left or right to tilt the motorbike and used the handlebars to accelerate and brake.

In the wake of its success Suzuki pushed the idea even further with 1986's *Out Run*, a car driving game inspired by his love of the film *The Cannonball Run*, a 1981 comedy about an illegal race across the US. Armed with a video recorder, camera and notepad, the sports car-loving Suzuki went on a two-week driving tour of Europe to gather information for his game. While visiting Monaco during this trip, he spotted the most desirable supercar of the 1980s: a Ferrari Testarossa. Then and there he decided that would be the car would be the focus of *Out Run*. Suzuki made *Out Run* a celebration of driving and an ode to '80s cool with its fast supercar, open road and a blonde-haired girlfriend in the passenger seat. There was even a virtual in-car stereo offering four synthesizer-heavy pop tunes that blared out as the player zoomed through scenery inspired by Suzuki's European tour. And while the need to reach your destination within a tight time limit meant players had to drive fast, *Out Run* was about the joy of driving rather than the racing.

Suzuki's final touch was a box-like cabinet designed to look like the inside of a car that used hydraulics to shake and move in line with the on-screen action. When *Out Run* was shown for the first time at a Japanese trade show in October 1986, the excited crowds left many struggling to get a glimpse. It went on to become one of the most popular driving games of the 1980s. Suzuki's experiments in cabinet design eventually reached their pinnacle in 1990 with *R360 – G-Loc Air Battle*, a gigantic arcade game that put players in a replica jet fighter cockpit capable of spinning and rotating through 360 degrees.

The developments in cabinet design and graphics technology helped keep people coming to the arcades, which stabilised at a turnover of between $2 billion and $3 billion a year in the US shortly after the crash. The leaps forward in technology also tempted Eugene Jarvis, the creator of early 1980s arcade hits such as *Defender*, back into the games business. "I could see that technology was evolving beyond the pixel-based artwork of the '70s and early '80s towards 3D animation, motion capture, digitization, you name it. The video game field was about to explode in a technological big bang enabling immersive and rich gaming as never before seen."

Inspired, Jarvis rejoined his former employer Williams and set to work on his comeback game: *Narc*, an ultra-violent ode to Reagan's war on drugs. "*Narc* was the war on drugs to the limit," said Jarvis. "The player characters Max Force and Hitman were on a mission to protect the innocent and punish the guilty. Although you could bust drug dealers and send them to jail, it was much more fun to blow them to bits with a rocket launcher or roast them in a glorious flamethrower BBQ. Let's face it, drug dealers enable the destruction of more lives than any disease in our society. It's about time they got their due instead of a suspended sentence." Aside from its brutal 'say no' message, *Narc* was also notable for being one of the first games to use digitization, a technique that allowed game designers to record film or sound and import it into their work allowing big improvements in visuals and audio. "The cool thing about digitized imagery is that you can work directly with actors, costumes, locations and directly digitize imagery into the game," said Jarvis. "It enables amazing photorealistic quality and you can capture all the nuances and lighting and skilled character acting and dialogue."

When *Narc* arrived in the arcades in 1988, however, the dark ages of interactive entertainment referred to by Electronic Arts' Gordon were over. And it was all thanks to a Japanese toy company that single-handedly brought video games back from the brink.

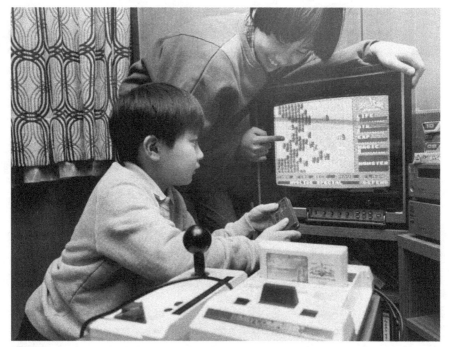

Family Computer: Two Japanese children play Nintendo
Reuters / Corbis

CHAPTER 12

A Tool To Sell Software

They made an odd couple. On one side of the desk sat Hiroshi Yamauchi, the 49-year-old chairman of Nintendo – a Japanese toy firm that started out in 1889 making playing cards. He was a hard-nosed businessman with a single-minded desire to turn the business he inherited from his grandfather into a global giant. His raw ambition was evident from the moment he took charge of Nintendo in 1949. His first act was to stamp his authority on the firm by firing his cousin, the only relative employed by Nintendo, before sacking every manager loyal to his grandfather.

On the opposite side of the desk was Shigeru Miyamoto, a banjo-playing daydreamer straight out of college whose shaggy shoulder-length hair was out of keeping with the neat style expected of Japanese businessmen. As a child Miyamoto had dreamt of becoming an entertainer, a puppeteer or a cartoonist and he shuddered at the thought of becoming a 'salaryman', one of corporate Japan's downtrodden drones. He hoped life at a toy company would be better. Nintendo's stern boss told Miyamoto, who landed the interview because his father knew a friend of Yamauchi's, to go away and return with an idea for toy. Miyamoto returned brandishing a colourful coat hanger made out of soft wood and decorated with pictures of animals drawn in bright acrylic paint. Metal hangers could hurt a child, explained the young designer. Impressed, Yamauchi gave Miyamoto a job as a graphic designer. It was 1977 and, although neither man yet knew it, together they would redefine both the business and the content of video games. Miyamoto joined Nintendo at a crucial moment. Under Yamauchi's direction the company was in midst of its second attempt to enter the video game business. Nintendo's first attempt had ended in failure. In 1975 it released *EVR Race*, a coin-op horse race betting game designed by Genyo Takeda, only to find the game's use of videotape footage meant the machine was prone to breaking down and needed regular maintenance.

Now Yamauchi hoped to achieve video game success by releasing two home *Pong* consoles: the Color TV Game 6 and Color TV Game 15. While the US had caught the home *Pong* bug back in late 1975, home video games were still new to Japan. Prior to 1977 the only game console on the market was toy firm Epoch's TV Tennis, a repackaged version of the Magnavox Odyssey.

Nintendo's timing could not of been better: 1977 saw a succession of home *Pong* consoles launched by companies such as Bandai, Hitachi and Epoch that got Japan excited by the chance to play video games at home. Nintendo was one of the victors in the Japan's home *Pong* wars. The Kyoto toy firm sold more than a million of its Color TV Game consoles making it a high-profile player in Japan's emerging video game industry. At the time Japan's game business was largely focused on the domestic market, even the big guns – Taito and Sega – concentrated on winning over Japanese arcade-goers rather than reaching a global audience. Namco, the country's other big coin-op video game company, didn't even have a development team of its own. Instead it concentrated on reaping the rewards of having the rights to bring Atari's coin-op games into Japan.

There was little reason to think Japan's video game industry might become a serious challenger to US companies such as Atari and Bally Midway. Post-war Japan may have gone from vanquished enemy to a successful exporter of manufactured goods, but its cultural influence had lagged far behind its growing economic power. For North Americans and Europeans, Japanese cinema meant *Godzilla*, which despite a cult following was regarded as something of a joke. And while the work of director Akira Kurosawa had directly inspired the Hollywood films *Star Wars* and *The Magnificent Seven*, few people knew about his influence. Manga comics and anime, meanwhile, were rarely – if ever – seen outside Asia.

But then came Taito's *Space Invaders*, a made-in-Japan global phenomenon that gave the nation's game companies a newfound sense of self-confidence. Suddenly the idea of competing on the global stage seemed a real possibility rather than a pipedream. Namco responded by starting to design its own games, Nintendo re-entered the coin-op game business and Taito opened a US office. *Space Invaders'* enormous earnings confirmed Yamauchi's belief that Nintendo's future lay in video games. He ordered his staff to throw out their old ideas and concentrate on devising bold new video game products that would give Nintendo the edge over its rivals.

Gunpei Yokoi was first to respond. Yokoi was Nintendo's toy-maker-in-chief, a creative genius whose inventive toys had sustained the company through the 1970s.[1] One evening on his commute home, he saw a bored businessman passing the time by

---

1. Yokoi joined Nintendo in 1965 and maintained its playing card production line until being asked by Yamauchi to create a toy for Christmas 1969 market. He came up with the Ultra Hand, a hand on a long plastic stick that grasped when a handle at the opposite end was pulled. More than a million were sold.

playing with a pocket calculator that had a liquid crystal display (LCD). A portable video game, Yokoi figured, would be a much more enjoyable way to pass the time. Inspired, he set about designing such a device. Yokoi's approach to creating his portable game followed his personal design philosophy: the "lateral thinking of withered technology". The concept, he explained, involved shunning the latest technology and finding new uses for "mature technology that can be mass-produced cheaply".

The result was the Game & Watch, a low-cost handheld LCD game that doubled as a digital watch. Nintendo released the first Game & Watch, the juggling game *Ball*, in April 1980 to instant success. Over the next 11 years, Nintendo would sell more than 30 million of the dozens of Game & Watch titles it released. The year after *Ball*'s release, Miyamoto's debut game *Donkey Kong* confirmed Nintendo's new status as a member of Japan's video game elite alongside Taito, Sega and Namco.

Yamauchi's desire for business success was far from satisfied, however. Although he had managed to turn Nintendo into one of the world's biggest video game companies in just a few years, Yamauchi wanted to dominate the whole industry. And to achieve this, he decided that Nintendo would build a cartridge-based video game console for the Japanese market in an attempt to replicate the massive success the Atari VCS 2600 had had in North America. In 1982, the year Yamauchi set his sights on Atari-esque levels of success, Japanese sales of home *Pong* consoles had melted away but no cartridge-based console had managed to electrify the nation's video gamers. Bandai had just abandoned its SuperVision 8000, Japan's first cartridge-based console, in order to launch the Japanese version of Mattel's Intellivision. Bandai's main rival, Epoch was touting the Cassette Vision, which launched in 1981 with tree-felling action game *Kikori no Yosuka* as its flagship title, but sales were slow.

While the console market had slowed to a crawl, sales of home computers produced by the likes of NEC and Fujitsu were accelerating, encouraging the formation of Japan's first video game publishers. Among the first was Koei, a software company founded by the husband and wife team of Yoichi and Keiko Erikawa. Koei kicked off their move into game publishing in 1981 with the release of *The Battles of Kawanakajima*, a historical strategy game set in feudal Japan. Koei would go on to make its name with its games based on Chinese and Japanese history, particularly the *Nobunaga's Ambition* strategy series that began in 1983, but its most influential release was 1982's *Night Life*.

Night Life wasn't even a game. It was a computerised sex guide for the NEC PC-8801 computer offering advice on sexual positions complete with primitive black and white line drawings and a menstrual cycle tracker. Its commercial success, however, spawned a whole new genre of video game: bishojo gemu.

Literally translated as 'pretty young girl games', bishojo games emerged when several companies sought to capitalise on the success of *Night Life* by fusing the interactivity of video games with the stories and art of bishojo manga comics. These comics, which are widely available in Japan and openly read in public, focused on young school-age girls and, while not always, are often pornographic in nature.[2] At its most extreme, bishojo manga offered readers access to unrestricted sexual fantasies including violent rape and paedophilia. The games at the pornographic end of bishojo, known as eroge, were no less explicit. Within a year of *Night Life*'s launch, Japanese game publishers were releasing games such as *Lolita Syndrome*, a game released by Enix that included a section where players throw knives to tear the clothes off pre-pubescent girls.

As a video game genre, bishojo games are defined primarily by their sexual or romantic content rather than play style, but none-the-less a number of connected game genres have evolved from them. There are yaoi, or boys' love, games that involve sexual or romantic encounters between teenage boys that are aimed at a female audience and are distinct from the bara, or men's love, games that cater for gay men. Dating simulations, meanwhile, challenged players to develop their characters' attractiveness or social skills in order to woo virtual girlfriends or boyfriends.[3] Sometimes sex was the end goal, but many focus only on romance such as Konami's massively popular high-school romance series *Tokimeki Memorial*. Another branch were the life simulations where players mentor a girl or group of girls, such as the 1991 game *Princess Maker*, where players dress and raise an orphan girl while training her for a future career that could be anything from prostitute to prime minister.

For the most part the Japanese authorities paid little attention to the explicit sexual content of bishojo games, but on occasion these games have tested the limits of the nation's tolerance. The first to do so was *177*, a 1986 game where the goal was to rape teenage girls. It provoked condemnation in the National Diet of Japan, the country's parliament, prompting its publisher to remove it from sale.

Bishojo games, manga and anime came under further scrutiny in 1989 following the arrest of Tsutomu Miyazaki, a serial killer who had kidnapped, tortured and murdered four girls aged four to seven. Before his capture, Miyazaki had taunted his victims' families by sending letters detailing his crimes and, in one case, leaving the bones of one victim on her parents' doorstep. After his arrest police discovered that Miyazaki had a collection of more than 5,000 videos, including many pornographic anime films.

---

2. Japan's tolerance of such comics and games stems, at least in part, from Shinto, the country's former state religion, which has a non-judgemental attitude towards sex.

3. Elf's 1992 pornographic *Dokyusei* kicked off the dating sim genre.

The discovery earned Miyazaki the nickname the Otaku Murderer, in reference to Japan's otaku youth subculture. The otaku tend to be obsessed with anime, manga and video games, to the extent that some dress up and behave like their favourite characters in public. Miyazaki's hideous crimes provoked national outrage and stirred up a wider moral panic about the otaku. In response to the outcry bishojo game publishers formed a regulator, the Computer Software Rinri Kiko, to police the content of their games. The Computer Software Rinri Kiko's first act was to restrict games featuring incest, bestiality and paedophilia.

Further controversy erupted in 2009 when New York-based women's rights group Equality Now launched an international campaign against the rapist sim *RapeLay*. The campaign sparked global condemnation that resulted in an embarrassed Japanese government threatening a legal ban. Game publishers quickly agreed to stop making such games in the future.

Due to their sexual content, bishojo games rarely got released outside of Japan but within the country it has been a popular genre, especially on home computers, with best sellers achieving sales of half-a-million copies or more. The genre's popularity in Japan was such that its influence has seeped into several mainstream games that have been released internationally. The most overt example is Tecmo's fighting game series *Dead or Alive*. Its obsessive focus on the jiggling breasts of the game's female fighters has a clear debt to bishojo. The series' bishojo connections, however, were shown most clearly in its 2003 spin-off *Dead or Alive Xtreme Beach Volleyball*, where players could dress the game's female fighters in a range of skimpy bikinis and buy them presents.

Life simulations and dating games also influenced Yasuhiro Wada, the designer of the gentile farming sim series *Harvest Moon*, which first appeared in 1996 on the Super NES. "When I first started out in game design there was a life simulation called *Princess Maker*, where you 'grow' a princess. Due to this influence, I created a very minor game for the PC Engine called *Metal Angel*," said Wada.[4] But while the more popular *Harvest Moon* features elements of dating games, gardening was the stronger influence on the series. "I have never had a job connected to farming but my hobby is to grow plants. I find taking care of living things very comforting, it also teaches you to treasure life," said Wada.

Bishojo games often take the form of visual novels, another uniquely Japanese form of video game. Visual novels grew out of the text adventure genre, which first reached Japan when the publisher Micro Cabin released Sierra's murder mystery game *Mystery House* in 1982. *Mystery House* inspired a spate of Japanese adventure games that

4. Released in 1993, *Metal Angel* is a life sim where the player manages a group of female superheroes.

featured manga-style artwork including the 1983 detective game *Portopia Serial Murder Case* and 1984's *Princess Tomato and the Salad Kingdom*, a game about anthropomorphic vegetables, which was designed to encourage children to eat their greens.

But as Japan moved towards consoles rather than computers in the mid-1980s, Japanese game designers began to simplify their adventure games reducing the commands to short lists of options to choose from. This reductionist approach eventually led to Chunsoft's 1992 game *Otogirisou*, an influential horror game regarded as the original visual novel. *Otogirisou* reduced the player's actions to a small number of choices that would influence how the story would progress. Although most visual novels only get released in Japan, one notable crossover hit was the 2001 courtroom drama *Phoenix Wright: Ace Attorney*.

While Japanese home computer game makers set about devising new video game genres, Nintendo was gearing up for the launch of its new console. Yamauchi had ordered its designer Masayuki Uemura to create a console that was not only a year ahead of the competition in technology but also a third of the price of the Epoch Cassette Vision. Uemura's original design for what he called the Family Computer, or Famicom for short, came with a modem, a keyboard and a disk drive, but in order to meet Yamauchi's price point demands he was forced to throw away most of the features. The end result was a simple cartridge-based console with a controller that took its cues from the cross-shaped directional pad used in the 1982 Game & Watch incarnation of *Donkey Kong*.

Once Uemura completed the console, Yamauchi picked three of Miyamoto's games – *Donkey Kong*, *Donkey Kong Jr.* and *Popeye* – as the console's launch titles before heading out to proselytise the benefits of the Famicom to Japan's retailers. The goal, he explained, was not to make money from selling the Famicom itself but from selling games over and over again to those who bought it. "It is really just a tool to sell software," he told retailers, before highlighting the attractive profit margins to made from selling Famicom game cartridges.

Yamauchi's instinct that the Famicom could replicate Atari's North American success in Japan was quickly proved right. Within two months of the Famicom's July 1983 launch around 500,000 had been sold. By the end of the year sales had topped the million mark. The Bandai Intellivision and Epoch's counter-console the Super Cassette Vision faded from view fast.[5]

---

5. The Famicom's success also overshadowed the MSX line of home computers that launched in June 1983. Designed by Microsoft Japan and the ASCII Corporation as a common format for home computing, the MSX was produced by a number of manufacturers including Yamaha and Sony. It was popular enough in Japan to become the one of the leading home computer formats for games in the mid to late 1980s and also gained a strong following in Spain and South America. But MSX sales lagged far behind those of the Famicom.

But success brought its own headaches. The Famicom's success had created a customer base of hundreds of thousands of people all desperate for the chance to buy more video games for their exciting new console. The problem was Nintendo just couldn't produce games fast enough. Yamauchi's solution revolutionised the video game business. Instead of trying to expand rapidly to meet consumer demand, he opened up the Famicom to other game publishers. In return for allowing them to make games for the lucrative, game-starved captive audience it had built up, Nintendo wanted cash upfront to manufacture the cartridges, a cut of the profits from sales and the right to veto the release any game.[6]

Many baulked at Nintendo's demands, but the lure of fast-growing Famicom audience was just too enticing. Hudson Soft, the makers of *Bomber Man* and *Princess Tomato and the Salad Kingdom*, were among the first to sign up.[7] Used to selling 10,000 copies of its games on home computers, Hudson Soft watched their debut Famicom release – *Roadrunner*, a Japanese version of the US platform game *Lode Runner* – make its way into a million Japanese homes. With those kind of sales it didn't take long for Japan's other leading game publishers to agree to Nintendo's conditions. By 1985 Nintendo had given 17 companies licences to make games for the Famicom.

One of these companies was Bullet-Proof Software, a game publisher founded by an American called Henk Rogers, who moved to Japan in the mid-1970s to work in the gem business with his father. In early 1983 as the sales of home computers in Japan began to rise, Rogers decided to enter the video game business to cater for the growing number of computer users. "I went to Akihabara, Tokyo's electronic district, to find out who was going to win that battle in Japan and it was obvious it was NEC's PC-8001," he said. In 1983 NEC dominated the Japanese home computer scene, holding an estimated 45 per cent share of the market and there was already a wealth of software available. "There were a lot of games," he said. "There were platform games, shoot 'em up games, there were adventures games, puzzle games. Pretty much everything you could find – even the strategy war games. Every genre that you could think of was there, except role-playing games. There were no role-playing games in Japan."

For Rogers, an avid *Dungeons & Dragons* fan who spent many hours as a University of Hawaii student playing the fantasy game, this was a glaring omission. Role-playing games were huge in the US. *Ultima* and *Wizardry* were best selling video games and

6. As well as vetoing poor-quality games, Nintendo also prohibited the release of pornographic bishojo games on its console.

7. *Bomber Man* has proved to be one of the most enduring video games ever made with more than 40 versions released in the 25 years since its 1983 debut. The original was a single-player game where players explore a maze by planting bombs to clear blockages while dodging killer balloons, but it rapidly evolved into a manic multiplayer game where players tried to blow each other up with the bombs.

Dungeons & Dragons was a cultural phenomenon. Yet Japan knew nothing of role-playing as the pen, paper and dice games had failed to connect with the Japanese. Rogers figured he would fill the gap: "It was my naivety that made me think that I could actually sell a role-playing game in Japan when in fact I had no idea what I was doing."

Drawing heavily on *Ultima* and *Wizardry*, Rogers created *The Black Onyx* – a straightforward role-playing game about dungeon exploration and monster killing. Rogers launched it in Christmas 1983 and watched it sink. His distributor broke its promise to order 3,000 copies and instead bought just 600. Japan's game-playing public, meanwhile, simply ignored it. "I had blown what little money I had on a couple of pages of advertising that was totally ineffective because people didn't understand what the hell it was about," he said. "So, in January 1984, I had burnt through my $50,000 start-up fund. I thought I was dead in the water." Desperate, Rogers hired an interpreter and started visiting every Japanese video game magazine to explain the concept of role-playing games and how to play *The Black Onyx*. It saved his company. "I'd create characters for them and say this is what you do and so on and so forth," he said. "A couple of months later all the magazine reviews came out and they were raving about it. They came out in March. In April we had orders for 10,000 copies, it was like 10,000 copies a month for the rest of the year."

The Black Onyx's success sparked a surge in interest in this new form of game and encouraged many Japanese players to try out *Ultima* and *Wizardry*, which were still only available in English at the time. Having familiarised themselves with the genre, Japanese game makers reacted to *The Black Onyx* in much the same way as they reacted to *Mystery House*: they absorbed the ideas and remade them in their own way. "From time to time Japan picks up on things that are not part of its culture, like rap, and they adopt it and they love it and it becomes part of their culture," said Rogers. "The role-playing game was one of those things."

The question of Japanese uniqueness was particularly high on the Japanese political agenda around the time *The Black Onyx* was released. The nation's bestseller lists regularly featured Nihonjinron ('theories of Japaneseness') books that advocated a belief in Japan's cultural uniqueness and, in some cases, superiority. These books often promoted the idea that one of Japan's unique traits was its ability to absorb the cultures of other countries and make them part of Japanese culture. Nihonjinron writers could well have cited the way Japanese game designers' absorbed and reconfigured the role-playing game as evidence of their theories.

Enix game designer Yuji Horii, who created the *Portopia Serial Murder Case* adventure game, led Japan's reinvention of the role-playing game with his 1986 game *Dragon Quest*. Unlike the designers of the first wave of Japanese role-playing games, which adhered rigidly to the *Wizardry* template, Horii wanted to make something different. Something Japanese. He rejected *Wizardry*'s attempts at realistic visuals and hired Akira Toriyama, an artist who had worked on the popular *Dragon Ball* anime series, to give his game colourful visuals that were more in keeping with the manga and kawaii artwork that was popular in Japan. He also recruited TV show theme tune composer Koichi Sugiyama to give the game's music a more Japanese feel. Rather than using the music to reinforce on-screen action as is common in American and European film, TV and games, Sugiyama created a continuous score that changed to reflect the overarching atmosphere of the game.

For the in-game text, Horii drew on the rhythms of haiku, the Japanese poetry that focuses on economy with words, to give the text a jaunty sound when read aloud. Finally Horii tailored his game to the Famicom, making it easier to play by reducing the reliance on statistics and complex controls of American role-playing games. The biggest problem with US role-playing titles, he opined, was that they were "very unkind toward the player". The final result bore only a passing resemblance to the US games that inspired it. The game dripped with Japanese influence and focused on the character development and resource management aspects of role-playing. Horii had created the first truly Japanese role-playing game.

Dragon Quest marked a major fork in the evolution of role-playing games, creating a divide between Japanese and North American visions of the genre.[8] If anything that divide has only widened over time as Japanese developers increased their emphasis on story and team management while Americans and Europeans sought to free players from the constraints of pre-defined narratives.

Dragon Quest became a sensation in Japan; a video game with a cultural impact comparable to a major Hollywood blockbuster with more than two million copies of Horii's fantasy epic flying off the shelves. Famicom sales soared as people bought Nintendo's console just to play Horii's adventure.

Dragon Quest formed the blueprint for the rush of Japanese role-playing games that followed. Among the first developers to explore similar territory to *Dragon Quest* was Square's Hironobu Sakaguchi. Like Horii, Sakaguchi was introduced to role-playing via US games. "My first experience of role-playing games came when I played the English version of *Wizardry* and *Ultima* on the Apple II," he said. "I was not attracted by the story of the early *Wizardry*, but I liked the system and worldview." His response to *Dragon Quest* was *Final Fantasy*, a darker game with an undercurrent of angst as opposed to the more light-hearted adventuring of Horii's game. "I brought the life of people and the

8. European role-playing games largely followed the North American model.

colliding of people's passion into the games," said Sakaguchi. "I wanted people to feel these passions more actively by playing roles in the games than by audio-visual works and novels, in which you feel them passively." Released in 1987, *Final Fantasy* also became a commercial and critical success.

Soon every Japanese publisher was producing a role-playing game. Nintendo released *Mother*, a role-playing game set in suburban America that replaced swords with baseball bats and healing potions with hamburgers. Sega came up with *Phantasy Star*, which took place in a world that blurred the lines between science fiction and fantasy, for its Master System console.

But not everyone in the Japanese game business was impressed by the creations of Horii and Sakaguchi. Nintendo, despite its involvement with *Mother*, was particularly lukewarm to the genre. Yamauchi described role-playing games as being for "depressed gamers" who sit in dark rooms, while Miyamoto professed to a "fundamental dislike" of their emphasis on pre-defined stories and level-based advancement. Not that this stopped Miyamoto from raiding the genre for inspiration. His 1986 game *The Legend of Zelda* adapted many of the conventions of role-playing into an action game format. The concept of character development, for example, was reinvented as extra abilities or weapons that players obtained as they progressed through the game.

The Legend of Zelda's biggest influence, however, was Miyamoto's own childhood. As a child he lived in Sonobe, a village 39 miles outside of Kyoto, and would spend his spare time exploring the local countryside. With *The Legend of Zelda* he sought to recapture the excitement and awe he felt as a child while wandering the countryside and never knowing what he would discover. He wanted players to experience the freedom of exploration and to encounter "amazing things". The result was a world on a micro-chip where players could take different routes to complete their quest, discover hidden passageways behind waterfalls and encounter strange creatures. Miyamoto's game was also a coming-of-age story, a tale of how an ordinary boy overcomes his fears to defeat an evil army. It was a motif that would become standard in many Japanese role-playing games.

Japanese role-playing games would become the country's most popular type of video game, a popularity exemplified by the pandemonium that greeted the launch of *Dragon Quest III* on the 10th February 1988. Nearly a million copies were sold on launch day as fans of Horii's series clamoured to lay their hands on a copy of the latest instalment. In an outrageous affront to Japan's disciplinarian school system, many children skipped school to buy their copy. Police took 392 truanting children and teenagers into custody, in what a National Police Agency spokesman described as "a national disgrace". Elsewhere there were reports of children stealing copies from their classmates and instances of attempted shoplifting. The disorder prompted its publisher Enix to promise the Japanese government that it would only release *Dragon Quest* games on Sundays or bank holidays in future.

The huge success of *Dragon Quest III* confirmed Nintendo's dominance of the Japanese video game business. The home computers that once seemed to be creeping ahead were now relegated to the sidelines, destined to be dominated by the bishojo titles that Nintendo would not allow on its console. The Japanese public had also ignored the Master System despite Sega's hopes that its popular coin-op games, such as the ninja-themed *Shinobi*, might help it challenge Nintendo.

Nintendo had turned the Japanese game industry into a client state. Its licensees were willing slaves to Nintendo's will: told how many games they could release, when they could release them and required to hand over a cut of the money they made from every game. Yamauchi's insatiable desire for business expansion and Miyamoto's wide-eyed creations had turned Nintendo into Japan's most powerful video game company. And Yamauchi now had the multi-million-dollar war chest he needed for his next and most ambitious goal: to repeat the Famicom's Japanese success in the US.

Mario man: Shigeru Miyamoto shows off his banjo skills
Nintendo UK

CHAPTER 13

I Could Have Sworn It Was 1983

Nintendo was a laughing stock. It was 1984 and the company was touting the Advanced Video System, the US version of the Famicom, at the trade shows. To avoid associations with Atari, Nintendo tried to distance its unreleased system from the consoles of the past by encasing it in video-recorder grey and showing off a keyboard attachment that turned it into a home computer. Nobody was fooled. This Japanese company was trying to bring back the home console just months after the whole video game business went down the tubes. The businesspeople, retailers, analysts and journalists at the trade shows laughed, pointed, teased and rolled their eyes, amazed by the audacity of these Japanese no-hopers.

Nintendo realised its plans had to be binned. Again. The previous year, just as the Famicom launched in Japan, Nintendo had sought to persuade Atari to bring its console to the US market. The marriage of Atari's brand and Nintendo's games would be a winning combination, the Japanese company thought. It was not to be. "Nintendo came to us and said do you want to make the Famicom?," said Manny Gerard, the Warner Communications executive in charge of Atari. "They wanted us to distribute and make the Famicom. We had enough crises of our own at that point and we couldn't deal with it." It looked hopeless for Nintendo. Atari had become the video game ground zero and had convinced the world that game consoles were finished; a historical footnote, an electronic Hula-Hoop. This was not a view that Hiroshi Yamauchi, Nintendo's chairman, was prepared to accept. The Famicom had succeeded in Japan, so why not in the US? He insisted that his son-in-law Minoru Arakawa, the president of Nintendo of America, find a solution.

Arakawa re-examined the state of the video game industry. Sales were collapsing fast. Game publishers were scrambling onto the home computer life rafts. Retailers burned by the Atari fallout wanted nothing to do with video game consoles ever again. Everything seemed stacked against Nintendo, but then Arakawa noticed something that all the business analysts with their heads buried in profit-and-loss accounts had paid little attention to: the kids of America were still playing video games. They were playing them on home computers and they were still pumping quarters into coin-op game machines. The players, he concluded, were not bored of games as a concept, just the average or substandard ones. He ordered a second redesign of the Famicom but this time as an unashamed, out and proud games console: the Nintendo Entertainment System or NES for short. Given his conclusion that the quality of the games would make or break for the system, Arakawa embraced the Japanese licence system for the Famicom and added a security chip to the NES so Nintendo could dictate what was available on its console. Any company that released an NES game without a licence, Arakawa decided, would risk legal action. Not that anybody outside Japan was interested in making games for the NES. "Numerous consoles had failed prior to the NES," said Trip Hawkins, co-founder of Electronic Arts, which was more focused on the latest home computers such as the Commodore Amiga and Apple Macintosh. "The entire US game industry thought the NES was a big step backwards. Worse, the licence agreement was completely draconian and reduced a publisher to being a captive developer with no control over its business."

Nintendo did not care what the US publishers thought. It had dozens of great games from Japan that it could bring to the US, many already well known through the arcades. Nintendo's real problem was the retailers. If they refused to stock the NES, all was lost. Arakawa decided the NES needed some gimmicks to distance it from earlier systems. Nintendo came up with two – the Zapper light gun and a 24-centimetre high robot called R.O.B. or Robotic Operating Buddy. The Zapper had already proved popular in Japan thanks to shooting gallery-style games *Duck Hunt* and *Hogan's Alley*. In Japan the Zapper was designed to look like a real firearm, but fearing criticism from anti-gun campaigners, Nintendo redesigned it as a laser gun for the North American market. R.O.B., meanwhile, helped people play by watching the action on screen and moving physical objects around that affected the game such as moving blocks to open doors in the dynamite-collecting game *Gyromite*.

Nintendo made the Zapper and R.O.B. the core of its NES marketing efforts in the hope of thawing the frosty attitude towards its games machine. But retailers attending the June 1985 Consumer Electronics Show remained unmoved and showed no interest in stocking the NES. The redesign also went down badly with the children invited to Nintendo's focus groups. Arakawa called Yamauchi and told him it was time to admit defeat. The US was simply not interested. Yamauchi refused. He rejected the verdict of the focus groups, declaring market research to be a waste of time and money. He told Arakawa to focus on making the NES a success in one American city before going national. Nintendo chose New York City, which was seen as the toughest city to crack in US. The reasoning was that if Nintendo could sell the NES to New Yorkers, it could sell it to anyone. Yamauchi gave Arakawa and his team $50 million to bankroll their assault on the Big Apple. The key staff from Nintendo's Seattle headquarters packed their bags and moved to New Jersey to work around the clock and make the NES a Christmas 1985 success story. Nintendo offered money-back guarantees to retailers, spent millions on advertising and showed off the Zapper and R.O.B. to shoppers in malls across the city. By Christmas Eve the NES was on sale in more than 500 New York stores. The push worked. That Christmas New Yorkers bought 90,000 NESs. The majority of the retailers recruited thanks to the money-back guarantee, agreed to continue stocking the console and its games. Nintendo then set about doing the same in Los Angeles, then Chicago, then San Francisco, then Texas before finally launching the NES throughout the US.

And when Nintendo released Shigeru Miyamoto and Takashi Tezuka's game *Super Mario Bros* in the US in March 1986, the NES went supernova. *Super Mario Bros* marked the return of Mario, the barrel-jumping hero of *Donkey Kong*, and transported players into a Dr Seussian cartoon world of secret rooms, cuddly enemies and day-glo land-scapes. Along with 1984's *Pac-Man* spin-off *Pac-Land*, *Super Mario Bros* heralded a new era for platform games. Instead of confining the action to a single screen, *Super Mario Bros* offered the thrills of exploration in a virtual playground far larger than players' TV sets and where there was always some unexpected delight around the next corner. It could be a castle with a moat of lava, a beanstalk stretching into the sky to climb or an *Alice in Wonderland*-inspired magic mushroom that turned Mario into the giant Super Mario.

For Miyamoto it was a game that, like *The Legend of Zelda*, recreated the joy he felt as a child exploring the countryside around Sonobe. And for a generation of American and Japanese children whose freedom to wander, explore and play outside was being curtailed by urbanisation, it was a virtual substitute. *Super Mario Bros* became a global phenomenon; millions of copies were sold alongside millions of the NES consoles needed to play it. Nintendo and its Japanese licensees added to the momentum of *Super Mario Bros* with a steady supply of quality games such as Miyamoto's *The Legend of Zelda*, Gunpei Yokoi's sci-fi action game *Metroid*, the vampire-themed adventuring of *Castlevania* and versions of arcade hits such as the street-fighting smash *Double Dragon*. Soon Wall Street analysts were throwing their weight behind Nintendo and the retailers who once dismissed the NES changed their minds.

By summer 1987's Consumer Electronics Show, it was clear that the video game console was back. In between the digital guitars, prototype CD-Video players and black-and-white video phones on display were enough video games to prompt *Popular*

Mechanics' correspondent to remark: "I could have sworn it was 1983." Other reporters agreed. *The Milwaukee Sentinel* described it as "one of the biggest comebacks ever". *Fortune* credited Nintendo with "single-handedly" reviving the games business. Few in the business would have disagreed. "Nintendo came out of nowhere. They were ballsy and they rebuilt the videogame console market," said Michael Katz, the former Coleco executive turned Atari Corporation vice-president. By December 1987 American children were snubbing traditional favourites such as Barbie and G.I. Joe dolls and asking for a Nintendo instead. Nintendo, however, was keeping a tight control on the supply of its system. Partly because it could not keep up with the demand it had created, but also because it was haunted by the collapse of Atari in the early 1980s.

Nintendo worried that retailers would slash the price of any excess stock, leading to the kind of discounting that helped destroy Atari. The tight control of supply also gave Nintendo more influence over retailers and generated a buzz about the NES because it sold out the moment it arrived in shops. Egged on by limited supplies and huge demand, the NES became the hottest toy of Christmas 1987. Parents desperate to please their children with a gift-wrapped Nintendo under the tree trawled the shops, hunting for the elusive present. As Christmas approached they became increasingly feral. "They're upset that we don't have it," one sales clerk told *The Milwaukee Sentinel*. "Now that it's getting closer to Christmas, they are getting ruder."

Supplies of Nintendo products may have been strictly controlled, but enough were sold that Christmas to make the NES the US's number one selling toy of 1987. The game publishers who once laughed at Nintendo were now begging for a licence to make NES games. Those that signed up were subject to exacting rules and controls that placed Nintendo in a position of incredible power over the video game business. Licensees had to pay Nintendo to manufacture their game cartridges so even if the game sold badly Nintendo made a profit. Nintendo also took a cut of every NES game sold, dictated when the game could be released, told licensees how many games they could release every year, and got to decide whether a game was good enough to be released.

Attempts to bypass Nintendo's rules were stamped on fast. Australian game developer Beam Software was one of those who incurred the wrath of Nintendo. Shortly after the NES appeared, Beam worked out a way to get around Nintendo's security protection and used this knowledge to create a development system that it hoped to sell to NES game publishers. "Our systems were much more user-friendly than the Nintendo ones and certainly less expensive upfront. We managed to sign-up one publisher before word of this reached Nintendo," said Alfred Milgrom, the co-founder of Beam. "It didn't take long for the heavy arm of Nintendo to come down on us.

The word was that any publisher who signed up with Beam for development systems would lose their licence." And with no publisher willing to lose access to the millions of NES owners, Beam was in trouble. "It was crucial to us – if we could not resolve this problem, Beam was out of business," said Milgrom. "The situation was horrific. That's when we really knew that Nintendo had enormous power, because they made it clear that if we didn't withdraw our system we'd never be able to do anything in games ever again." Panicked, Milgrom made frantic calls to Nintendo's Seattle offices trying to speak to Arakawa about the situation. "I kept phoning him every half an hour saying 'can I come over and can I talk to you'," he said. "After a while he rang the president of our publisher Acclaim and said: 'Why is this guy pestering me and calling every half an hour?'. He said: 'Well you did say you were going to put him out of business'."

Beam saved itself by agreeing to become a Nintendo licensee. "Once we were in the family the reality was different," he said. "Nintendo's a big paternalistic company, it's a lot like a dictatorship. Nintendo say 'here you are in our family, but you're going to have to obey the rules of the family'. It's quite strict and obviously Japanese."

Nintendo's standards were exacting. "In terms of game testing they revolutionised the concept," said Milgrom. "They said zero defects – we will not allow you to release a game that has any bugs in it whatsoever. Now zero defects was an unheard of concept in any other software or on any other gaming platform. Nintendo knew if they were going to sell it in the supermarkets and sell it to mums and dads it had to work off the shelf and had to be flawless. They didn't want returns. We had to change our programming attitude and the way we developed games, which was brilliant. It was really hard work. If you had a bug in your final version you could miss Christmas because it would take a month for them to go through the testing of the title."

Nintendo's attention to detail became clear when Beam submitted *Aussie Rules Footy*, a NES game aimed at the Australian market, for the Japanese giant's approval. "One of the quirks about Australian rules football is that in the real game you can keep the ball in play if the ball is inside the line, even if the player holding the ball is over the line," said Milgrom. "Nintendo actually picked up a situation where the player went over the line but wasn't called out. So we said 'well, here's the rule', but you wouldn't expect this to come up as a bug. It was just so meticulous."

Nintendo's hawk-like examination of its licensees' games didn't stop there. Keen to avoid controversy or another *Custer's Revenge*, Nintendo produced an extensive list detailing what game publishers could not put in a NES games. The rules echoed both the Hays Code, which policed the films of Hollywood from 1934 to 1968, and the 1954 Comics Code. The Hays Code emerged in response to a spate of scandals in the 1920s that earned Hollywood the nickname 'Sin City'. Enacted by Will Hays, the head of the movie business trade association and a campaign manager for US President Warren Harding, the code was actually written by the Catholic Father Daniel Lord. The Hays Code banned sex, drug use, nudity, swearing, positive portrayals of criminals and the ridicule of religion. Its rules encouraged Hollywood to spend three decades creating innocent fantasies or moralistic parables where the bad guys were always punished for their crimes.

The Comics Code was born out of a political row about adverts for porn, drug paraphernalia and weapons in comics that coincided with increasing public disquiet about gore, violence and sexual content of comic books. The Comics Code took its cues from The Hays Code. It banned cannibalism, zombies, torture, sex and werewolves. It required that "in every instance good shall triumph over evil and the criminal be punished for his misdeeds" and demanded that judges, government, police and other respected institutions were not treated in a negative way. Many of Nintendo's restrictions could have been lifted directly from the Hays Code. Nintendo prohibited graphic depictions of death, the Hays Code barred studios from showing brutal killings in detail. Both barred sex, nudity, random or gratuitous violence, criticism of religion and illegal drug use. Nintendo also banned games from featuring tobacco and alcohol, and prohibited sexist and racist content. The NES remake of the ultra-violent, anti-drugs *Narc* played down the drug references and removed the blood featured in the original coin-op game. "The game was watered down to almost unrecognisable levels," said the arcade version's creator Eugene Jarvis. Jaleco was forced to remove nude Greek statues from its NES version of Lucasfilm Games' *Rocky Horror Show*-inspired adventure game *Maniac Mansion*.[1] Even Miyamoto couldn't escape the censors. His 1984 *Pac-Man* clone *Devil World* was refused a US release because it featured demons, Bibles and crucifixes – a breach of the rules of the treatment of religious imagery.

But Nintendo's code differed from the Hays Code and Comics Code in its motivations if not content. Nintendo's rules did not emerge in response to public or political pressure but more from an expectation of controversy at a later stage. It was also a unilateral censorship code rather than one agreed to by the video game industry as a whole – as was the case with the movie and comic industries with the Hays Code and Comics Code.

Few game publishers minded. Most were happy to trade creative and business freedom for the huge profits to be made from NES games. As in Japan, much of the US video game industry was now subject to the will of one company alone. Nintendo's

1. Nintendo did, however, miss the option for players to microwave a hamster in the game. Only after the first 250,000 copies had gone out to retailers did Nintendo spot it. Jaleco was told to remove the option if it wanted to have any additional copies manufactured.

competitors Atari Corporation and Sega watched their 7800 ProSystem and Master System consoles get savaged in the retail market and starved of games as Nintendo licensees decided it was safer not to jeopardise their relationship with the increasingly monolithic Nintendo by flirting with the opposition. "I couldn't get any arcade exclusives on the 7800 because Nintendo had exclusive agreements – formalised or not," said Katz. "We couldn't get the hot arcade games for the 7800. So I thought I should try and get the hottest titles from the old computer game companies. It was the only strategy we could employ."

By 1989 Nintendo products accounted for 23 per cent of all toys sold in the US. Macy's and Toys R Us devoted whole sections of their stores to Nintendo, shrines to the new messiah of video gaming. *Nintendo Power*, Nintendo's promotional magazine, became the US's biggest-selling children's magazine with a monthly circulation in the region of five million copies. Nintendo's premium rate helpline for players wanting tips on how to beat games was taking more than 50,000 calls a week. And there was enough Nintendo merchandising for fans to literally eat, sleep, drink, dress and study shrouded in Nintendo branding. There were Mario cereals, Zelda trunks, Nintendo notebooks, wallpaper, bed sheets, rulers, tennis shoes, birthday cake pans, portable radios and soft toys.

Nintendo's success did, however, make it a target for Americans upset about the growing influence of Japan on the US – a concern that peaked at the end of the 1980s when the NES was at its zenith. After defeating it in the Second World War, the US set about turning Japan into a democratic free market outpost in Asia. The US bankrolled improved infrastructure, helped Japan gain membership of international trade associations and encouraged US companies to share technology with the Japanese. It also sought to make it easier for Japanese companies to sell their products in the US by reducing trade barriers and agreeing a fixed exchange rate between the yen and the dollar. Japan, meanwhile, introduced protectionist laws that kept foreign firms out of Japan and helped kill off Atari Japan.

By the end of the 1970s Japan was being described as an "economic miracle" and, aided by low wages, Japanese corporations were making massive in-roads into the US market at the expense of American manufacturers. Many Americans hated this. They saw Japan's protectionism and the US's openness as an unequal arrangement that was destroying American companies and jobs.[2] But despite the domestic pressure, the strategic importance of Japan as a Cold War ally often caused attempts to address the issue

---

2. US companies also contributed to their own downfall. Many were dismissive of the Japanese. They believed that the Japanese could not match the technology and innovation of Americans. They were utterly wrong. By 1987, 95 per cent of the world's 100 million video recorders were Japanese despite the video recorder being a US invention.

to be sidelined. Measures such as US President Richard Nixon's decision to restrict Japanese TV imports only made matters worse by encouraging Japanese companies to open US factories or buy out US businesses. Nowhere was the row about Japanese influence more fraught than in the debate about the car industry. The US car industry was more than a collection of businesses to Americans; it was a symbol of national economic virility. So as Japanese companies made in-roads at the expense of iconic companies such as General Motors and Ford, the anger boiled over. There were incidents of people smashing up Toyotas, while others made patriotic appeals for people to buy American motors.[3] Wild-eyed commentators compared Japan's economic success in the US to a "second Pearl Harbour".

To some, Nintendo's success was just another example how American business was being crushed by Japan's economic steamroller. For Nintendo, however, the anti-Japanese feeling only came to a head when Washington State senator Salde Gorton asked if the company would buy the Seattle Mariners baseball team to prevent it moving out to Florida. Yamauchi, hoping to give something back to the country that had made Nintendo so huge, used $60 million of his own money to buy a majority stake in the club. Yamauchi didn't even like baseball. "Baseball has never really interested me," he told reporters at the time.

He explained the purchase as returning a favour to the country that helped make Nintendo one of the world's most profitable companies. Furious baseball fans did not see it that way, however: they saw it as another example of the Japanese buying America.[4] Nintendo found itself the latest focus of anti-Japanese sentiment. A poll conducted at the time found 61 per cent of Americans wanted the Japanese out of Major League Baseball. When news of the row reached Japan, Yamauchi found himself being criticised by the Japanese for inflaming American ill feeling. Japan was well aware of anti-Japanese sentiment in America at the time. One Japanese company, SystemSoft, even responded with a video game called *Japan Bashing*, where, as the Americans, the player's goal is to change Japan by trying to make the Japanese eat wheat or to stop hunting whales. The computer-controlled Japan, meanwhile, fights back by trying to turn hamburgers on the US coast into sushi.

Nintendo faced other sources of criticism as well. Its huge success prompted accusations that the company was engaging in monopolistic practices that stifled competition. However, the attempts to challenge Nintendo on these grounds in the courts came

---

3. The car-smashing incident was unintentionally reflected in the Japanese fighting game *Final Fight,* where players get to smash up a car with a Toyota-esque badge between levels.

4. This came not long after Sony's 1989 buy out of Columbia Pictures and Matsushita's purchase of MCA in 1990.

to nothing. Health campaigners, meanwhile, blamed Nintendo for making American kids fat. The National Coalition on Television Violence released figures in November 1988 suggesting 83 per cent of NES games were violent in nature.

More troubling for Nintendo was the work of Eugene Provenzo Jr., the professor of education at the University of Miami. Provenzo had become fascinated by Nintendo's huge success and noticed there was very little research into video games from a cultural perspective. He decided to conduct one of the earliest studies of video games examining the portrayal of gender and violence within the 47 highest-selling NES games. "My colleagues thought I was a lunatic and crazy to be working in this area," he said. "I got kind of stubborn and really got fascinated with it and thought this was a very important emerging phenomenon. They sort of humoured me."

Provenzo started looking for funding. The Harry Frank Guggenheim Foundation, which funds studies into new media and violence, was his first port of call. "They kinda laughed me out of the place. I went to the Spencer Foundation, the US Department of Education. They all basically said 'charming, but who would possibly care about this?'. Bear in mind at that, at the time, the notion of looking at film sources, new media and popular culture as an area of serious research wasn't there yet, least of all looking at it in terms of kids' culture."

Provenzo finished his book *Video Kids: Making Sense of Nintendo* off his own back and sent it to Harvard University Press. "I sent it in on a Thursday and I got a telephone call, which was very unusual, on Monday morning," he said. "The editor basically said this is a breakthrough book." In his book Provenzo accused many of the NES games he examined of promoting aggression and containing racist and sexist stereotypes. His work marked the start of academic study of video games, but it was not the kind of conclusion the game industry, and Nintendo especially, wanted to hear. "The game industry was extremely hostile afterwards," he said. "Nintendo of America's publicity office and legal office were intimidating enough that the original title of the book was going to be *The World According to Nintendo*, which is the Nintendo motto, but the publisher decided they didn't want to take the risk."

The criticism, however, did little to detract from Nintendo's success and, with North America and Japan wrapped up, Nintendo turned its attentions to Europe. The NES started to arrive in Europe in 1986, but Nintendo's lack of a European office and patchy distribution, meant the console only really started to appear in most countries during 1987. By then, however, European game players were aspiring to own one of the new home computers such as the Commodore Amiga or Atari ST – machines that offered visuals no NES game could match. The four-year-old NES just looked old hat compared to the work of game developers such as the UK's Bitmap Brothers.[5] The NES and its games were also expensive for a continent weaned on that budget games costing as little as £1.99 and dirt-cheap home computers. "When Nintendo originally exhibited the NES nobody was very impressed with it compared to the Amiga and stuff like that," said David Darling, co-founder of British budget game publisher Codemasters. "It looked really quite old technology, the cartridges were expensive and no-one in the industry could foresee it as being a success."

The terms of Nintendo's NES licences also shocked the freewheeling European game industry. "It was restriction of trade. I still find it totally amazing," said Geoff Brown, the founder of US Gold – one of Europe's largest publishers at the time. "It not only told you how many products you could put on that format, it also says Nintendo has to see it in advance. They approved the release of them when they are your biggest competitor. It was incredibly expensive to manufacture and you couldn't manufacture it yourself. It was a total lock down of the format. I just thought it was outrageous. If I want to make a game for a console and it's terrible and I want to spend the money it's my problem, not theirs. I wasn't alone in this, there was a negativity amongst publishers – they didn't want to support it." Others saw it as a creative affront. "It would have been impossible for us to develop a creation such as *Captain Blood* on a console because we would never have obtained the concept approval," said Philippe Ulrich, the founder of French game publisher Ere Informatique. Not that Nintendo cared. It already had the cream of Japanese and American games for its console and saw little need to pander to the game publishers of Europe. "They couldn't care less," said Brown. "They didn't really need regular UK publishers. They had brilliant games of their own. We didn't go to them and they didn't come to us."

Nintendo also found itself faced with stronger competition from Sega. While Nintendo's distribution arrangements were patchy, Sega had struck deals with leading European video game distributors such as the UK's Mastertronic and Ariolasoft in West Germany. Sega's Master System went on to outsell the NES in Europe, although neither came close to weaning European game players off their home computers. There were also a few abortive attempts by European companies to challenge the Japanese systems. British electronics firm Amstrad had enjoyed huge success with its CPC home computers and its founder Alan Sugar, who like fellow British computing pioneer Clive Sinclair later received a knighthood, thought his company could be a

---

5. The Bitmap Brothers was a London-based team of game designers that presented themselves as leather-jacketed and shade-wearing rock stars and gained a strong following in Europe. They took advantage of the extra power of the Amiga and Atari ST to serve up flashy, rave music-enhanced, visual feasts such as *Speedball 2: Brutal Deluxe*, a violent futuristic sports game, and the Bomb the Bass-soundtracked shoot 'em up *Xenon 2: Megablast*.

European answer to Nintendo. Amstrad repackaged the CPC as the GX4000 console, which Sugar presented to UK publishers as an alternative to the draconian licensee terms of Nintendo and Sega. "All we did was pay lip service to it, because the Amstrad CPC had been incredibly successful. We thought we'd better do a game just to keep him happy," said Brown. "We went to a meeting with Sugar where he said: 'We've got a driving game'. I said: 'Yeah, but you haven't got *Out Run*'. He says: 'What the bloody hell's *Out Run*?'. I said: 'It's just a brilliant, brilliant driving game'. He says: 'But we've got a driving game, mate, that's all we need'. What he couldn't understand was that a driving game is not just a driving game. *Out Run* was *Out Run*. For me it was a significant point in Alan Sugar's understanding of the video game market: he was great at moving boxes, but he was a seller of hardware rather than software."

The GX4000 sank without a trace, selling little more than 10,000 units compared to the two million CPC computers sold throughout Europe. The NES performed far, far better, but still came second to Master System and neither format achieved the level of support that the Amiga and Atari ST would in Europe. Even Luther De Gale, the former UK head of Konami – one of Nintendo's closest Japanese partners, who was hired as a consultant to help try and save the NES in Europe, admitted to the press that Nintendo had failed to win over European consumers.

But while European publishers resisted getting involved with Nintendo at home, they were more than happy to try and crack the US NES market. Especially when they realised the popularity of the NES on the other side of the Atlantic. British game designer Philip Oliver was one half of the Oliver Twins, a game development duo consisting of him and his fraternal twin brother Andrew that had gained success making cheap and cheerful budget games for Codemasters such as *Fruit Machine Simulator*, *Grand Prix Simulator* and *Dizzy*, an arcade adventure starring an anthropomorphic egg that became the UK's answer to Mario. He was shocked at the scale of the NES market in the US compared to the UK industry:

"We went to America to a show in Las Vegas and just couldn't believe the size of the show and the size of Nintendo's exhibition and the number of games. On the ST and Amiga the sales numbers of the best games would be a few hundred thousand, then you looked at the NES in America and the average game sales were about a million a piece and some of the better games like *Super Mario Bros* was like 28 million. You were just going 'oh my god'. We just thought this is what we need to be doing."

The Oliver Twins' publisher Codemasters had come to the same conclusion. It developed a device called the Game Genie that plugged into the NES and allowed players to cheat at games by giving themselves extra lives or unlimited ammunition. "We licensed it to a toy company in Canada and they set up a licence for an American toy company," said Darling. "The American toy company took it to Nintendo and they couldn't get a licence to sell it. We looked into the whole legal side of it and there was no reason why we needed a licence. So with the toy company and the lawyers we decided to go ahead and market it anyway." Codemasters also decided to release games on the NES without Nintendo's approval. Its first release was the Oliver Twins' 1991 game *The Fantastic Adventures of Dizzy*, a NES-only addition to the *Dizzy* series. Code-masters hoped would bring the eggy hero to a new legion of fans. It barely registered in the huge NES market. "I don't think it was massively popular, but since America is so big it still sold a lot," said Oliver. "I wouldn't be surprised if it sold 100,000 or 200,000 units, but it didn't reach a level of saturation."

More successful were Chris and Tim Stamper, the founders of Ultimate Play The Game – the iconic Spectrum publisher that had created the groundbreaking *Knight Lore*. They turned their back on the UK market and reinvented themselves as Rare – a NES developer for hire. "The Stamper brothers did this very clever thing," said Geoff Heath, managing director of the UK's Sega Master System distributor Mastertronic in the late 1980s. "They reverse engineered the Nintendo and then went to Japan, saw Nintendo and said 'hey, what do you think of these games?'. Nintendo went 'wait a minute we don't understand what's going on. You don't have a licence, how have you done this?'. They said we just reversed engineered all your technology and here are the games. Nintendo were smart enough to say: 'Since you're that brilliant, we'd better give you a special deal'. They got a very, very sweetheart deal from Nintendo." Rare devoted itself to pumping out games designed to appeal to the US audience. They created TV and film tie-ins (*Wheel of Fortune, Who Framed Roger Rabbit?*), converted arcade games (*Narc, Marble Madness*) and developed a smattering of original titles (*R.C. Pro-Am, Battle-toads*). Unlike the Oliver Twins' effort, Rare's NES titles sold millions, turning the Leic-estershire siblings into the UK's most successful game designers of the late 1980s.

For Nintendo, the failure to conquer Europe made little difference, as the incred-ible success of *Super Mario Bros 3* underlined. Nintendo marketed the game in much the same way as a movie studio might nurture hype about its latest blockbuster film, spending months building consumers' anticipation to fever pitch. The pre-publicity effort included the 1989 feature film *The Wizard*, a Universal Studios picture about three kids who go to California to take part in a video game tournament. Essential-ly a 100-minute-long Nintendo advert, *The Wizard* gave *Super Mario Bros 3* a starring role. Nintendo also joined forces with McDonald's to offer Mario Happy Meals in its US stores to coincide with the game's February 1990 launch. It was a marketing operation of a scale unheard of for a single game. *Super Mario Bros 3* also marked something of a creative comeback for Miyamoto and Tezuka, who had ended up working on different visions for *Super Mario Bros 2*. Tezuka's *Super Mario Bros 2* was darker version of the original featuring levels designed to challenge the very best players. While it came out in Japan, Nintendo felt it was too hard to bring to the US market and instead asked Miyamoto to rework his Japanese NES title *Doki Doki Panic* into a *Super Mario Bros 2* for the US market.[6] Neither game matched the brilliance of Miyamoto and Tezuka's original effort. *Super Mario Bros 3*, however, revived the sense of wonder that made *Super Mario Bros* so special with features such as chain chomps, black balls with gnashing teeth constrained by a chain that were inspired by Miyamoto's memories of his neighbours' aggressive dog, and a range of ability-changing costumes for Mario to wear. *Super Mario Bros 3* became, both critically and commercially, the culmination of Nintendo's journey from unknown Japanese toy maker to global video game giant.

The game sold more than 17 million copies worldwide, grossing around $550 million – more than Steven Spielberg's film *E.T. The Extra-Terrestrial*. In 1990 the Q Score survey measuring the popularity of celebrities and brands reported that Mario was now more famous and popular than Mickey Mouse. Miyamoto became a world-famous game designer, even attracting the attention of former Beatle Paul McCartney and Spielberg, both of whom travelled to Kyoto to meet him. Nintendo, meanwhile, became the focus of business acclaim and anxiety. In 1989 the *Japanese Economic Journal* named Nintendo as Japan's most profitable company ahead of both Toyota and Honda. Nintendo's average employee was earning the company $1.5 million of profit a year. Apple Computer's president Michael Spindler went so far as to name Nintendo as the company he feared most during the 1990s. The business acumen of Yamauchi and the creative abilities of Miyamoto had turned the laughing stock of 1984 into one of the world's most formidable companies.

Nintendo's success reconfigured the games industry on a global level. It brought consoles back from the dead with its licensee model, which became the business blueprint for every subsequent console system. It revitalised the US games industry, turning it from a $100 million business in 1986 to a $4 billion one in 1991. Nintendo's zero tolerance of bugs forced major improvements in quality and professionalism, while its content restrictions discouraged the development of violent or controversial games. The NES also put Japanese games at the centre of the world's video game industry. Japan was seen as having the best game makers and instead of looking to California, game players started looking to the Japanese archipelago for the next amazing game.

---

6. Nintendo eventually released it outside Japan in 1993 as *Super Mario Bros: The Lost Levels*. The US *Super Mario Bros 2* was released in Japan as *Super Mario USA* in 1992.

Rotoscoped: Jordan Mechner turns footage of his brother David into frames of animation for Prince of Persia
Courtesy of Jordan Mechner, www.jordanmechner.com

CHAPTER 14

Interactive Movies

Mitchy saw it all. She watched Atari grow from scrappy pioneer into corporate behemoth. She witnessed the birth of the Atari VCS 2600 and the hours spent designing the innards of the Atari 400 and 800 home computers. Now she was a spectator to the creation of a next-generation home computer that would help shape of the future of video games. Mitchy didn't know what was going on, she was, after all, just a dog. But her owner Jay Miner, the computer engineer behind the developments Mitchy witnessed, took his pet cockapoo everywhere. As Miner slaved over schematics and toyed with microchip designs, Mitchy would sit patiently by his side waiting for attention.

Miner had quit Atari in 1982 after the company refused to fund his dream of creating an advanced home computer based on Motorola's 68000 microprocessor. Miner took his ideas to the Amiga Corporation, a Californian start-up bankrolled to the tune of millions by a group of dentists from Florida.[1] The company embraced Miner's vision and gave him the money and resources he needed to turn it into reality.

As a huge fan of flight simulations, Miner made it his goal to build a system that would be home to the cream of the genre. With Mitchy at his side, Miner designed graphics chips that could simultaneously display thousands of colours, at a time when 16 colours on screen at once was a serious challenge. His graphics technology could also update the display independently of the microprocessor, a feature far beyond the abilities of the era's best arcade machines and that meant his computer would not get bogged down by the demands of processing its advanced visual capabilities. Miner also built a sound chip that put the sonic capabilities of every other home computer available at the time to shame.

Impressed by Miner's work, his former employers Atari snapped up the rights to the still-in-development computer on 21st November 1983. When the Amiga Corporation

---

1. Activision co-founder Larry Kaplan started the Amiga Corporation as Hi-Toro in 1982 and recruited Miner. Kaplan, however, quit shortly after its formation and, by the end of 1982, the company had been renamed the Amiga Corporation.

began to show off Miner's computer, the Amiga, in public it was met with a mixture of disbelief and barely contained excitement among game designers. Some believed it was a fraud while others salivated at the very thought of what they could create using Miner's audio-visual powerhouse. Atari, however, was in no state to enjoy the moment and by the summer of 1984 Commodore founder Jack Tramiel was poised to take over the ailing firm's computer division.

The thought of working for Tramiel horrified the Amiga Corporation; it had already gained first-hand experience of Tramiel's abrasive business style during talks about selling the rights to Miner's computer and wanted nothing to do with him. Desperate to escape its Atari deal, the company formed an alliance with Commodore, which had ousted Tramiel at the start of the year. Commodore bought Amiga out of its deal with Atari just days before Tramiel's takeover was completed in July 1984. The Atari Amiga would now become the Commodore Amiga. Bob Jacob, an agent representing game and software designers, was one of the first people to see the finished version of the Amiga, a few months prior to its official release in 1985. "It was 1984, I got a call from a company called Island Graphics that had a contract to develop three graphics programs for the Commodore Amiga," he said. "This company and Commodore had a falling out, so Island wanted to place the project elsewhere. I went up to see them and I had never seen an Amiga before. It was really cool. After seeing the Amiga I figured things were going to be different and I wanted to take a more direct approach to game development."

Inspired, Jacob and his wife Phyllis formed Master Designer Software in January 1986 with the intention of using the Amiga to "rethink what a computer game could be". Jacob decided his game studio would not look to existing video games for inspiration but to the hills of Hollywood. "I wanted to tell stories. I wanted to give people a movie-like experience," said Jacob. "I became obsessed by the idea of trying to create games that had the mood-altering quality of an arcade game but had a story and some minor role-playing game aspects. What I really liked about arcade games was that when I was playing them I couldn't think about anything else, I couldn't think about my problems – it took up all my attention. It definitely became a mood-altering experience. At that time I thought computer games were very crude. They were really slow. A lot of them had keyboard interfaces, ugly graphics; a whole host of elements that served to really kick you out of the experience." Jacob wanted his company to address these flaws of video games without sacrificing the emotional power of action games. "If I had a breakthrough creatively it was the idea that I wanted action but I didn't want action by itself," he said. "I wanted the action element of success or failure to branch the story, to move things along. Action for a purpose. I wanted to create a different feeling."

Jacob homed in on the idea of using movies as the basis for his new breed of video game and decided Master Designer Software would release its games under the name Cinemaware. Cinemaware's first release was November 1986's *Defender of the Crown*, a 'knights in shining armour' game set in medieval England. Its action scenes paid homage to the 1952 film *Ivanhoe* and its strategy elements drew on the world conquest board game *Risk*, but it was the game's lush visuals and cinematic presentation that made it stand out. "*Defender of the Crown* was a phenomenon," said Jacob. "It was the first game that showed the power of the Amiga graphics. It was beautiful. At the time the Amiga had a lot of games that were essentially Commodore 64 ports that really didn't show off the ability of the hardware. Literally every person who owned an Amiga at that point bought that game, we had almost 100 per cent sell through." Cinemaware followed *Defender of the Crown* with further attempts to explore Jacob's fusion of cinematic storytelling and video game action with titles such as gangster film-inspired *King of Chicago*, the '40s sci-fi serial pastiche *Rocket Ranger* and the '50s b-movie ode *It Came from the Desert*. "Bob Jacob really wanted to put characters into the games," said Ken Melville, who wrote the script for *It Came from the Desert*. "So you saw these real breakthrough notions like *King of Chicago* and *Defender of the Crown* where characters were right on the screen talking to you. Cinemaware was really the first to bring actual characters and story elements into direct interaction with the player." Cinemaware's movie influences ran deeper than just surface presentation and storytelling however. Hollywood's movie development processes would also inform its approach to game development. "We would have story meetings, we would flow chart the game and come out with storyboards," said Jacob. "The games we were doing were different to the games other people were doing at the time. We really had to figure out where we were going with the game. We were doing games that had storytelling and role-playing and action and this and that and the other thing. If we didn't know where we were going it would be a disaster and that forced us into a level of oversight that was rare at the time."

Cinemaware blazed the trail for the concept of 'interactive movies' – narrative-driven games where cinematic storytelling was as important as play – but others were not far behind. Across the world game designers, empowered by new machines such as the Amiga, were starting to examine what they could learn from the craft of filmmaking. The games of aspiring screenwriter Jordan Mechner, for example, drew heavily on the visual language of cinema. His 1984 debut, the martial arts title *Karateka*, imported the camera techniques of silent film using rotoscoping, cross-cutting and tracking shots to convey its simple girlfriend-rescuing story without resorting to text. *Karateka* became a hit, but Mechner was unsure whether to continue with video games, torn between his desire to be a filmmaker and his success in the game industry. "There's no guarantee the new game will be as successful as *Karateka* or that there will even be a computer games market a couple of years from now," he wrote in his journal in July 1985. Although plagued with doubt, Mechner eventually settled on completing his second video game: the *Arabian Nights*-themed action title *Prince of Persia*. Mechner once again raided the toolkit of cinema. He bought a video camera and filmed his brother David running and climbing around a New York City parking lot so he could make the player's character move in a realistic way. He spent hours watching the duel between Errol Flynn and Basil Rathbone in the 1938 movie *The Adventures of Robin Hood* to work out how the in-game swordfights should look and he drew on the storytelling techniques of silent movies to explain the story through the on-screen action and character movement. Released in 1989, *Prince of Persia*'s cinematic presentation and attention to visual detail turned it into a major video game series that was still going strong 20 years after its debut.

Game designers were not the only ones exploring what video games could learn from the movies. Film director George Lucas had also started to explore the intersection between movie and video game. The *Star Wars* director created a game development studio within his Lucasfilm production company in 1982 after Atari gave him $1 million in return for having first refusal on releasing whatever he did with the money. Instead of creating games based on *Star Wars* or *Indiana Jones* as Atari probably hoped, Lucas used the money to create a studio whose mission was to develop original game franchises of its own. But while encouraged to find its own creative voice, the game studio's output reflected many of the production values of Lucasfilm the movie studio. In keeping with Lucasfilm's reputation for special effects, Lucasfilm Games sought to mirror the high standards of Lucas' movie work by creating games that excelled both visually and, unusually for the time, sonically. "The importance of music and sound effects had been overlooked in earlier video games as it was in movies until productions such as *Star Wars* and *Raiders of the Lost Ark* signalled a new artistic awareness in the industry," said Peter Langston, the head of Lucasfilm Games, at the 1984 launch of the firm's debut games *Ballblazer* and *Rescue on Fractulus!*. "We're very satisfied with the successful role that music and special effects play in making both games a total sensory experience for the player." Lucasfilm Games also devoted a huge amount of effort to detailing the game worlds it built. During the making of *Rescue on Fractulus!*, a sci-fi game set in fractal-generated canyons that players had to fly around in order to rescue stranded spacemen, Lucasfilm built life-sized spaceship models and even selected a colour for the uniform of the game's unseen hero in order to deepen their understanding of how the game world should feel.

Lucasfilm Games' cinematic parentage really came to the fore when it started making adventure games. Lucasfilm's entry into the adventure genre began when one of its programmers, Ron Gilbert, came up with an alternative to text input that fused the mouse-based interface of ICOM Simulations' *Déjà Vu: A Nightmare Comes True* with the animated visuals of Sierra's text adventures *King's Quest* and *Leisure Suit Larry*. Gilbert and artist Gary Winnick used the format to create 1987's typing-free adventure game *Maniac Mansion*, a parody of horror b-movies where groups of teenagers end up in a dangerous location and become separated before being killed one by one. Lucasfilm followed *Maniac Mansion* with a run of adventure games that used the same approach from slapstick swashbuckler *The Secret of Monkey Island*, movie tie-in *Indiana Jones & The Last Crusade* and the ethereal fantasy of *Loom* – a game set in a dream-like world inspired by Tchaikovsky's *Swan Lake* and Disney's *Sleeping Beauty*. "There wasn't any distinct message I was trying to communicate with *Loom*," said its creator Brian Moriarty. "It was more like a mood I was trying to sustain, a dreamy, melancholy feeling like that evoked by the music for Tchaikovsky's *Swan Lake* ballet." As its roster of adventure games grew, Lucasfilm became more and more interested in applying the techniques of cinematography, moving from the largely static scenes of *Maniac Mansion* to the panning cameras and close ups of *The Secret of Monkey Island*. As Doug Glen, the general manager of Lucasfilm Games, said in 1991: "There's a whole area of cinematography – cutting, panning, zooming and so on – yet to be properly exploited in games."

The integration of movie techniques into games was not just a US trend. Delphine Software, the Parisian game studio set up by French music label Delphine Records, was also exploring how to make games more cinematic. After some initial success making adventure games such as *Cruise for a Corpse*, which used a similar interface to that created by Lucasfilm, it came out with *Another World*, a superb action game about a computer programmer transported to an alien world. Created by Eric Chahi, *Another World* echoed the *Prince of Persia* model, using character animation to tell the story visually. Where Chahi differed from Mechner was his focus on capturing the pace of movies in his game. "What I really learned with this game, and what was a lot more important than the cinematographic aspect, was to create a game rhythm, moments of relaxation, of tension," he said. "I wanted to bring into a game an immersive, cinematic feeling. I think it succeeded because there was a good balance between the cut scenes, which were punctuations rather than sequences, just a plan inserted at the right moment. It was not like what they do now [2009] where they feed in sections of film unconnected with the game play."

The game's opening where the player is chasing through the alien world by a large black beast demonstrated the approach. The beast is seen lurking in the background as the player starts to explore the alien landscape. Then, suddenly, the game cuts to a brief scene showing the roaring beast jumping into front of the player's character, before switching back to the action and the ensuing chase. Chahi also brought *Another World*'s distinctive flat sharp-edged visual style, inspired by comic book artwork, to the fore by removing the kind of information that usually littered game screens. There was no score, no lives, nothing but the game itself. "I was fed up of score because it was nonsense," he explained. "It was in conflict with the universe of *Another World*. I wanted a visceral implication of the player, no distraction other than the world itself. No artificial motivation, which score is. Score's a capitalistic view of game play, no?"

The cinematic exploration of Chahi, Mechner, Lucasfilm and Cinemaware were dwarfed in size, however, by the work being carried out by Hasbro and Axlon in the late 1980s. Axlon was the latest business venture of Atari founder Nolan Bushnell.[2] Formed in 1988 with Tom Zito, Axlon's big idea was to make a console that ran games stored on VHS videocassettes rather than cartridges. Toy makers Hasbro embraced the idea and teamed up with Axlon to develop the system, which they codenamed NEMO.[3] The NEMO team read like a video gaming who's who. There was Bushnell, *Spacewar!* co-creator Steve Russell, Imagic's Rob Fulop, Cinemaware's Melville, Activision's David Crane and a gaggle of former Atari coin-op employees including Steve Bristow and Owen Rubin. The decision to use VHS as a storage medium was based on technology that allowed videotape to be divided into several tracks, only one of which would be shown on screen at any one time. The approach gave game makers the chance to use real-life film footage in their titles rather than computer-generated graphics, although the system lacked the ability to rewind or pause the tape during play. "The tape was always running, so you can't sit there and wait for the user to decide to go left or right," said Fulop. "That worked on laserdisc, a laser can always jump, but this thing was always running and that's why we made a game like *Night Trap* – there's a house with cameras all over the place and you jump between each one.[4] That was designed to work on tape and the system worked; the story's moving and you would just move where you viewed the action from."

The need to use film footage in the NEMO's games required game developers and filmmakers to work in unison. "We had a $4 million budget for *Sewer Shark*, which was unheard of at the time," said Melville.[5] "We built huge, elaborate sets, had a big film crew, actors, sound stages, location shoots – the whole ball game. We hired John Dykstra, the guy who had literally made *Star Wars* via his insane special effects, and he had his top guys build all the tunnels and rats and everything – all practical, motion-control work. Meticulous, expensive, serious Hollywood sci-fi movie production values. This was the real first

---

2. Bushnell first returned to the video game industry in 1983 with Sente, a producer of cartridge-based coin-op machines. The appeal of the approach was that when players grew tired of a game, arcades could just insert a new cartridge to change the game rather than buying a whole new machine. It was an idea that would later become widespread, but Bushnell was forced to sell Sente after his Chuck E Cheese pizza restaurant chain ran into financial trouble. Senate's buyer Midway failed to exploit the full potential of the idea, which Japanese coin-op makers eventually adopted to great success.

3. Short for 'Never Ever Mention Outside'.

4. *Night Trap* was a horror game where the player has to protect a slumber party of teenage girls from vampires.

5. *Sewer Shark* was a futuristic shoot 'em up that was to be the NEMO's flagship title.

marriage of Hollywood and gaming." Not that Hollywood and Silicon Valley necessarily gelled. "Since game and movie production had never before been merged, we had to blaze all the contractual, licensing and ego trails," said Melville. "Hollywood folks had a huge attitude about us Silicon Valley geeks. They treated us at first like commercial clients – shoot our Bubble Yum advert spot, that kind of thing." As well as shooting the footage, the NEMO team also faced the challenge of getting every scene to line up to ensure continuity. "There was a timeline drawing on the walls that spanned the entire hallway, top to bottom, that showed all the connections and scenes and how they matched up," said Rubin. "It was incredibly complex."

Hasbro lavished $20 million on developing its VHS games console and hopes were high amongst the team that it would be a success. "The idea of an actual guy talking smack directly to you based on your performance was, for the times, mind-bending," said Melville. "Our main competitor, Nintendo, was producing blocky little cubes at the time. We would have lit 'em up like a sewer rat. We did a focus test once and after a little kid had played some of our games and various characters were talking to him face-to-face, he turned to our host and asked in deep wonder: 'How do it know?'." The NEMO was set to launch in January 1989 as the Control-Vision but, just three months before its debut, Hasbro pulled the plug. Ultimately Hasbro could not get the figures to add up. The NEMO would have cost $299, far more than rival consoles and, with its games costing millions to develop, Hasbro concluded it could never make enough profit to justify the venture. "It was too expensive to make these things. You had to make a movie," said Fulop. "Look at *Sewer Shark*, we shot that in Hawaii. The console cost a lot too, it was a loss-making thing, would break a lot and the games, while they worked fine, they weren't that replayable." The NEMO was a write-off, but it did test out the idea of Hollywood and Silicon Valley working together, a practice that would become increasingly common from the 1990s onwards, and furthered game designers' understanding of how video games could absorb ideas from film. "In interactive you have to write very minimalistically. You have to get to the next decision point fast, you had to edit for that lightning pace, the player has no interest in waiting around," said Melville. "For instance, I had John Dykstra editing the stuff we shot in Hawaii, but it was too slow. It constantly needed to be upcut. I took over and finished it to get that fast cut edge that John, as a movie guy, couldn't possibly understand not being a gamer. Hasbro was really the first time gaming notions, the real needs of gamers, were given full shrift in a Hollywood production environment. We invented the system of merging the two highly disparate cultures."

Yet while many game developers began pushing for more cinematic, story-driven experiences, a rival camp were pulling in the opposite direction and seeking to free players from the constraints of designers' visions.

The Entrepreneur: British game designer Peter Molyneux

CHAPTER 15

Ah! You Must Be A God

In 1984 Will Wright and Peter Molyneux made their debuts as game designers. In time, their work would define an alternative vision of video gaming to the narrative-driven cinematics of Lucasfilm Games and Cinemaware. But in 1984 the two designers were worlds apart.

Born on 20th January 1960, Will Wright was an American raised in Atlanta, Georgia, and Baton Rogue, Louisiana. He was a voracious reader obsessed with model making and robots, a love that led him to video games. "What got me into games were robots," he explained. "I was building robots as a teenager, weird mechanical things out of random parts. I bought my first Apple II computer to connect to my robots, to control them. Some of the very first computer games were coming out and they were basically simulations. I got fascinated with simulations as a form of modelling, so I started writing simple simulations of my robots and got fascinated with artificial intelligence and simulation."

One piece of software in particular captured Wright's imagination: *Life*. Created in 1970 by the British mathematician John Conway on a PDP-7 computer, *Life* sought to demonstrate how complexity could emerge from very simple rules. The program displayed a screen divided into a grid of cells. At the start of the 'game', the user set the initial conditions by deciding how many cells to bring to life. After that the program ran itself by following three simple rules: (1) live cells with 2 or 3 live neighbours survive; (2) live cells with fewer than 2 neighbours die, as do those with 4 or more neighbours; (3) 'dead' cells with 3 live neighbours come to life.

Despite their simplistic nature, these three rules created often-mesmerising animated patterns that generally moved towards some kind of stable system over time. Conway's creation fascinated Wright: "It's so extraordinary because the rules behind it are so simple but the behaviours so complex. It's like Go in that a lot of people I know lost major chunks of their life in both these endeavours. There is some underlying aspect that they capture about reality and complex systems in that they arise from incredibly simple rules and interactions. It became a major design approach: put together simple rules to create complex behaviour. That was a huge inspiration for me."

While fascinated with *Life*, Wright's debut video game – *Raid on Bungeling Bay* – showed little of its influence. *Raid on Bungeling Bay* was a Commodore 64 shoot 'em up where players controlled a helicopter as it flew around an archipelago trying to bomb military factories. "I was trying to find things on the Commodore you couldn't do on the Apple II," said Wright. "The Commodore 64 had these graphic features that were advanced for what it was and you could have this big scrolling window on a larger world. I also always loved helicopters. I designed that game around the technology, around what you couldn't do on the Apple II." Released by Brøderbund, a game publisher based in San Rafael, California, *Raid on Bungeling Bay* sold poorly on the Commodore 64. "There was a lot of piracy. Everyone had a copy but I only sold 20,000 to 30,000 copies," said Wright. "Luckily for me it was one of the first American games licensed into the Japanese market on the Nintendo Entertainment System. It sold about a million units in Japan and, back then, the terms you got from publishers were pretty generous. I earned enough from that game to live on for several years."

Molyneux's first taste of the video game business could not have been more different. Born on 5th May 1959 in the English town of Guildford, just to the south west of London, Molyneux hated school and had a vague dream of becoming a successful businessman. His first attempt at running his own company was a disaster. "It was based on this ridiculous business idea of selling floppy disks to schools. We put software on the floppy disks and sold it to them, the unique thing is you get this free software," he said. "Of course the floppy disks were quite expensive because people could buy them enormously cheap from Taiwan or something and they only needed the software once. A school would order 10 disks to get the software when, to make the business work, I needed them to order 10,000."

For his next business venture, Molyneux decided to join the early 1980s boom in British video games with a business simulation called *The Entrepreneur*. Confident that his mail order game would be a huge success, Molyneux took out adverts in game magazines and warned the Royal Mail to expect a flood of orders.

"I thought it was going to be the greatest success ever," he said. "In those days I didn't think in any sensible way. I very much floated around in the stream of life going 'hey let's do business stuff, let's do *The Entrepreneur*'. If I'd thought about it properly I wouldn't have made a business sim, I would have made a *Space Invaders* clone like everybody else. My contemporaries were working on what people really wanted and I was doing what people didn't want." Molyneux sold just two copies of *The Entrepre-*

neur and his business collapsed. He decided to give up on games and focus on making business software.

While Molyneux was watching his businesses turn to dust, some 5,300 miles to the west Wright was using his earnings from *Raid on Bungeling Bay* to explore the potential he saw in that game's level editor. "While building *Raid on Bungeling Bay* I had to build lots of other programs to help me," he said. "One let me scroll around the world and place buildings and roads on these little islands. I had more fun with that than flying the helicopter around it." Free from the immediate pressure of having make a living and intrigued with his world-building tool, Wright began experimenting. "At first it was just a toy for me, I was making my editor more and more elaborate and thought it would be cool for the world to come to life, so I started researching books on urban dynamics, traffic and things like that," he said. "I came across Jay Forrester, who was kind of the father of system dynamics."

Forrester was an electrical engineer who helped build some of the first computers in the early 1950s at Massachusetts Institute of Technology. In 1956 he became a professor at MIT's management school and began exploring how he could apply his knowledge of electrical systems to other kinds of systems. His work gave birth to the field of system dynamics, where computer models or simulations are used to examine social systems and predict the implications of changes to complex systems such as cities. "He was one of the first people who simulated a city on a computer, except in his simulation there was no map it was just numbers – population level, number of jobs – kind of a spreadsheet," said Wright. Wright combined Forrester's theories with the living system of *Life* to bring the conurbations he built using his enhanced world-building tool to life.

Another influence was Wright's brief spell at a Montessori school as a young child: "Montessori is part of what's referred to as constructivist education, using toys that are creative to help kids learn geometry and math. Things I took from it were (a) rather than educate someone, I'd rather inspire and (b) I think that self-directed learning is much more powerful than if you lead someone on a leash."

Finally Wright added an interface based on the Macintosh operating system and Apple's computer art package *MacPaint* in particular. "Probably the biggest inspiration was *MacPaint* – you have your tools, your canvas and you grab the tools and draw them. I always thought of *MacPaint* as the underlying architecture for it," said Wright. Eventually he began to think his experimental plaything could work as a game. "I found when I was reading all these books that simulating it on a computer was more interesting than reading it in a book. It bought the whole subject to life for me," he said. "I started to think other people might like it. I didn't think it would have a broad appeal, maybe architects or city planner types but not average people."

Despite feeling his audience would be planning and social science experts, he avoided making his game too deep a simulation: "It's a caricature of how a city works. We're really emphasising things such as gentrification. You elect these truisms in city planning." Wright called the game that was evolving out of his experiments *Micropolis* and approached Brøderbund to see if they would be interested in releasing it. They weren't. 'How do you win or lose?', the baffled publisher asked. 'You don't', replied Wright, 'you just build and manage your city and see what happens'. Until then games were all about winning and losing. A video game with no goal, no purpose and no recognition of success or failure was just unthinkable.

"They were expecting more of a traditional game out of it. I wanted it to be more open-ended, more of a toy," said Wright. "Because we weren't formally defining success, the first thing players had to think about was what kind of city did they want to create. What is success? Is it a big city? Low crime? Low traffic? High-land value? This puts more meaning into that possibility space."

Wright found other publishers thought the same about his unusual Commodore 64 game, but decided to continue refining his virtual town planning game anyway. Then in 1987 Wright met Jeff Braun at a party. Braun wanted to get into the games business and was looking for games to publish. Wright told him about *Micropolis* and invited him to come and see it. Braun loved it. The pair formed Maxis and decided to make *Micropolis* its second release.[1] But with the Commodore 64 losing ground in the home computer market, they decided to convert the game to the Macintosh, Atari ST and Amiga. And at the suggestion of a friend they changed its name to *Sim City*.

Maxis asked Brøderbund to be its distributor. Don Daglow, the Brøderbund producer who handled the talks with Maxis, saw the appeal of *Sim City*, not least because he had ventured into similar territory with 1982's *Utopia* while working as a game designer for the Mattel Intellivision console. "*Utopia* was a counterpoint to the action and sports games," said Daglow. "We thought, let's do something that has educational value, that is covert rather than overt: a strategy game. I pitched it and, bless them, Mattel's marketers and executives approved it. There was a tradition of simulation games and playing simulations that weren't games that existed on the mainframes in the '70s. You ran them and a bunch of numbers appeared and that was what influenced me on *Utopia*." *Utopia*, which was a two-player game, put players in charge of an island and challenged them to score points by keeping their citizens happy by

1. Braun already had *SkyChase*, a wireframe 3D flight sim for two players, lined up as Maxis' debut release.

building of hospitals and developing industry. To try and sabotage their rival, players could also fund rebel activities on their island. The game sold less well than Mattel's arcade and sports games, but it did gain widespread press attention. "We went to the Consumer Electronics Show in 1982 and I got a call from my wife in the hotel on the second day," said Daglow. "She said the morning paper had a front-page article about the new video games with *Utopia* as the lead story in the article. It was the game that got the press attention." So when Maxis approached him with *Sim City*, Daglow saw the appeal and gave them a distribution deal. "Clearly *Utopia* and *Sim City* come from the same spirit. They're both city sims. I signed the distribution deal for *Sim City* for Brøderbund. The fact that I did is probably no coincidence." It had taken five years for Wright to turn his world-building tool into *Sim City*, but by 1989 the game was finally ready for release.

* * *

While Wright spent 1984 to 1989 crafting his city simulator, Molyneux spent his time battling to make ends meet. His early optimism about his entrepreneurial ideas had evaporated. Everything he touched, it seemed, went wrong. His business software company was struggling and housed in dismal offices. "The office was so shit," said Molyneux. "We were using the sink as a toilet, it was unbelievably scummy." Then, out of the blue, he got a call from Commodore, who wanted to give his business software company – Taurus – a bunch of Amiga computers. It transpired that Commodore was actually trying to get hold of another software company called Torus, but for Molyneux Commodore's erroneous gift was a rare flash of good fortune. Molyneux and his business partner Les Edgar initially created a database product for the Amiga, but shortly afterwards they decided to move into video game development and renamed their company Bullfrog. "My daughter Louise had a thing about frogs and Bullfrog seemed like the sort of daft name a games company might have," said Edgar. "At the time we were still reeling from the shock of the cost of supporting our Amiga database product, and with increased competition from other, larger publishers in that space, we had to do something. The game arena seemed attractive because development times were short, costs were low, support was not necessary and results were fairly instantaneous. If you screwed up, you could start another game without massive losses."

Bullfrog managed to get some work converting Firebird's *Gauntlet* clone *Druid II: The Enlightenment* to the Amiga and then persuaded Electronic Arts to publish its first original game: the 1988 vertically scrolling shoot 'em up *Fusion*. Bullfrog's game became one of the first to be published by Electronic Arts' new European operation that had opened in Slough, a London satellite town not far from Guildford. "Europe was still in the earlier 'cottage industry' or 'import' phase because there had never been an indigenous phenomenon like Atari or Nintendo, which had educated Japanese and American consumers and retailers on what to do," said Electronic Arts' founder Trip Hawkins of the company's expansion into Europe. "Retail promotion was lagging in sophistication and needed more direct support from major publishers. There were unfair trade practices in many countries. In Germany you could only get into half the stores because two competing distributors had divided up inappropriate retail exclusives and would then insist on game exclusives as well, so your game was either in the first or second distributor's half of the stores and the consumer could only see half the games in their store."

Electronic Arts set out to bring its professional corporate style to Europe's unwashed video game industry. Geoff Heath, who was the managing director of Melbourne House when Electronic Arts opened its European arm, remembered the US company's arrival as a wake up call to the UK industry. "The company that really brought professionalism to the games industry was Electronic Arts," he said. "They ran it like a business whereas prior to that we all ran these companies that made money half the time, didn't know why we were making money and had a load of fun. Then along came Electronic Arts run by savvy MBAs who knew their subject."

Fusion sold poorly, but the small advance Electronic Arts gave Bullfrog to pay for its creation, provided the studio with enough time to make some headway with its next project, *Populous*. *Populous* emerged out of an isometric pictograph created by Bullfrog programmer and artist Glenn Corpes. "I thought it looked fascinating and said shall we put little people on it as if these isometric blocks were mountains," said Molyneux. "Then there was the thing about how are people going to move around on this land?"

Molyneux set about bringing the miniature citizens of this isometric land to life and immediately ran into problems. "The fact that I programmed it meant some of the fundamental things that programmers can do in their sleep I couldn't do," he said. A particular problem was getting the virtual people to navigate around walls rather than getting stuck. "I didn't know how to do that. I tried to do it, tried to invent it myself and couldn't and I thought 'oh fuck it, I'll just get the player to solve the problem for me by raising and lowering the land'. That became the game's fundamental mechanic. Pure and utter luck. Suddenly you're raising and lowering land with little people, 'Ah! You must be a god'."

Molyneux turned Corpes' isometric drawing into a game that put players into the role of a deity seeking to gain more worshippers using Old Testament-style powers to change the landscape, cause earthquakes and create volcanoes. The ultimate aim was to wipe out a rival group of people who were followers of another god. It was, said Molyneux, a game "written, designed and everything for one person: me".

Britain's game publishers agreed. "Publishers kind of looked at it and went okay… thank you very much, can you get those little people shooting each other? Er, no. Can you get the mouse cursor to fire? No. Oh well, next!," said Molyneux. One executive from Mirrorsoft, the UK game publisher owned by media mogul Robert Maxwell, exclaimed 'who would want to play at being god?'. "I seem to remember he didn't believe in the internet too," said Edgar. Like *Sim City*, *Populous*' lack of shooting, lack of pre-defined solutions and megalomaniac experience was so unusual no one knew what to make of it. Eventually Electronic Arts agreed to publish it. "They didn't have any games for Christmas 1989 so they said 'ok, we'll sign that *Populous* thing'," said Molyneux. "It didn't really feel like they wanted it, it felt like it just filled a gap for them."

With *Populous* complete and Electronic Arts on board, Molyneux braced himself for another disappointment: "I thought it would be a flop because everything in my life that I had touched up to that point had just gone awfully badly. School had gone awfully badly, then my first business, *The Entrepreneur* and selling disks to schools. There really wasn't an idea of success."

The only hint that Molyneux's pessimism might have been misplaced came when Electronic Arts arranged for Bob Wade, a journalist from the UK games magazine *Advanced Computer Entertainment*, to visit Bullfrog for a preview of the game. "I didn't know what to do with journalists, so I thought we'll just go out and get drunk," said Molyneux. "We got drunk and I was dying to ask him what he thought of the game, although I was also terrified to because I was pretty sure it was going to be like everything else in my life. Eventually, after god knows how many pints, I said what did you think of *Populous*? He said it's one of the best games I've ever played. My first thought was he must never play *Populous* again because he must have been on some other planet when he played it."

* * *

By the time *Populous* was ready for release in late 1989, *Sim City* had been on sale for several months. Although sales were initially slow, the game ended up in the hands of a *Time* magazine reporter who wrote a full-page review of Wright's groundbreaking Macintosh game.

The review prompted a massive surge in sales and, as *Sim City* started appearing on other computers, the sales just kept rising.

"Typically in the game market you released the game and 80 per cent of sales were in the first six months," said Wright. "*Sim City* was a totally different profile. The first year did well; the second year sold a lot more, the third even more. *Sim City* paid for a lot of mistakes, which was great because we made a lot of mistakes with our company."

As well as appealing to many video game fans, *Sim City* also connected with an audience who would normally shun the bouncy platform, martial arts and fierce shoot 'em up games that typified the games available on the NES and in arcades. Millions became hooked on the joy of growing their own cities, although some felt Wright's simulation was biased – particularly when it came to traffic management. "Any simulation is a set of assumptions. A lot of people thought we were really biased towards mass transit, others thought we were biased against it," said Wright. "The interesting thing is a model like that gives you something to reflect against and, in fact, when people start to argue with the model, that's when I think it's been successful. When they're playing a game like *Sim City*, which is really one set of assumptions, and they start arguing with those assumptions, the game has crystallised an internal model to the point where they can now argue against that model. In some sense that's the point of it."

By the end of the year *Populous* had joined *Sim City* in becoming a surprise international hit. The magazine reviews lavished praise on the game and, to Bullfrog's surprise, it seemed as if they might get some royalties from the game. "Given the reception by the press in the UK, I was expecting that we would make some money, but when the first royalty cheque came in – £13,000 as I recall – I didn't imagine we would get much more," said Edgar. "The second cheque was significantly larger and they just kept coming. Then the Japanese arrived and things really took off." Molyneux was shocked: "Electronic Arts couldn't manufacture enough to satisfy the demand and then it was released all over the world and everybody was playing it. The publisher, David Gardener, phoned up and said 'you are a millionaire now' and it was like 'my god!'. It was just this amazing moment of thinking 'Christ, what have I done?'. I expected to sell four copies, instead it was getting close to a million within a matter of weeks."

Unusually, both *Sim City* and *Populous* became popular in Japan, a country that usually ignored the smattering of North American and European games that reached its shores. But it was only when Molyneux and Edgar went to Japan to do some publicity work that the scale of their success in Japan became clear. "When we arrived at the airport in Tokyo there was a TV crew there to meet us and an infinite number of magazine interviews lined up," said Edgar. "The reaction from the press was staggering – they had all played the PC and Amiga versions."

Back in the UK, a bemused Molyneux became a star game designer: "The journalists at the time made this genre up. They called it a god game because they had no other way of describing it and suddenly I was attributed with creating this whole genre. I didn't create the genre, the journalists did – I just created a game that allowed them to describe the genre. We had these round tables started by the press. David Braben was there, Jez San, Archer MacLean and there was me.[2] They were all talking about assembler and machine code and how they can get the blitter to interact with the copper. I was sitting there thinking 'shit, what are they talking about?'. It was all double dutch to me."

Together, *Populous* and *Sim City* gave form to a disparate game development movement that rejected the controlled, confined and directed experiences of their cinema-worshipping peers in favour of more open-ended experiences that were closer to toys than board games in concept and embraced creation and construction as a play mechanic. It was an approach to video game design that had been circulating for some time, from the nation management of *Utopia* and the freedom of Braben's *Elite*, but the success of Wright and Molyneux took these games into the mainstream of game design and encouraged other designers to start investigating the possibilities.

One of those Wright and Molyneux inspired was Sid Meier, the co-founder of US game publisher Microprose. The Maryland publisher had carved out a lucrative niche producing military-themed simulations such as *F-15 Strike Eagle*, *Silent Service* and *Gunship*. But as the 1980s drew to a close, Meier started exploring ideas outside Microprose's military comfort zone, thanks partly to the influence of Dani Bunten Berry's *Seven Cities of Gold*. Inspired by the Spanish Conquistadors, *Seven Cities of Gold* was a game of exploration that condensed the history of the discovery of the New World into a video game and sought to convey the panic of being lost in the uncharted wilderness. *Seven Cities of Gold* inspired Meier to create *Pirates!*, a open-ended game about adventure, trade, robbery and romance on the high seas in the 17th Century, and got him thinking more generally about how to make games that simplified complex scenarios.

Shortly after completing *Pirates!*, Meier discovered both *Sim City* and *Populous*. For Meier, these games demonstrated that creating something could be just as fun

---

2. Braben was the co-creator of *Elite*. San had designed the then visually impressive 3D space combat game *Starglider*. MacLean had created *IK+*, a three-player martial arts game featuring a number of cheats for players to use against their opponents that were designed to be part of the fun.

and compelling as the destruction usually peddled by video games. Meier's response was to devise a game that put players in control of a civilization's journey through history.

His grand vision became *Civilization*, an epic turn-based strategy game that offered a beguiling concoction of military conflict, diplomacy, exploration, city building, history lesson and resource management. The goal was to take a tiny tribe, ignorant of the world, and turn it into a great world power shaped by the choices and decisions of the player themselves. Core to *Civilization*'s appeal was its ability to make players feel as if they were writing history as they went, with centuries marked by war and instability giving way to golden ages of scientific progress before going back to high-stakes clashes with other large nations. It was an aspect that gave *Civilization* – more than *Sim City* or *Populous* – a narrative quality, albeit one defined by the player rather than Meier.

Civilization would have a significant impact on strategy games, which had not moved too far beyond their tabletop origins since first appearing on home computers. *Civilization*'s branching technology tree, where each discovery opened up more and more options for scientific research, was a big influence on *UFO: Enemy Unknown*, a 1993 strategy game where players defended the earth from aliens by unlocking the secrets of alien technology and using it against them (the game was known as *X-COM: UFO Defense* in North America).

The same feature also inspired Las Vegas game developer Westwood Studios, which was working on a video game version of Frank Herbert's sci-fi novel *Dune*. After an argument with one of the vice-presidents of strategy game specialists SSI about the future of strategy games, Westwood's co-founder Brett Sperry decided to use *Dune* to reinvent the genre. As well as *Civilization*, Sperry and his team drew on ideas from the Japanese console game *Herzog Zwei*, a strategy game where the battles happened in real-time, and the Macintosh operating system.

The result, 1992's *Dune II: Building of a Dynasty*, fused action with strategic thinking and opened up the strategy genre to a new audience. People who once associated strategy games with slow and turgid bouts of number crunching now equated it with high-pressure action that rewarded quick, strategic thinking. The marketing description of *Dune II* as a "real-time strategy game" was soon adopted as the byword for the rush of similar titles that followed in *Dune II*'s footsteps. The real-time strategy games quickly displaced the turn-based strategy games of old and by the end of the 1990s *Civilization* was the only big name strategy game series that had not gone real-time.

Video games had evolved into three broad philosophical movements. There were the action games that were forged in the cash-hungry environment of the arcades and served up instant thrills. Then there were the narrative-based, cinema-worshipping works that saw video gaming as a story-telling medium enhanced by interactivity. Finally, there were the sandbox games; the simulations that sought to entertain by giving players the freedom to experiment and the ability to create.

These movements were not discrete, their ideas would repeatedly cross-pollinate, but each pulled games in different directions, enriching the medium with new ideas and constantly pushing back the barriers of what a video game could be.

East meets West: Alexey Pajitnov (left) and Henk Rogers in Moscow, February 1989
Courtesy of The Tetris Company

CHAPTER 16

A Plane To Moscow

Alexey Pajitnov lived an unassuming existence in the early 1980s. As a mathematician in the computing department at the Moscow Academy of Science, Pajitnov spent his days programming computers fuelled by copious amounts of black coffee and cigarettes.

"I was just a regular programmer and researcher. My lifestyle was regular for every young hacker in the world at that time," said Pajitnov. "I worked about 11 hours a day, started relatively late and stopped around midnight or half-past midnight. Every day was no different. All the young programmers are workaholics, mostly, and I was just one of them. Nothing special, very usual."

His job was to research serious computing applications such as speech recognition, but when he got a spare moment he would create mathematical puzzle games on the academy's computers for his own amusement. "I came to the game from the puzzle. From the board games, the regular board games, wooden castles and tasks and riddles," he said. "I always considered the game more like a puzzle or mental challenge. I was fascinated by puzzles and mathematical riddles from the time I was a kid. As a kid I participated in all kinds of contests and I was in a special programme at school, so I was addicted to mathematics and mathematical tasks and puzzles. I liked the challenge."

One day in 1984 Pajitnov wrote a game based on Pentominoes, a puzzle he found in a Moscow toy store. Pentominoes consisted of a collection of flat plastic shapes, each built from five squares of equal size arranged in different ways. The goal was to take them out of the box and then fit them back together within the box, rather like a jigsaw. Pajitnov thought Pentominoes would be more interesting as a computer game where the pieces rained down from the top of the screen into the play field, his virtual equivalent of the Pentominoes box. While the academy boasted some of the most advanced computers in the Soviet Union, they were primitive compared to the home computers available outside the communist world. Pajitnov's work computer was an Elektronika 60, a Soviet computer based on the PDP-11 that first appeared in the US in 1970. "We were way behind," he said. "Our best machine repeated the level of the rest of the western world about five to eight years late. The Elektronika 60 was one of the very first micro machines in Russia, almost desktop. Not very powerful but really convenient."

The Elektronika 60 had no graphics so Pajitnov had to construct his digital Pentominoes pieces, which he formed out of four rather than five squares[1], using punctuation marks. "I thought Pentominoes might be a very good basis for two-player games," said Pajitnov. "My original idea was that you just put the pieces on top of the field and your task is to put as much as possible in the play field. If you put in more than the other player then you win the game. But when my first prototype started working I realised that the game ended in something like 10 seconds and you didn't have too much fun out of it."

Pajitnov concluded that the size of the play area was spoiling his game. "One solution was to make a really, really long play field and scroll it through, but it was technically difficult at that time," said Pajitnov, who also regarded play areas that did not fit on a single screen as annoying. He then noticed that once a player filled up a horizontal line in the play field it became redundant and blocked access to empty space below.

"It just took up space and was doing nothing, so I decided to get rid of them and give more space for the game to continue," said Pajitnov, who also dropped the two-player mode in favour of a single-player experience. Now whenever a horizontal line was filled it vanished allowing the game to continue indefinitely provided the player kept completing horizontal lines to open up space in the play area. And since the 'pent' of Pentominoes referred to the Greek word for five, Pajitnov called his game *Tetris* after tetra, the Greek word for four.

The result was a captivating battle to keep the play area clear of pieces. "As soon as the finished prototype started working, I couldn't stop playing. I understood the game was very addictive. I realised it was something special," said Pajitnov. Pajitnov was not the only one entranced by his compelling creation, his colleagues were also hooked. So when the academy got its first IBM-compatible PC, shortly after the creation of *Tetris*, Pajitnov's co-worker Vadim Gerasimov rewrote it on the new computer, adding proper graphics and a score counter. "As soon as this conversion was done, we gave it to our friends and it self-distributed itself very, very quickly. Like a virus," said Pajitnov. *Tetris* spread like wildfire across the computer-equipped offices of Moscow,

1. Pajitnov also reduced the number of different shapes from Pentominoes' 12 to just seven.

infuriating managers who watched their workforce spending their time playing Pajitnov's game rather than being productive. Copies of the game began to seep beyond the city limits and spread to offices across the USSR and then into the communist nations of Eastern Europe.

Tetris was clearly something special, but trying to sell it was out of the question. In 1984 Mikhail Gorbachev had yet to become the leader of the USSR and his perestroika reforms, a programme of economic liberalisation that allowed Soviet citizens to form their own businesses and would help end communist rule, were still three years away. Soviet law strictly prohibited the formation of businesses and rejected the concept of copyrights or intellectual property – nothing could be done for personal gain, everything was owned by the state.

But even if the legal system had allowed Pajitnov to sell *Tetris*, there was no one to buy it. While cheap mass-market computers could be found in homes in many industrialised democracies, such technology was only accessible to a small cabal of computer researchers in the USSR and the Warsaw Pact nations of Eastern Europe. The few people who did have home computers usually obtained them at great expense and personal risk on the black market. The Hungarian authorities, for example, estimated that there were 30,000 home computers in Hungary by early 1985, 90 per cent of which arrived through unofficial channels. Czechoslovakian Marek Spanel got his introduction to computers thanks to the black market. His first computer, a Texas Instruments TI-99/4, was smuggled into the country and bought by his parents in 1985 – six years after the computer's US launch and four years after it went out of production. "My brother Ondrej and I were probably the only users of that type of computer in the entire country at the time," said Spanel, who went on to form Prague-based game developers Bohemia Interactive.

There were no shops to buy games from, so people had to rely on illegal copies of Western games shared among a clandestine network of home computer owners or, like Spanel, learn how to make their own. "In the mid-'80s computers were rare and people wrote games for fun and distributed them for free as selling them was forbidden," recalled Serge Orlovsky, president of Nival Interactive, a Moscow-based game development studio that formed in 1996. While home computers were rare, there were a few coin-op video games that had started to appear in the arcades of communist countries around the time that Pajitnov created *Tetris*. East Germany's *Poly Play* was one of the earliest and is believed to have first appeared in 1985[2]. *Poly Play* was

---

2. No official records confirming exactly when the *Poly Play* machine launched appear to have survived the fall of communism in East Germany.

created by VEB Polytechnik, a state-owned electronics company based in Chemnitz, which was then known as Karl-Marx-Stadt. Encased in functional wood panelling, it looked more bookcase than an arcade game and offered players a choice of 10 games including slalom skiing game *Abfahrtslauf*, deer hunting title *Hirschjagd* and *Hase und Wolf* – a maze chase featuring the wolf and rabbit characters from the popular Soviet TV cartoon *Nu, pogodi!*. *Poly Play* was mainly seen in youth hostels and the exclusive holiday homes run by East Germany's powerful trade union federation, the Freier Deutscher Gewerkschafsbund. It was often set up so it was free to play. *Poly Play*'s technology, however, was years behind that of the arcade games available across the border in West Germany. Its visuals looked closer to the first colour arcade games that appeared in the West in 1979.

The handful of arcade video games created in the USSR also reflected the technology gap with primitive offerings such as the overhead-view racing game *Magistral*.[3] Soviet coin-op games were also expensive, costing as much as 15 kopeks a go, enough to buy a small meal. The ideology of communism tended to influence every aspect of life behind the Iron Curtain and these Soviet arcade games were no exception. The operators' manuals made it clear that these machines were not just entertainment, defining their purpose as "entertainment and active leisure, as well as the development of visual-estimation abilities". Unlike US, Japanese and Western European games, there were no high scores, largely because such a feature would have been seen as encouraging a culture of competition rather than co-operation between players. Games involving the shooting of people rather than targets or animals were also shunned by the authorities.

When perestroika came into effect, the Eastern Bloc's arcades began to catch up. Towards the end of the 1980s, the Vinnytsia-based Extrema-Ukraine teamed up with Ukrainian manufacturer Terminal to produce more advanced arcade games using its custom hardware. The most notable game these companies created was 1988's *Konek-Gorbunok*, a platform game that echoed Activision's VCS game *Pitfall!* and was based on Pyotr Pavlovich Yershov's 1834 fairytale of the same name. Yershov's fairytale was a satire of pre-communist Russia's feudal government that told of an honest peasant boy who marries a princess after meeting the unreasonable demands of the evil tsar with the help of magical horse. The Russian government banned it for 20 years for making the tsar look foolish, a decision that turned it into a favourite of Soviet communists. Extrema-Ukraine's involvement in games was, however, short lived: it

---

3. Again there is no clarity about the exact year of release, although it is certainly some point in the 1980s prior to the collapse of communism in Eastern Europe.

abandoned arcade games in 1991 to concentrate on producing gambling and strip poker machines instead.

With no prospect of selling the game commercially, Pajitnov was happy to give *Tetris* away and get on with his artificial intelligence research. Then, out of the blue, he received a Telex message from a London businessman called Robert Stein.

Stein was the founder of Andromeda, a company specialising in selling Hungarian-made software to western publishers. Stein grew up in Hungary but fled the country aged 22 in the wake of the failed 1956 uprising against communist rule. He eventually claimed asylum in the UK. After spending the 1970s selling pocket calculators and digital watches he realised there was money to make in home computer software and formed Andromeda.

While most of the Eastern Bloc was strictly opposed to capitalism in any form, the 1956 revolt had prompted Hungarian leader János Kádár to try and quell discontent with limited economic and social reforms that allowed people to form businesses that exported goods to the West and granted some additional human rights. Kádár's fusion of communism, free enterprise and human rights was nicknamed Goulash Communism by Soviet leader Nikita Khrushchev in reference to the Hungarian national dish. By the time Stein formed Andromeda in 1982, Kádár's reforms had already allowed Ernö Rubik to export his Rubik's Cube puzzle toy to the West, where it became a phenomenon and a pop culture icon of the '80s. Stein saw Hungary's willingness to trade as an opportunity to access a supply of cheap software to sell on to western publishers. He teamed up with two Budapest companies – the state-funded Novotrade and a small private business called Mikromatix – and started bringing their creations to the West.[4]

"Communism was pretty soft by that time, though the really visible falling apart was a few years later," said Pál Balog, the Novotrade programmer who worked on 1984's *Bird Mother*, a Commodore 64 game where players played a bird raising a family of chicks. "Business was not forbidden with the West, it was just restricted by several factors. Trade with a net income of hard currency was quite welcome and what Novotrade did clearly fit that description."

Stein sold Novotrade and Mikromatix's games to a number of UK publishers. Media mogul Robert Maxwell's game company Mirrorsoft chose Mikromatix's mouse-catching game *Caesar the Cat* to be one of its earliest releases and *Eureka!*, a text adventure designed by *Fighting Fantasy* author Ian Livingstone and programmed by

---

4. Novotrade was headed by Gábor Rényi, the son of Peter Rényi, who was the deputy editor of the Hungarian Socialist Workers' Party newspaper *Népszabadság*.

Novotrade, became the debut release of UK publisher Domark. As well as Hungary's embryonic game industry, Stein had links with a Budapest computer science research centre called SZKI.

In early 1986 he paid a visit to SZKI to see if there was anything new he could take back to the UK. While there he saw someone playing *Tetris*. Stein was no fan of video games but he too found Pajitnov's game a compulsive experience and asked to buy the rights to it. SZKI gave him Pajitnov's details and, on his return to London, Stein sent the Russian computer scientist a Telex message in English expressing his interest in buying the rights. "My English wasn't very good," said Pajitnov. "He sent the proposal of the licence and I tried to answer him that I'm very glad he is interested and I like his proposal and we need to start negotiation, which I assumed was a necessary stage for any agreement. But somehow he interprets my answer as an agreement to his terms."

Taking Pajitnov's response as a yes, Stein began selling the rights to *Tetris* to game publishers across the world. First in line were Mirrorsoft and its US sister company Spectrum Holobyte which bought the rights to the computer versions of Pajitnov's game. The Maxwell-owned firms were keen to push *Tetris* as the first game from behind the Iron Curtain, conveniently ignoring the Hungarian titles that had already made that journey. To emphasise the Soviet connection they packaged the game in a red box adorned with a illustration of St Basil's Cathedral, a hammer and sickle, and a logo that spelt *Tetris* with a backwards r.

And when Stein indicated he could also get the home console and coin-op game rights, Mirrorsoft bought them and promptly sold them on. Soon five companies spread across the world were developing versions of *Tetris*. All thought they owned the rights, but Stein had yet to get a contract signed. By December 1987 Stein was getting worried but could not get an answer to his enquiries out of Moscow. In January 1988, with the rights situation still murky, Mirrorsoft and Spectrum Holobyte launched their computer version of *Tetris*. Pajitnov's hypnotic game became a sensation made all the more exciting by the exotic novelty of being a video game that originated in the Soviet Union. Around 100,000 copies were sold in 1988 alone.

Back in Moscow, Gorbachev's perestroika reforms were starting to take hold and Pajitnov had decided to see if his latest creation, a psychological assessment program called *Biographer*, could be sold commercially. He arranged a meeting with Electronorgtechnica, a recently formed government agency responsible for software exports that was known as Elorg for short.

The meeting was going well until Pajitnov mentioned that *Tetris* was on sale in the West. Suddenly the atmosphere turned frosty. The agency's director Alexander

Alexinko informed Pajitnov that only Elorg had the authority to make such deals. Elorg immediately set out to regain control of the situation. The agency contacted Stein and told him the deal was off. Stein's panicked objections eventually managed to persuade Elorg not to cancel the home computer rights deal, but Stein knew that getting the console and arcade rights he had already sold was now going to be tricky.

To add to the pressure on Stein, an American called Henk Rogers who ran the Japan-based game publisher Bullet-Proof Software was nagging him for the handheld rights. Rogers had already bought the Japanese computer, console and arcade rights for *Tetris*. He had sold the coin-op rights to Sega and was making a fortune with his Famicom version of Pajitnov's game. Rogers now hoped to secure the handheld rights for *Tetris* as he thought it would be perfect for the Game Boy, Nintendo's soon-to-be-released portable games console.

The Game Boy was the latest invention of Gunpei Yokoi, the Nintendo engineer who designed the company's Game & Watch line of handheld games. The Game Boy was pure Yokoi, adhering to his design philosophy of the 'lateral thinking of withered technology'. While Nintendo's competitors focused on engineering flashy handheld consoles with colour graphics and impressive sound capabilities such as the Atari Lynx and Sega's Game Gear, Yokoi opted for a monochrome screen and a tinny speaker. But what the Game Boy lacked in hi-tech oomph it made up for with its unrivalled 10-hour battery life and lower retail price. Nintendo had planned to make *Super Mario Land*, a Game Boy spin-off of its popular *Super Mario Bros* series, the lead game for the machine when Rogers suggested *Tetris* as an alternative.

"The Game Boy was just getting started and I thought *Tetris* was the perfect product for it," said Rogers. "I spoke to Nintendo of America's president Minoru Arakawa and said 'I can get the perfect game for you to pack in with the Game Boy'. I had already released it in Japan and the game was starting to make waves within Nintendo. He said: 'Well, we're going to pack in Mario'. I said: 'Well if you want little boys to buy your machine then pack in Mario, but if you want everybody to play your game, pack in *Tetris*'. I convinced him."

With Nintendo on side, Rogers contacted Mirrorsoft and asked to buy the rights. Mirrorsoft, however, was evasive so Rogers decided to skip the middleman and went straight to Stein, who said he would strike a deal with the Russians on his behalf. Weeks passed with no word from Stein, who was desperately trying to get Elorg to sell him the rights he had already sold on. Eventually Stein informed an increasingly suspicious Rogers that he was going to Moscow and suggested he joined him there in a week's time. Rogers decided it was time to bypass Stein entirely and deal with Elorg directly. "Two days later I was on a plane to Moscow," said Rogers.

Rogers and Stein were not the only people in Moscow hoping to strike a deal with Elorg that week. Unknown to Rogers and Stein, Kevin Maxwell, the son of Robert and the ultimate boss of Mirrorsoft, had also decided to deal with the Russians directly rather than wait for Stein to confirm the deal.

Rogers arrived in a cold, drab and paranoid Moscow with no idea where to find Elorg and no prior appointment with the agency. "It was the tail end, but it was still communism," he recalled. "Moscow was like being in a prison. No colour, no advertising. People didn't seem to smile. I tried to win people over with my personality and smile but people thought I was crazy. The first night I couldn't get any food. I'd ask and they said sorry the restaurant's just closed. I said ok, where can I get food? They'd go I don't know. I went to the hotel room, turned on the TV and sparks flew out. I had to unplug it from the wall to get it to stop. Maybe 12 little squares of toilet tissue in the bathroom. Everything about it was strange."Rogers spent two days hunting for Elorg's offices only to find, after hiring an interpreter, it was just around the corner from his hotel. On the 21st February 1989, the day before Maxwell and Stein's appointments, Rogers strolled into the Elorg offices.

"My interpreter wouldn't come in with me because she hadn't been invited. In their mind if you're not invited you cannot go in – it's unthinkable," said Rogers. The first person he encountered was Elorg's new director Evgeni Nikolaevich Belikov, who joined the agency after working at the Communist Party's headquarters. "I didn't know he was the important guy and said I'd like to talk to someone tomorrow about the rights to *Tetris*," Rogers said. Seeing Rogers' unexpected appearance as further leverage to use in his talks with Maxwell and Stein, Belikov arranged a meeting with Rogers. The next day all three would convene on the Elorg headquarters hoping to leave with the rights to Pajitnov's lucrative game. Rogers was first through the door. He was shown into Elorg's large conference room, a gigantic space with 15-foot high ceilings, long drapes hanging off the walls and an enormous table. "It was like a mansion they had converted into a ministry," said Rogers. "Some of the meetings we had were in dining rooms that would fit comfortably 40 people on these long tables. They sat on opposite sides on this long table."

Amid the army of Elorg bureaucrats at the table was Pajitnov. Rogers explained he wanted the handheld rights and then showed them one of the *Tetris* Famicom cartridges his company was selling in Japan. Belikov said no console rights had been granted and demanded to know what Rogers was doing selling their game without permission. Suddenly Rogers realised Stein did not have the rights in the first place and envisaged his whole business imploding: "I was thinking 'oh my god' because I had 200,000 cartridges in production and that involved hawking all the money from my in-laws. Everything was on the line and here I am in another country with someone telling me I don't have the rights."

Rogers explained the web of *Tetris* deals to Belikov, who then showed him the contract agreed with Stein that gave him only the rights to the home computer versions. "It was a horrible contract," said Rogers. "It was like spanking a baby or something like that. Elorg was getting a percentage of a percentage of a percentage, so by the time it got to Moscow it was nothing."

Rogers sensed an opportunity and told Belikov he wanted to put things right. He handed Elorg a $40,712 royalty cheque for the 130,000 *Tetris* cartridges already sold in Japan and said he wanted to buy the console rights. Expecting legal action from Mirrorsoft and Atari Games if he got the deal, Rogers asked Belikov to hold off from making any deals until he returned with backing from Nintendo.

While Rogers got on with bringing in Nintendo, Stein faced an angry reception at Elorg. Belikov shoved an amended version of the home computer contract under his nose and told him if he did not sign it immediately, the arcade rights deal was off. Desperate to get the coin-op deal, Stein homed in on a clause about fines for late royalty payments and negotiated on that. After signing it, Belikov sold him the coin-op rights. Only later did Stein notice the crucial change that the late payments clause was designed to distract him from: a clause defining a home computer in such a way that it was completely clear he did not have the home console rights.

Maxwell, meanwhile, was hoping to use the influence of the Maxwell Communications empire to win all the rights to *Tetris*. But when Belikov handed him Rogers' Japanese *Tetris* cartridge and demanded to know what it was, Maxwell stumbled. Unfamiliar with the details of Mirrorsoft's dealings, Maxwell said it must be an illegal copy. Eventually Elorg offered Mirrorsoft the option to bid for home console rights in return for the rights to publish Maxwell Communications' reference books in the USSR.

Keen to get the lucrative *Tetris* rights, Maxwell agreed and signed over the reference book rights. And when Rogers and Nintendo returned with their offer for the console rights, Belikov gave Mirrorsoft 24 hours to make a better offer. Mirrorsoft couldn't respond fast enough and the next day Nintendo and Rogers walked away with the deal. Atari Games, which thought it had the US home console rights and had just released its NES version of *Tetris*, suddenly found itself served with a court injunction and left with a warehouse full of games it was barred from selling much to Nintendo's delight.[5]

Pajitnov's game proved crucial to the success of the Game Boy. It became the must-have game that drove the handheld's massive sales across the world. More than 40 million Game Boys with copies of *Tetris* were sold worldwide. It made Rogers rich

5. Atari Games had refused to sign up to Nintendo's NES licensee system and the two firms were engaged in a number of tit-for-tat legal spats about Nintendo's control of the NES market. Atari Games wasted millions promoting, developing and manufacturing its version of *Tetris*, which was on sale for about a month.

and let Nintendo gain a level of dominance in the handheld games market that was even greater than its hold on the home console business.

Maxwell and his father were furious when he discovered how Rogers and Nintendo had grabbed the rights they thought were theirs. The Maxwells were well connected with the Gorbachev regime and decided to use political influence to try and scupper Nintendo's deal. "They put pressure on the Politburo. That pressure came down on the ministry and Belikov, who had made the decision to go with me rather than Kevin Maxwell," said Rogers. "Belikov thought he was a dead man. Nobody talked to him. It was like he had a disease or something. He was like an untouchable."

Government officials turned up at Belikov's offices unannounced and rifled through Elorg's files searching for evidence to use against him. The Elorg director feared he was under KGB surveillance. As a former Communist Party official he knew full well what he was up against: "I was afraid because I understood that it was like trying to stop a tank with your hands. They had switched on that soulless state machine, which is totally uninterested in any reasons."

Luckily for Belikov, communism was fading fast and Gorbachev's reforms were already changing attitudes in Moscow. Eventually the Kremlin decided that ministries such as Elorg should be free to make decisions independently and the soulless state machine was switched off.

For Pajitnov, the success of *Tetris* brought him little wealth, but it did give him and his family a new future. In 1991 he moved to Seattle to open his own video game company and later joined Microsoft to make puzzle games such as *Hexic*. And when Elorg's rights to *Tetris* expired in 1996 he teamed up with Rogers to form The Tetris Company, a business dedicated to managing the global rights to Pajitnov's game and exploiting its on-going appeal. In particular *Tetris* has reaped the benefits of the growth of mobile phone games, selling more than a 100 million copies on myriad models. "It's amazing and very strange for me. I didn't expect it to live that long," said Pajitnov. "But then the game did not change too much and the human brain remained the same, so I don't see any reason why the game should be less popular."

By the time Pajitnov left Moscow for Seattle, the communist regimes of Eastern Europe had come crashing down; their demise hastened by Gorbachev's economic and political reforms. Although these countries were now free of communism, its legacy proved hard to shake off. It was a legacy that would have profound implications for video game development in the former Eastern Bloc. Decades of communist rule had left the economies of the Eastern Europe in tatters and these nations spent the 1990s making an often-traumatic metamorphosis into free market economies. Those that tried to make a living creating games faced not only a hostile economic climate, but also rampant piracy and a lack of home computer owners to sell to. In communist times piracy was the only way computer owners could get hold of games, but in the new capitalist world this cultural inheritance only undermined efforts to build a video game industry. "There was nothing in the way of software on sale under communism and 99 per cent of games were pirated," said Spanel, who formed Czech game company Bohemia Interactive in 1999. "It is still a big problem and there is still the mentality that games are free."

The economic problems facing Eastern Europe and the high levels of piracy also dissuaded western game companies from investing in the region. "Our countries were seen as part of the socialist bloc and, after the Iron Curtain fell, time had to pass before people realised the bad image was no longer true," said Gábor Fehér, managing director of Hungarian game studio Digital Reality. Console manufacturers made no attempt to expand into the new democracies, figuring that the impoverished population could not afford to buy expensive luxuries such as video game consoles. As a result, Eastern Europeans saw little of the video game consoles that were common in Japan and the West beyond a few shops where people could play Sega Megadrive and Super Nintendo games on a pay-per-minute basis. The only consoles on sale were illegal clones such Steepler Company's Dendy, a Russian copy of Nintendo's Famicom. The absence of consoles meant home computers became the de facto game platform for the former communist nations of Eastern Europeans. At first low-cost ZX Spectrum clones that ran an operating system called TR-DOS proved popular, but by the mid-1990s people began switching over to PCs as price reductions and increased wages made these affordable.

These factors, along with a lack of people with the skills needed to make video games, slowed the growth of Eastern European game studios to a crawl. It would take until the late 1990s before a games industry of any significance formed. Only a few of the video games created during the immediate post-communism period stood out. *Filler*, a 1991 abstract colour-matching game created by Russian programmer Dmitry Pashkov, was one of the few to reach an international audience after French publisher Infogrames released it as *7 Colours* in Western Europe.

But as home computer sales in the former communist bloc increased towards the end of the 1990s, more and more game studios began to form. These ambitious new companies wanted to make their mark on the international stage, not just in their homelands. They wanted to show the world that the former communist bloc had more to offer than *Tetris*. Their emergence coincided with a change in attitude towards Eastern Europe among western game publishers. Although the growing sales of games in the region helped change attitudes, saving money was the primary motive.

By the turn of the century, the cost of creating video games was soaring into the millions and big-league game publishers hoped Eastern Europe's pool of cheap talent could help cut costs. The interest from western game firms further fuelled the formation of Eastern European game studios as more and more amateur programmers went professional hoping to make it big in the video game industry.

And when Czech game studio Illusion Softworks' military action game *Hidden & Dangerous* became a million-seller in 1999, any lingering doubts about Eastern European studios' ability to deliver vanished. Illusion Softworks' breakthrough was followed by a succession of hit games from Eastern Europe including Ukrainian strategy game *Cossacks: European Wars* and Bohemia's intense military combat simulation *Operation Flashpoint: Cold War Crisis*. But while these games appealed globally, the legacy of communism still lingered within the subject matter of the games forged in Eastern Europe and Russia.

Operation Flashpoint drew on designer Spanel's experience of communism. It cast the player as a frontline soldier who is part of a NATO attempt to counter a fictional military rebellion against Gorbachev shortly after his rise to power in 1985. "We've experienced communism from the inside and now live outside it and so we know what both are like and this influenced the story of *Operation Flashpoint*," said Spanel. Unlike most military games, *Operation Flashpoint* showed warfare without the glamour of Hollywood heroism. The full-on assaults that typified other military games would almost always result in death in *Operation Flashpoint*; planning, discipline and patience were a necessity. Spanel's game also left players in the dark about the overall state of the conflict, instead there were simply orders to follow and need-to-know information provided from on high. It was war at its most unheroic, paranoid, confusing and dangerous. There was no room for Rambos in this game.

Spanel was not alone in drawing on his nation's history for inspiration. *IL-2 Sturmovik*, a flight simulation designed by Russian game studio 1C: Maddox Games, recreated the air battles between Soviet and Nazi forces in the Second World War that are usually overlooked by western flight sims. Ukrainian studio GSC Game World, meanwhile, looked to the fenced-off zone surrounding the exploded Chernobyl nuclear reactor that lies 110 kilometres from their Kiev offices. They used this unpopulated zone and the ghostly abandoned city of Pripyat, which lies within the area, as the template for the irradiated world of its 2007 first-person shooter *S.T.A.L.K.E.R.: Shadow of Chernobyl*.

By the mid-2000s, Eastern European studios had completed their journey from communism to globally successful video game creators. But while the move to making games for a global audience resulted in British and French game designers playing down their national influences, Eastern Europe's games made few concessions to US audiences. Having fought for years to free themselves from communism, Eastern Europe's game developers were not about to hide their cultural distinctiveness. That desire to ensure Eastern European culture was not extinguished was the real legacy of communist attempts to erase the national identities of the countries once under its control.

Sonic boom: Sega's blue hedgehog captivated millions
Pitchal Frederic / Corbis Sygma

CHAPTER 17

Sega Does What Nintendon't

Sega's prospects looked bleak. The arcade game giant's attempts to challenge Nintendo with its Master System console had ended in disaster. And now its latest home console, the Megadrive, had bombed in Japan.

When it launched in October 1988, Sega was hopeful that the Megadrive, which was based on the technology used in the company's coin-op games, would make serious inroads into Nintendo's dominance of the Japanese market. Despite boasting conversions of popular Sega arcade titles such as the fighting game *Golden Axe* and the second instalment of Yuji Naka's role-playing game series *Phantasy Star*, the Japanese had snubbed the Megadrive. Embarrassingly even NEC's PC Engine, a NES rival launched in 1987, was outselling Sega's flashy new system.[1]

Confident that its superior hardware would win out in the end, Sega shrugged off Japan's apathy towards its new console and pressed ahead with the 1989 launch of the Megadrive in North America, where the system would be called the Genesis. Few analysts believed Sega could succeed in North America. NEC, despite mild success in Japan, had come unstuck in the US when it attempted to challenge Nintendo's NES with the TurboGrafx-16 console, the North American version of the PC Engine. The TurboGrafx-16 sold so little in the US that NEC abandoned its plans to bring the console to Europe. NEC's low sales were compounded by the lack of games being developed for the TurboGrafx-16 outside Japan. As a result, many of the games on NEC's console were titles designed for a Japanese audience and were ill-suited to the American market, such as *Kato-chan and Ken-chan*, a *Super Mario Bros*-esque romp starring a popular Japanese comedy duo who farted, defecated and urinated their way through the game.[2]

Sega, at least, had its popular coin-op games to fall back on, but getting other

---

1. Sega had been so dismissive of the PC Engine that it even allowed its arcade shoot 'em ups *Space Harrier* and *Fantasy Zone* to be remade on NEC's console.
2. The US version – released as *J.J. and Jeff* – removed the game's toilet humour.

companies to go against Nintendo and back the Genesis was an uphill struggle. Even a licensee deal offering more favourable terms failed to persuade big-name publishers to support the Genesis. Michael Katz, the head of Sega of America whose primary goal was to make the Genesis popular in the US, had faced similar problems before.

Prior to joining Sega, Katz had been in charge of marketing Atari Corporation's 7800 ProSystem, its unsuccessful attempt to challenge the NES. "Atari couldn't get the hot arcade titles," said Katz. To try and plug the gap, Katz arranged for the most popular home computer game titles to be converted to the 7800 ProSystem. The approach failed. Katz decided to try a different approach for the Genesis. "I thought we should go after personality licences, especially in sports," said Katz. "It was very hard to get support from third-party publishers for the Genesis. That's one of the reasons we needed strong personality licences, because then we could woo guys into doing a football or baseball or basketball game because they knew that if we had a good personality licence attached that would give it good volume and they would know we were going to put good amounts of TV behind it."

First on Katz's shopping list was Joe Montana, the quarterback for the San Francisco 49ers and one of the most valuable American football players at the time. Sega paid $1.7 million for the rights to create American football video games bearing the sports star's name for the next five years and hired Mediagenic, the company formerly known as Activision, to create its *Joe Montana Football* game. It was to be Sega's flagship Christmas game for the Genesis. "Sega needed a football game and fast. We saw the beginning of what looked like one at Mediagenic so I asked if they could do it for us and we paid them a lot of money," said Katz.

With *Joe Montana Football* under way, Katz formulated a marketing plan designed to take Nintendo head on. He decided to position the Genesis as a console for teenage boys, figuring that the children who grew up playing cheery and cute Nintendo games would want something more edgy now they were entering puberty. The Genesis would, he decided, be pitched as the console Nintendo owners "graduated" to, an argument Sega's line-up of arcade hits and sports games was well placed to reinforce. Katz then decided to ram the message home with an advertising campaign that attacked Nintendo directly. "The Japanese would never do competitive commercials," said Katz. "They thought they were in bad taste in terms of business ethics, but we convinced them that was what we needed since we were against Nintendo. So it became 'Sega does what Nintendon't'."

But within a month of the Genesis' August 1989 launch, Sega discovered its crucial Christmas game was in trouble. "Each month we were checking on the game, but we weren't doing a very good job of checking it and/or we were being deceived," said

Katz. "Mediagenic weren't nearly as far ahead on the game as we thought they were. We found out in September that the game wasn't going to be finished. I was in a bind. I owed Joe Montana $1.7 million and we were counting on a game for Christmas."

With time running out, Katz turned to Electronic Arts for help. Since its formation back in 1982, Electronic Arts founder Trip Hawkins had aspired to make the company a leading producer of sports games. "My personal desire to make authentic sports simulations was the primary reason that I founded Electronic Arts in the first place," said Hawkins, whose interest in sports games stemmed from his love of Strat-O-Matic's dice-based sports games.[3] "I would watch games on TV and then want to go outside and run around and pretend to be my sports heroes. Then I wanted to be them in Strat-O-Matic, but I couldn't get that many of my friends excited about it because it was too complicated. When I saw my first computer I realised I could put all the computation and administrative stuff in the computer and just put nice graphics on a TV screen. I figured the more we made it look like TV the more people would be able to relate to it."

Sports games had come a long way since the early 1970s, when text-only statistical sims and *Pong* clones disguised as sports titles ruled. By the early 1980s, advances in graphics technology had encouraged game designers to explore new ways of representing sport in video games. Some experimented with views from the stands, such as Texas Instruments' *Indoor Soccer*, or bird's eye views of the pitch as in *Atari Football*, but more game designers took their inspiration from TV sports coverage.

Don Daglow and Eddie Dombrower were among the first to follow the example of TV with their 1983 Intellivision game *World Series Baseball*, which fused the statistical backbone of Daglow's text-only *Baseball* with action viewed from TV-inspired angles.

"Sports simulations started with no graphics, so we started to get the maths and the simulation part right first because that was all we had," said Daglow. "Then game designers started integrating graphics and had to explore the trade-off between mathematical accuracy and graphical display. On *World Series Baseball* we started imitating TV coverage. That game came out of wanting to mimic the way television covered baseball and watching a baseball game one day and realising I knew how to make the Intellivision do that."

The viewpoints used in TV coverage had significant advantages, said Daglow: "TV coverage has always experimented with trying to find the best camera angle that

3. Strat-O-Matic's sports games first appeared in 1961 and used a combination of dice rolls and player statistics to simulate sports matches.

gives you the best close up, but still lets you follow the action, because the players are recognisable rather than specks in the field. It's the best trade off between showing the action and portraying human beings."

As the 1980s progressed, game designers continued to explore the fusion of mathematical simulation and TV presentation pioneered by Daglow and Dombrower. And, as game makers sought to increase the realism of their sports games, they began to include real-life sports stars, starting with Electronic Arts' 1983 basketball game *Julius Erving and Larry Bird Go One-on-One*, and more managerial elements such as training your virtual sportspeople.[4] These ideas and more came together in Daglow and Dombrower's *Earl Weaver Baseball*. As well as bearing the name of and design input from the former Baltimore Orioles manager, the 1987 Amiga game simulated baseball in incredible statistical detail while pushing TV-style coverage to new heights with slow-motion instant replays and computerised commentary. *"Earl Weaver* was a case where machines had become more powerful," said Daglow. "It was originally conceived for the Amiga and we now had the power to do split screen so we could show the batter and the pitcher on one side of the screen and the field on the other, so you could actually see the players in detail."

The leap in visual capabilities also meant Daglow and Dombrower had to pay attention to the skin colour of the baseball players. "When you had four colours to choose from everybody is going to look the same and no-one's going to think anything about it," said Daglow. "But when you get to the point when you've got that many colours and that big a human figure, you can't have an African-American pitcher and a white pitcher look the same. At the time I was concerned because I felt anything else was going to be disrespectful and that there could be negative feedback. Ironically, we ended up getting tremendous support from the community precisely for having acknowledged race in a game. At the time I felt we were taking chances, but I ended up feeling very proud of the responses we got."

The following year Cinemaware took the union of TV coverage with sports games to its logical conclusion with *TV Sports Football*, a title that offered all the razzmatazz associated with broadcasts of American football matches. "I saw that people related to sports through television and that the way to do it was to emulate the TV broadcasts," said Cinemaware founder Bob Jacob. "We had scores, the half-time show, we had marching bands, adverts. We had everything."

In the same year that *TV Sports Football* came out, Hawkins tried once again to

4. Sports game specialist GameStar's 1985 game *Barry McGuigan World Championship Boxing* (*Star Rank Boxing* in the US) was one of the earliest titles to include training sessions that allowed players to enhance the abilities of their virtual boxer in a manner that echoed the character development approach of role-playing games.

realise his dream of making the best sports video games with *John Madden Football*, an American football title endorsed by the former Oakland Raiders coach turned sports commentator. Electronic Arts' first effort, created on the Apple II, was a dud. Compared to the showbiz trappings of *TV Sports Football* and the depth of *Earl Weaver Baseball* it was an underwhelming effort that only offered full-size teams because Madden himself intervened when he discovered that the developers planned to make it a six- or seven-a-side game. "The first version was exceptionally crude," admitted Roger Hector, the Electronic Arts producer who oversaw the game once Hawkins decided to concentrate on other projects. Scott Orr, the founder of GameStar, felt it was dull. "It emphasised strategy over action," he said. "Unfortunately, the graphics – even by Apple II standards – were unimpressive and the game play was boring. Not surprisingly sales were disappointing too."

"Madden became known around Electronic Arts as 'Trip's folly'," said Hawkins. "Everyone other than me thought the project should be killed. The accountants insisted that all the money, including Madden's advance, be written off as unrecoupable. But I am a determined fellow and eventually got it right."

At Hawkins' insistence Electronic Arts tried again with *John Madden Football*. To help reboot the series it handed control of the project over to Rich Hilleman, who had just finished producing the company's 3D racing simulation *Indianapolis 500: The Simulation*. Hilleman immediately brought in veteran sports game designer Orr, who had sold GameStar to Activision in 1988, as the lead designer. While Hilleman and Orr set about rebooting the game, Sega's Katz called Hawkins pleading for help. "I called him up and asked him did he have any back up Madden football games we could use under the Montana name," said Katz.

Up to that point Electronic Arts had focused on home computers. It distrusted Nintendo's dominance of the console market and had made minimal effort to expand into the NES game business. "Part of Trip Hawkins' original founding vision was that the future of gaming would be on PCs, not consoles," said Hector. "At the time I joined Electronic Arts, they were committed to this strategy and to speak otherwise was heresy." But the Sega Genesis had caught Hawkins' attention. "When I heard that Sega would introduce a $189 console based on the 68000 microprocessor in late 1988, it was a revelation," he said. "Electronic Arts had significant experience and software assets for the 68000, which had been in the Lisa, Macintosh, Amiga, Atari ST and also in many coin-op arcade systems. I was also excited that Sega might give us an alternative to Nintendo."

Sega was in trouble and Hawkins, knowing he had the upper hand in negotiations, pushed for a licence to make Genesis games on his terms. "Since he was doing us a favour and it was critical he took the opportunity to say 'I'll do that but I want a break on the royalty that we have to pay you guys for each unit of Sega software that's sold'," said Katz, who readily agreed, seeing the advantage of having a major publisher like Electronic Arts supporting the Genesis. Despite using a stripped-down version of its still-in-development Madden game, Electronic Arts missed Sega's Christmas deadline and *Joe Montana Football* eventually came out in January 1990. But for Electronic Arts the deal had given it a chance to bring its new version of *John Madden Football* to the Sega Genesis.

Orr and Hilleman's reboot had changed the game significantly. "The Genesis and subsequent versions of Madden had nothing in common with the Apple II version other than John's name," said Orr. "Rich wanted an action game with realistic strategy. I remember several testy meetings between Rich and Trip regarding the direction of the game. Trip wanted a more computer-oriented feature set while Rich insisted on keeping our shared vision for a realistic arcade action game designed for console gamers. Fortunately, Rich prevailed."

Orr and Hilleman also sought to get Madden more involved in the game's design. Madden's input into the original title came mainly from discussions with Hawkins that took place on a two-day train journey, but now the team wanted to get more input from the football star. Madden's belief that American football games are won or lost on the results of individual confrontations on the field became crucial to Orr's thinking. "The biggest innovation was one-on-one confrontations between the on-field players," said Orr. "Another innovation was a significant use of individual player skill ratings to create more sophisticated dynamics and play results. The final innovation was the end zone's 3D look and pop-up windows showing close-up views of the receivers. Up to then most football games used 2D side-scrolling or top-down views."

The Genesis version of *John Madden Football* would become a defining moment in sports games – a shift away from the overt statistical modelling of *Earl Weaver Baseball* towards a more action-based experience where the mathematical elements subtly enhanced the action rather than dominated it. *John Madden Football*'s marriage of action and simulation became the template for the majority of sports titles that followed it.

By the time the Genesis version of *John Madden Football* was released in 1990, Katz's daring head-on assault on Nintendo seemed to be working. Half a million Genesis consoles had been sold in North America and Nintendo's new Super NES was still a year away from launch. But Sega knew it had to do better. "My mantra was to sell a million units. I was supposed to do this chant every morning: 'hyaku man dai' – one million – to represent the units we wanted to sell," said Katz. Relations between Katz and Sega's Japanese bosses were worsening. They wanted to know why Katz had not broken the million sales mark already and they felt he wasn't doing enough to keep them informed on his progress. "I wasn't on the phone every day telling them how great I had been doing since that wasn't me, which probably would have been good to do if I wanted to stay in the favour of the Japanese, but I do my own thing and people either like it or they don't. I only did half a 'hyaku man dai' and that wasn't good enough."

Sega's Japanese bosses decided to fire Katz. "The Japanese didn't really understand that you don't destroy a 90 per cent franchise, which is what Nintendo had, in one year. It takes a while," said Katz. Katz's replacement Tom Kalinske, a former president of Mattel, decided Sega needed to use the time it had before the arrival of the Super NES to secure its position in the market. He ramped up the anti-Nintendo adverts, slashed the price of the Genesis and ditched the Greek mythology-influenced *Altered Beast* as the game sold with the console because he feared its imagery would be equated with "devil worship" in the US's bible belt. To replace *Altered Beast* he picked *Sonic the Hedgehog*, the latest game from *Phantasy Star* creator Naka.

Sonic the Hedgehog was the result of an internal competition to create a character-led game that would provide Sega with a new mascot to replace Alex Kidd, the platform game hero that Sega touted as its answer to Mario during the late 1980s. Naka wanted his game to address his own dissatisfaction with Nintendo's *Super Mario Bros* series. "When I played Mario I was really frustrated that even though I became better at it, the first level took the same amount of time," he said. "I wanted that time to get shorter and my play to get faster as I got better at the game, so I came up with Sonic."

Naka made speed the overriding quality of his game and created worlds that were a cross between a pinball table and a rollercoaster ride. There were pinball bumpers to rebound off, rollercoaster-style loop-the-loops that allowed fast-moving players to travel upside down for brief moments and steep slopes for the game's character to slide or roll down. By comparison *Super Mario Bros* dawdled. "The way I see it, we tried our best to think like Sonic and come up with the kind of world he would like to run through," said Naka. The game also needed a distinctive character suited to the velocity of Naka's world. He initially decided on a rabbit. "Sonic was supposed to be a long-eared rabbit, but a rabbit needs to stop before they can bring down an enemy, so we went for the hedgehog character that can keep running while bringing down its foe with its spines," said Naka.

Sega wasn't too keen on Sonic being a hedgehog, primarily because few people in Japan or North America were familiar with the prickly nocturnal mammal. While Sega didn't pick Sonic as its mascot, it realised that Naka's exhilarating platform game was one of the best titles created for the Genesis. *Sonic the Hedgehog* arrived in the US in June 1991, a few months ahead of Nintendo's Super NES. It became an instant success, giving Genesis owners a Mario of their own. Genesis sales rocketed both in North America and Europe ensuring there was no way Nintendo could recapture the level of dominance it had enjoyed in the days of NES. "Nintendo introduced the Super NES too late and allowed Sega to get in," said Katz. "They let Sega develop a very strong sports line. I guess we were willing to take Nintendo on and spend the money necessary to do a decent launch for a period of 18 months. *Sonic* turned out to be lucky."

Nintendo had underestimated Sega. Many of the video game publishers who once refused to work with Sega for fear of upsetting Nintendo, now signed up to make games for both machines. Nintendo's total control of the home games business was over. But Nintendo was far from finished. It sold 4 million Super NES consoles within a year of its November 1990 Japanese launch and still had millions of loyal fans in the US. By the end of 1991 – just over three months after the Super NES reached North America – market analysts NPD Group estimated that the Super NES had already taken 45 per cent of the post-NES market compared to Sega's 55 per cent. NPD didn't even bother to measure how NEC's TurboGrafx-16 was doing. The next few years would see Nintendo and Sega engaged in a bitter battle for consumer attention, hoping to grab that extra slice of market share. Exclusive games were their weapons.

To challenge Nintendo on the exclusives front, Sega set up a new game studio in San Francisco called the Sega Technical Institute, which was tasked with developing games for the US market. "The Sega Technical Institute was a specially created isolated research and development centre, where the best Japanese development talent could work with American developers in an ideal environment that would foster creativity and produce internationally successful games," said Hector, who replaced former Atari coin-op game designer Mark Cerny as its general manager in 1992. The core members of the original *Sonic the Hedgehog* team, including Naka, were among those transplanted to West Coast of the US. "I believe that the influence of living in the US had a profound impact on the games and was a huge contributing factor to the success of the games, so much so that when I returned to Japan I wanted the creative team for the *Sonic* games to be based in the US and so we set up Sonic Team USA and continued to develop *Sonic* there," said Naka. "Having American children nearby and being able to observe their reaction to the game influenced the final product greatly." The studio produced many of Sega's flagship games during the first half of the 1990s including the sequels to *Sonic the Hedgehog* and *Comix Zone*, an action game that took place within the panels of a comic book brought to life.

Nintendo's arsenal of exclusives included a number of games based on its existing franchises including *The Legend of Zelda: A Link to the Past*, *Super Metroid*, *Super Mario World* and *Super Mario Kart*, a go-kart racing game for up to four players starring the company's best-known characters that spawned the kart-racing genre. Both companies also arranged exclusives with their licensees. Crucially, Nintendo's relationship with Japanese arcade game manufacturer Capcom gave it two of the arcades' most popular martial arts-based fighting games: *Final Fight* and *Street Fighter II*, both of which were designed by Yoshiki Okamoto.

Although fighting games can be traced back as far as 1976's Sega coin-op *Heavyweight Champ*, the genre really took off in 1984 with the arrival of the genre-defining *Karate Champ* and *Kung-Fu Master*. Both of these Japanese arcade games took their cues from the kinetic kung-fu movies produced in Hong Kong during the 1970s, such as Raymond Chow's 1971 film *The Big Boss*, which turned Bruce Lee into a international star. Irem's *Kung-Fu Master* envisaged the fighting games as a journey, challenging players to battle through hordes of attackers to reach an end goal. Data East's *Karate Champ*, meanwhile, drew on martial arts tradition, pitting players against a computer or second person in a bout of one-on-one combat. Released within weeks of each other, these two games provided the blueprint for almost every subsequent fighting game.

Final Fight was a descendant of *Kung-Fu Master*. Set in a New York-inspired city riddled with crime, it let up to three players join forces to punch, kick and bash their way through the streets in order to take out the city's crime lords. While *Final Fight* was popular, the groundbreaking *Street Fighter II* marked the most significant leap forward for fighting games since *Kung-Fu Master* and *Karate Champ*. At heart it was a *Karate Champ*-style one-on-one fighting contest with a range of fighters for players to choose from. But its big breakthrough was its use of powerful secret moves that players could unleash by moving the joystick and pressing buttons in the correct order. Okamoto also divided the types of attacks players could do into fast-but-weak and slow-but-strong moves to give the game a tactical edge where players needed to decide if a making a slower-but-more-powerful punch would leave them too exposed to a counter-attack. It was an approach that countless fighting games would follow. *Street Fighter II* became one of the biggest arcade successes since the collapse of Atari in the early 1980s, with more than 60,000 machines sold worldwide. Nintendo's exclusive deal with Capcom proved to be a major coup that prevented Sega's console from carrying two of the most popular arcade games of the early 1990s.

Sega's licensees responded with games such as *Ecco the Dolphin*, a 1992 game developed by Hungary's Novotrade where players controlled a dolphin in search of the rest of his pod, boutique Japanese developer Treasure's shoot 'em up *Gunstar Heroes* and Virgin Interactive's *McDonald's Global Gladiators*, a 1992 tie-in with the fast-food chain programmed by British game designer David Perry. "I got a call from the head of Virgin USA saying would you like to come to the US and help us get a game done in six months," said Perry. "He said whatever you're earning now we'll pay it with royalties on top, we'll give you a car and an apartment. How could I say no? I went out to California and the game was for McDonald's. We didn't do it the way McDonald's wanted it, we did it the way we wanted it. It got a game of the year award."

The clash between Nintendo and Sega, together with the more advanced technology of the new 16-bit consoles, raised consumers' expectations of video game quality and drove up the cost of game development. And as development costs grew, the game industry began to think more carefully about what types of games would sell, rather than giving game developers free reign. "Back in the 8-bit days, literally anything you thought of you'd just do it because there was such little cost – you could make a game for $3,000," said Perry. "But then, when you start to get into 16-bit, prices went up because the development was more expensive: there's more work to be done, more graphics to draw and that started to get people more serious about the whole thing, a bit more careful. You didn't quite go with the crazy ideas anymore, you were thinking about what's actually going to sell."

Perry's 1995 platform game *Earthworm Jim 2* reflected the changing nature of the games business with its ISO 9000 level, named after the international set of management standards. "Virgin was becoming very big and very powerful, and so they hired middle managers to take care of us," said Perry. "I was basically given a boss and he had little books on how to manage people and what he kept talking was about ISO 9000. I was quite enraged that I had to deal with this guy. He had no clue what he was doing there."

Perry's frustration with his new boss and the long hours he and the development team were spending creating *Earthworm Jim 2* eventually came to a head. "I would literally sleep in my car when I worked at Virgin because I was working so hard and then in came the middle manager asking me to detail out everything that was going to be done in the game," said Perry. "There was a watershed moment where I made up a whole load of rubbish; I just made up fake stuff that didn't make any sense. I presented that as the plan and because he didn't understand any of it anyway he was like 'ok, this sounds good'."

"I lost all my respect for him. We were rebelling against it, the whole let's make video game companies into corporations, into ISO 9000 corporations where you pretend to follow those standards." The team's run-ins with the middle manager turned into a level built out of mountains of folders and files where Earthworm Jim battled angry filing cabinets that spewed out paperwork and masked lawyers that jumped out when least expected.

Earthworm Jim 2's defiant level would do little to challenge the direction of travel, however. The video game industry, fuelled by the quality arms race between Sega and Nintendo, was growing up.

Under fire: Senator Joseph Lieberman wields a gun controller during the Senate inquiry into video game violence
AP Photo / John Duricka

CHAPTER 18

Mortal Kombat

On Wednesday 1st December 1994, the Washington press corps gathered for a news conference called by Joseph Lieberman, the Democrat Senator for Connecticut. Next to Lieberman on the stage was Bob Keeshan, aka Captain Kangaroo – the US's favourite Saturday morning kids' TV presenter.

Once the assorted reporters had taken their seats, they were shown footage from two of the latest video games to reach the market. In one scene a digital image of an actor playing a martial arts fighter ripped the still-beating heart out of his opponent's chest. The next scene showed film footage of a young woman in a nightdress being dragged off-camera by vampires before the sound of a high-pitched drill being used to extract the victim's blood is heard. The games in question were *Mortal Kombat* and *Night Trap*, and, like many adults in the US at the time, most of the people in the room were unaware of these games, let alone their gory content. "We're not talking *Pac-Man* or *Space Invaders* anymore," Lieberman told the stunned journalists. "We're talking about video games that glorify violence and teach children to enjoy inflicting the most gruesome forms of cruelty imaginable."

Captain Kangaroo told reporters he could "not believe anybody could go that far" and said the nation's children were being exposed to such material in the name of greed. "Violent video games may become the Cabbage Patch dolls of the 1993 holiday season," Lieberman added. "But Cabbage Patch dolls never oozed blood and kids weren't taught to rip off their heads." Video game makers were in the dock facing charges of corrupting the nation's children.

* * *

The industry had, of course, been here before. There were the bomb threats that followed the hit-and-run driving game *Death Race*, the outrage over the pornographic *Custer's Revenge* and the 1985 protests about the release of *Raid Over Moscow* in the UK.

"In *Raid Over Moscow* you had to break through and bomb the Kremlin," said Geoff Brown, founder of the game's UK publisher US Gold. "It was number one in the charts, hundreds of thousands of copies being sold to kids and CND took it upon themselves to picket our offices.[1] I thought it was fantastic. I mean how could it get any better? It was in every newspaper. They were there every day. We used to give them coffee and they used to walk around with banners like 'ban the bomb'. I kept saying you'll be better off if you didn't keep coming here, it would actually sell less."

But in late 1994 the game industry wasn't laughing off the outrage. There had been a growing sense that the video game industry would eventually end up facing the wrath of Washington's politicians and the early 1990s had already seen some minor skirmishes with lawmakers. In June 1990 a Democrat Party member of the California Assembly called Sally Tanner tabled a law to ban games that featured alcohol or tobacco. Among the games under threat of being outlawed was Roberta Williams' 1987 children's nursery rhyme-themed game *Mixed-Up Mother Goose*, which featured a pipe-smoking Old King Cole. The game industry managed to get the law dropped after pointing out that Tanner's application of the ban to video games and not other media was unfair.

Tobacco had also caused another headache for the game business in January 1990 when controversy erupted about Sega's arcade racing games *Hang-On* and *Super Monaco GP*. Unknown to tobacco manufacturers, both featured barely disguised adverts for cigarette brands such as Marlboro, which the game's developers had included because it was in keeping with the tobacco-dominated advertising boards that adorned racetracks at the time. When the row broke, tobacco companies such as Philip Morris threatened to sue, conscious of being accused of trying to peddle cigarettes to children as a result. Sega agreed to remove them. But the intervention of Lieberman and Captain Kangaroo took video game controversy to a whole new level. Lieberman openly declared that what he really wanted was an outright ban, although the US constitution would probably not allow it. Instead he, together with fellow Democrat Senator Herbert Kohl, organised a public inquiry to investigate the problem of violent games and called some of the leading lights of the game business to appear before it for questioning.

* * *

The events that led to the Senate's inquiry began back in 1991 when artist John Tobias and programmer Ed Boon, who worked for Chicago-based arcade game

---

1. CND, the Campaign for Nuclear Disarmament, were a UK pressure group. *Raid Over Moscow* also caused controversy in Finland, where a member of the country's parliament prompted a debate about whether its sale should be allowed.

manufacturer Midway, started thinking about their next project. *Street Fighter II* had just become the biggest arcade hit for years and Midway wanted a fighting game of its own. "We were fans of head-to-head arcade games and their appeal to the arcade crowd," said Tobias. "I was also a huge geek for Hong Kong martial arts cinema and was looking for an excuse to use some of those influences."

Having agreed to create Midway's answer to *Street Fighter II*, the pair started debating how they could make their game – *Mortal Kombat* – stand out from Capcom's landmark game. "A big goal of mine was to differentiate the visual qualities of *Mortal Kombat* from any of the other 'fighting' products out there. It was important for players to look at *Mortal Kombat* and know immediately that it was, at the very least, different," said Tobias. The arcades were already overflowing with fighting games and some arcade owners were already tiring of them. "We purchased fighting games when they first came out, but after a while we stopped buying them," said Bob Lawton, founder of New Hampshire arcade Funspot. "The sequels were coming out at a pace we couldn't keep up with and as soon as a new version was released, no-one wanted to play the older one."

Tobias hit on the idea of using digitized footage of real-life actors, an approach that had already been used in a small number of coin-op games, such as Williams' *Narc* and Atari Games' fighting game *Pit-Fighter*. "We thought that by using the digitizing technique we could achieve a high level of detail, given the size of the on-screen characters in an expedited amount of time," he explained. The team hired actors to play out the roles of their virtual fighters in the game and, after some touching up, imported their images into the game.

Aside from the digitized characters, *Mortal Kombat* did not stray far from the *Street Fighter II* formula, sticking to Capcom's combination of secret moves, two-player action and tactical brawling. Until they started testing it that is. "There was an anti-climactic moment at the end that created the opportunity to do something cool," said Tobias. "We wanted to put a big exclamation point at the end by letting the winner really rub his victory in the face of the loser. Once we saw the player reaction, the fact that they enjoyed it and were having fun, we knew it was a good idea."

The 'exclamation points' Tobias and Boon came up with were a selection of gory takedowns that could be enacted using secret button and joystick combinations when the game urges the player to 'finish' their defeated opponent. Tobias and Boon called these gruesome finishing moves Fatalities. "We certainly weren't out to cause controversy. We were out to serve the needs of our players and make sure that they enjoyed themselves while playing – that was our number one goal all of the time," said Tobias.

And enjoy themselves they did. *Mortal Kombat* became the hottest game in the arcades since *Street Fighter II* as players fought each other in the hope of delivering a brutal fatality move to their crushed opponent. With *Mortal Kombat* taking the arcades by storm it was only a matter of time before it arrived on the home consoles. Acclaim Entertainment snapped up the rights and converted the game to both the Sega Genesis and Super NES. Sega approved the game complete with the violence from the arcade original, but Nintendo was more squeamish and insisted the fatalities were removed. Acclaim told the Japanese giant that the fatalities were the selling point and removing them would give Sega the superior game. Nintendo refused to change its mind and Acclaim grudgingly cut the gore from the Super NES edition.

With the home console versions of Midway's arcade smash ready, Acclaim prepared one of the biggest game launches ever seen at that point. The company declared that the launch day, Monday 13th September 1993, was 'Mortal Monday' and lavished $10 million on TV advertising to ram home *Mortal Kombat*'s arrival in US homes. Children and teenagers across the nation were inspired to start urging their parents to buy them a copy. One of them was the nine-year-old son of Bill Anderson, the chief of staff for Lieberman. Anderson was shocked at the violence in the game and told Lieberman about his son's request for a copy. Lieberman decided to check it out for himself. He too was shocked at a game that he saw as rewarding players for violent acts. He delved further into the video games on sale in the shops and found *Night Trap*, one of the first games released on CD-ROM.

Night Trap began life back in the late 1980s as part of Axlon and Hasbro's abandoned VHS videocassette console the NEMO. After the NEMO project collapsed, Axlon co-founder Tom Zito managed to recover the rights to the game and formed a development studio called Digital Pictures to bring the game out on the Sega CD, an add-on for the Sega Genesis that hosted some of the first games to be released on compact disc. The goal of the game was to protect a group of teenage girls at a slumber party by setting traps for vampires. Lieberman felt the scenes where the partygoers were dragged off by the vampires were sexist; particularly a scene where a woman dressed in a nightgown is attacked in the bathroom.

After finding out that most console owners were under 16 and that most of the parents in Connecticut who he spoke to knew nothing about the content of the video games their children played, he decided to act. The violent games Lieberman had homed in on were exceptions, but they justified the unease about video games that many parents felt. It was a distrust that reflected the historical pattern of new forms of media or entertainment being viewed – at least initially – with suspicion. Such reactions to new forms of media could be seen in Greek philosopher Plato's criticisms of theatre and Hollywood's 'sin city' image during the late 1920s and 1930s. As British psychologist Dr Tanya Bryon noted in *Safer Children in a Digital World*, her 2008 report for the UK government: "The current debates on the 'harms' of video games and the internet are the latest manifestations of a long tradition of concerns relating to the introduction of many new forms of media."

It was a suspicion that had even caused games such as *Lemmings*, a 1991 puzzle game created by Scottish game studio DMA Design, to face criticism. The game was a play on the popular myth that the rodents regularly commit mass suicide by jumping off cliffs. It charged players with trying to save the creatures by leading them to safety while bouncy renditions of tunes such as *How much is that doggy in the window?* played in the background. But the inclusion of levels where the lemmings had to avoid rivers of lava and negotiate volcanic rocks before jumping into the fiery mouth of a demon to escape prompted one southern US TV station to call for the game to be banned for its satanic imagery.

But such criticism was not just about video games. The early 1990s were a time when fears about violence in society were particularly high on the US political agenda. Congress was debating whether to restrict violent TV programming and a new gun control law, the Brady Handgun Violence Prevention Act, was about to be signed into law by President Bill Clinton.

Other nations also shared America's unease about *Mortal Kombat*. Germany's youth media watchdog the Bundesprüfstelle für Jugendgefährdende Medien got the game banned outright in 1994 for its extreme violence. Until then the only games outlawed in this way were a bunch of free neo-Nazi propaganda games that started circulating around Germany and Austria at the end of the 1980s.[2]

* * *

The first day of the Senate inquiry into video games was set to take place on the 9th December 1993 just eight days after Liberman's press conference with Captain Kangaroo. Ahead of the hearings Nintendo, Sega and the other video game publishers in the firing line suddenly realised just how poorly connected they were in Washington, D.C.. While many of them belonged to the Software Publishers' Association, its focus was on supporting business software companies such as Lotus and Microsoft rather

---

2. These racist and anti-Semitic games were not released commercially, but distributed for free and, apparently, quite widely. One newspaper poll of Austrian students reported that 22 per cent of students had encountered these games, which included concentration camp management games and quizzes testing how Aryan the player was. The Bundesprüfstelle für Jugendgefährdende Medien successfully sought bans on seven such games between 1987 and 1990.

than video game publishers. With the political pressure mounting, Sega and Nintendo found themselves forced together to try and come up with a strategy for tackling the hearings. The two harboured conflicting views of what should happen. Sega, with its more teenage audience, wanted an age rating system so it could carry on publishing games featuring violence. Nintendo, on the other hand, saw little need for an age rating system as it had its own family friendly policies that it applied to all games released on its console. But with the pressure on, the US's leading game companies agreed to back an age-rating system that would be managed by the industry itself in the hope of defusing the row.

This opening gambit took some of the heat out of the situation, but the proponents of video game restrictions at the hearings still tried their best to land some blows on the game makers. Marilyn Droz, vice-president of the National Coalition on Television Violence, asked Senators how they would feel if their teenage daughter went on a date with someone who had just played *Night Trap*. Eugene Provenzo, the University of Miami professor who had already annoyed the game business with his critical 1991 book *Video Kids: Making Sense of Nintendo*, also made the case for intervention. Drawing on his research, Provenzo told the hearings that 40 of the 47 games he examined for his book were violent. Video games were also "overwhelmingly" racist, sexist and violent, he added.

There was some truth in Provenzo's claim. Female and black video game heroes were a rarity and this absence was still pronounced years later as a 2001 report by US children's charity Children Now discovered. Children Now's report revealed that women accounted for just 16 per cent of playable human characters in the 10 most popular games of 2000. It also reported that 58 per cent of playable male characters were white but that if they had excluded sports games the proportion would have been much higher. "Video games do seem to do worse than other mediums, particularly when it comes to the representation of women," said Patti Miller, director of the charity's children and the media programme. "The lack of racial diversity in video games seems to be on a similar level to that of US TV."

It wasn't just a matter of ignoring women and black people. A few Japanese games had descended into racist stereotyping of black people, partly because the country's more homogenous racial make-up meant such racism was rarely confronted. For example, the 1989 role-playing game *Square's Tom Sawyer*, based – somewhat ironically – on the Mark Twain's novel *The Adventurers of Tom Sawyer*, portrayed black characters as 'blackface'-style caricatures with giant lips. It was never released outside Japan. Many Japanese games did, however, seek to avoid portrayals of race altogether, through the use of 'mukokuseki' characters that are drawn in such a way to obscure their racial origin, so that while they could viewed as white or Asian, they were neither.

In the West the racism in video games was more subtle and behind-closed-doors, with some game publishers pushing designers to 'whiten' the skin tone of black characters or remove them entirely. "It all boils down to money," Shahid Ahmad, a British Asian game developer, told *Edge* magazine in 2002. "Publishers believe that games with black or Asian characters could lose them money, although they won't openly say it."

Homosexuality, meanwhile, remained largely taboo in video games throughout the 1980s beyond Japan's yaoi, or boys' love, titles. The few that did mention the subject usually did so in a negative way. It took until the 1995 adventure game *The Orion Conspiracy* for the subject to be tackled in a less prejudiced way. Created by UK developer Divide by Zero and published by Domark, *The Orion Conspiracy* cast the player as a father who travels to a space station to investigate the death of his son.

While searching for the truth, it emerges that the son was gay. "When I first read the script I was quite surprised," said 'Tardie', an artist who worked on the game. "It was a very daring thing for Domark to do. The gay character was embedded into the game and you found out about it as you questioned people and discovered more about your son. It got quite charged as the father did not know, so you got to see how he handled it."

* * *

The real battle in the Senate on the day of the hearing, however, was not between the proponents and opponents of restrictions on video games but between Sega and Nintendo. It did not take long for the bitterness between the two video game giants to bubble to the surface and soon Washington's finest were watching Nintendo of America chairman Howard Lincoln and Sega of America vice-president Bill White Jr. engage in their own verbal equivalent of *Mortal Kombat*.

Lincoln used Nintendo's earlier decision to force Acclaim to remove the gore from the Super NES version of *Mortal Kombat*, as a stick to beat its commercial rival with. He said Nintendo had lost money by sanitising its version of *Mortal Kombat* and had even received angry letters and telephone calls from children demanding the violence before proudly adding that *Night Trap* would "never appear on a Nintendo system", ignoring how the lack of a CD drive meant the Super NES was technologically incapable of handling such a game. White countered that Sega had an older audience demographic than Nintendo, a claim echoing the underlying message of his company's 'Sega does what Nintendon't' advertising campaign: Sega is for cool teens, Nintendo is for children. He added that Sega had already introduced age ratings on its games voluntarily and that it wanted other companies to adopt its system.

Lincoln stuck the knife in. He dismissed Sega's age ratings system as a panic measure introduced when the controversy about *Night Trap* started and refuted White's assertions that the video game industry was now catering for adults as well as children. White started listing the violent games available on the Super NES and showed the Senators a Nintendo light gun. Lincoln described *Night Trap* as "outrageous". White defended it by pointing out that the player has to try and stop the vampires only to get slapped down by Lieberman who interjected: "You're going to have to go a long way to convince me that that game has any moral value." Lieberman watched the pair tearing chunks out of each other with amazement.

Nintendo and Sega weren't the only people in the games business divided by the controversy. The game developers who worked on *Mortal Kombat* and *Night Trap* were also divided over how to react to the row their creations had started. "I felt like we were being attacked by a bunch of people who were mostly ignorant of what they were attacking," said Tobias. "Watching the news coverage at the time, you'd think that *Mortal Kombat* was created by some evil corporation. Anyone who knew me or Ed personally knew that our intentions weren't anything other than ensuring our players were having fun."

Rob Fulop, who had designed *Night Trap* while it was still part of Hasbro's NEMO project, found the experience harder. "The scandal was kind of silly and it was also deeply embarrassing because friends of mine, my parents, and my girlfriend didn't really get games. All they knew was a game that I had made was on TV and Captain Kangaroo said it was bad for kids," he said. "I fell out with my girlfriend because I thought it was completely bullshit."

But while he thought the inquiry was nonsense, after nearly 15 years of game making, Fulop had begun to worry about the message video games were sending out to children. "I grew up in a generation where you watched TV, that's all we did. TV was basically 30-minute stories and always had a happy ending. Whatever the problem, in half an hour the problem was worked out," said Fulop. "You tell that story to kids 20 million times and they grow up believing everything will work out. That was my generation. You believed everything would work out because of TV. Now think of video games – the message is no matter what you do you always lose. You never ever, ever, ever, ever win. Once in a million you win, but most of the time you never win. Unless you can find the cheat, so what does that teach you? I think that's created a whole different culture – a very fatalistic 'what's the point?' attitude. It's a personal philosophy, I don't know if it's true or not."

The controversy surrounding *Night Trap* and the reaction of his family and friends inspired Fulop's next game: "I decided that the next game I made was going to be so cute and so adorable that no-one could ever, ever say it was bad for kids. It was sarcastic. It was like what's the cutest thing I could make? What's the most sissy game I could turn out?"

A shopping mall Santa Claus gave him the answer. "I would go at the end of the year to see Santa Claus at the stores and he would tell me exactly what kids wanted," said Fulop. "He knows better than anyone because his job all day is to ask them what they want. You want to know what's going to sell? Go talk to Santa. I did that every year. I went that year and he said the same thing that was popular every year was a puppy and has been for the last 50 years." Fulop's plan to make an adorable game and children's Christmas wishes for puppies came together to form 1995's *Dogz: Your Computer Pet*, one of the first virtual pet games.[3] *Dogz* installed an enthusiastic cartoon puppy on the player's computer that was housed within a playpen window but could be stroked or taught to do tricks. Fulop's goal was to make players grow attached to their virtual puppy: "The dogs follow your mouse and they can't wait to be petted, you do a mouse click and they come running over, it was their biggest excitement. People get attached to pets because they need you. You're needed. You come home and this thing is in your face…if you're not there you know it's going to be very unhappy or it could die. I don't think a virtual pet is any different."

Fulop's company PF Magic played on this attachment to encourage sales of the game. "It was sold the same way real puppies are sold," said Fulop. "We'd give it to you for 10 days and then ask for it back. You give a puppy to a kid and ask for it back five days later – see what he says. In sales this is called the puppy dog close. We gave you five days' worth of food and after five days you ran out of food and if you want more food you've got to call us and give us $20 and we'll give you a lifetime's supply of food, otherwise your puppy dies or you have to delete it. And who can delete it? It's cruel, it's a little puppy and you won't feed it."

Fulop's ploy worked like a dream and *Dogz* became a huge success, becoming the starting point of the long-running *Petz* series of virtual pets.

* * *

With evidence about the relationship between video game violence and real-life

---

3. Bandai's portable virtual pet the *Tamagotchi* came out shortly after *Dogz* in 1996. Created by Aki Maita, the electronic toy became a global sensation similar in scale to the Rubik's Cube. Tens of millions were sold worldwide.

behaviour inconclusive, the Senators closed their hearing by telling the video game industry to return to Washington on 4th March 1994 to report on its progress with creating an age rating system.[4]

In the three months between the hearings and the industry's return to the Senate, the US video game industry reorganised itself. The country's leading game companies quit the Software Publishers Association and formed the Interactive Digital Software Association, which was headed by veteran Washington lobbyist Douglas Lowenstein. After several weeks of rows, the industry also created the Entertainment Software Rating Board (ESRB). The ESRB's job was to manage the industry's new age-ratings system. Sega also stopped distributing the Sega CD version of *Night Trap* in January, although America's leading toy stores Toys R Us and Kay-Bee Toys had already stopped stocking it because of the Senate hearings.[5] The changes were enough to satisfy the Senate. The controversy was laid to rest and Lieberman and Kohl's threats of creating a state-run regulator with it.

The game industry came out of the controversy with more than just a lesson in how to handle politicians. It had also learned that violence and controversy sells. Sales of *Mortal Kombat* and *Night Trap* soared during the hearings. *Night Trap* in particular had gone from a poorly selling Sega CD title to a game that was selling 50,000 copies a week at the height of the row. When Zito's Digital Pictures released *Night Trap* on the PC to coincide with Halloween 1994, its advertising campaign embraced the scandal: "Some members of Congress tried to ban *Night Trap* for being sexist and offensive to women (Hey. They ought to know)."

Arcade game makers Strata, which also got a ticking off in the hearings, stuck its middle finger up to Washington within months of the hearings by launching *Blood Storm*, a *Mortal Kombat* clone featuring even more extreme violence and a hidden character that had Lieberman's head so players could beat up the Democrat Senator.

Lieberman's attempt to challenge video game violence failed. If anything the changes that resulted from his intervention made video game violence more acceptable, as the age ratings system would identify violent or controversial games as for adults not children, helping game publishers defend themselves against future accusations of peddling violence to children.

And with an age ratings system in place Nintendo no longer felt compelled to filter out the violence from games released on its consoles. When Acclaim launched

---

4. The evidence is still inconclusive. As Bryon's 2008 report for the UK government noted: "It would not be accurate to say that there is no evidence of harm but, equally, it is not appropriate to conclude that there is evidence of no harm." She added: "The research evidence for the beneficial effects of games is no more convincing than the work on harmful effects."

5. The bigger selling *Mortal Kombat*, however, remained on sale.

Mortal Kombat II on the Super NES, the fatalies and blood remained in place. "The inquiry didn't impact anything," said Tobias. "We were content with the M for mature on our packaging. Developers and publishers fell in line, accepted the ratings system and developed games according to the need of the product."

The Senate hearings had actually made it safer for video game developers to create violent games, not harder.

Backstage: An actor in full costume waits for filming to begin on the set of Phantasmagoria: A Puzzle of Flesh
Courtesy of Andy Hoyos

CHAPTER 19

A Library In A Fish's Mouth

Rand Miller got his first taste of video games when he and his junior high school classmates were given a tour of the University of New Mexico's computer centre in Albuquerque. After being shown around the facility, Rand and his classmates got to try some of the early text-only computer games using a terminal linked up to an IBM System/360 mainframe. His classmates paid little attention, but Rand was hooked. "I was the geeky kid in school. Although I played American football and musical instruments, I loved computer and science stuff. When I was young you couldn't have a computer at home, so I was really intrigued," he said. "We sat down for probably 15 or 20 minutes and we got to play whatever little games happened to be in the catalogue. I was a lot more intrigued than the other boys."

The university's computer centre was only a couple of blocks away from his school, so Rand kept returning, keen to play some more games. "I would get access to the computer by getting paper out of the trash cans where people had left their username and password printed out on the paper," he said. "I would steal their passwords and user names and change the passwords. I had my own login and catalogue area and began to get interested in writing my own programs. The first ones I wrote were games."

Years later as a married man living in Dallas, Texas, Rand decided to buy an Apple Macintosh and started looking for some software that would interest his young daughter. The search proved fruitless. "Games were beginning to be of a higher quality but children's software was like the dregs," recalled Rand, who was working as a programmer at a bank at the time. "It was like, if you couldn't make it making games, then you would just slop out crap for children. I remember having the distinct feeling that this did not mimic other markets – books in particular. A good children's book will actually be appealing to adults as well and that's what inspired my brother Robyn and I with *The Manhole.*" At Rand's suggestion, the brothers decided to create an interactive children's book for the Macintosh. As well as reading the story, they wanted children to be able to interact with the still pictures so that when they clicked on certain objects they would

hear a sound or something would happen. The first picture Robyn drew for their computerised book showed an image of a manhole and a fire hydrant. He programmed it
so that when users clicked on the manhole the cover slid open and a beanstalk grew
out of it. "I sat there looking at the beanstalk and the open manhole and I realised I
didn't want to turn the page," said Robyn. "I needed to know what was at the top of
the beanstalk and down inside the manhole or even inside the fire hydrant, which I
began to imagine being like a house."

The brothers never created the next page of the book. Instead they started creating
pictures of what was inside the fire hydrant, down the manhole and up the beanstalk.
"I think it came from a universal desire to explore. I think people love to explore –
people wonder what's around the next corner," said Rand. "The ideas were just streaming out of Robyn's mind, it was like: 'What's down the manhole cover?', 'How about
a boat and an island?', 'Okay', 'And on the island I'll stick a walrus and an elevator'."
Riding their wave of spontaneous creativity, the pair paid little attention to process. "I
just jumped in and went to work," said Robyn. "Here's another planet. What should it
be? How about a library in a fish's mouth?" They created worlds within worlds, moulding an unpredictable piece of software that served up a stream of constant surprises
that echoed the insanity of Lewis Carroll's 1865 children's novel *Alice's Adventures in
Wonderland*. Billed as "a fantasy exploration for children of all ages", *The Manhole* was
initially released via mail order in 1988 but, in 1989, video game publisher Activision
decided to re-release it on CD-ROM, a new storage medium being touted as the next
big thing in computing.

Although the compact disc had been developed in the 1970s and Sony had released
the first CD music player in 1982, its potential for storing data had gone unused by
video game companies largely due to the lack of home computers with CD drives. By
the late 1980s, however, the situation was changing. The music industry's efforts to get
people to buy audio CDs had brought down the cost of CD drives, which were now
becoming available for computers such as the Macintosh and IBM PC compatibles.[1] By
February 1989 the first CD-based computer, Fujitsu's FM Towns, was on sale in Japan.
CD had two big attractions for the video game industry. First, CDs were cheaper to
manufacture than microchip-based game cartridges. Second, CDs could store around
600 times as much data as a floppy disk and around 300 times as much as a cartridge.

1. The IBM PC was not a standard computer system. Keen to get a low-priced business computer on the market fast, IBM
built its PC using widely available hardware rather than creating its own technology. This meant other manufacturers could
reproduce the PC without fear of legal action by using the same hardware, which they started doing within a year of the
PC's 1981 launch. Microsoft, which wrote the PC-DOS operating system for IBM's PC, also helped the makers of clone
machines by allowing them to buy the same software under the MS-DOS name. Eventually the sales of IBM PC-compatible
computers would far outstrip the sales of IBM's own brand system.

It was a win-win situation, publishers lowered their production costs and developers could create bigger games and use audio and video recordings in their work.

At the time developers were finding the limited storage available on floppy disks a major frustration. "It was clear CD-ROM was going to be the future of games. People like Sierra were releasing games with 10 disks – it was getting crazy," said Bob Jacob, the co-founder of Cinemaware. The Miller brothers shared this sense of frustration. "Floppies had been insanely limiting," said Robyn. "We shipped the original version of *The Manhole* on five floppies, which meant people had to continually switch back and forth between disks. Very annoying. It was near luxurious working with that huge amount of space. More than anything, it allowed us to make the world as big as we wanted." Taking advantage of the format, the Millers added extra music, animations and scenes to the CD version of *The Manhole*.

The sonic potential of CD was particularly exciting for game audio programmers, who had spent most of the 1980s trying to squeeze out tunes and sound effects in machine code within miniscule amounts of memory. In the same year that *The Manhole* moved to CD, British game musician Rob Hubbard quit Newcastle-upon-Tyne for San Francisco to become Electronic Arts' sole audio person. He had spent the past four years becoming a minor celebrity on the UK gaming scene by writing video game music on computers such as the Commodore 64. "I did a game on my own and people thought the game sucked, but they thought the audio was really good," said Hubbard. "At that time, game music was really, really dreadful so I thought I should just try doing the audio. There wasn't much competition. I don't think it's anything to do with being particularly brilliant at anything because there was hardly anybody else doing it. All you had to do was something half-decent that made a bit of sense as opposed to some of the stuff back then, which was often really, really awful."

Electronic Arts' San Francisco studios was a world away from the ramshackle cottage industry Hubbard left behind in the UK: "They were streets ahead in what they were doing with their thinking and were talking about the optical devices as well. It seemed like they were on the bleeding edge of what was going on." CDs allowed video game musicians such as Hubbard the freedom to concentrate on the music rather than programming. Instead of having to write music in machine code they could now write complex scores, record real-life musicians, sample sound and use voice actors. In short, CD allowed them to stop fighting with microprocessors and focus on being creative. "CD was a radical transition for game audio," said Hubbard. "It opened the floodgates." CD also allowed video games to include music from popular pop and rock acts. "*Road Rash* on the 3DO console was one of the first games that exploited using licensed bands. Electronic Arts struck gold because they licensed a band called

Soundgarden when they were just a college band and, while they were making the game, Soundgarden became massive, really big. So they shipped *Road Rash* with these Soundgarden tunes. It basically started the whole idea of having a CD track playing while you have a game going."

There were limits to this promised land, however. "When CD-ROM first came out, the drives were so slow that you had to plan your content well to make it work as it would come off the drive very slowly," said Rand. "Everything had to come off the CD because the hard drives didn't have room for much stuff, so you had to stream your music from the CD in chunks large enough that when someone clicked to a new picture in *The Manhole*, you could get the new picture and then go back to where you were streaming the music from without getting a break in the tune."

CD also opened up new visual frontiers. As well as having the capacity to store more art, CDs could also hold short video clips that could be integrated into games. For Cinemaware, a video game studio that always wore its Hollywood aspirations on its sleeve, this was an especially appealing feature. In early 1990 NEC had decided to try and tap into the growing hype about CD games by launching the TurboGrafx-CD, a CD drive for its struggling TurboGrafx-16 console. But the company knew it needed some eye-catching games to generate interest in its CD add-on and turned to Cinemaware. In return for a CD-enhanced remake of *It Came from the Desert*, Cinemaware's *Them!*-inspired tale of a desert town terrorised by giant radioactive ants, NEC offered to buy 20 per cent of the company. Cinemaware was quick to say yes. The company was in dire straits at the time, suffering primarily from its decision to focus on the Amiga computer that had flopped in the US despite selling well in Europe. "They saw us as a high-quality developer and at the time I was like 'hey, getting some money for this would be good'," said Jacob. "We did the deal and obligated a significant proportion of the resources of the company to doing development for NEC. In retrospect this was the decision that killed Cinemaware."

Cinemaware sank $700,000 into the project, a huge investment at a time when a typical console video game cost around $150,000 to make. The development team was headed by David Riordan, the designer of the original Amiga version of *It Came from the Desert*, and also included programmer Mike Livesay and scriptwriter Ken Melville. "It was a dream project with all the filmic aspects and cool '50s sci-fi content," said Melville. The team filmed actors and imported them into the game where they were imposed on still photos of real-life locations to create some of the first live-action film footage seen in a video game. "I'm very proud to have been part of the team bringing motion video for the first time to a game," said Melville. "It looked like shit, of course, but there it was. Mike had to really bust his hump to extract video and game play out of those old systems. Guys like Mike were the original game heroes who made lemonade out of lemons."

Although the CD version had a relatively large budget, the team found the money did not stretch far when cameramen, film studios and actors had to be hired. "The *It Came from the Desert* shoots were very minimalist and low-budget," said Melville. "They were shot in a small studio and designed to pop up occasionally as exposition. Non-interactive stuff." The project's budget got even tighter when Cinemaware's financial problems finally dragged it under in 1991 before the game was finished. Under Riordan's leadership the team completed the game on a shoestring budget and it eventually reached the shops in 1992. But by then it was clear NEC's bid to revive its fortunes with the TurboGrafx-CD had failed miserably.

Cinemaware's collapse came as the video game industry was preparing for the leap to CD. Game companies started investing in recording and film studios, and exploring how the new technology could enhance their products. They adopted the language of Cinemaware, talking about interactive movies and of blurring the boundaries between Silicon Valley and Hollywood. UK publisher Psygnosis was one of the first to embrace the CD future. The Liverpool-based company, formed out of the wreckage of early 1980s game publisher Imagine, had developed a reputation for beautiful-looking games, such as *Shadow of the Beast* and *Agony*, that were packaged in luxurious boxes adorned with the fantastical artwork of Roger Dean, the artist famed for his album covers for 1970s prog rockers Yes. Psygnosis saw the audio-visual capabilities of CD as a chance to push its focus on presentation even further. It bought high-end computers that had once been the preserve of movie special effects teams to create stunning visuals that it then filmed so they could be shown on more primitive computers via CD's video capabilities. It also exploited its prog rock connections by hiring former Yes keyboard player Rick Wakeman to write and record a score for its flagship CD game *Microcosm*, a shoot 'em up set within the human body that was inspired by the 1966 film *Fantastic Voyage*. Sierra Online were even more ambitious. The company built a dedicated film studio close to its headquarters in Oakhurst, California, and created *Movie 256* – a custom-made software tool that allowed its developers to edit and import video footage for use their games. "It cost us around $1.5 million," said Ken Williams, co-founder of Sierra. "It was a true studio with all that goes with it: sound rooms; blue-screen stage; editing bays; server rooms; etc. It was very cool."

What the industry lacked, however, was a popular CD-based gaming system. The FM Towns had achieved cult status in Japan, but Fujitsu had no plans to bring it to North America or Europe. Commodore had come up with the $999 CDTV, an Amiga-based console that – in a break with console tradition – was designed to look like a

piece of hi-fi equipment. Created under the leadership of Atari founder Nolan Bushnell and touted as a consumer electronics product rather than a console, the CDTV
won support from several game publishers thanks to its Amiga-based technology, but
few people bought it. Philips' CD-i console did only marginally better.[2] Sega had plans
to bring out a CD drive for its Genesis console in 1992, but no one knew how well
it would do. Nintendo and Sony had teamed up to create a CD version of the Super
NES, tentatively titled the PlayStation, but by the summer of 1991 the two companies
had fallen out. With console manufacturers struggling to produce a CD system that
the video game industry could unite behind, it was the IBM PC-compatible computers
that came to the rescue.

During the 1980s the PC had been regarded as a video game backwater. It was a
popular business machine but lacked the kind of audio-visual capabilities game players
had come to expect. This began to change in the second half of the 1980s, starting
with the arrival of VGA graphic cards for PCs in 1987 and sound cards the year after.
More than any other game 1990's *Wing Commander* underlined the PC's transition from
dull business machine to gaming powerhouse. *Wing Commander* was the brainchild of
Chris Roberts, a British game designer who, like Hubbard, had quit the UK for the
US's larger games business, taking a job at Origin Systems in Austin, Texas. Inspired by
the Second World War air battles between the Japanese and American forces above the
Pacific Ocean, *Wing Commander* was a sci-fi epic that mixed non-interactive cinematic
storytelling scenes with fraught space battles. For many it was the first time they had
looked at a PC game and been impressed. "It was a groundbreaking product for PCs.
There is a running gag that it made the PC market because all the PC manufacturers
had to upgrade their product so people could play *Wing Commander*," said Geoff Heath,
the managing director of the game's distributor Mindscape International.

The PC's evolution into a popular system for video gaming was confirmed by an
agreement between the world's leading computer companies including Fujitsu, Microsoft, Philips and Tandy on the 8th October 1991. The agreement defined the Multimedia PC format – a set of standards for PCs that combined CD drives with graphics and
sound cards. The standard gave software developers clarity about the kind of hardware
they could expect on CD-equipped home PCs and, in turn, gave hardware manufacturers the confidence to mass produce equipment such as CD-ROM drives in volumes
large enough to bring down the cost significantly. Over the next year and a half CD

---

2. There were actually several types of CD that could be used for video games competing to become the standard format
in the early 1990s. The CD-ROM format focused on data storage while Philips' CD-i format was primarily designed for
storing video. To confuse matters further, some were even touting digital audio tape (DAT) as a superior alternative to CD
since it held twice as much data as a CD. CD-ROM won the day.

drive-equipped PCs made huge in-roads in the home computer market at the expense of the computers produced by Commodore, Atari and Apple. By 1993 the number of multimedia PC users had grown large enough to make CD-ROM games commercially viable. The game that made the sales breakthrough was *The 7th Guest*, the creation of Oregon studio Trilobyte. Part inspired by David Lynch's TV drama series *Twin Peaks*, the game challenged players to solve 21 puzzles in order to discover the secret of a haunted mansion. After each puzzle was solved, the game used video clips to develop the narrative before moving onto the next challenge. Released in April 1993, the game sold more than a million copies, a huge number for a computer game and the first CD game to achieve that level of popularity.

That same year LucasArts, the rebranded Lucasfilm Games, delivered *Star Wars: Rebel Assault*, another million-selling CD title for the PC. Designed by Vince Lee, *Star Wars: Rebel Assault* used footage and sound taken from the *Star Wars* movies to build graphics on high-end special effects computers that, as with Psygnosis, were then turned into video clips that could be played on standard home computers. The game had players flying in pre-set paths through movie-like scenes from the world of *Star Wars* while taking down Empire fighters. *Star Wars* director George Lucas later wrote to the team praising their work for taking his sci-fi story into a new medium.

The success of both these games paled, however, compared to the phenomenon that was *Myst*, the latest creation of Rand and Robyn Miller's game studio Cyan Worlds, which was based in the town of Mead near Spokane, Washington. Since *The Manhole*, the Millers' work had been slowly moving into more traditional video game territory. Their second game, *Cosmic Osmo and the Worlds Beyond the Mackerel*, added story and characters to the bizarre worlds conjured up by Robyn. *Myst* continued this journey, taking in lessons from adventure games such as *Zork!* and *Déjà Vu: A Nightmare Comes True*. The brothers, however, were at pains to distance themselves from video games. "It kind of seems silly now, but during these early products, we always rejected the term game," said Robyn. "It's understandable why. Our early stuff had no real goals and the only point was exploration. We really saw very little in common with games. We always called them 'interactive worlds'. Interviewers would refer to *Myst* as a game; we would politely correct them."

Myst's interactive world was a mysterious unpopulated island that players saw through a collection of still pictures that shunted into view like a slideshow when players moved or changed the direction they were facing. The game's beautiful, strange and lonely island proved to be its main selling point. "We heard from tons of people that they had this sense of really being there, on those islands," said Robyn. "People liked to turn down the lights, turn up the sound and lose themselves in the *Myst* world."

Myst challenged players to learn the secret of the island's past by solving a number of puzzles. "Our desire was to achieve something more story-like, even if it wasn't a very substantial story," said Robyn. "We wanted to build a narrative into the environment. That's what attracted us to this different kind of world. We never saw the puzzles as simple mindbenders; we saw them as extensions of the story."

But, while the Millers had incorporated more of the storytelling and puzzle elements of adventure games, they were keen to avoid many of the other trappings of video games. "We had played other games that had dead ends," said Rand. "I remember one that took place in a city where you go down the city streets and between buildings there were alleys. We wanted to go down the alleys, but you couldn't. So we tried to design the world so it was contained enough that you could go anywhere you felt like there was a place to go." And, following the example of the adventure games created by LucasArts, they also rejected the use of player death as a play mechanism. "Everybody continued to do 'die and start over', 'die and start over'," said Rand. "From our point of view we were trying to mimic the real world and we decided the real world doesn't do that. There are consequences from doing the wrong thing, but you don't always have to start over. We felt like the world would be large enough that we wouldn't have to send people back to the beginning and the puzzles were difficult enough that players felt they got enough game play without having to start over."

For the fast-growing number of PC owners searching for something to show off the capabilities of their new CD-ROM drive, *Myst*'s gentle exploratory play was ideal. "We were in the right place at the right time," said Rand. "We had come out with an application that was just primed for CD-ROM. It was a killer application because it was so safe. Anybody at any age could walk into a store and say I need some software for my 10-year-old son all the way up to my 80-year-old grandmother for my new CD-ROM drive, what do you recommend? *Myst* was safe." Prior to the game's release in September 1993, Rand and Robyn had discussed their hopes for the game. They had designed it with the mass market in mind and Rand hoped it would sell around 100,000 copies. Rand's hopes were surpassed within weeks. *Myst* topped a million sales. Then two million. Then three. Then four. And it just kept selling. "At some point the numbers stopped having meaning. It was just numbers on paper," said Rand. *Myst* became the biggest-selling PC game of all time.[3]

In the wake of the commercial breakthroughs of 1993, CD games flooded the market. Most concentrated on exploiting the video capabilities of the format, a focus that had video game studios grappling with the same challenge of fusing the creative

---

3. Until *The Sims* beat its record in 2002.

cultures of games and films that Hasbro had tried to solve during the development of its unreleased NEMO console. "It was kind of the 'best of times, worst of times'," said Jane Jensen, the designer of Sierra's *Gabriel Knight* series of southern gothic horror adventure CD-ROM games. The first game in the trilogy, 1993's *Gabriel Knight: Sins of the Father*, used voice actors to bring the dialogue to life, but the second, 1995's *The Beast Within: A Gabriel Knight Mystery*, was a production closer to a Hollywood movie than a video game, utilising an army of actors, camera operators, costume designers and make-up artists. "It was fantastic to work with live actors and get a whole other dimension to the work. But it was the first time I or the producer had done it and the learning curve was pretty astronomical. We had a 900-plus page script to shoot, which was also daunting for the young director and actors," said Jensen. *The Beast Within*'s gargantuan script reflected the challenge of turning films into interactive experiences. "It's long because you need to provide different scenes for what happens when the players approach something from different angles," said Jensen.

While the production challenge was huge, the game designers involved found working with filmmakers an exciting experience. "It was definitely a time of exploration and experimentation," said Robyn Miller. "Developers suddenly had all this space and didn't exactly know what to do with it." French game designer Muriel Tramis spent four months on the streets of Paris filming material for Coktel Vision's 1996 adventure game *Urban Runner*.[4] "I had some unforgettable moments of creativity with the director," she said. "Him from the world of cinema with all its codes and myself from the world of 3D images, we had to find common ground, we had to invent other tracks to introduce interactivity."

Others also took to filming games on location. "My feeling was why not film the game on sets and on location using Hollywood talent," said Andy Hoyos, director of Sierra's horror game *Phantasmagoria: A Puzzle of Flesh*. "A cinematographer, a practical special effects crew for on-set horror and make-up effects was what I felt was needed to push this kind of product to the next level. As long as the budget could support incorporating these kinds of technologies and approaches that's what I wanted to do. However, doing this meant learning the ropes of this kind of 'movie-making' approach, which was no easy task. Learning to make a game more like a film director was a tad painful in some ways and, consequently, the hours were long and arduous." Lorelei Shannon, the horror writer who designed and wrote the game, remembered the production as a terrific experience: "Pretty much everybody involved had a great time. We brought in an actual film crew and professional actors. The actors were a

---

4. A 1996 adventure game set in Paris in which the player took on the role of an American journalist framed for murder.

little bemused by having to do things like shooting 'walkers' – a generic loop of the character walking – but they were all pros about the whole thing. Of course due to the interactivity, it was a very long shoot and we had many hours of footage. Editing was a big job."

On its release in 1996, *Phantasmagoria: A Puzzle of the Flesh* immediately courted controversy due to its graphic violence, sex scenes and a gay kiss involving the player's bisexual lead character. It was content unthinkable prior to the creation of the US age ratings system introduced in the wake of the 1993 Senate inquiry into video game violence. "We all knew we were treading on unsteady ground with such controversial material, but the feeling was that we should go for it," said Hoyos. "I really wanted to make a 'splash' no matter what. Even something that might even be considered shocking. I really wanted to blow the lid off of the horror game genre." The game's sex and violence was enough to prompt bans in Australia and Singapore, something Shannon welcomed. "It just added to its notoriety," she said. "I think the main reason games get more negative press for sex and violence is because people hear the word 'game' and immediately associate games with children. But games are just another form of entertainment, like movies. Some movies are for kids and some are for adults."

By 1996, however, the fusion of game design and movie making ushered in by the arrival of CD was causing a different kind of controversy as the game industry began to question the merit of the approach. Game designers started to find that the once vast capacity that CD offered had been swallowed up by hours of movie footage. So much so that games such as *The Beast Within* spanned six CDs. "We filled it up right away," said game designer Rob Fulop, whose CD work included *Max Magic*, a virtual magic set for the CD-i. "The first time you go 'wow, it's a million times more storage' and then you go 'oh, we've run out of it'."

Another problem was the limitations video placed on game designers. "We couldn't experiment much because once you filmed these scenes you couldn't just go and create another one," said Fulop. "With video you can't go 'let's hire the actor back and get them to do a back flip'. We couldn't try many things like that."

The increasingly grand cinematic visions of game designers and the inability of many of the games to deliver on interactivity eventually sparked a backlash within the specialist gaming press. Critics talked about the 'interactive movie' in terms that echoed the late 1970s British punk rock movement that made its hatred of the pretensions of progressive rock acts such as Yes and Pink Floyd central to its philosophy. "Looking back on it now it seems we were a little too enamoured with this new process and didn't spend the time polishing the quality of the game play," said Hoyos. "But to be fair to the designer of *Phantasmagoria: A Puzzle of Flesh*, the executives of Sierra kept whittling away at the design of the game, pairing down the actual game play to save money, until what was left was little more than an interactive movie. How they felt this was going to square with the game-playing public I'm not sure."

By 1999 when Jensen's *Gabriel Knight 3: Blood of the Sacred, Blood of the Damned* reached the shops, the video game industry had largely abandoned its attempts to turn games into films. Instead of the video shoots of its pre-equal, *Gabriel Knight 3* used characters drawn using 3D graphics. "The company simply wasn't going to do full-motion video, period," said Jensen. "I was happy to give it a try and I think the game turned out really well. There are definitely some things we gained with 3D but a lot that we lost as well. It allowed for more variety of puzzles and a deeper sense of exploration but the drama and emotion of the story was harder to convey with the 3D characters, especially the love scene. I miss having the actors."

The rise of 3D graphics techniques would sweep away the era of the interactive movies much like the punks brought the heyday of the prog rockers to a sudden halt. And fittingly, it was a team of rebellious young game designers who plastered their back-to-basics action games in gore, swastikas and industrial metal that would deliver the killer blow.

Merchants of Doom: (left to right) John Carmack, Kevin Cloud, Adrian Carmack, John Romero, Tom Hall and Jay Wilbur
Courtesy of John Romero

CHAPTER 20

The Ultimate Display

In 1965 computer graphics pioneer Ivan Sutherland laid down an ambitious challenge to the computer scientists at the annual International Federation for Information Processing Congress. He outlined an elaborate vision of the future, a time when computers would create "the ultimate display". The computer display of tomorrow, he ventured, would not just look like the real world, it would feel, respond and sound like reality too. Creating that virtual reality, he argued, is what computer researchers should see as their ultimate goal.

Sutherland's bold vision fired the imagination of computer scientists.[1] They threw themselves enthusiastically into trying to create "the ultimate display". They built head-mounted displays, helmets with computer screens for each eye that consumed the wearers' field of vision like a pair of hi-tech binoculars. They figured out how to construct virtual objects out of coloured polygon shapes, usually triangles, to create an illusion of 3D on 2D computer screens.[2] They built electronic gloves to let people interact with these 3D worlds using their hands and designed haptic feedback devices that conveyed the sensation of touch with their vibrations.

Although some of these breakthroughs seeped out of the research labs in the form of flight simulators for pilot training, for the first two decades following Sutherland's landmark speech the work of virtual reality researchers went, for the most part, unnoticed.

But as the 1990s began the wider world finally latched onto the idea of virtual reality, partly as a result of talk about the development of the internet and the connected world it would usher in. The virtual realities these researchers sought to create

---

1. Sutherland's vision of computer-generated worlds also inspired countless authors, film directors and game developers resulting in novels such as William Gibson's *Neuromancer*, *The Matrix* movie trilogy and the 1985 computer role-playing game *Alternate Reality: The City*.

2. Prior to the use of polygons, 3D objects had usually been represented by wireframe graphics that used thin outlines of objects that resembled the steel frames of under-construction buildings.

provided an easily understandable and visual representation of the global network the internet promised. "Virtual reality was a great symbol of how the internet would take over our lives," said Jonathan Waldren, the founder of British virtual reality research company W. Industries, which later renamed itself Virtuality. "There was a lot of hype, people were being told by analysts that the internet was going to be a paradigm shift for everything in life and for a lot of people that was really bewildering."

Virtual reality may have been a separate idea from the internet, but for a world trying to get its head around the abstract idea of what a networked world would be like, it brought the concept to life. After 20 years of being ignored, virtual reality became one of the most talked about areas of computer research. Investors pumped millions of dollars into research projects and virtual reality start-ups hoping to cash in on the new world. Journalists flocked to see the latest developments before returning with reports about how in the future we could be spending as much time in the virtual world as in the real world. TV documentaries excitedly discussed the possibility of cybersex in the virtual reality worlds that, at the time, seemed to be just around the corner. To a lot of people the polygon 3D realities being promoted by those trying to engineer these digital worlds looked a lot like a video game.

As it happened, video game makers had also been wrestling with the same problem as virtual reality researchers, namely how to make believable computer-generated 3D worlds. But while the virtual reality set focused on engineering hardware devices that could make the digital real, game developers were battling to deliver Sutherland's immersive vision through software that would work within the limitations of mass-market computer technology.

The earliest 3D video games, such as *Tailgunner* and *Battlezone*, had relied on wire-frame visuals to create an illusion of depth. It was an elegant solution given the technology available, but game developers knew the polygon 3D visuals that virtual reality researchers had been playing with would let them create far more believable worlds.

Atari became the first company to bring the technique to video games with 1983's coin-op game *I, Robot*. Devised by Dave Theurer, the creator of *Tempest* and *Missile Command*, the game mixed shooting with platform jumping in a world that looked as if it was built out of Lego bricks. *I, Robot* not only introduced the visual approach to game players, but demonstrated the big advantages of 3D visuals by allowing players to see the action from a range of viewpoints.

Unlike the 2D graphics that were widespread at the time, 3D graphics did not need to be drawn in advance. Instead they were generated using mathematical equations and so all a game had to do to alter the player's perspective was recalculate the position and size of each polygon relative to the player's location in the virtual world. And by using more polygons you could create more elaborate and realistic looking objects. It was easy in theory, but the more polygons you had the more calculations were needed and as the number of calculations increased so did the time it took for computers to carry out the instructions. "The maths for 3D graphics is very simple," said David Braben, the creator of *Zarch*, a 1987 shoot 'em up for the Acorn Archimedes set above a 3D patchwork landscape of agricultural fields created out of polygons.[3] "The main problem was making the graphics go fast enough for the motion to look smooth."

Not that the challenge stopped game developers from trying. Flight simulation developers led the way, keen to inject more realism into their work, but polygon graphics soon began being used in other types of games, from 1989 racing sim *Indianapolis 500: The Simulation* to French game designer Christophe de Dinechin's *Alpha Waves*, a 3D platform game for the Atari ST promoted by its publisher Infogrames as a dream-like relaxation experience. "*Alpha Waves* was the first full 3D platformer game," said Frédérick Raynal, who converted the 1990 game to the PC. "Everything was moving fast on the screen and the game play was very challenging. I think Infogrames made a mistake trying sell it as a New Age brain motivating experience instead of an efficient and modern platformer."

Others built whole worlds out of polygons, most notably *Midwinter*, Mike Singleton's 1989 game of guerrilla warfare that took place on a large snow-covered island made up of light blue polygons. Atari Games' 1989 coin-op driving sim *Hard Drivin'* went a step further than most in its plundering of ideas from virtual reality, combining 3D polygon visuals with variable wheel resistance and force feedback technology that had its origins in the haptic feedback hardware engineered by those chasing Sutherland's Holy Grail.

The intertwined worlds of virtual reality and video games finally came together when Virtuality decided to bring virtual reality to the masses. Virtuality began life in the English city of Leicester where it engineered virtual reality equipment for corporate customers. "We used to work in a tiny little place in Leicester that was underground; under an old shoe factory," said Waldren. "It was a government-run place – they were trying to get more IT companies in and we hired some rooms. It was cheap space – no cost. The idea was to build these virtual reality systems and sell

---

3. Released in June 1987, the Archimedes was the UK's answer to the Atari ST, Commodore Amiga and Apple Macintosh computers. It is notable for being the first computer to use Acorn's ARM microprocessor technology, which thanks to its low power use went on to feature in almost every mobile phone and handheld device in circulation by the late 2000s.

them to professional development organisations." After doing some work for British Telecom, Virtuality joined forces with Leading Leisure, a UK firm that had created a mini-flight simulator called *The Venturer* for use in funfairs and large arcades.

"The idea was that we'd build a little person simulator using virtual reality technology and link it to the flight simulator," said Waldren. "The grand vision was we could have these simulation centres where people could be in an aircraft or exoskeleton in some kind of game environment."

The grand vision never happened, but Virtuality figured there was potential in making coin-operated virtual reality game machines for arcades. Virtuality's team designed sit-down and stand-up arcade machines that boasted the head-mounted displays, 3D joysticks and glove-like controllers that symbolised virtual reality in the public mind. In October 1991 Virtuality unveiled its first game: *Dactyl Nightmare*. *Dactyl Nightmare* allowed up to four players, each using a separate Virtuality machine, to fight each other with guns and rocket launchers in a polygon 3D world while dodging aggressive pterodactyls that flew menacingly above the play area's surrealistic chequerboard floors.

"People were just lost in that thing; they were really, really immersed," said Waldren. "They just went over that line where they just forgot about the rest of the world." But with each machine costing $65,000, the onus was on getting people in and out of the game as fast as possible to make it profitable for arcade operators. "That really cramped our ability to do anything of depth," said Waldren. "Our goal was to get people in there, give them a very highly intense, intuitive experience for three to four minutes, but then you had to change over because typically there was a queue and, obviously, the operator has to get the next people in there."

Virtuality's games gave many people their first taste of a virtual reality experience, fuelling the belief that the future envisaged by Sutherland was almost here. Nothing could have been further from the truth. While Virtuality continued to make games until the late 1990s, including 1996's *Pac-Man VR*, the hype surrounding virtual reality waned fast. "The equipment was too expensive and there was no good business model beyond training," said Brenda Laurel, the former Atari Research employee who got involved in virtual reality work in the early 1990s.

Jaron Lanier, a virtual reality researcher who also started out exploring the technology during a stint at Atari's research arm, feels in retrospect that Virtuality's attempts to take the technology to the public may have done more harm than good.[4]

4. "One crazy thing I was working on at Atari Research was a force-feedback broomstick – a sort of simulator for a witch's broom," said Lanier. "I thought it would make a nice arcade game. At some point someone actually tried that, but it didn't catch on."

"On one hand they had gone quite a bit further than anyone had at creating something that was a manufactured product and they had really worked on a business plan – this $1-a-minute experience – and that was all good," he said. "The problem is that they were a little unrealistic about whether they were ready for primetime. I think people spent their dollar and were disappointed with the level of graphics that were possible at the time."

This gap between the rhetoric and actuality of virtual reality was something video game designers were about to really expose. In November 1991, one month after Virtuality unveiled *Dactyl Nightmare*, a Texan game developer called Id Software released a PC game called *Catacomb 3-D* that marked a major breakthrough in video game 3D graphics. Id began as a tight-knit group of game-loving colleagues who worked for Softdisk, a disk magazine publisher based in Shreveport, Louisiana. Popular in the 1980s, disk magazines provided subscribers with a floppy disk packed with articles, adverts and software for their home computers. Softdisk started out in 1981 with its monthly Apple II magazine *Softdisk Magazine* and by 1987 around 100,000 subscribers were paying $9.95 a month for the company's disk magazines, which were now available for a variety of computer systems.

One of the programmers churning out software to tight deadlines for the firm was John Romero, an energetic and ambitious game designer who dreamed of achieving wealth and fame through his creations. He had started out selling his games to disk magazines such as *UpTime* before joining Origin Systems, the publisher of the *Ultima* series of role-playing games, in 1987. Origin was supposed to be Romero's big break, but it turned sour.

His first project at the then New Hampshire-based firm was cancelled before it was finished and he then decided to quit to join his former boss at Origin who had left to form his own game studio, which went bust almost as soon as it started. Romero was left broke and so when Jay Wilbur, the former editor of *UpTime* who had joined Softdisk, called and offered him a job he jumped at the chance. "I could count on John. He was a machine," said Wilbur. "When I worked at *UpTime*, he was putting out a game per month, which for me as someone buying games to put on a periodical was perfect. He'd make these great games and each one of them had double consonant titles, so it was like Wacky Wizard or Deep Dungeon or something like that."

Shreveport was a long way away from video gaming's spiritual heartland of Silicon Valley. The city was built on the back of oil and gas, but in the mid-1980s the industry collapsed leaving behind a legacy of racial tension, double-digit unemployment rates, high levels of homelessness and gang crime. The offices in the downtown skyscrapers that once symbolised Shreveport's wealth sat empty and dis-used buildings in the city centre were boarded up. Despite being told he would be able to make games at Softdisk, Romero spent his time converting the company's existing programs so they could be put on its IBM PC-compatible magazine *Big Blue Disk* in a monotonous nine-to-five working environment. One day in 1990 Romero cracked and confronted Softdisk's owner. Romero told him that Softdisk should make a game subscription disk for PC owners and, if the company didn't, he would quit.

"I was tired of writing programs for *Big Blue Disk* and wanted to make games full time," said Romero. Softdisk's boss agreed and *Gamer's Edge*, a disk magazine for PC gamers that came out once every two months, was born, with Romero at the helm. To supplement the *Gamer's Edge* team, Romero brought in Adrian Carmack, a Softdisk artist fond of drawing dark and disturbing images part-inspired by the pictures of injured and diseased people he saw while working in the photo archives of a Shreveport hospital, and the unrelated John Carmack, a fiercely intelligent and talented programmer recommended to Romero by Wilbur, who was charged with overseeing the business end of the *Gamer's Edge* project.

"John Carmack submitted an overhead-view dungeon crawl game to me to buy for Softdisk," said Wilbur. "It was magnificent. I would have loved to have purchased it but couldn't because of the parameters of what I could do on the disk. So I said: 'Hey John, this is awesome but it's too large. Can you do something small?'. He came back with this tennis program that had accurate physics – *Pong* with physics and an isometric view. It was brilliant. It was like 'oh my god, this guy's got something going on here that's beyond what we can do'. He is clearly up there as one of the best. He is a monster at what he does."

Another Softdisk employee, Tom Hall, was the last member of the gang. Hall was not part of the *Gamer's Edge* team but he loved games and would regularly contribute ideas. On the evening of 19th September 1990, Hall and John Carmack stayed late at work fooling around with game ideas and working out how to make a PC game scroll smoothly. That night Carmack cracked the problem. "Over that night they developed a perfect – and when I'm talking perfect, I mean pixel-by-pixel – rendition of the first level of *Super Mario Bros 3*," said Wilbur.

The pair stayed at work until 5am recreating Nintendo's hit game before head-ing home. On Romero's desk they left a floppy disk containing their creation, which they named *Dangerous Dave in Copyright Infringement* after the video game character created by Romero that they had used in place of Mario, with a note saying 'load me'. Romero was stunned, as were the rest of the *Gamer's Edge* crew. Romero saw the potential immediately: this was their ticket out of Softdisk. "We started piling in the next morning and they showed it to us and we were like 'my god, that's perfect, this is *Mario 3*'," said Wilbur, who got in touch with Nintendo to see if they would be interested in a PC version of their game. They were not. "I contacted my friends at the legal end of Nintendo who said: 'We have no desire to exploit this property outside of Nintendo's hardware'."

As luck would have it, around the time that Nintendo turned them down, a man called Scott Miller got in touch with Romero. Miller was the founder of Apogee, a Dallas-based game publisher specialising in selling titles via shareware. Shareware was an alternative to the mail order or retail distribution methods used by most software publishers, and was dreamt up by Andrew Fluegelman, the founding editor of *PC World* magazine. In 1982 Fluegelman created a communications software program called *PC-Talk* but, instead of seeking a publisher, he decided to use it for an economic experiment.

He gave it away and asked people to send him a cheque if they liked it. Despite having the option of not paying, hundreds of users paid up leaving Fluegelman swamped by cheques. His trust-based experiment inspired a movement. By 1988 the estimated turnover of the shareware software market was somewhere between $10 million and $20 million in the US alone, even though, on average, only one in 10 users paid up. Miller liked the shareware concept, but noticed that while people paid for applications or utilities they rarely paid for games. He started wondering why and whether there was a way to solve the problem.

"Shareware games did not make money before I formed Apogee," he said. "The reason was because shareware game authors – and there weren't many – made the mistake of releasing their full game as shareware, giving no incentive for players to send them money. I decided to try a new method: release an episode rather than a full game and then sell the remaining episodes." To test his idea he created *Kingdom of Kroz*, a 1987 maze game divided into three episodes, and gave away the first part for free. If players wanted the remaining two episodes they would have to buy them from Apogee.

"The first few months were slow, just a trickle of mail-in orders," said Miller. "But it picked up and soon I was getting $100 to $200 a day in orders. On some days as much as $500. This spurred me on to create several more episodes of *Kroz* – seven in total. In 1989 I made around $100,000 and in 1990 I decided to quit my $30,000 per year day job and focus all my time on Apogee."

Miller wanted more games to release via Apogee and started looking for developers who could help him grow the company's product line. Miller had been impressed by Romero's games and decided to contact him. Miller hoped Romero would be interested in remaking *Pyramids of Egypt*, a maze game he wrote for *UpTime* in 1987. Romero instead showed him the work Hall and Carmack had created. Miller was also blown away. "The technology was well ahead of anything I'd yet seen on a PC, so I made a pitch: I'd fund a game if they presented me with a design," said Miller. "A week later I had a compelling design created by Romero, Carmack and Hall.

The two-paragraph design was for a game titled *Commander Keen in Invasion of the Vorticons*. I loved it." In keeping with their Nintendo-inspired demo, *Commander Keen* was the kind of colourful action game common on home consoles but a rare sight on PCs. Miller sent the team a cheque for $3,000 and they started work on *Commander Keen* under the name Ideas from the Deep. The team wrote *Commander Keen*, a game about an eight-year-old boy who lives a double life as a planet-saving space hero, in their spare time while still working for Softdisk and, on the 14th December 1990, Apogee released it in three parts. The first part was given away free via bulletin board systems (BBSs), while the second two episodes could be bought by mail order for $30. "It set a new standard on the PC for scrolling platform games," said Miller. "Not even the big publishers had anything as good. From day one Id was a technology leader and this alone generated a lot of buzz and attention. On top of that, *Commander Keen* was a damn fun and often funny game."

Apogee's monthly sales quadrupled and *Commander Keen* became the talk of the BBS world. The team's first royalty cheque for *Commander Keen* topped the $10,000 mark prompting the *Gamer's Edge* team to quit Softdisk on the 1st February 1991 to form Id Software. After relocating briefly to Madison, Wisconsin, the team moved to Dallas. Id planned to focus on producing more console-style games on the PC, but everything changed when Romero heard about a game being put together by Blue Sky Productions, the new game studio formed by his former Origin colleague Paul Neurath.

Neurath envisaged the Massachusetts-based studio, which would rename itself Looking Glass Studios in 1992, as a game design think tank, a place where pushing back the boundaries of game design and the exploration of new video game concepts were the order of the day. To fulfil Neurath's grand vision, the studio used its proximity to the Massachusetts Institute of Technology (MIT) to recruit some of the US's brightest graduates who would help the company foster a rigorously intellectual approach to game design. It was a place where every day felt like a game design symposium.

"It was like being in university," said game designer Ken Levine, who joined Looking Glass in 1995. "Their focus was building up principles of game design that would turn games from being something where we sort of guessed what would be fun to putting a theory on it and then building on those theories to move forward with confidence. They all came from MIT, a place where they could all define a problem set. They brought that thinking to games."

During its lifespan, Looking Glass continually broke new ground in game design.[5] When the studio decided that flight simulations suffered from a lack of realism it created *Flight Unlimited*, which incorporated satellite imagery of real-life terrain to give players a more authentic experience. And when it decided that audio was underused as a game play device it produced 1998's *Thief: The Dark Project*, where players had to listen out for footsteps and use noise to distract guards so they could successfully carry out their thieving. "We started working on the game system for *Thief* writing documents about the stealth mechanic and comparing it to the stealth fighter and submarine games where you have all these tools like noise makers to throw off detection," said Levine, who helped develop the game's initial concepts. "The noise maker in *Thief* is right out of submarines, they have noise makers and that's where the idea came from."

But the studio's first breakthrough was its most significant. In 1990 one of the company's programmers, Chris Green, devised a revolutionary 3D graphics engine that was not only fast, but also allowed wallpaper-like 'textures' to be attached to each polygon, turning them from solid blocks of colour to patterned images. After Neurath told him about Green's breakthrough, Romero promptly informed John Carmack. Id had just put together a texture-less polygon 3D game of its own called *Hovercraft 3D*, where players drove around a maze shooting enemies and rescuing hostages, and Carmack soon worked out how to do what Green had done. And in November 1991, just a few months after Neurath's conversation with Romero, Apogee started selling Id's first texture-mapped game – *Catacomb 3-D*. Viewed through the eyes of the player's character, the fantasy-themed game was set in maze-like dungeons filled with monsters that could be zapped with fireballs shot from the player's virtual hand, which stuck out from the bottom middle of the screen. It wasn't until March 1992 that Blue Sky Productions got its texture-mapped game, *Ultima Underworlds: The Stygian Abyss*, on the shelves.[6]

Encouraged by the positive reception for *Catacomb 3-D* and Carmack's ideas for improving his 3D graphics engine, Id halted work on its console-style games to focus

5. Looking Glass closed in 2000 due to cash flow problems.

6. *Ultima Underworlds* was partially inspired by *Dungeon Master*, a 1987 role-playing game for the Atari ST that reshaped the genre with its use of real-time rather than turn-based combat.

on its third 3D game. At Romero's suggestion they based it on *Castle Wolfenstein*, a 1981 game for Apple II where players sneaked around a Nazi castle bumping off guards quietly in order to steal secret war plans. The original idea was to remake the Apple II game in 3D, but the Id team found it was much more fun mowing down the Nazi soldiers with a machine gun rather than sneaking past them. They began reducing the game down, stripping it back to a one-man assault on Hitler's underground bunker. The result, 1992's *Wolfenstein 3D*, was brutal.

Opening to the strains of the Nazi Party anthem *Horst Wessel Lied*, it married John Carmack's cutting-edge 3D visuals with gory artwork created by Adrian Carmack and Id's latest recruit Kevin Cloud to create a trigger-happy first-person shooter set in a maze-like bunker adorned with swastika-bearing flags. The result shocked and excited in equal measure. Its action was, at heart, not far removed from 2D overhead-view games such as *Gauntlet* where players run around a maze shooting swarms of enemies, but the 3D visuals gave the experience a whole new level of intensity and realism.

Wolfenstein 3D became a huge success. By the end of 1993 more than 100,000 copies had been sold, making it the biggest-selling shareware game released at that point. Its inclusion of Nazi imagery earned it a ban in Germany and objections from the US pressure group the Anti-Defamation League. It also raised the standard for 3D visuals to a new level, forcing other game designers to rethink their work. Overnight, Id became the hottest game developer around. Nintendo paid them to bring the game to the Super NES provided they cut the blood and replaced the dogs of the PC original with rats. Then Id dropped a bombshell. It was willing to let other game developers buy a licence to use John Carmack's revolutionary 3D technology. Until then game developers had treated their in-house technology like their most valued possession; secret weapons that could give their games the edge over the competition. Even saying you should let your rivals use your technology was viewed as heresy.

The idea to let others use the *Wolfenstein 3D* engine came direct from Carmack and won support from most of Id's creative team. "Carmack, Romero and the bulk of the creative team wanted to do that," said Wilbur, Id's chief executive. "They were like 'Oh, this is great – it's going to be a lot of fun, it's going to open up the market'. To be honest I was the blood-sucking business and financial guy and I was a little apprehensive, thinking we were going to open up a whole can of worms."

Despite Wilbur's doubts, Id pushed ahead with the plan and began selling licences to other game studios such as Raven Software, who used it to make 1993's *Shadowcaster*. It was a revolutionary business move that would reshape the way games were made. Before *Wolfenstein 3D* game developers would build the systems and tools they needed to make their games themselves. After *Wolfenstein 3D* they had a choice. Instead of creating their own game engine they could buy Id's technology, and focus on being creative. By 2005 the idea of buying in technology would be so commonplace in game development that dozens of companies, specialising in making software that did everything from 3D graphics to generating the leaves on virtual trees, had emerged. Even the developers of top-selling, big-budget games such as *Grand Theft Auto: Vice City* would use the software written by these 'middleware' providers to help make their games.

One licensee of Id's technology was Christian game publisher Wisdom Tree, who used it to make *Super 3D Noah's Ark* for the Super NES. Wisdom Tree was formed by Color Dreams, a game publisher that started out by making games for the NES console without Nintendo's approval before deciding to start making games for the Christian market. The first Wisdom Tree release, 1991's *Bible Adventures*, sold more than 350,000 copies.

"Our main goal was to provide scripturally correct games that offered families an alternative to the violent and sexually oriented games in the secular market," said Brenda Huff, the company's sales supervisor who would buy out the company in 1997. "Wisdom Tree's games were sold in Christian bookstores, ministry premiums and other Christian venues." Wisdom Tree also got its games promoted in the magazines produced by Focus on the Family, one of the organisations at the heart of the US's conservative Christian movement, which rose to prominence in the 1980s. "Focus on the Family offers premium items for sale in its magazines and the approval process for inclusion in these publications is a very structured and involved process," said Huff. "Founder James Dobson had talked about the dangers of video games a few months before we presented *Bible Adventures* to them. In April 1991 our game was featured in the *Focus* magazine. It was the equivalent of the *Good Housekeeping* Seal of Approval."

It is doubtful that Focus on the Family would have approved of Id's next game – 1993's *Doom*, where Id took the violent intensity of *Wolfenstein 3D* to a whole new level. *Doom* would prove to be a landmark release that would shake up the entire video game industry. From the outset Id aimed to create a game that would make *Wolfenstein 3D* look like yesterday's news. John Carmack rewrote his 3D engine, determined to once again demolish the competition. His revised code allowed Id to build rooms of any height, create curved walls and use new lighting effects such as flickering ceiling lights. Adrian Carmack also upped the shock value of the artwork, creating a nightmarish carousel of twisted monsters and twitching bodies impaled on spikes. They sampled animal noises to provide the game's demonic opponents with intimidating growls and roars and added a pulsing, clanging musical soundtrack inspired by the work of industrial metal acts such as Ministry and Nine Inch Nails. And, in a total rejection of the approach of the movie-inspired CD full-motion video epics flooding onto the market at the time, Id abandoned the idea of story almost entirely. "The full-motion video games were trying to look as good as possible without the hard programming necessary to do a true 3D game like we were making," said Romero. "I saw those games as the remnants of a dying niche. The future belonged to the 3D programmers."

Carmack summed up Id's thinking on game narrative by comparing video game stories to those of porn movies – expected but unnecessary. *Doom*'s scenario consisted of no more than telling the player they were on Mars and that demons from hell were attacking. Your job was to kill them. All of them. To help players carry out their bloody task, Id supplied players with a selection of vicious weapons ranging from shotguns and rocket launchers to mini-guns and chainsaws.[7] *Doom* was about one thing and one thing only: survival. It was terrifying, exhilarating, primal and angry. "With *Wolfenstein 3D* we wanted to shock people with the speed of the engine and the violence," said Romero. "With *Doom* we wanted to shock people with everything. It was the best game ever."

Doom's action and 3D visuals alone would have been enough to send shockwaves through the games business, but the Texan studio's game also ripped up the rules of the industry. Having already explored the idea of letting outsiders use its technology by licensing the *Wolfenstein 3D* engine, Id decided it was time the players also had access. The team drew inspiration from the *Wolfenstein 3D* fans who had hacked into the game and created new versions featuring different new graphics and levels. This practice was called modding, as in modification, and most video game companies were fiercely opposed to it, regarding it as a breach of copyright. But rather than frowning on its mod-making fans, Id embraced them. *Doom* came with the tools fans needed to redesign the game as they wished and share their work. The idea of letting players create their own levels dated back to early 1980s games such as *Pinball Construction Set*, but players' level of access was usually tightly controlled. Id's pro-modding stance went a step further, giving fans unprecedented access to the code that made *Doom* tick. "It gave users the opportunity to literally touch the tools that

---

7. *Doom*'s grisly action led to it being linked to the spate of high school shootings that shocked the US during the mid- to late-1990s, particularly the 1999 massacre at Columbine High School in Littleton, Colorado. The claims of a connection prompted both an FBI and Senate inquiry into video gaming's influence on the perpetrators of the crimes. The Senate inquiry went nowhere and the FBI's report dismissed the idea that playing games like *Doom* was a distinguishing trait of high-school shooters.

we used for the games we make, allowing them to turn themselves from an amateur developer into a professional developer," said Wilbur. It also brought business advantages, he added: "It gives the game legs, so a game that might exhaust its time in the marketplace in six to 12 months might get an additional 12 or 18 months or more depending on how popular it is because users are creating more content."

Doom also let players fight each other by connecting their computers together. Romero named these player-versus-player battles 'death matches'. The idea had been tried before. *Maze*, the original first-person shooter, let up to eight players battle each other and games such as the 1987 Atari ST game *MIDI Maze* let up to 16 people play together by connecting up their computers with cables. But the need to have each player's computer in the same room meant these features were rarely used. *Doom*'s arrival, however, coincided with the opening up of the internet, which meant computers could be connected via phone lines rather than direct cables. Soon after *Doom*'s release fans modified the game so it could be played over the internet as well as on PCs connected by cables. Soon thousands of people were spending their nights and days playing *Doom* death matches online.

Id's marketing for *Doom* was just as revolutionary. Having decided to publish *Doom* itself as shareware rather than go through Apogee, Id started using viral marketing years before the term had even been invented. "It was more how do we make the most impact based on what we had, which was zero dollars," said Wilbur. "We'd had some impact on the consumer with *Wolfenstein 3D*, we were talking about what we were going to do next and everybody was listening, so we just used that." Id kept its increasingly rabid fan base in a state of heightened excitement with a constant drip, drip, drip of information about the cool features that would be in *Doom*. By the time the team was getting close to completing the game, some fans were so excited they began phoning Id's offices demanding that they hurry up and get the game out.

By the launch day – the 10th December 1993 – the fans were at fever pitch. So many fans were sat waiting on the website where the free shareware version of *Doom* would first appear that it took Id hours to upload the game. The stampede to download the game as soon as it appeared caused the website's servers to crash several times. Within five months of its launch, the free demo of *Doom* had been downloaded more than 1.3 million times and Id was raking in $100,000 a day as fans began buying access to the rest of the game. Romero had achieved his dream of becoming a video game superstar.

"When *Doom* was released, I knew we were absolutely number one in the industry. Without a doubt," said Romero. Id was the hottest video game company in the world and the outgoing Romero, with his long black hair, was the perfect front man.

"We made a conscious decision," said Wilbur. "Romero wanted to be the rock star and he started to dress in the role. At the time the industry needed a rock star. The company and the software happened to be there and we had a guy out in front willing to take the mic and play the lead singer, so we pushed him out there."

Romero readily agreed: "I was the only person the press could talk to that knew every single aspect of what we were doing – design, art, coding, publishing, playing. I loved talking to people about our games, how we made them, why we made them, what was upcoming and everything else about Id Software. I was perfectly suited for the job. I wasn't just a mouthpiece – I made these games."

Video games were never quite the same after *Doom*. It was to games what The Beatles' *Sgt Pepper's Lonely Hearts Club Band* was to pop: a paradigm shift. Its most immediate impact was to be the catalyst for video gaming's shift from 2D to 3D visuals. *Doom* inspired a flood of first-person shooters that fuelled demand among PC gamers for 3D graphic cards. These hardware add-ons contained graphics processing units (GPUs): fast microprocessors dedicated to doing the maths needed to create realistic 3D visuals that would further accelerate gaming's move into three dimensions. Ironically, GPUs were a by-product of the virtual reality research that had faded from popular interest by the end of the 1990s. "Cheap 3D graphics hardware is what enabled the current explosion of video games," said Warren Robinett, the former Atari game designer who later moved into a virtual reality research.

Id also changed video games in more subtle but equally revolutionary ways. Id's willingness to let other game companies licence its technology was the starting point of a culture shift that has made the exchange of technology between developers widespread in North America and Europe.[8] The company's embrace of modding was another break with tradition and one that has evolved into a multi-pronged movement that has spawned hit games and acted as a training ground for hundreds, possibly thousands, of game developers.

By the end of the 1990s, however, the Id team that started out as friends back in the humidity of Shreveport, Louisiana, was no more. Relations between Romero and John Carmack broke down during the development of *Quake* and, shortly after that first-person shooter's 1996 release, Romero left to form his own studio with Hall, who had quit Id during the development of *Doom*.

Wilbur also quit after his five-year-old son asked him why he never went to see him play baseball when all his friends' dads did. "Things changed at Id. Early on we

8. Japanese companies tended to stick more to the traditional approach of developing their own technology and not allowing others to use it.

were a rag-tag bunch of friends then at some point in time we got successful and money started rolling in and the dynamic changed," said Wilbur. "It stopped being that crazy, fun place to work and started to get a little more serious."

Quake's release may have marked the end of the Romero-era Id but, in those short five years, he and his colleagues had an impact on video games that was still being felt more than 15 years after *Doom*'s release.

Grrl gaming: The PMS Clan (co-founder Amber Dalton, centre) attend the 2006 E3 trade show in Los Angeles
Associated Press / Reed Saxon

CHAPTER 21

We Take Pride In Ripping Them To Shreds

It should have been a coup. On the 28th May 1991, at the Consumer Electronics Show in Chicago, Sony proudly revealed that it was working with Nintendo to create a version of the Super NES with an in-built CD drive. The two Japanese companies had been working together in secret on the project, tentatively titled the Nintendo PlayStation, since 1989 and with the hype about CD-ROM reaching fever pitch, Sony's announcement should have been a highlight of the trade show.

But behind the scenes all was not well. Since agreeing to the alliance, Nintendo had become increasingly nervous about Sony's intentions, fearing that it wanted to use the project to muscle in on the game business. Nintendo's paranoia was justified. Ken Kutaragi, the Sony engineer who initiated the whole project, saw the partnership as the first step to achieving his dream of getting Sony to start making game consoles. Suspecting as much, Nintendo decided to strike first. The day after Sony gave its announcement, Nintendo announced it was dropping Sony and was now working with its Dutch rival Philips instead. Sony was shocked at the public humiliation Nintendo had inflicted on it. But if Nintendo had hoped to push Sony out of the game business the move backfired. Sony's president Nario Ohga was furious and, egged on by Kutaragi, decided to seek revenge by creating Sony Computer Entertainment, a new division headed by Kutaragi that would take Sony into the game console business. The result was the Sony PlayStation, a console that married the two biggest technological developments for video games in the 1990s: CD storage and state-of-the-art 3D graphics.[1]

In retrospect, the coming together of CD and 3D seemed like a logical next step for consoles but, when Sony first began trying to get game makers interested in its 3D system in 1993, opinion in the video game business was divided. At the time most of the breakthroughs in 3D graphics had yet to happen. *Doom* wouldn't be released until

---

1. The PlayStation would also come with a controller that rejected the flat plastic slabs of earlier console controllers in favour of a moulded plastic design that fitted comfortably in players' hands – an approach followed by almost every subsequent game console.

the end of the year and, even then, the monsters within Id's 3D world were created using 2D images. The first 3D graphics cards for PC had yet to appear and even in the arcades – the natural home for the latest in video game technology – game developers were only just starting to explore the visual approach.

Many concluded that Sony's promises of advanced 3D hardware within the Play-Station would simply mean the company's console would be expensive and, therefore, unpopular. The price problem would only be emphasised later that year when, in October 1993, Panasonic launched the first game console based on Trip Hawkins' new 3DO hardware standards. The 3DO was born from Hawkins' belief that the dominance and power of console manufacturers such as Sega and Nintendo was bad for the video game industry. "I looked out towards the mid-1990s with the concern that the industry could not advance with cartridge systems and restrictive licenses," said Hawkins. "The PC of the time was not a good alternative and none of the console companies had a vision that was constructive for consumers or developers."

He concluded that there needed to be a VHS of video games: a common hardware platform for developers to create games on that would be manufactured by a range of companies rather than one corporation. And in 1991 he quit Electronic Arts, the game publisher he founded in 1982, to make the idea real by forming The 3DO Company. "The purpose of the 3DO was to advance the game industry through 3D graphics, multimedia capabilities, optical disc mass storage and liberal licensing models," said Hawkins. "The 3DO was trying to push media back in the direction of more open and democratic licenses, where the costs are very low and nobody tells you what kind of music you can or cannot make."

Panasonic was the first company to buy the manufacturing rights to Hawkins' system and in October 1993 it launched its 3DO console, the FZ-1, amid widespread media fanfare. Consumers, however, baulked at the $699.95 price tag that resulted from its advanced technology. "The realisation that it was not going to be a success came in stages," said Hawkins. "The first arrow was poor 1993 holiday sales. In 1994 it was a big set back that all the developers who had welcomed our charitable $3 license fee were flocking to competitors with bad licenses and $10 fees. It was like the collapse of a union. If developers had been able to stick together they would have had collective bargaining power and could have permanently shifted the value chain."

One of the competitors developers were flocking to was Sony, which was finally starting to persuade the industry that it really could deliver on its promise of an advanced 3D games console with an acceptable price tag.

Ironically, one of the big reasons for the shift in developer opinion was a game developed by rival console maker Sega: Yu Suzuki's 1993 coin-op game *Virtua Fighter*.

Having pushed the use of hydraulics in coin-op video games to the max with his 1990 360°-motion air combat game *R360 - G-Loc Air Battle*, Suzuki began exploring 3D visuals after seeing Atari Games' *Hard Drivin'*. He decided to create a 3D driving game of his own but, rather than recreating the contemplative pace of Atari's driving simulator, he set out to make an exhilarating Formula 1-style racing game.

The result, 1992's *Virtua Racing*, used an expensive graphics microprocessor created by military contractors Lockheed Martin to move its Lego-like polygons around the screen at breakneck speed. It was an impressive and popular evolution of *Hard Drivin'*, but Suzuki had bigger plans for his next 3D project. One of the key objections to 3D graphics that developers had been raising with Sony was that while polygons worked fine for inanimate objects such as racing cars, 2D images were superior when it came to animating people or other characters. *Virtua Fighter*, Suzuki's follow-up to *Virtua Racing*, was a direct riposte to such thinking. Released in November 1993, it featured fighters built out of polygons. The characters may have resembled artists' mannequins but their lifelike movement turned Suzuki's game into a huge success that exploded claims that game characters couldn't be done successfully in 3D or that people would not warm to them.

But while it was Sega that had proven the full potential of 3D, it was its newest rival – Sony – that capitalised on it. Sega had been wary of basing its successor to the Megadrive console, the Sega Saturn, on 3D graphics and instead built a system that offered some 3D capabilities but was primarily a 2D graphics powerhouse. Sega's caution meant that Sony had the upper hand when it came to 3D visuals. Teruhisa Tokunaka, chief executive officer of Sony Computer Entertainment, even went so far as to thank Sega for creating *Virtua Fighter* and transforming developers' attitudes.

Soon Sony was winning over developers across the world. Japanese arcade firm Namco became Sony's first big-name supporter when it decided to abandon Nintendo and throw its weight behind the PlayStation. Namco's reproduction of its 3D arcade racing game *Ridge Racer* became one of the PlayStation's flagship games when it launched in Japan in December 1994.

None of this, however, was enough to put Sega out of the race. Sega still had its bank of popular arcade hits at its disposal and was an established player in the console business. Indeed, when both the Saturn and PlayStation launched in Japan in late 1994, it was Sega who took the sales lead thanks to the home version of *Virtua Fighter*. Within months of its launch the Saturn was fast becoming Sega's most successful console in Japan. But as 1995 continued and the battle between the two firms expanded into North America and Europe, Sega's early lead began to slip, not least because it couldn't match Sony's mammoth $2 billion worldwide marketing budget for the PlayStation.

Sony was also reaping the benefits of its relationship with Namco, who gave it *Tekken*, a fighting game to rival *Virtua Fighter*, and its 1994 decision to buy UK game publisher Psygnosis. Sony used the Liverpool publisher's 1995 amphetamine-driven futuristic racing game *WipEout* to connect the PlayStation with Europe's house music scene. *WipEout*'s soundtrack brought together some of the continent's biggest dance acts, including The Chemical Brothers, Leftfield and Orbital, and Sony's European marketing team deliberately aligned the PlayStation with club culture by installing Play-Station demo pods with the game in nightclubs. *WipEout* would sell more than 1.5 million copies worldwide and helped establish the PlayStation as the first console to take Europe by storm.[2]

Psygnosis also helped attract more game developers to the PlayStation by creat-ing software that made it easier for other developers to create games for the system. In comparison, those working on the Saturn found themselves faced with little help in creating games for Sega's complex machine. "Many factors caused the Saturn to do poorly, but they started with the complex system architecture of the Saturn," said Roger Hector, general manager of Sega's US development studio the Sega Technical Institute at the time. "It was not easy to learn to program and there were very few tools or documentation available when it came out."

Soon game companies once loyal to Sega began to question their support for the Saturn. One of those companies was Core Design, a UK developer based in Derby that was part of the CentreGold, a video game conglomerate that had evolved out of US Gold. "Sega had been very good to me," said Jeremy Heath Smith, the founder of Core Design. "They had transformed my company and I had made a lot of money out of Sega." Core Design was working on four games for the Saturn and PlayStation at the time, one of which was *Tomb Raider*, a 3D action game designed by Toby Gard. "Toby was working on *Chuck Rock Racing* – this crazy *Flintstones*-esque racing game – and he said 'I've had this idea for doing an Egyptian tomb-raiding type game'," said Heath Smith. "We loved the concept because there was always going to be so much material to use. So I said let's get *Chuck Rock Racing* out of the way and do it. We started in pretty broad strokes, which was to have a world that was within tombs that went into a fantasy world because nobody really knows what is in tombs and pyramids and things. The concept did not actually change, but month-by-month as the engine and the technology we were working on got more and more powerful, the game designers were just going bloody mad and we were able to do some amazing things with it."

2. Germany, however, remained impervious to the appeal of consoles. A 2005 report revealed that just 15 out of every 100 German households owned a console compared to around 60 out of every 100 UK homes and 30 out of 100 in France.

Initially Gard designed a male character for the game. "I went down to look at what he was doing and on the screen was Indiana Jones," said Heath Smith. "I said 'what the bloody hell's that?'. He says 'that's going to be our character'. There was no way we could use that, we'd just be sued from here to the Moon and back again. So he went away and literally two weeks later he says 'what about this one?'. I said 'well it's a girl, what use is that to anybody?'."

Gard's new creation, Lara Croft, was Indiana Jones reimagined as a busty English aristocrat whose hot pants hugged her wasp waist. She was more adolescent fantasy than feminist icon, but for video games it was a huge departure from the norm. Video games were still largely a male pursuit and asking a male audience to play as a woman still seemed like a daring move that bordered on commercial suicide.

"At that stage all game characters were still big and burly men," said Heath Smith. "There had never been a popular female character, but she did have something about her. Toby said 'let's just run with it and see. If it doesn't work we'll swap it to a guy'. It was just as ridiculous as that."

As the game got closer to completion, Core Design knew it had created something special. Lara Croft had evolved into a distinctive hero who had potential to gain publicity and the lavish 3D world she had to explore took in caverns with vertigo-inducing drops, tropical jungles and ruined temples. The inclusion of a giant Tyrannosaurus Rex dinosaur to fight also added to the game's wow factor. Heath Smith decided the game should debut on the Saturn first. "I just felt I could give something back to Sega by giving them a three-month exclusive and hoped it would help them with their hardware sales," he said.

Tomb Raider became incredibly popular, but it was on the PlayStation rather than the Saturn that it really found its audience. "Sony came out all guns blasting and ripped Sega apart," said Heath Smith. "Everyone saw how great *Tomb Raider* was on the Saturn and was waiting for it to come out on the PlayStation. We sold seven or eight million copies on the PlayStation and its hardware sales went through the roof."

Tomb Raider's action heroine also tapped into the buzz about the 'girl power' movement that emerged in 1997. Although the phrase was used to help promote the late 1990s pop group the Spice Girls, girl power had its origins in the underground 'riot grrls' feminist movement that rejected the collectivism of early feminists in favour of a more individual and assertive vision of feminity that embraced popular culture. Lara Croft's position as an interloper into the usually male-dominated world of video games made her a natural fit for the movement, despite her origins in male fantasy.

"At the time *Tomb Raider* came out, that whole girl power movement was really kicking off," said Heath Smith. "You've got *Tank Girl*, you've got movies with girls really taking leading roles. Women were coming far more into the forefront and rightly so. All that was happening when some of the early press appeared in the magazines and they just loved it. They picked up on the whole girl power coming to video gaming angle."

* * *

Tomb Raider's release coincided with an increasing awareness within the game industry that female players were growing in number and becoming more vocal. Until the mid-1990s, few game designers had attempted to appeal beyond the industry's core audience of young males. "They didn't care to and just assumed that girls wouldn't play games because they didn't play the ones that were out there as much as boys did," said Brenda Laurel, who in 1996 formed Purple Moon – a publisher dedicated to creating games for girls. "The video game business was totally vertically integrated around a male demographic – from designers and programmers to marketers and distributors to retailers and customers."

There was also a cultural perception that home consoles and computers were off-limits to women and girls. "Technology has been gendered male for a long time. Girls were trained to believe that they might break something if they touched it without understanding it," said Laurel. "The fear was not so much of technology as of humiliation or exclusion – the idea that a real woman wouldn't mess with that stuff. I was the only girl in chemistry class in my high school and it became clear to me that my choice was between having dates and taking the class. I dropped the class. These taboos were common in the '50s and '60s and they die hard."

Video games had fallen into line with such thinking and had become an overwhelmingly male activity. In 1987 just 14 per cent of players were thought to be female. But in the mid-1990s this began to change when a wave of female game designers emerged hoping to address the gender imbalance.

Leading the way was Theresa Duncan and Monica Gesue's 1995 game *Chop Suey*, a vivid interactive story narrated by novelist David Sedaris that told the story of two teenage girls who pig out on Chinese food before going on a trippy and wistful journey through a Midwestern town. Described by Duncan as an attempt to capture "the mystery and beauty of just hanging out at the picnic table for an afternoon", *Chop Suey*'s non-directed play and emphasis on discovering the unusual echoed Cyan's eclectic interactive children's book *The Manhole*. While not a sales success, *Chop Suey* was a critical victory that *Entertainment Weekly* named as the best CD-ROM release of 1995.

The sales breakthrough came soon after, when toy company Mattel decided it knew how to make games appeal to girls. "Mattel had been studying girls' play patterns for many years and understood what would appeal to them. They also understood that their powerful brands could create a huge shift in the market," said Nancie Martin, director of girls' software development at Mattel Media. "At that time, games designed to appeal to girls and women didn't have strong brands behind them and very few of those involved women in the design or production process or took the time to research what might truly be of interest to them."

Mattel, of course, had no shortage of brands, most famously its iconic Barbie dolls. Mattel Media made Barbie the focus of its attempts to bring video games to girls. One of its first releases was 1996's *Barbie Fashion Designer.* "Andy Rifkin, who'd been a toy inventor for years and headed development for Mattel Media, had an eight-year-old daughter named E.J. who wanted to be able to design clothes for Barbie on her computer," said Martin. "He came up with the idea of the fashion show and the printed clothing; my team made the production work in a girl-friendly way."

Barbie Fashion Designer allowed players to design new clothes for their Barbie dolls and print them out on specially made fabric paper that they could colour in using marker pens. "While it had several outcomes, including a virtual fashion show and the clothes you could print out for your doll, it was not competitive as traditional games of any kind generally are, but it certainly provided its players with a tremendous sense of accomplishment," said Martin.

Mattel's game proved to be the breakthrough hit for games aimed at girls. Thanks to the Barbie brand, it encountered little of the retailer scepticism that had made games such as *Chop Suey* hard to find and its play, finely honed through rigorous testing with children, ensured it became a huge seller. "I believe it sold several million copies all told," said Martin. "I was not at all surprised by *Barbie Fashion Designer*'s success; I'd been saying for years that the reason girls weren't playing video and computer games was that no one was making games they were likely to find appealing. For many girls it was the first time they'd ever played with anything on a computer."

Laurel agreed: "It showed games could appeal to girls and still be successful. The problem with it was, quite simply, that it perpetuated a version of femininity that was fundamentally lame."

The games produced by Laurel's company Purple Moon attempted to push the emerging girl game movement further by focusing on relationships between the characters and an underlying goal of trying to get its target market of eight- to 12-year-olds girls using computers. "I was sick of the industry giving nothing to girls in those days," said Laurel. "Girls were generally afraid of the technology. Boys had the advantage of gaming to get them involved in the world of computing. I wanted to create the same sort of bridge for girls, using forms and content that would engage them and get them over the hump of putting their hands on the computer." And, thanks in part to, change in retailer attitudes after the commercial breakthrough of *Barbie Fashion Designer*, Purple Moon's debut releases *Rockett's New School* and *Secret Paths in the Forest* both sold well. *Barbie Fashion Designer*'s success also changed game developer and publisher attitudes to female players. The shift could be seen at the 1997 Computer Game Developers' Conference in Santa Clara, where game industry delegates packed out five sessions on designing games for girls. In one case a group of more than 20 delegates who were refused entry to a packed session forced their way in, desperate to learn how to sell their products to this previously ignored audience.

But it wasn't just those making the games who were transforming attitudes. Female players of *Doom* death matches were taking girl power into cyberspace with the aid of virtual rocket launchers and shotguns. Many of the women who played *Doom* and similar games online found themselves confronted with rampant sexism from some male players who accused them of being transvestites, claimed that they were inferior players because of their gender or offered comments such as "girls who play online do it because they can't get a man" or "I'll give you a rocket to ride".

Fed up with the abuse, female players retaliated by organising themselves into women-only teams influenced by the riot grrl feminist punk movement of the early 1990s. These all-women teams embraced a fierce brand of feminism to challenge the chauvinism they encountered, adopting team names such as Clan Psycho Men Slayers and Crack Whores and forming websites such as Grrl Gamer. They set out to counter the sexism in the only way *Doom* allowed – by blasting the chauvinists into a shower of bloody lumps. As one grrl gamer, Street Fightin' Mona of the Crack Whores, put it: "We take pride in ripping them to sorry little shreds." "The game grrls really took on sexism and gave it a good punch in the nose in the hardcore gaming community," said Laurel.

By the middle of the first decade of the 2000s, the sexism faced by those early clans was on the back foot and later female clans, such as PMS Clan, were becoming more about community than retaliation.[3] "The importance of PMS Clan is that it went past the 'grrl movement' and female teams and into establishing a complete 'gamer movement'," said PMS co-founder Amber Dalton, who played under the name Athena Twin. "Back then we needed to do it on our own as females. We needed to prove we could battle against the best of them. The environment was thick with harassment and prejudice for a long time against women players, but we stuck through it all. Now,

---

3. When PMS Clan formed in 2002 its initials stood for Psychotic Man Slayers – no relation to the Psycho Man Slayerz group that had broken up long before – but in 2005 it changed the meaning to Pandora's Mighty Soldiers in reaction to its increasing mainstream appeal.

women are much more normalised in games since our numbers have thankfully grown and some of the players now have wives and daughters that play as well. People realise we have made our home and no matter what they throw at us we are still going to be here, probably long after they are gone. There is still harassment of course, and always will as it is the nature of online spaces, but at least people are used to us now in the gaming world and many show respect and even admiration."

By 2010 PMS Clan had expanded to include men, who play under the H2O banner, and become one of the best-known gaming clans around with more than 2,000 members of both sexes. "We were one of the first gaming groups ever to get mainstream media with features in outlets like *Entertainment Weekly*, *Forbes*, *ABC News* and *Fox News*," said Dalton. "That did not just help our organisation, it helped expose competitive gaming to an entire new mainstream audience and grow online gaming."

The girl games movement led by the grrl gamers, Mattel, Duncan and Laurel, had forced a significant change in attitudes within the video game industry and among players themselves. Ironically, having helped change attitudes the games produced by the Purple Moon and its peers ended up being regarded as patronising relics. "I get a lot of crap from both women and men who don't understand the social context in which Purple Moon and its sister companies came to be," said Laurel. "They don't remember the time that girls were afraid of computers, boys dominated computer labs in elementary school and girls thought that tech was not gender-appropriate for them. The conditions that we were trying to address when we started Purple Moon no longer exist."

* * *

Meanwhile, *Tomb Raider* had become an international pop culture phenomenon. Lara Croft had become the face of post-Nintendo gaming, the socially acceptable Mario; a representation of Sony's desire to present games as cool. Fashion house Gucci gave *Tomb Raider* publisher Eidos $30,000 to have the virtual icon model its clothes. Eidos then hired real-life models, including Nell McAndrew, to become the 'official' embodiment of Croft and help publicise the game around the world. Health drink Lucozade used the character on its adverts to reposition itself as an energy drink. In 2001 the game became a blockbuster film, *Lara Croft: Tomb Raider*, which earned more than $250 million.

Croft's success as a pop culture icon helped changed attitudes to video games. "Suddenly video gaming was an acceptable media – it was talked about around dinner tables," said Heath Smith. "We had got characters like Lara that virtually everybody had heard of, it was talked about and there was no embarrassment about it. There was a massive shift from playing a computer game being deemed as a thing spotty geeky kids did in their bedroom into being a very acceptable part of entertainment."

The maturation of video games from toys to home entertainment was aided by game developers' increasing attempts to cater for older teenagers and adults. These attempts were only partially a response to the popularity of the PlayStation. The US age ratings system introduced following the Senate hearings of 1993 had given publishers the confidence to produce games for older players without fear of retribution. And it was Capcom's 1996 horror game *Resident Evil* that was the most significant of all these adult-orientated games.

Until the early 1990s game developers rarely tried to produce horror games, largely because the limited technology often made the task of scaring players difficult. One of the early attempts was Five Ways Software's 1985 adaptation of *The Rats*, James Herbert's gory horror novel where a plague of mutant rats terrorise London. The game used text adventure scenes where players experienced the attacks of rats from the victims' perspective, but with the twist that these encounters happened in real-time rather than the usual turn-based approach. Five Ways Software's subversion of the adventure game format, which is ill suited to real-time action, helped create a sense of panic when the rats attacked as players struggled to survive. The game also had images of rats that would 'burst' out of the on-screen page obscuring the text to add to panic. It was an interesting experiment in trying to invoke fear in players but it proved to be a one-off.

Seven years after *The Rats*, French game designer Frédérick Raynal created the modern horror game with *Alone in the Dark*, a nerve-racking 3D horror game for the PC set in a mansion in 1920. Raynal hit on the idea while working on the PC version of the 1990 3D platform game *Alpha Waves*. "It made me think in 3D," said Raynal. "While making it I was thinking about what could be the future of adventure games with 3D computed animation and skinned characters."

Being a big fan of 1970s horror movies such as the zombie films directed by George A. Romero, Raynal decided to create a horror game: "I was very attracted to those movies. Almost all of them were survival horror movies with just one survivor at the end. I had wanted to do a game with that simple principle since I started using computers: your goal is to survive and exit the house."

Aware that the primitive polygon 3D visuals available in 1992 were unlikely to scare players, Raynal focused on using the unexpected and the anticipation of danger to instil fear in the player. "The main idea was to implement inevitable death from what the player does all the time – playing and controlling the character," he said. "When the floor of the first corridor fell out from under you, then you were always afraid of walking. One of the first doors you opened unveiled a monster just behind, so then you asked yourself what will happen with every door."

Alone in the Dark became a huge seller for its publisher Infogrames and turned Raynal into one of France's foremost game designers, but it would take *Resident Evil* to really establish horror as a distinct type of video game. While it had many similarities with *Alone in the Dark*, the Japanese team that made *Resident Evil* was in fact trying to update *Sweet Home*, a 1989 role-playing game for the Nintendo Famicom based on a Japanese horror film of the same name. "It was more influenced by the basic structure of *Sweet Home*," said Jun Takeuchi, who worked as an animator on *Resident Evil* before going on to produce later editions of the series. "The fact that a western-style house is where the game takes place, there are traps and problem-solving sections in the game and other such components."

But while *Sweet Home*'s simple 2D visuals struggled to convey a sense of horror, *Resident Evil*'s tale of zombies created by a genetically engineered virus made fear and frights its raison d'etre. "Movies were a big influence in the area of creating the atmosphere of terror. We learned a lot from the skills and techniques that had been cultivated by the movie makers of yesteryear," said Takeuchi. As well as absorbing horror film techniques, such as claustrophobic camera angles and surprise attacks, Capcom's team made the need to avoid combat and a sense of vulnerability core to the experience by strictly rationing the amount of ammunition available for the player's gun and only allowing the game to be saved at a few spread out locations. "The game was designed to create fear via movement within the game itself, therefore being able to save only at specific points in the game heightens the sense of fear," said Takeuchi. "When a player is unable to invoke the special privilege of 'continue' the player is psychologically destabilised and therefore in a more vulnerable position and more susceptible to the feeling of fear. In other words, constraining the save option to certain parts of the game is one of the tools for enhancing the feeling of fear."

Resident Evil's nail-biting action and global success encouraged the arrival of more horror games. Many stuck to the b-movie feel of *Resident Evil*, but others sought to move beyond the Romero zombie movie template. Konami's *Silent Hill* was a particularly significant development. Instead of taking its inspiration from the gory shock movies of the US, its Japanese developers absorbed ideas from the new wave of Japanese horror films, which tend to concentrate on psychological fear rather than shocking images and the crossover between natural and supernatural worlds.

Silent Hill cast players as Harry Mason, a father searching for his adopted daughter in the fog-shrouded and abandoned Midwestern town of Silent Hill. The town itself reflected the emotional turmoil of the lead character with its decaying buildings, horrific scenes and disturbing monsters that echoed the unsettling visions of painter Francis Bacon. And while the player did have opportunities to fight monsters, *Silent Hill* was primarily a game where the horror manifested itself in the town's oppressive and haunting atmosphere, which was enhanced by the screeching radio interference that warned players of approaching creatures.

From these two templates the horror game became an increasingly important and diverse genre on video game consoles. Japanese publisher Irem's 2002 PlayStation 2 release *Zettai Zetsumei Toshi* was an especially innovative take on the genre, swapping the threat of monsters for the dangers of natural disasters. "Since I grew up on the novel and movie *Japan Sinks* and the cartoon *Survival*, I wanted to make the game with a theme of disaster," said Kazuma Kujo, producer of the game.[4] "Additionally, I had a chance to hear about the horror of disaster from my seniors and friends during the big 1995 earthquake at Kobe. A combination of those things inspired me to make a game about surviving a disaster."

While fear remained paramount in the game, the earthquake theme meant that the monsters or enemies that were usually the causes of those fears were absent. "Since the disaster, especially the actual earthquake, is not visible, it required some creative thinking to express it through the screen," said Kujo. "For example, we combined the small shakes and big shakes to express the approaching earthquake."

While Irem explored the fear of acts of god, British game developers Rockstar North took the horror genre in a darker direction with 2003's *Manhunt*, which put the player in the shoes of a convicted murderer who is forced to kill by the director of a snuff movie. *Manhunt* revolved around its murders, which ranged from suffocating people with plastic bags to clumsy beheadings with meat cleavers, and rewarded players for the brutality of the executions they carried out. The graphic deception of murder and the player's complicity in the carnage made *Manhunt* the one of the most controversial games of the first decade of the 21st century, earning bans in Australia, Brazil, Germany and New Zealand. In the UK, national newspapers linked *Manhunt* to the 2004 murder of 14-year-old Stefan Pakeerah in a Leicester park by his 17-year-old friend Warren Leblanc. The police and the courts dismissed the link, not least because it was Pakeerah not Leblanc who owned the game. Leblanc was jailed for a minimum term of 13 years for the murder.

The biggest development in horror games during the first decade of the 2000s, however, was 2005's *Resident Evil 4*. *Resident Evil 4* saw Capcom reassess the whole horror genre and take it in a more action-based direction. The lumbering zombies of the

---

4. *Japan Sinks* was an award-winning 1973 disaster novel written by Sakyo Komatsu.

original games were replaced with fast-moving villagers under the control of a para-sitic organism. "Familiarity is the enemy of evoking the sense of fear and terror," said Takeuchi. "We need to keep reinventing the fear aspect of the game, so sometimes we need to go against the grain of the *Resident Evil* concept itself to keep this sense of the unknown, ergo fear, alive. I think *Resident Evil 4* has cleared many of these problematic issues of familiarity and loss of the sense of fear in very eloquent ways. We managed to destabilise the expectations of the players by changing the speed of the zombies, but were able to retain the essence of *Resident Evil* throughout."

Resident Evil 4 re-energised the horror genre leading to a spate of more action-ori-entated games, such as 2006's *Dead Rising* and 2008's *Left 4 Dead*, a multiplayer-focused game where players work together to try and escape from an onslaught of fast-moving zombies. *Resident Evil 4*'s viewpoint, similar to having a camera perched over the right shoulder of the player's character, was also influential, providing a way to include the player's character on screen without blocking players' view of the action. The camera angle quickly resurfaced in titles such as 2006's *Gears of War*.

The idea of player vulnerability as a play mechanic was also explored in another genre that the PlayStation helped establish as part of the video game lexicon: the stealth game. Although early games such as 1981's *Castle Wolfenstein* and submarine simulations such as *Silent Service* revolved around stealth, the emergence of the PC game *Thief: The Dark Project* and the PlayStation titles *Metal Gear Solid* and *Tenchu: Stealth Assassins* in 1998 developed the concept enough to cement it as a distinct genre. Although the feu-dal Japan-based *Tenchu: Stealth Assassins* was first to be released, it was Hideo Kojima's *Metal Gear Solid* that really became the focal point for the genre's birth.

Envisaged as a 3D update of his late 1980s *Metal Gear* series of games, *Metal Gear Solid* drew on a cornucopia of influences ranging from childhood games of hide and seek to Hollywood movies such as *Escape from New York*, Cold War politics and tales of Tokyo in wartime. The player had to infiltrate a radioactive waste facility to prevent terrorists launching a nuclear weapon, armed with nothing more than a pair of bin-oculars, a radar that detects guard movement and a packet of cigarettes that could be smoked to help spot infrared trip wires. Kojima's non-confrontational, anti-nuclear parable became a huge seller and would spawn numerous sequels. The first, 2001's *Metal Gear Solid 2: Sons of Liberty*, saw Kojima bring his love of films to the fore through the inclusion of non-interactive story-telling scenes that could last up to 40 minutes. It was a divisive approach that repelled as many as it attracted.

Metal Gear Solid became one of the five best-selling PlayStation games, but the biggest seller of all was 1997's *Gran Turismo*, an ode to the joy of driving and car ownership. Created by Tokyo game developer Polyphony Digital, *Gran Turismo* was a petrolhead's dream. It combined real-life car models with the opportunity to build up a virtual garage packed with cars to drive. More than 10 million copies were sold worldwide.

The likes of *Tomb Raider*, *WipEout*, *Metal Gear Solid*, *Gran Turismo* and *Silent Hill* were perfect reflections of the image Sony sought to give the PlayStation – that it was not a kids' toy but a desirable piece of consumer electronics. Sega's resistance crumbled fast and even Nintendo found the going tough when it finally came up with its follow-up to the Super NES, the Nintendo 64, in 1996. The Nintendo 64's release had suffered numerous delays, partly because Nintendo was keen to give its leading game designer Shigeru Miyamoto the time he wanted to perfect *Super Mario 64* – the first 3D Mario game. While a number of games had tried to create platform games in 3D on PlayStation, not least Sony's own *Crash Bandicoot*, they offered none of the freedom of movement that had once been the defining feature of platform games such as *Super Mario Bros*. Miyamoto was scathing about such attempts describing them as attempts to "fool people" into thinking it's a 3D experience. He spent months figuring out how to bring Mario into a truly 3D environment, spending days working out how the virtual in-game camera should move around in response to the players' actions. He also spent weeks perfecting the areas Mario would explore in his 3D masterpiece; it was a process he compared to designing a theme park. Even the Nintendo 64's controller was built around the demands of Miyamoto's game.[5]

When the Nintendo 64 and *Super Mario 64* finally appeared in the shops in June 1996, Miyamoto's game was hailed as a Mario's best adventure yet and proof that the platform game could adapt to the 3D era. While Sony's *Crash Bandicoot* steered players down tight paths, *Super Mario 64* provided an open 3D playground to explore and interact with.

But while Nintendo's console had 3D graphics that could match those of the PlayStation, the company decided to stick with cartridges rather than adopt CD due to its concerns about long load times, durability and illegal copying of CD games. Nintendo's rejection of CD gave Sony another advantage in addition to its already strong lead over Nintendo: Square, the publisher of the *Final Fantasy* series of Japanese role-playing games.

Since its debut in 1987, the *Final Fantasy* series had only appeared on Nintendo's consoles but its creator Hironobu Sakaguchi wanted the seventh game in the series

---

5. The Nintendo 64 controller featured a thumb-sized version of the pressure-sensitive analogue joysticks mostly usually used in conjunction with PC flight simulator games and an expansion port for a 'rumble pak' that allowed games to make the controller vibrate. Sega and Sony quickly followed Nintendo's example and both features are now standard in game console controllers.

to take full advantage of extra capacity of CD. So when Nintendo decided to stick to cartridges on the Nintendo 64, Sakaguchi decided that rather than dilute his vision he would move the multi-million selling series to the PlayStation. It was a major blow for Nintendo as *Final Fantasy VII* proved to be the most popular in the series to date. *Final Fantasy VII* ushered in a new cinematic style to the series using the audio and 3D capabilities of the PlayStation to create an epic tale of fantasy eco-warriors out to stop the ruthless Shinra Energy Power Company from mining the energy of life itself. "I wanted people to be aware that planets are alive too," said Sakaguchi. "All living creatures are in the circulatory system, which is connected with planets."

Sakaguchi's ambitious epic, spread over three CDs, became the first Japanese role-playing game to really make it big outside Japan. It sold more than eight million copies across the world and provided one of the most iconic moments in game history: the unavoidable murder of Aerith, a key female character who helps the player in the game.[6]

Nintendo's console never came close to matching the sales of the PlayStation, despite Nintendo of America chairman Howard Lincoln's assertions that customers would favour Nintendo's more expensive Cadillac over Sony's cheaper Chevrolet. Within a year of launching the Nintendo 64, the company's chairman Hiroshi Yamauchi even admitted to Japanese newspapers that Sony now dominated the market and that Nintendo had lost its edge. But Nintendo still had popular brands and a loyal following. Releases such as *Super Mario 64*, *The Legend of Zelda: Ocarina of Time* and *GoldenEye 007*, a first-person shooter based on the James Bond film that was created by British game developers Rare, helped Nintendo sell around 30 million Nintendo 64s, but these sales were far below the 100 million plus sales of the PlayStation. It was Sega who took the brunt of the damage. By February 1998 it had stopped making the Saturn.

With Nintendo's position as the console top dog ended, Sony had got its revenge. The video game business now revolved around the decisions and actions of Sony and Nintendo looked like yesterday's men. As well as changing the balance of power in the game industry, Sony had reshaped society's attitudes to video games with its efforts to reach a more mature audience. Video games now had a popular culture relevance that seemed unimaginable just a few years earlier. Sony's realignment of the video games industry would also have big implications for the coin-op game business.

6. Deaths and tragedies that affect the player are common in Japanese role-playing games compared to North American and European role-playing games. Even as far back as in 1987, when Sega's *Phantasy Star* included a scene where a deranged father kills his daughter, who is one of the player's team of heroes.

Dance Dance Revolution: Rhythm action in Tokyo
Haruyoshi Yamaguchi / Sygma / Corbis

CHAPTER 22

Beatmania

It's the late 1980s and Psy•S are one of Japan's biggest pop groups. Their brand of synthesizer-driven pop had spawned a succession of hit singles, including the dreamy *Lemon no Yuuki* and the zesty *Angel Night*. Their music, reminiscent of early Depeche Mode, featured on popular anime TV series such as *City Hunter* and their tours packed out the nation's concert halls. "We were a kind of popular band in Japan, we played on tour, we performed in several places in huge halls," said Masaya Matsuura, who founded the group with Osaka jazz club singer Chaka in 1983.

Matsuura had become hooked on the musical possibilities of synthesizers as a teenager growing up in Osaka. He loved the way the technology allowed him to manipulate sound and had long been interested in the interface between computers and music. In 1980, aged 19, he became fascinated by *Kaleidoscope*, a 1979 demo created by Applesoft to show off the colour graphics of the Apple II home computer. Thinking that *Kaleidoscope* would be much better if it also had music, he wrote a soundtrack to accompany the demo's bright, pulsing visuals. His early experiment with computer music on the Apple II came back to him when he began wondering what else computers could bring to his musical work in the early 1990s.

"I was thinking music expression doesn't have to be a record," he said. "I had been using a computer for the composing and I would construct the data to play the music. Every time I did this I thought why does the sequencer always look the same or why can't I release the sequence data to the public instead of making the record? These kinds of thing made me think of a new way of expressing music."

In 1993, keen to explore the possibilities of computer music, he founded software firm NanaOn-Sha and began releasing multimedia CDs for the Macintosh that fused sound, graphics and interactivity. The more he did with NanaOn-Sha, the more Matsuura became enthused by the possibilities of computerised music. So when Psy•S broke up in 1996 he decided to concentrate on his experiments in software full time. By then he had begun thinking about making a video game for the PlayStation that enabled players to create music. "In the '80s I always used sampling and the sampling synthesizer is a very strange kind of machine," said Matsuura. "The most exciting part of the sampling machine for me was sampling the voice so you could play it on the keyboard. It is a very toy-type of thing and I thought if we had phrases it would be much more fun to play."

Matsuura's ideas resulted in *PaRappa the Rapper*, a PlayStation game where players had to repeat lines rapped by a selection of colourful characters by pressing buttons on the controller in time with the rhythm of the music. To give the game a sense of purpose Matsuura and his team cast the player as PaRappa, a floppy-eared dog out to woo a female sunflower with his rapping skills who seeks help from various teachers.

Matsuura hired New York City artist Rodney Greenblat to bring the world of PaRappa to life. Greenblat created an eccentric world packed with bizarre characters ranging from a chicken that teaches PaRappa how to make seafood cake by rapping the recipe on a TV cooking show to a Rastafarian frog called Master Prince Fleaswallow. Rather than creating fully-fledged 3D characters, Greenblat designed them as paper-thin cartoon characters that looked and moved as if they were paper cut outs brought to life. "Rodney and I spent a lot of time debating the characters of the game and the paper look," said Matsuura. "We felt with the hardware that maybe polygons were too blocky to express this in 3D."

As well as straightforward repeat-after-me play, the 1996 game enabled players to be creative with their rapping, provided they stuck to the overall rhythm. "*PaRappa* had a performance function so if you got a cool rating the teacher disappeared and you could play as you liked," said Matsuura. "This is very like music. I'm still very proud of the scoring system, which rates players on following the song and improvisation. This kind of improvisation feeling is a basic idea for music."

Prior to *PaRappa the Rapper*, the concept of a game that let players create music had rarely been explored. The most significant attempt prior to Matsuura's game was *Moondust*, a 1983 game for the Commodore 64 created by virtual reality pioneer Jaron Lanier. "The play was very weird for its time and if it was introduced now it might be more trendy," said Lanier. "There were six spaceships and one little spaceman, who bounced around the screen. You hit the button on the joystick and the spaceman drops a glowing seed. You have to the get the spaceships, which you coordinate all at once, to run over the seed to spread it towards the target, which was this sort of writhing, organic, vaguely female-symbolic astral phenomenon."

And as the player moved the spaceships, *Moondust* created music to reflect their actions. "It had an interesting harmonic model internally where you could improvise on it," said Lanier. "It always sounded normal, it didn't go into some weird atonal thing, but at the same time it didn't have a particular harmonic base. It achieved some intermediate level of harmonic complexity, which is not that easy to do whether writing a computer program or even writing real music."

Despite its strangeness, *Moondust* became one of the most popular Commodore 64 games of 1983. But while it earned Lanier enough money to start his own virtual reality research company, the music-based play of *Moondust* failed to inspire other game designers to explore the format. The arrival of Matsuura's *PaRappa the Rapper* 13 years later, however, did just that. It became a big seller worldwide and in Japan the game's rapping hero even became the face of the PlayStation brand for a short time. Its success encouraged other game companies to start making music-based games. And it was in the arcades that the first games to adopt Matsuura's approach appeared.

The arcades in 1996 were in trouble. Coin-operated games had always relied on having the best visuals, the best sound and the best technology, but the arrival of the PlayStation, Nintendo 64 and Saturn had narrowed the gap between arcade and home considerably. The arcades no longer looked like the natural location for the latest in game technology. Their role as social gathering places was also fading thanks to the emergence of the internet. "Arcades in the past were not only a fun place to play games – there was the social interaction between the game players and their friends," said Howell Ivy, executive vice-president of Sega Enterprises USA. "This social aspect has been taken over by the instant access to the internet. Social websites fulfil most of the needs the amusement arcade visitors craved in the past."

The losses of the arcade's social and technological advantages were compounded by the restrictive nature of coin-op game design. Unlike the rich, complex and involving worlds of the home, arcade game designers were still serving up snack-like experiences. "The arcade scene was limited by its economics of short play time," said Eugene Jarvis, who formed arcade game makers Raw Thrills in 2001. "A short intense experience is the only way for an arcade game to pay its way, but there are so many other ways to play games much more cheaply and more conveniently with many varied formats not subject to intense time pressure. The arcade, like a primordial rat, remained in its primitive evolutionary niche while the rainbow of today's gaming experience took shape." And as interest waned in arcade video games so did arcade operator profits. In 1994, when the PlayStation came out in Japan, the annual sales of coin-op games in the US were $1,570 million. In 1998 this had dropped to $1,129 million and in 2002 it was just $523 million.[1]

---

1. The popularity of arcades declined most significantly in Europe and North America. Japanese arcades, while declining steeply in number, did better thanks in part to the huge popularity of sticker photo booths. In mainland Asia and other nations, including Afghanistan, where home computers and consoles are rare, arcade video games are still successful and in some cases growing in number.

Another development that was tearing customers away from the arcades was the rapid growth of mobile phone ownership that began in the late 1990s. Now players could play games wherever they were, grabbing a quick gaming fix at the bus stop or while waiting for a friend to turn up. The first mobile phone game to make a major impact was Finnish phone manufacturer Nokia's 1997 game *Snake*, a simple remake of the 1976 arcade game *Blockade* for the Nokia 6610 phone.

Snake challenged players to steer a constantly moving snake around the screen so it could eat apples that would cause its tail to grow. Crashing into the edge of the screen or your tail resulted in instant death. It was a simple but compelling game designed to be played in short bursts, much like traditional coin-op games, and it introduced tens of millions of people to the idea of playing games on their mobile phones. In many ways the mobile phone game took over the niche once held by arcade games by providing short gaming snacks but with the added advantage of communication features and mobility. This combination of bite-sized games, communications and portability was something that game makers quickly latched onto with early mobile games such as *Alien Fish Exchange*, a fish-breeding game released in 2000 where players raised virtual fish and used their internet-connected phones to swap their fish with other people.

Arcade operators responded to the declining interest in coin-op video games by investing in different types of machines. Gambling machines, claw grabbers and photo booths – which all offered experiences players could not get at home – began to displace video games, much like the video game pushed electro-mechanical games out of the arcades in the 1970s. Only a few video games such as Sega's enduringly popular *Daytona USA*, a 1993 driving game that allowed people on different machines to race each other, bucked the trend. "*Daytona* was one of the first real simulators that gave the player a first-person view and truly placed the player inside the game," said Ivy. "It appealed to the full breath of game players, both men and women."

Coin-op game manufacturers were not, however, about to sit by as the arcades abandoned them. They began hunting for new ways to distinguish their titles from the games on offer in people's homes and pockets. Many tried creating unique controllers to capture players' interest, from the train control panels in train-driver sim *Densha De Go!* and the football to kick in Spanish game maker Gaelco's *Football Power* to leashes in dog-walking game *Inu no Osanpo*.[2] Most bizarre of all was the South Korean title *Boong-Ga Boong-Ga*, which challenged players to insert the pointing finger of a plastic hand into the anus of a

2. Galeco was one of the last remnants of Spain's once thriving video game industry. While Spain was a major European producer of games in the 1980s, the move to more advanced home computers at the end of that decade proved traumatic for the country's game publishers. By the middle of the 1990s almost all of Spain's leading game companies had shut down with only Dinamic surviving thanks to its popular budget soccer game *PC Fútbol*. While Spain still had its fair share of game developers, the early promise of the country's game business was never fulfilled.

plastic posterior to punish on-screen characters that ranged from ex-girlfriends to child molesters. Saved games, stored on memory cards or provided in the form of passwords, were also used in games such as Sega's multiplayer racehorse management game *Derby Owners Club* to try and keep players coming back to the arcades.

A few focused on appealing to loyal arcade-going video game players by producing ever more challenging and intense variants on existing game genres. One of the fruits of these attempts to appeal to the dedicated arcade gamer was the emergence of the 'bullet hell' shooter, a sub-genre of 2D shoot 'em ups that threw players into ever more hectic fire fights where the screen was often covered in a blizzard of enemy bullets. Arcade game makers Toaplan had paved the way for the bullet hell shooters with its fierce 1993 game *Batsugun*, but the genre really took shape when a group of former Toaplan staff founded Cave.

Cave took the challenge of *Batsugun* to new heights with its debut release, 1995's *DonPachi*, which pitted players against a daunting storm of enemy fire that could only be survived if players memorised the pre-set patterns of the bullets so they could steer through the maze of enemy fire. For Japanese shoot 'em up aficionados in need of ever tougher games the ability to beat bullet hell games became a source of pride. The sub-genre became so popular in Japan that VHS tapes demonstrating how to complete these games became big sellers.

But of all the coin-op game makers' efforts to keep players coming to the arcades it was the music games that followed in the footsteps of *PaRappa the Rapper* that were most successful. Konami became the first company to reinvent the music game for the arcades with 1997's *Beatmania*, which gave players a DJ turntable and mixer and challenged them to play in time to various techno tunes. Instead of following *PaRappa's* call-and-response action, *Beatmania* required players to time their actions to coincide with the beats that appeared on the screen as lines falling down towards a marker that indicated when the beat would be played. *Beatmania* lacked the free-styling improvisation of *PaRappa the Rapper*, but its booming speakers and its ability to make players feel as if they were playing the music made it a big draw in Japanese arcades. Konami followed *Beatmania* with a spate of similar games that used different instruments, including 1999's *Guitar Freaks* and the following year's *Keyboardmania*.

Konami's music game efforts also resulted in 1998's *Dance Dance Revolution*, where players had to dance in time to on-screen prompts by stepping on large metal buttons.[3] As well as bringing a physical element to the game, Konami sought to make

---

3. The idea had been tried before with *Dance Aerobics*, a 1987 game for the NES that worked with Bandai's Power Pad accessory – a plastic sheet with 12 buttons that were controlled by stepping on them.

the game enjoyable for spectators as well as players, equipping the machine with loud speakers, flashing lights, giant screens and an almost stage-like platform for the player to strut their stuff on. The very best *Dance Dance Revolution* players could even throw in their own dance moves on top of those required by the game. *Dance Dance Revolution* became a huge success in Japan and Europe, sending Konami's income soaring by 260 per cent between the 1997/98 and 1998/99 financial years. It even caught the attention of West Virginia's health officials who put the machines into schools in the hope of tackling the state's position as the US's obesity capital.

But the edge that music games gave the arcades was short-lived. Konami released a PlayStation version of *Guitar Freaks*, complete with a sold-separately guitar controller, in 1999 and console-only titles such as the improvisational duelling of *Gitaroo Man* and the synaesthesia-inspired trance experience of *Rez* quickly reclaimed music games for the home.[4]

But most of these music games, with their Japanese visuals and pop, had failed to connect with US consumers, who tended to prefer traditional hard rocking stars to cartoon dogs with a taste for ragga or techno. As a result music games remained on the fringes of video games in North America. The breakthrough that took the music game into the US mainstream came from Harmonix Music Systems, a game studio based in Cambridge, Massachusetts, that wanted to find ways to let non-musicians play music. Harmonix specialised in music games but, despite critical acclaim, its abstract music titles *Amplitude* and *Frequency* had been snubbed by consumers. Figuring that its earlier efforts were just too cerebral to be popular, Harmonix decided to embrace the mainstream.

It started by teaming up with Konami to produce *Karaoke Revolution*, a PlayStation 2 game that rated players on their pitch and octave as they sang along to a selection of popular hits. *Karaoke Revolution* was a groundbreaking step, paving the way for later karaoke music games such as Sony's *SingStar* series and Harmonix's own *Rock Band* series, where up to four players could experience performing as a band complete with singer. After completing *Karaoke Revolution*, Harmonix decided to revisit the guitar controller ideas of *Guitar Freaks* and *Gitaroo Man*. But instead of the jazzy Japanese pop of *Gitaroo Man*, Harmonix married the guitar game to western rock music to create *Guitar Hero*, a game that let non-musicians live out their dreams of rock stardom.

"In November 2005 Harmonix founder Alex Rigopulos sent me a copy of *Guitar*

---

4. Synaesthesia is a neurological condition where the stimulation of one sense prompts a person to involuntarily experience another sense. For example, a person with the condition might see a blue sky and that provokes the taste of apples in his or her mouth. Russian abstract painter Wassily Kandinsky's exploration of the condition in his art inspired *Rez's* creator Tetsuya Mizuguchi to synchronise the game's sounds, graphics and the vibrations of its specially made 'trance vibrator' device.

Hero," said Manny Gerard, the former Warner Communications executive who invested in Harmonix in the late 1990s. "I have four grandchildren aged 11 to 14. That Thanksgiving I stuck them in a room with a PlayStation 2 and handed them this game. Three hours later we had to pry them out of the room, if we'd left them alone they'd still be in there playing this game."

Guitar Hero became a multi-million selling success that later evolved into a music platform that allowed players to download new tracks to play. It even had enough influence to break new artists and introduce old acts to a new generation. British heavy metal band Dragonforce were just one example. After its song *Through the Fire and Flames* appeared in 2007's *Guitar Hero III: Legends of Rock*, sales of the track soared from 55,000 to 624,000 copies. Soon *Guitar Hero*'s publisher, Activision Blizzard, was talking about charging record labels to have their artists' songs featured in the game rather than paying for the rights to include them.

For arcade game producers the music game genre offered only temporary salvation from decline. The on-the-move gaming of mobile phones, the depth of home games and the social networking of the internet meant the arcades were increasingly irrelevant for those after video gaming thrills. Arcades began turfing out video games in favour of more profitable snack dispensers and gambling games. And as the arcade operators turned their backs on video games, many of the companies that had forged the game industry did the same to the arcades. Midway, which had also bought its arcade-making rival Atari Games in 1996, stopped making arcade games altogether in 2000. Data East sold off its arcades to Sega before going under in 2003 while Taito shut down its US operation.

New Hampshire's Funspot, the world's largest arcade centre, refocused on bingo, bowling, mini-golf and food. The centre's video games ended up being used to form its American Classic Arcade Museum section – a nostalgic not-for-profit throwback to the glory days of the coin-operated video game. "The museum is a labour of love," said Lawton. "Most of us were around when these games were new and being a part of the preservation process is a rewarding experience and it is fun to watch the players who come into relive these great classic games."

The video game arcade – the birthplace of video gaming – had been reduced to rose-tinted nostalgia and fading memories.

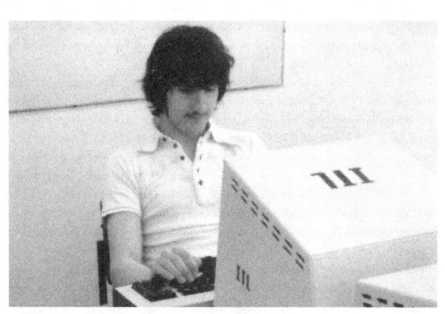

Making MUD: Roy Trubshaw working on MUD, circa late 1979/early 1980
Courtsey of Richard Bartle

CHAPTER 23

You Haven't Lived Until You've Died In MUD

Just after midnight on the 9th August 1997, game designer Richard Garriott logged onto his latest and most ambitious creation *Ultima Online*, an online game that allowed thousands to play together in his fantasy realm of Britannia. Controlling his in-game persona Lord British he headed to the ramparts of his castle in the town of Trinsic with Lord Blackthorn, the alter ego of the game's associate producer Starr Long, to address the crowd gathered in the castle grounds below.

The courtyard teemed with people. They had gathered because today was the end of the world – the pre-release 'beta' testing of *Ultima Online* was about to end. Over the past few months a select few fans of Garriott's games had been living a second life within the virtual world of *Ultima Online*, helping its creators spot any problems that would need to be fixed before its formal launch in September.

The seconds to the end of the test were counting down as Lord British began to address the crowd. Soon the Britannia they had come to know would be no more. Lord British thanked the crowd for their contribution to *Ultima Online* and began to explain the details of the commercial version that would launch a few weeks later.

"This day was the finale of the beta," said Garriott. "We had everyone who was playing on the same server. Everybody knew that at a certain hour the servers would be turned off and in the hour or so that preceded literally turning off the switch, me and Starr teleported from city to city to thank people for their participation."

The atmosphere as Garriott and Long teleported through the world they had created on their way to Trinsic was jovial. "People would do really funny things," said Garriott. "In the city of Moonglow when we arrived people, as a joke, mooned us – took off their pants leaving them in underwear really and faced away from us and bowed as a mass of people. It was hilarious to see everybody in Moonglow mooning us."

One of the people gathered in Trinsic to hear Garriott's farewell address was Ali Shahrooz, an internet consultant from Indianapolis whose *Ultima Online* persona was a thief called Rainz. And he was in a mischievous mood. Shahrooz decided to pickpocket some of the people in the crowd and came across a firefield spell that could create a wall of fire. For fun he promptly cast the spell onto the battlements where Lord British and Lord Blackthorn were standing. Shahrooz expected nothing to happen. It was well known that both Lord British and Lord Blackthorn were invincible characters. "Both Richard and I responded 'ha, ha, ha, don't you know our powers? You can't destroy us, we are much stronger than your puny firefield'," said Long. To emphasise the point Garriott made Lord British walk back into the flames: "My first reaction because it was a fire was to step out, however then I thought I don't have to worry, my character's immortal. So I just walked back into the fire and then fell over dead."

Garriott, unlike Long, had failed to 'switch on' his character's immortality. Lord British was dead. Pandemonium ensued. Rich Vogel, the producer of *Ultima Online*, was among the game developers at the scene: "We had something like 15 game masters surrounding him like the secret service.[1] It was like when someone shoots the president – everyone runs around and protects him. We immediately surrounded him and teleported him out of the area." Long, meanwhile, had decided to deal with the culprit. "We were all just shocked," said Long. "Staying in character, I became enraged and started summoning demons into the crowd and summoning lightning."

Chaos broke out in the castle grounds as giant demons slaughtered the crowd indiscriminately and the defenceless fled in panic. "All of the game masters started summoning demons and devils and dragons – all of the biggest, most horrible monsters they could," said Garriott. "The creatures they summoned were far, far more powerful than any human in the vicinity so in no time they wiped out everyone in the square. People are shocked and horrified and screaming, running for their lives, so to speak, and the ones that are killed are trying to get themselves resurrected and back into the game before the servers go off."

As the scene filled with monsters and death-dealing spells, the action ground into slow motion as the computer server running the game struggled to cope with the sheer volume of information it had to take in and send out to the players sat on their computers across the world. Eventually the server buckled and *Ultima Online*'s beta test ended abruptly in the middle of a chaotic bloodbath played out on hundreds of computer monitors across the world. "We had this initial reaction of being really angry at this guy for doing it but then we realised it's kind of awesome really, it shows that this game is really about what the players want – they're really more in control than we are," said Long.

---

1. Game masters are people appointed by the developers of massively multiplayer online games such as *Ultima Online* to police the virtual world on their behalf. The term originates from the name given to the person who would oversee games of *Dungeons & Dragons*.

Ultima Online marked the fruition of a 20-year quest to create a virtual world that could be inhabited by hundreds of players on separate computers connected via telephone lines. The journey that ended in Lord British's assassination began in 1977 when Don Woods released his reworking of Will Crowther's *Adventure*, the first text adventure game. Among the many early computer users who played *Adventure* was Roy Trubshaw, a British computer science student at the University of Essex. Trubshaw thought its text-based input method would make a good interface for his pet project to build a virtual world that users of different computers could explore together. "I liked the idea of multiplayer games – wandering around the locations in an *Adventure*-like environment and doing stuff to or with other folks in the same game as you was an unutterably cool idea," said Trubshaw.

Trubshaw invested his time in figuring out how to get the computers that the various players would use to communicate with each other, and then began designing a virtual world for people to log into, explore and interact with. There would be no goals, no game as such, just a world other than our own to discover. He called his virtual world *MUD* – short for *Multi-User Dungeon* in a nod to the early name for the text adventure *Zork!*. To help him fill this world with content, Trubshaw turned to his friend Richard Bartle, another student at the university. "We first met when he joined the Essex Computer Society," said Trubshaw. "This society was created ostensibly to allow students to study all things computer-like in their own time but in reality allowed us to gain more access to the main computer and play with it. It soon became apparent that Richard was (a) a genius and (b) interested in games of all kinds but especially games of strategy and cunning."

Bartle had been designing games for his own entertainment since his early teens, mainly pen-and-paper role-playing games and choose-your-own-adventure story books. "One that had quite a big influence on my views on *MUD* was a single-player role-playing game that I started when I was 12 or something," said Bartle. "Role-playing games hadn't formally been invented back then – there wasn't even a name for that particular type of game. I started off creating a continent like mid-1800s Africa. I had an explorer and what happened is I would write the diary of the explorer and explore the continent I had created. I wouldn't know exactly where I was going to go or what was going to happen. It was a big influence on my ideas of what an imaginary world was, how to build a world and how to make it real."

From the off Bartle started pushing for Trubshaw to turn *MUD* into more of a game and, since his attempt to create a fully functioning virtual world was proving impossible given the computing power available, Trubshaw agreed. "Roy wanted to build a world, a place that was separate from the real world. It wasn't puzzle-driven, it wasn't a game world at all," said Bartle, who took control of the game's development when Trubshaw stepped back from the project to focus on completing his degree. "I wanted there to be game aspects to it. One of the obvious things to do was to put in puzzles, but they don't really work in multiplayer games, so most of the puzzles were chainings of goals. So I want to get some treasure, the treasure is behind the door, to get through the door I need a particular key which is held by a particular monster, so I need to kill that monster, which is behind a portcullis, and to get past the portcullis I need someone to help me open it. That's chaining of goals."

Having used these kind of treasure-hunting goals to give *MUD* a gaming foundation, Bartle incorporated other game concepts such as the character levels from *Dungeons & Dragons*. He figured that levels were a good way to motivate players and a useful means of conveying each individual's knowledge of the game to others: "I wanted people to have some goal that was attainable and I wanted people to be able to tell, just by looking, how good another player was. I also named the characters by their level. I didn't say Freddy level 6, each level had a name, so it would be Freddy the Warrior or Freddy the Necromancer."

Bartle's completed version of *MUD*, which went live in 1980, offered people the chance to adopt a new persona through which they could live, converse, fight, fall in love and explore a virtual fantasy world. It was not the first game to let users of different computers play together, but never before had a multiplayer game brought them together in an alternative reality where socialising and freedom of choice was as important a part of the experience as the goals of the game itself.[2] "Right from the beginning we knew this was something that was always going to be a wondrous place that was different," said Bartle. "*MUD* was always about freedom. We always wanted to make a virtual world – a place where you could be and become yourself free from the constraints of the real world. What we were trying to express through the design was a statement about freedom."

The social interaction and player freedom offered within *MUD* took video games into uncharted territory. Before it, multiplayer games limited players to following the rules laid down in the computer code. *MUD*'s virtual world, however, allowed players to define the shape and atmosphere of the world through their actions. In theory Bartle still had ultimate control, but in practice the players also had power. "The people who run the game are effectively gods – they can do anything they want, they can change the

2. 1974's *Maze* (see chapter 5) was not the only multiplayer game created during the 1970s. In 1973 John Daleske created an eight-player space combat game called *Empire*, which ran on terminals connected to the University of Illinois' education-focused PLATO network. Empire underwent several updates, including 1976's *Empire IV*, which allowed up to 80 PLATO users to play at once. PLATO's online communications capabilities also made it host to some of the world's first message boards, chat rooms, instant messaging and emoticons.

physics of the world," said Bartle. "However, if they do anything they like and the players don't like it, the players can leave and you end up a god without worshippers. Both players and developers have the ultimate weapon, the developer can do anything they want and the players can walk away."

This tug of war between players and developers made *MUD* a social experiment as well as a video game. And within its digital walls, a new culture would be born. Players and developers debated how and when disruptive players should be punished and coined new words to describe the actions and types of players within *MUD*. The disruptive or abusive players, who hounded and deliberately annoyed other visitors to the world of *MUD*, became known as 'griefers'. New players unfamiliar with the world and its mores were called 'newbies' or 'noobs'. The exact origins of the term 'newbies' is unclear. "It's one of those words that has probably been invented before – maybe private schoolboys used it or the British Army. But I'm fairly confident that 'newbie' is one I made up and since I've never been to public school and never been in the army, I can't of imported it from that," said Bartle. Whatever its origins, the word 'newbie' caught on in *MUD* and then spread onto the bulletin board systems frequented by *MUD* players. And when the internet arrived, the term spread across cyberspace until it eventually crept into everyday language.

Another cultural legacy of *MUD* was an acceptance that male players could become female characters inside virtual worlds. "When we first had *MUD* running, all our players were men because it was a computer science department in 1978," said Bartle. "But they weren't really role-playing because they hadn't got the idea that they were playing as someone else, not themselves, so they didn't create female characters and all that."

Bartle solved the problem by creating Polly, the first female character in a virtual world, and venturing into the world of *MUD*. "So here's me, who everyone knew was male, playing this character who is female," said Bartle. The stunned players who witnessed Polly's arrival were unsure how to react – are you gay, some asked, or are you a transvestite? Bartle's response was straightforward: "I don't care what you think. This is Polly and this is me. Of course I'm not the same as Polly – Polly's part of a computer program."

And once the gender divide had been crossed, attitudes changed. "I'd kind of given them permission to play as female characters and they did," said Bartle. "Most would try it, some didn't because they felt it was a slur on their sexuality, but most did have a female character. It became accepted fact that you could have a female character. If *MUD* had been invented somewhere other than the UK people would still play characters of the opposite gender, but it might not have taken root because there might have been social reasons that prevented the original players from doing it."

Gender swapping would become widespread in the virtual worlds that followed. A 2008 study by the UK's Nottingham Trent University found that 54 per cent of men and 68 per cent of women who played multiplayer online games such as *Ultima Online* had characters of the opposite sex.

But *MUD* could have easily slid into obscurity if it wasn't for two crucial decisions by the University of Essex around the time of the game's creation. The first was its decision to test a new computer communications system developed by British Telecom. "There wasn't an internet back then, but there was a system called EPSS – the Experimental Packet Switching System – that British Telecom had implemented," said Bartle. "Only a very few universities in Britain had it – maybe two or three. We were one of the universities being used to trial this. You could use EPSS to connect to what would now be called the internet, which was back then ARPAnet."

Using EPSS, the pair connected to places such as Xerox PARC, Stanford University and MIT and opened accounts so computer users in these institutions could access *MUD*. "To get an account you had to say something that described what you were doing," said Bartle. "I had to think of some words to put and came up with 'You haven't lived until you've died in *MUD*'. That was our calling card."

The University of Essex's second crucial decision was to let computer users from outside its campus log onto and use its computer systems. "That was an important decision," said Bartle. "They allowed people to access the computers during off-peak times when they would otherwise be idle. That meant other people could play the game – we called them externals, the people who weren't in the university. They could play it and because they could play it they saw the proof of concept and some thought 'I'll write my own'."

Those inspired to create their own virtual worlds were also aided by Bartle and Trubshaw's belief that *MUD* should not be a for-profit project but a gift to other computer users. "Back then the mentality of programmers was that software was meant to be free and available to everybody," said Bartle. "So if people asked me for a copy then as long as they weren't going to charge money for it then I would give them a copy."

Programmers inspired by *MUD* soon began to hack the code of Bartle and Trubshaw's game to make *MUD*-style worlds of their own. Soon *MUD* became a byword for text-based multiplayer virtual worlds ranging from *Shades*, a for-profit fantasy-themed *MUD* developed by British Telecom, to *Rock*, a free-to-play *MUD* inspired by Jim Henderson's children's TV series *Fraggle Rock*. The most popular by some way was *AberMUD*, the 1987 creation of four students at the University of Aberystwyth in Wales. "*AberMUD* wasn't a great *MUD* – there were better ones around, *MUD* itself was better – but it ran under the Unix operating system," said Bartle. "Most of the American universities were running Unix and we didn't run Unix at the University of Essex, we had a DECSystem-10 back then. So once people got hold of *AberMUD* it just spread across US universities like wildfire. A thousand copies of the game were running in America within six months to a year of its creation."

AberMUD initiated an explosion in the number of *MUD*s. By the end of the 1980s there were around 20 or so *MUD*s, but by 1992 an estimated 20,000 people were living second lives in around 170 *MUD*s. However, while *MUD* had sought to strike an even balance between game and social network, *AberMUD* emphasised the game aspect, much to the disappointment of those who preferred the more social side of the experience. "The social players who'd been playing these games before because they liked socialising with people began to feel left out, so in 1989/1990 there was this schism where the social players broke off," said Bartle.

The social players found a new home in 1989's *TinyMUD*, which deliberately snubbed the game elements in favour of a purist social experience that echoed Trubshaw's original non-game vision for *MUD*. Its creator James Aspnes, a student at Pittsburgh's Carnegie-Mellon University at the time, rejected the scoring and levelling up of *MUD* and sought to build a world where simply being part of it and interacting with other people was what mattered. *TinyMUD* and its followers eventually evolved into a new genre of multiplayer game – the *MOO* or *MUD Object Orientated*. And as socially orientated *MUD* fans gravitated towards *TinyMUD* and *MOO*s such as *LambdaMOO*, a pioneering text world where players could use computer code to create interactive objects within the game world, those who lusted for the traditional gaming thrills of competition and adventure started to weed out the social elements that they no longer felt obliged to maintain. The most influential of these new game-focused *MUD*s was 1991's *DikuMUD* – the creation of a group of Danish students from the Datalogisk Institut Københavns Universitet (DIKU), the computer science department of the University of Copenhagen. "They took the basic *MUD* concept from *AberMUD* and super-gamified it," said Bartle. "They deliberately added some concepts from *Dungeons & Dragons* – things like classes and races – and honed the game play. *DikuMUD* swept the whole *MUD* thing to one side – if you were encountering a *MUD* for the first time you're going to go for the one with the game play that is most compelling rather than the most cerebral."

In parallel to the spread and evolution of *MUD*s, a number of similar virtual worlds also emerged independently during the 1980s. The first was *Scepter of Goth*, a 1983 game created by Minnesota programmer Alan Klietz. *Scepter of Goth* had a lot in common with *MUD*, but its author knew nothing of Bartle and Trubshaw's game at the time and, rather than giving away his game, Klietz charged players to play. Other American game designers followed in his footsteps hoping to profit from home computer owners who owned modems.

But while a few companies and individuals made money from these games, none achieved the level of influence of *MUD*, largely because of Bartle and Trubshaw's decision to give away their game. "Because they were trying to be commercial all the expertise was kept in-house," said Bartle. "And because we let people have the code of *MUD* and we didn't object to people writing their own games – we encouraged them, in fact – we got this large number of people who were skilled in creating *MUD*s. So when virtual worlds finally did take off and people needed to recruit large numbers of experienced online multiplayer game designers, there were a thousand *MUD* makers for every one that came from these commercial games. That's why today's massively multiplayer online games are descended from *MUD*s rather than *Scepter of Goth*."

Trying to make a profitable online game during the 1980s was, however, an uphill struggle. Relatively few people owned a home computer and even fewer had one with the dial-up modem needed to connect to pre-internet computer networks. And even for those with modems, the cost of playing games online was prohibitive. First players had to buy access to an online network such as The Source, which demanded a $100 set-up fee and charged users $10 for every hour they were connected to it. These networks were separate from each other and users could only access the games, software and information that their network operator provided, making them more like a single website than the internet.[3] Once connected, early online gamers then faced additional charges for playing commercial online games such as *Heroic Fantasy*, a 1982 turn-based role-playing game played through email that charged $2.50 per move.

Given such high charges and the considerable expensive of providing such services, networks such as The Source, CompuServe and Quantum Link went to great lengths to keep players online for as long as possible. And multiplayer video games with their social interaction and compelling game experience were seen as especially good at keeping people online. Quantum Link, a network for Commodore 64 owners, was particularly successful at using games to keep computer owners racking up the bills. Its most popular effort was *RabbitJack's Casino*, an online gambling game created by former Imagic

3. Modem owners could also connect to bulletin board systems or BBSs. These had their own specific telephone number that the modem had to call in order to connect to the BBS, which sometimes limited the number of users at any one time to just one person. The content of BBSs ranged widely but included news updates, email facilities, the ability to exchange software and – in some cases – games to play, such as computerised versions of card games like Blackjack. Incidentally, the pre-internet situation in France was different. There the government-owned France Telecom launched the Minitel system in 1985. Minitel worked like a nationwide time-sharing computer. French citizens could get a Minitel terminal for free, plug it into their phone line and connect to Minitel in order to check weather reports and train times, read newspapers and play games such as a version of *Des Chiffres et des Lettres* – the French TV game show known as *Countdown* in the UK. Ironically, the popularity of its early exposure to online communications meant France was relatively slow to embrace the internet.

game designer Rob Fulop in 1987. "I remember meeting the president of Quantum Link at a trade show and saying they should do a casino," said Fulop. "They were new then and they were charging $4 an hour and wanted things to entertain people. It wasn't mass market then, it was a pretty niche business."

RabbitJack's Casino was designed with one commercial goal in mind: keeping people online. "We built the slot machines to be very generous. Unlike a casino slot machine, *RabbitJack's* was designed to give you chips. You win a lot and so you're like 'Wow, this is the greatest casino in the world!'," said Fulop.

While *RabbitJack's Casino* allowed up to five people to play Poker together, most of its gambling games were single-player experiences, such as the slot machines and Bingo, that pushed the social element of online gaming to the fore. "In the single-player games you could chat, so in Bingo you could all chat and play Bingo together," said Fulop. "Bingo's a very easy game, you don't have to do anything – it's a brain-dead simple game to play. Also the big advantage of Bingo is scalability – you can have 10, 100, 100,000 people playing together, all excited, waiting for P4. There's not many games like that. Good luck trying to figure out a game where you can entertain a million people in five minutes. You can't beat Bingo."

RabbitJack's Casino became the single most popular game on Quantum Link with around 15,000 regular players who ate up 3 per cent of the network's capacity to handle traffic. But the most ambitious and innovative title to appear on Quantum Link in the 1980s was a graphical virtual world called *Habitat* that was created by Chip Morningstar, a game designer at Lucasfilm Games.

Habitat evolved out of Lucasfilm's belief that by creating a visual virtual world it could extend the appeal of online gaming far beyond those who enjoyed text-based *MUD*s. While a few online games had used basic graphics before, *Habitat* sought to create a fully-fledged visual world with animated characters that brought the players' digital alter ego to life. *Habitat*'s goal was to create a place where players really could experience an alternative reality. Set in a vaguely modern day world, Morningstar designed a persistent virtual world built out of 20,000 single-screen locations and filled it with numerous interactive objects for players' avatars to use.[4]

To ensure players felt connected to the world of *Habitat*, the game allowed them to customise the looks of their digital self, decorate their virtual homes and adopt computerised pets. Even the customer service aspects of the game became part of the game world. Players could contact the system administrators with their problems via *Habitat*'s

---

4. Most virtual worlds at this time, *MUD*s included, were not permanent. In *MUD*, for example, once players had located all the treasure in the game, the game would reset the world back to its initial state with the treasures placed in new locations. *Habitat*'s world by comparison never stopped, it was built to last.

Bureaucrat in a Box, use Pawn Machines to sell the objects their digital selves owned and check the state of their virtual finances using ATMs. And to give players a sense of purpose, Lucasfilm created dozens of fun activities to participate in ranging from road rallies to games of Chess and treasure hunts.

The idea of these activities, explained Morningstar, was to ensure no matter what a player's personal interests were there would be something within *Habitat* that appealed to them. As if this wasn't ambitious enough, Lucasfilm also intended to allow up to 20,000 people to play in *Habitat* at once. It was an unheard of figure. *MUD*s rarely topped the 100-player mark and they didn't even try to represent their worlds in graphics. Given the volume of traffic and the unpredictability of player behaviour, Lucasfilm and Quantum Link decided to do an initial pre-release 'beta' test of the game in 1986 with 500 players. Morningstar and his team started out believing that *Habitat* should function in a way similar to a theme park, creating new activities designed to keep the visitors to its virtual world entertained.

The developers spent weeks honing some of *Habitat*'s more involved quests and introduced them to the game world confident that it would keep players occupied for weeks on end. Players solved them in just a few hours. Deciding that their original centrally planned approach was doomed to failure, Lucasfilm did a u-turn and embraced a more libertarian, free-market attitude where they would be facilitators not directors. They gave players weapons and allowed them to kill each other in the hope of encouraging them to make their own fun.

That too backfired. *Habitat* became a lawless world as players gunned each other down in the streets and stole each other's items. One player, a Greek Orthodox priest in the real world, responded by forming the Order of the Holy Walnut, a popular virtual religion that preached non-violence. In the *Habitat* town of Populopolis, players fed up with the lawlessness formed a virtual town council and elected a sheriff before launching a campaign to get Lucasfilm to grant their law enforcer of choice special powers. As a result Lucasfilm agreed to hold a referendum among players to decide what powers *Habitat*'s first police officer should have.

But before the referendum could be held, *Habitat* was shut down. *Habitat* had become a victim of its own success. Despite the challenges, *Habitat* had largely delivered on its vision but the beta test had revealed a major problem that no one foresaw. While just 500 people had access to the trial version of the game, they were playing it so much that the beta test version of *Habitat* swallowed up 1 per cent of Quantum Link's network capacity.

Quantum Link realised that if *Habitat* matched the success of *RabbitJack's Casino* and people played it as often as the first 500 players, its network would be unable to cope.

And since increasing the network's capacity was seriously expensive at the time, Quantum Link and Lucasfilm cancelled the game's full release.[5]

Habitat wasn't the only game to appear on Quantum Link that sought to push back the boundaries of video games. In 1988 the network teamed up with a writer called Tracy Reed to create *The Quantum Link Serial*, an experimental interactive fiction project that foreshadowed the fact-and-fiction-blurring alternate-reality games that would gain prominence thanks to 2001's *The Beast*, a promotional game for Steven Spielberg's movie *A.I.*[6] Included as part of the subscription to Quantum Link, *The Quantum Link Serial* combined storytelling with online chat and email. Reed would write the story in weekly instalments after asking readers to suggest how they could be included in the tale. *The Quantum Link Serial* became one of Quantum Link's most popular features, not least because the readers themselves had become part of Reed's fictional world.

In 1989 Quantum Link also released the first of two games it had commissioned from Don Daglow's Stormfront Studios. The pioneering game developer, who had created 1971's *Baseball* and the Intellivision game *Utopia*, started working with the network after a visit from its executive vice-president Steve Case. "Steve did a tour of all the game companies looking for ways to hook up with them and get games on Quantum Link," said Daglow. "I was at Brøderbund and he showed me what they were doing. I privately thought the Apple II was starting to fade and Brøderbund was the leading Apple II game company. I did a two-title deal with Steve for a token payment because they didn't have any cash at that point – Quantum had 40 employees and was operating on a shoestring – to get us in the door."

Stormfront's first creation for Quantum Link was 1989's *Quantum Space*, a text-only turn-based strategy game released just before Quantum Link renamed itself America On-Line (AOL). *Quantum Space* relied on the network's internal email service. "You would get your turn reports and then you'd email back a message that had what you wanted to do in a form," said Daglow. "The system would turn it into a string of data that would go to us and get processed."

Stormfront's second online game, *Neverwinter Nights*, was much more ambitious. It took the format of game publisher SSI's series of official *Dungeons & Dragons* computer games and created an online role-playing game that, unlike the various *MUDs*, used 2D

---

5. Quantum Link later launched a heavily stripped-down version of *Habitat* called *Club Caribe* in 1989, which managed to attract 15,000 players at the height of its popularity. The following year *Fujitsu Habitat*, a remake of the original *Habitat* with a new look, appeared on the Japanese FM Towns computer.

6. Alternate reality games tend to take the form of elaborate puzzles where players are given a few cryptic clues and usually use websites – some real, some created just for the game – and other media (such as newspaper adverts or answer phone messages) to try and solve it.

graphics rather than text. Costing between $4 to $8 an hour to play, it became one of AOL's most profitable games as players became absorbed in its visual representation of the *Dungeons & Dragons* world, spending hours socialising, fighting and adventuring. As with *MUD* and *Habitat* before it, *Neverwinter Nights* once again showed how the bringing together of players often led to unexpected results, including player-organised comedy nights and poem readings as well as virtual and real marriages between players.

Daglow was not the only video game pioneer attracted to online gaming. In 1988 Dani Bunten Berry, the creator of the multiplayer strategy and trading game *M.U.L.E.*, had also begun to explore online as part of her career-long interest in multiplayer games. Bunten had always been interested in games that brought people together, an interest she put down to a childhood where playing board games with all of her family gathered around the table were some of her happiest childhood memories. After dabbling in single-player games with *Seven Cities of Gold* and *Heart of Africa* she became an outspoken advocate for multiplayer games. At the 1990 Computer Game Developers Conference she summarised her philosophy in one snappy soundbite: "No-one on their deathbed ever said 'I wish I had spent more time alone with my computer'." Her first venture into the online space was *Modem Wars*, a robot-themed war game released in 1988. Inspired by playing soldiers with her brothers as a kid, *Modem Wars* set out to avoid the complexity, pensive and lengthy experience of most strategy games with its action-based interpretation of war gaming. It foreshadowed the real-time strategy games that became popular following the release of *Dune II*, but its focus on action alienated traditional war game fans and, since few people owned modems at the time, sales were poor.

Sierra Online's founder Ken Williams also got the online bug and in late 1991 his company began exploring the frontiers of cyberspace with The Sierra Network. "The original mission statement for The Sierra Network came from me trying to think of something my grandma could do from home," said Williams. "I came up with a product that would be card and board games for seniors. They would be able to pick up a game and chat, 24 hours a day, seven days a week. This was pre-internet, but we made it happen and it became the basis for The Sierra Network."

Sierra's ambition initially got the better of it. Al Lowe, the creator of the company's popular *Leisure Suit Larry* comedy adventure games, was commissioned to create *Leisure Suit Larry 4* as a multiplayer game.[7] "It was a tough road to follow because we didn't know anything about online games," said Lowe. "Ken hired a guy who had done low-

---

7. *Leisure Suit Larry* began life as a graphical version of *Softporn Adventure*, a crude and lewd text adventure released by Sierra in 1981. But post-AIDS, Lowe felt *Softporn Adventure* was so out of date it should be wearing a leisure suit and turned it into a humorous adventure game about a leisure suit-wearing sleaze ball trying to charm the ladies. The first *Leisure Suit Larry* game came out in 1987.

level communications software and said I want you to write what was basically a server – we didn't call it that, we didn't know those terms at the time."

After several months of struggling to figure out how to make an online *Leisure Suit Larry*, Sierra scaled back its plans. It refocused on the simpler games, such as Chess, Checkers, Bridge and Backgammon, that Williams originally envisaged and started constructing the infrastructure for the system. "At the time there was no internet," said Williams. "We had to deploy our own national network and servers."

The Sierra Network managed to attract around 30,000 users who paid $2 an hour to access its mix of parlour games and online chat, but the expense of running such a network meant the venture lost Sierra millions. "We figured it would take 50,000 people to make it successful," said Lowe. "Back then 50,000 was a huge number. Not that many people owned modems back then. We had to write the code so it would deal with a 1,200-baud modem and that was pretty state of the art – it was not cheap, it cost hundreds of dollars to get a modem like that."

Sierra eventually sold off The Sierra Network to telecoms firm AT&T in 1994. By then, however, the days of online networks such as Quantum Link were about to end. Back in 1974 a group of computer researchers working on APRAnet, the computer communications system developed back in the 1960s using military funding, began talking about creating the internet – a unified global communications network that all computer users could use. The concept became an ambition for those working in computer communications, who began creating the systems and software that could make the internet a reality. Over the next decade and a half, APRAnet evolved into the embryonic internet as communications standards were adopted and email services were connected. In 1988 the internet was opened up to the business world, allowing the formation of the first internet service providers and the following year British computer scientist Tim Berners-Lee developed the concept of the world wide web, a hypertext-based system that would make the internet easy to navigate and pave the way for websites and web browsers. The final restrictions on internet use were ditched on the 6th August 1991 when it was opened to the public for the first time. One by one, online networks such as Prodigy, AOL and CompuServe connected their systems to the internet, vastly increasing its capacity and user base. By 1994 the internet was poised to go mainstream, ushering in a communications and computing revolution that would reshape society on a global level.

One of those excited by the possibilities of internet games was Starr Long, a project manager in Origin Systems' headquarters in Austin, Texas. "One of the technical managers, Ken Demarest, and I started mucking about with multiplayer games and looking at everything that was going on in the market," said Long.

The pair's journey into the online realm spanned the whole multiplayer game world. They explored AOL's collection of online games, went on quests in the realms of *MUD*s and slaughtered people they had never met in *Doom* death matches. They talked about how exciting it would be if you could play Origin's flagship role-playing game series *Ultima* with your friends. "We were affectionally calling it *Multima* and I became extremely passionate about multiplayer. I had this feeling that multiplayer games was what games in general were really about from the beginning," said Long. "If you look at the oldest games, such as dice, they really were social experiences and the games provided a framework for a social experience. The internet was just beginning, it really was this new frontier – it was like 'wow, people don't even physically have to be in the same space anymore, they could get social experiences through the internet'."

Long pitched the idea for an online *Ultima* to Origin's founder Richard Garriott, who saw the potential. Garriott asked Electronic Arts, who had bought Origin in 1992, to bankroll the project. "We went through this period of time where we tried to convince everybody that this would work, but since there was never previously a successful online game, people's sales projections were basically zero," said Garriott. "It took us a long time before we managed to convince Electronic Arts to give us some money to do a prototype and that was $250,000. We were already in an era where games cost many millions, so it was useless or almost useless."

Long and Garriott built a prototype using the isometric-view graphics of 1992's *Ultima VII: The Black Gate*. The team set out from day one to turn the world of *Ultima* that Garriott had been designing since the late 1970s into a living, breathing world. "The idea of simulation as the foundation for the game was always there," said Long. In keeping with the simulation approach, the team decided that player freedom was crucial to the game and set out to remove anything that prevented players from playing as they pleased. They created numerous professions and jobs for players to do in their virtual world, from warrior and wizard to baker and glassmaker. They designed the trappings of a virtual economic system and built towns with shops and bars for people to gather in. They even built population dynamics into their digital realm. "We wanted to create this virtual ecosystem where the grass on the ground was a resource that the rabbits would consume and the wolves ate the rabbits, so if there weren't any rabbits the wolves died off," said Long.

Despite the miniscule budget, Origin had a working version of their game world ready by 1996 and decided to hold a pre-launch beta test with a select few *Ultima* fans to give themselves time to fix any unseen problems. "We put up a single web page that said 'hey we're the *Ultima* development team and we're doing *Ultima Online* and we'd love you to help us test it'," said Garriott. To make sure those who signed up would actually play the game, they asked fans to pay $2 for the CD they needed to run the game. Since the most popular online game at the time had attracted, at most, 30,000 players, they expected hardly anyone to sign up. They were in for a surprise. "Within two or three days, 50,000 people had signed up to pay," said Garriott. "That was the day the future changed. That was the day that this game no one knew or cared about became the most important game currently being developed at Electronic Arts or Origin. Immediately not only were the coffers opened up, but so was management oversight, much to our chagrin." Since the release of *Neverwinter Nights*, the business model for online games had changed radically. The internet's explosive growth had sent bandwidth costs plummeting, slashing the overheads involved in running online games. On top of that a game called *Meridian 59* had provided an alternative business model for online role-playing games that eschewed the expensive hourly charges of old in favour of a set monthly subscription that allowed players to spend as long as they wanted playing.

The brainchild of Archetype Interactive, a game studio formed by two pairs of brothers spread across the US, *Meridian 59* sought to reinvent *MUD* using the kind of 3D graphics seen in *Doom*. A test run of the game in early 1996 had already caused a minor stir, attracting the interest of around 10,000 players and prompting Trip Hawkins' The 3DO Company, which was reinventing itself as a game publisher after the failure of its 3DO console, to buy Archetype before the game had even officially launched. "It was essentially a visual *DikuMUD*," said Rich Vogel, the senior producer of *Meridian 59*. "It was the first game that was actually internet accessed. It wasn't accessed by a propriety network like AOL, CompuServe or GEnie. It was the first one where if you had a web browser you could login and register. That game was such a trailblazer, we were excited to have as many people as we had on it because it was the first time anybody had done this and because the internet in 1996 was just kind of getting there then."

But while its 3D visuals were a first for an online role-playing game, it was *Meridian 59*'s payment system that was truly revolutionary. "It seemed like a reasonable idea at the time because role-playing game fans wanted to play a huge number of hours per month and with a flat monthly fee it was only pennies per hour," said Hawkins. "Ironically, the users complained that its $9.95 a month subscription was too high."

Ultima Online would, Origin decided, adopt the same subscription-based business model. And since *Meridian 59* attracted around 25,000 players at its peak – just half of the number of people who wanted to take part in the test of *Ultima Online* – Electronic Arts were convinced they had a major commercial success in the making.

With *Ultima Online* set to become the biggest and most complex online game ever launched, Origin appointed a community manager to help manage the player relations in the hope of avoiding the headaches of player management that Lucasfilm had experienced with *Habitat*. The idea of a full-time community manager stemmed from one of Long's finds during his early exploration of online gaming: *Air Warrior*. First released in 1987 on the GEnie network, *Air Warrior* evolved out of a multiplayer text-only flight simulation developed by University of Virginia physics students Kelton Flinn and John Taylor back in 1977. The game recreated the air battles of the Second World War and charged players $10 to $12 an hour for a dose of dog-fighting fun. Despite the price tag, *Air Warrior* gained a loyal following among flight simulation enthusiasts thanks to its realistic physics and social interaction.

By the time Long investigated the game, its 30,000-odd players had become a thriving community that would hold long debates on player etiquette and chivalry – berating cowardly players who would quit the game when they were about to get killed. "*Air Warrior* is a flight simulator, but in every other definition it was a massively multiplayer online game," said Long. "Everybody was in the same space simultaneously playing together in real-time, communicating live. It was in *Air Warrior* that we started to realise that this was really a community and that it wasn't just a bunch of people playing together. *Ultima Online* was the first game that had an official community manager as a full-time job. Before that people had facilitated communities but they were also a game designer or a programmer. *Ultima Online* was first to say this is a full-time job. *Air Warrior*'s probably the biggest contributor to that realisation for us, so it had a big influence."

But even with their dedicated community manager, *Ultima Online* was soon riddled with the kind of social problems and surprises that Lucasfilm encountered when it made *Habitat*. And the first victim was Origin's carefully built virtual ecosystem. "Once we introduced the players into the equation they did what happens in the real world and wiped out everything," said Long. "They killed all the rabbits because they were easy to kill, so all the wolves died off and then there was nothing to kill. Frighteningly like the real world, but not very fun for the new player coming in who had nothing to hunt, so we had to scale back a lot on our original ambitions for creating ecosystems."

The biggest problem, however, was the general lawlessness of the online version of Britannia. "We left it open for players to attack each other," said Garriott. "We thought that's just part of reality. People are going to have grievances that they are going to want to fight over. I didn't have a problem with people fighting each other, but we didn't at all anticipate the PK'ing – the player killing."

Within weeks of the test launch, opportunistic players were rampaging through the game world slaughtering those weaker than themselves and looting whatever items their virtual victims were carrying. Thieves lurked outside the game's towns to rob new players as they took their first steps into Origin's virtual world. Criminal gangs would gather at the entrance to mines and wait for players who had spent hours breaking virtual rocks for gold to emerge, before pouncing on them and stealing their treasure. Angered players formed vigilante gangs that prowled the world looking for criminals to meat out mob justice. Others resorted to stripping their characters of items and clothes and wandering the world in their underpants hoping the obvious lack of possessions would keep them safe. "The problem we had was that we didn't have enough tuning time before we released it and one of the things that needed tuning was player-versus-player combat," said Vogel, who became *Ultima Online*'s producer after working on *Meridian 59*. "We never realised how bad it would get. It took about three months to notice."

The extent of the lawlessness varied depending on the server that players were logged into.[8] "It's interesting how they developed differently," said Vogel. "We had servers near the North East US, which were very bad servers, and then we had servers in the Midwest, which were calm and nice. The Pacific coast and the one in the North East were our worst. The way they grew up was like the broken window syndrome, because if you get a bunch of bad people in one area you're screwed. Everything from extorting people for their money, holding them captive and teleporting them to islands to steal their money. It was just amazing. People were scared to leave the cities."

The players of *Ultima Online* soon started developing their own slang to describe the situation, which soon spread out onto the web, seeping into the language of other online games and eventually into everyday conversation. "Almost all the terminology in use today came out of *Ultima Online* – griefing, nerfing, killer dudes, raids – all this kind of stuff really developed out of *Ultima Online*," said Vogel.

Even direct intervention from the game's creators did little to stop the problems. Garriott once encountered a thief robbing a new player while wandering around the game as Lord British. He caught the thief and told him not to do it again and returned the items to the victim. The thief promised not to do it again and promptly broke his word. Garriott intervened a second time only for the thief to strike a third time. "I said ok that's it, I've warned you twice, you did it three times in a row so I'm about to ban you from the game forever," said Garriott. "The thief then drops character and goes 'Ok Richard Garriott, if that's who you really are, I'll have you know that I'm only playing the role as you defined it in the game. I'm playing a thief and I'm using the thieving skill that you put in the game and if you are a thief and the king of the land comes and tells you not to steal of course you're going to tell him you won't, before going somewhere else and getting back to thieving because that's what you do."

Garriott was dumbfounded. This was his world: the murders, the violence, the

---

8. While *Ultima Online* had 250,000 subscribers, they could not all play online in the same world at the same time due to technical constraints. As a result players' characters exist on copies of the game stored on entirely separate servers.

chaos. It was all his and his team's doing and his game was no longer under his full control. "I went 'wow, that guy's right'," said Garriott. "So I said 'ok, you make a very good point' and teleported him all the way to the other side of the world where he couldn't mess with this woman. I went off to rethink the rules and think about the fact that people are just gaming the system you provide. You can't really blame the player killers, you can't blame the people stealing stuff from each other, you can only blame the vision and rules and structure that you put into play. So we began to take much more care in the development of our inter-personal systems."

Fearing the lawlessness would cause many of *Ultima Online*'s 250,000 players to cancel their subscriptions, Origin became embroiled in a desperate battle with the player killers, criminals, thieves, griefers and vigilantes who doubled as their customers. "There's many people who say the danger is part of what made the game very, very exciting, but for many it was very challenging or a big turn off, especially when they felt like it was being abused," said Long. "When the strong kill the weak over and over there's nothing to be gained from that – it just humiliates the weaker character and becomes not very fun for the weaker player."

Over the course of 1998, Origin began a crackdown on the griefers. It turned cities into safety zones where players could not attack each other, introduced reputation scores for players so troublemakers could be spotted and avoided, and created virtual jails to lock up problem players for periods of time as punishment.

The game industry took two important lessons away from *Ultima Online*. First, that the internet had finally made online games commercially viable. Second, that giving players as much freedom as possible was a recipe for disaster. The first game to apply those lessons was Sony Online Entertainment's 1999 fantasy role-playing game *EverQuest*, which encouraged players to team up with each other rather than fight. It attracted more than twice as many subscribers as *Ultima Online*.

The open, anything-goes vision of online games such as *Ultima Online* was rapidly replaced with more directed, entertainment-driven online games. Only a select few, such as Icelandic developer CCP's *Elite*-inspired *EVE Online*, embraced the risky and unpredictable path that Origin explored. "There is a little bit of scariness about open-ended virtual worlds because of *Ultima Online*'s problems because, frankly, when you give people a lot of power they abuse it," said Vogel.

Long agreed: "It's been shown rather clearly that there are larger audiences for more structured, linear experiences than there are for a truly open space. *EverQuest* did better than *Ultima Online* and was very structured. *World of WarCraft*, which is also very structured, is better than everything else ever. The audiences voted with their feet about what kind of game they preferred."

Released in 2004, Blizzard's *World of WarCraft*, a spin-off of the US company's fantasy-themed strategy series *WarCraft*, was a deliberate attempt to avoid some of the problems that had beset *Ultima Online* and other online role-playing games. "It was very similar to what we've done with some other genres," said Rob Pardo, executive vice-president of game design at Blizzard. "We were playing lots of *Ultima Online* and *EverQuest* and really did see the potential of that genre. We felt that once you got into those games, once you got past the really steep learning curve and some of the rules that made them unappealing to a more casual core audience, they had so much immersion and stickiness. The ability to share your game socially with people had so much potential, there was a huge opportunity there."

Blizzard toned down the player killing of *Ultima Online* and *EverQuest*, by requiring players to be willing participants in fights with other players. It also reduced the need to keep accruing experience by killing monsters and other creatures to become more powerful. "*Ultima Online* and *EverQuest* had really long level curves – you had to work really long and hard to gain a level and there wasn't a lot of direction, so you would have to go out and starting killing beetles or wolves or something like that until you got to a new level and could kill slightly tougher creatures," said Pardo. "The biggest thing we did to make these games more accessible was to have a quest-driven experience from the very beginning so you would always know what you could do next and have some choices of what quest you wanted to take on. We were the first massively multiplayer online game where, from the first level all the way through to the maximum level, you would be on a quest and have things to do."

The popularity of *World of WarCraft* confirmed Blizzard's suspicions that the difficulty of earlier online role-playing games had alienated players. By 2009 the game had racked up more than 10 million subscribers across the world. The sheer size of Blizzard's *World of WarCraft* operation was mind-boggling. Some 20,000 computers were needed to run the game. Blizzard's customer support team consisted of 2,396 people and there were a total of 13,250 copies of the game running on servers that containing in excess of 1.3 petabytes of data. In addition, Blizzard had 451 people employed full-time who constantly built new quests, art, music and sounds for the game as well as two historians who catalogued its virtual history.

The virtual world envisaged by Bartle and Trubshaw on a primitive computer lurking on the campus of the University of Essex back in the days when home computers were still new had grown up to become a globe-straddling Goliath that brought millions of people together.

China online: Gamer at a internet cafe in Taiyuan, Shanxi province
Stringer Shanghai / Reuters / Corbis

CHAPTER 24

Second Lives

Japan annexed Korea on the 22nd August 1910. The fast-rising imperial power had already defeated China in the First Sino-Japanese War in 1895 and done the same to the Russians in the Russo-Japanese War in 1905. These military victories earned Japan the respect of other imperial powers, even prompting the formation of a 'learn from Japan' movement in Britain.

Japan's victories and new diplomatic standing gave it unopposed influence over Korea, a country that had only gained independence from China in 1897. In the five years that followed the Russo-Japanese War, Japan used its influence to dissolve the Korean army, hand itself a veto over the nation's laws and take control of its foreign policy. And in August 1910, Japan forced the Korean government to sign an annexation treaty that officially ended Korea's independence.

The Japanese were harsh rulers. Those suspected of aiding rebel fighters faced execution or enslavement. In some cases, whole villages were rounded up and locked in public buildings that were then set alight as a punishment for helping dissidents. Japan also set out to erase Korea's culture. The country's history books were burned, the Korean language suppressed and most of the Gyeongbokgung, the Korean palace built in 1394, was demolished to make way for the Japanese governor's headquarters.

After Japan's defeat in the Second World War, Korea was divided into US and USSR occupied zones. The US zone became South Korea and the Soviet Union's zone became North Korea. One of the first laws South Korea's government introduced was a ban on the import of Japanese goods and cultural items, a legal reflection of the understandable fury and bitterness Koreans felt towards their former occupiers. This law, which remained largely unchanged until the late 1990s, would have a significant influence on the video game business both in Korea and across the world. Its most noticeable impact was that it severely limited the supply of video games to South Korea in the 1980s. Japanese companies could not export their games or consoles to the country and since American and European companies were only just starting to expand into other developed nations they saw little incentive to head to the relatively poor nation of South Korea. And since home computer ownership was rare in South Korea, few Koreans attempted to build a game industry of their own.

Those wanting to play video games had little choice but to resort to buying illegal copies of games. "Japanese movies, music, comics and games were all banned from being imported," said Jake Song, a future game designer who got his first taste of games on a school computer. "But for the few who enjoyed games, they played illegal copies of Japanese games. It was fairly easy to buy Taiwanese clones of the Nintendo Super NES and the UFO, a device that enabled the use of illegally copied discs instead of cartridges, and it was fairly easy to get pirated software from PC bulletin board systems or the computer-related shopping districts in Seoul."

None of this was out of the ordinary for a mainland East Asian country in the late 1980s and early 1990s. Few game companies had any presence in the region and so the black market ruled supreme. But then in the early 1990s South Korea entered a period of rapid growth that greatly improved living standards. Sales of PCs began to rise, encouraging a few Korean companies to start releasing video games. Nearly every single one sold in low numbers. Koreans were too used to cheap or free pirated games to consider paying out for an official version. One after another, Korean publishers of retail games would give up.

Then in 1994 the IT firm Samjung Data Service hit on a way to get Koreans paying to play while trying to think of a product that would appeal to the small-but-growing number of Koreans buying internet-ready PCs. Samjung's founders were a group of graduates from the Korea Advanced Institute of Science and Technology in the city of Daejeon. During their time at the institute they discovered *MUD*, the text-only online world created by Richard Bartle and Roy Trubshaw back in the late 1970s. The founders realised that people couldn't easily copy an online game and, even if they managed it, they would lack the infrastructure necessary to run it. This, they concluded, was the way to sidestep the piracy problem. They put together *Jurassic Park*, a *MUD* of their own, and began charging players for the time they spent playing it. *Jurassic Park* did not become the success Samjung hoped, largely because few Koreans owned an internet-connected PC at that point, but it had showed the country's would-be game makers the way forward. All that was needed was for internet access to become more widespread and, as luck would have it, the Korean government was about to make that happen. In 1995 the South Korean government announced its intention to create a "knowledge-based society" founded on widespread broadband internet access. It was a crazily ambitious project. The internet was still a new concept, slow dial-up connections were still cutting edge and here was a relatively minor Asian economy pledging to plough hundreds of millions of US dollars into building the world's best computer communications network. But for Korean game companies the announcement was a gift. Armed with the promise of fast broadband internet access for all and the business model of *Jurassic Park*, Korea's game industry gave up on retail games and prepared for a cyberspace future.

One of the first to embrace this future was the Nexon Corporation, a company co-founded by Kim Jung-Ju in 1994 after he saw *Jurassic Park*. Kim's plan was to release an online multiplayer game that used graphics rather than text. His business partner was Song, who had ended up programming business software for a living but also created text-only *MUDs* in his spare time. While games such as AOL's *Neverwinter Nights* were already offering US game players graphics-based online games by the time Nexon formed, South Korea's isolation meant Song and Kim were in uncharted territory as far as they were concerned. "They weren't any terms for this genre, so we called it graphic *MUD* or *MUG*," said Song.

The result of Nexon's work was Song's 1996 game *The Kingdom of the Winds / Baram Eui Nara*. *The Kingdom of the Winds* emphasised its Korean roots. Its virtual world was based on the country's history and mythology and its art style drew its inspiration from Korean 'manhwa' comic books. "Nationalism was part of what I was taught since birth and throughout my education," explained Song. "So I think I had been compelled to make a game with a Korean theme. But after *The Kingdom of the Winds* I was relieved of that kind of imperative. Because I am Korean, I think even if I make a medieval European fantasy game, Korean sentiments will naturally melt into it."

The Kingdom of the Winds was a turning point for Korean games. Its use of graphics raised the bar for every other Korean game maker – text-based *MUDs* would no longer cut it. It also became the first Korean game to achieve mass appeal. At its height, nearly one million people were paying a monthly subscription fee in order to play. In one fell swoop, Kim and Song had shown that South Korea was a viable games market and that Korean developers could compete on equal terms to Japanese and American companies without compromising their culture.

But just as *The Kingdom of the Winds* started to take off, South Korea and the rest of mainland Asia was engulfed in economic turmoil. A loss of confidence in the Thai currency, the baht, in July 1997 sparked off a wave of investor panic that sent East Asia's economies into freefall. The South Korean economy shrivelled by a third in the space of a year and the national debt doubled. Soon Korean workers were being laid off in their thousands.

In theory, the crash and the resulting surge in unemployment should have strangled the country's barely formed game industry. But rather than destroy it, the economic woe actually boosted it. Some of the newly unemployed hit on the idea of opening 'PC bangs', internet cafés where people could play online games. It was an appealing business prospect. The government, which accelerated its broadband drive in response to the crisis, was offering generous subsidies to internet-related business and, thanks to *The Kingdom of the Winds*' success, there was a huge demand for online games. Not to mention a lot of out-of-work people with time to kill. PC bangs charged customers around $1 an hour to play and became immensely popular, encouraging more and more to spring up across the country. And as the number of people hanging out in PC bangs swelled so did consumer demand for online games. By 1998 around 3,000 PC bangs had opened across South Korea, giving millions of potential game players access to high-powered PC gaming on the cheap, and more were opening all the time. By the following year the number had risen fivefold to 15,150 and it was all thanks to two 1998 games: *StarCraft* and *Lineage*.

StarCraft was an American real-time strategy game with a sci-fi theme that revolved around frantic battles between three distinct alien races, each of which had their own weaknesses and strengths. The game's creators Blizzard Entertainment expected little success in South Korea. "Korea was already a market for us, but not a very big one," said Rob Pardo, the executive vice-president of game design at Blizzard. "We didn't localise it – you actually play the English language version over there. There were lots of things going on in Korea around the time that *StarCraft* came out: the growth of PC game rooms; the growth of internet connections. *StarCraft* came out in that time frame and captured a lot of the Korean imagination. They really like science fiction, they really enjoy that fast-paced strategic thinking and they have a very competitive gaming culture."

Koreans became the world's biggest *StarCraft* fans. Of the 9.5 million copies of the game sold worldwide, 4.5 million were sold in South Korea. Blizzard's video game was so popular it became a national sport. TV stations began broadcasting matches between the best players. One channel, iTV Game, even dedicated itself to broadcasting *StarCraft* matches to an eager audience. The professional gaming scene that *StarCraft* spawned in South Korea made the low-key US gaming competitions based on *Doom*-style death matches and 1980s arcade games look like a joke by comparison. The US's Cyberathlete Professional League got excited when 30,000 people decided to watch the event online. But South Korea's World Cyber Games could attract 50,000 spectators to the arena in which it took place plus hundreds of thousands more on TV and the internet, enough to persuade multinational companies, including Samsung, to pay huge amounts to sponsor the event. The World Cyber Games also had generous backing from the South Korean government, which contributed to the 2002 contest's $350,000 prize money. The government's support even included giving video gaming official recognition as a sport.

StarCraft's mammoth success in Korea was mirrored by another 1998 game: Song's *Lineage*. In stark contrast to *The Kingdoms of the Winds*, the NCSoft-released *Lineage* embraced a more European vision of fantasy. "When I was creating *The Kingdom of the Winds* it was very difficult to find all sorts of reference data because it was set in ancient Korea," said Song. "So I set out to make a conventional medieval European fantasy and the comic book *Lineage* happened to be published serially in one of the magazines I was reading at the time. I felt online games should provide a playground and that the players create the stories themselves, so I didn't really make it to relay the storyline of the comic and used only the basics of the world, for example the names of kingdoms or cities."

Song sought to make his new role-playing game more accessible than *The Kingdom of the Winds* by creating simple one-click controls and focusing the game on grandiose battles where dozens of players would co-operate to defend or storm castles in order to gain territory and the resulting tax revenue. "I wanted to create a structure where the users naturally created the content amongst themselves and it made sense when it was continuously repeated," said Song. "That's how the concept of players warring over castles came about."

These large-scale battles made *Lineage* a compelling social experience. "These events required intense co-operation between large groups of people and that encouraged people to play as they could build up a guild and then go siege a castle," said Starr Long, who became a producer at NCSoft in 2001. "Western massively multiplayer games are, a lot of the time, like a single-player game, where you just happen to be with a group of other people – they are built to allow you to play by yourself for a large amount of time. Korean games were about groups."

Lineage became a blockbuster-sized hit in Korea. Within a year of its release, more than 500,000 people were subscribing to the game making it the world's biggest online role-playing game. By 2003, at its height of popularity, *Lineage* had more than 3 million players. The Korean-made game's success took American, European and Japanese game publishers by surprise.

"We did not believe the numbers," said Long. "We were very, very sceptical. We started playing it and it was super simple. We were asking ourselves how come it's doing so well? Why are people so crazy about it? *Lineage* was huge. Electronic Arts even thought about purchasing NCSoft at one point."

For Song one of the biggest surprises about *Lineage* was how its popularity result-ed in players selling virtual items from the game, such as swords or even whole characters, for real money. "When I first heard of it I couldn't believe it and said 'no way'," said Song.

Lineage was not unusual in this respect. Across the world online game players, especially those playing multiplayer role-playing games, were putting virtual items up for sale and earning real money from doing so. "People were offering real money for objects in the game that were not real," said Rich Vogel, the producer of *Ultima Online*. "In *Ultima Online*, houses and castles were worth real money. If you had a house in a good area of the game you could get $10,000 for it."

The belief that items within virtual worlds have real value gained further weight when economist Edward Castronova published *Virtual Worlds: A first-hand account of market and society on the cyberian frontier*. His 2001 paper was a study of the economics of *EverQuest*, then the US's most popular online role-playing game. Castronova con-cluded that Norrath, the fictional world of *EverQuest*, had a gross domestic product per capita of $2,266 – more than China and India and about the same as Russia. The California State University professor noted that, if it was a real place, Norrath would be the world's 77th richest nation based on these figures.

By 2004 it was estimated that the global trade in virtual game items was worth at least $100 million. And with serious amounts of money to be made, a new industry was born – gold farming. Primarily operating out of low-wage countries such China and Mexico, gold farming businesses employed people to spend their days in video games gathering items that could then be sold for real money. While the scale of gold farming is hard to quantify, some of the better guesses suggest that by 2008 the industry employed 400,000 people who each earned an average of $145 a month. Gold farming was a controversial development. Game companies worried that these players' relentless harvesting of virtual gold and equipment could cause deflation in their virtual worlds as once rare items become widely available or inflation as prices in the game were pushed up by the growth in the amount of digital money in circulation.

"Managing economics in games is very hard and you have to think about infla-tion and deflation," said Vogel. "Gold farmers do affect this and that's why you try to look out for those people and ban them when you find them. Most people who buy from the gold farmers buy money. Gold farmers cut the time people have to play – that's why people buy."

Concerned, online game makers started hiring full-time economists to help them manage the macroeconomics of their virtual worlds. Another worry for game companies was that if gold or virtual items became plentiful thanks to the gold farm-ers, there would no longer be any incentive or challenge to keep people playing.

These concerns led to numerous attempts to stop or limit gold farming. *Ever-Quest* publisher Sony Online Entertainment created an official real-to-virtual money exchange system to try and control the market. NCSoft threatened to sue gold farm-ers for violating their intellectual property rights, since they legally own all the virtual items in their games. Blizzard began throwing suspected gold farmers out of *World of WarCraft*. "Some of the companies steal people's accounts, go onto their accounts, strip them of all their gold and then sell that," said Pardo. "That is really harmful to the game and that's something we work very diligently to stamp out."

The game industry also portrayed gold farming businesses as sweatshops, although the evidence to support such claims was patchy at best. In China, where most of the gold farming was taking place, the pay of gold farmers is around the same as the average wage and the working conditions are often better than those experienced by most Chinese employees.

Not every virtual world developer, however, has viewed the sale of its digital currency with suspicion. San Francisco's Linden Lab made such exchanges central to 2003's *Second Life* – a virtual equivalent of the Burning Man Festival that takes place each year in Black Rock, Nevada, where thousands of people spend eight days build-ing a fantasy town on the dry bed of an ancient lake.

Linden Lab's founder Philip Rosedale saw *Second Life* as a place where people could connect, do business, socialise and express themselves through their creativity. "I thought there would be a lot of advantages over the real world," said Rosedale. "You can build things in a digital space that are imaginable but unbuildable in the real world – that could be architecture or more metaphorical things like the ability to move around very fast or to recreate your own identity."

Second Life was in many ways the spiritual successor to the text-based social worlds that evolved out of *MUD*, such as *LambdaMOO*, which allowed players to create interactive virtual objects. "The text-based virtual worlds that were relatively open-ended and empowered people to create things were definitely precursors to *Second Life*," said Rosedale. In keeping with Rosedale's vision for empowering the users of his virtual planet, *Second Life* gave players the copyright over their creations and allowed them to sell their creations for real and virtual money.[1] "We just felt that anything you created in *Second Life* would be analogous to creating your own web

---

1. Unlike most online games, *Second Life* is a unified whole. All players exist in the same world rather than copies of the world spread across different servers, as is the case with most online game worlds including *World of WarCraft*.

page. If you wanted a lot of content in there and you wanted people to be excited and entrepreneurially invested in it, it couldn't possibly make any sense to lay any claim to the content."

The creations of *Second Life*'s population ranged from clothes for people's virtual selves to interactive objects that entertained and surprised. "A lot of the creations are at the intersection of art and design, where people have taken real world ideas and gone totally berserk with them," said Rosedale. "There was this amazing artist who made a magic wand and if you had it in your hand when you chatted with people, all kinds of remarkable things would happen that were somewhat consistent with things you were saying. So if you said Santa Claus, Santa Claus would fly by with his reindeer. If you were standing next to a person who had one of these things then you were perpetually entertained. The guy sold them for about US$50 each and it was one of the most coveted things people could have."

While the population of *Second Life* could profit from their creations, Linden Lab earned its keep by selling virtual real estate and running the Lindex, a currency exchange that converted US dollars into *Second Life*'s Linden dollars and back. "The Lindex trades about $300,000 a day," said Rosedale. "You can put money in and take money out. It's a completely open market – the price is not regulated one bit. We actually sell new currency into that market with the goal that the exchange rate remains roughly constant. It's monetary policy: increasing the money supply to match GDP growth essentially."

* * *

Back in South Korea the explosion in the popularity of video games and PC bangs fuelled by *StarCraft* and *Lineage*, coupled with additional government investment had encouraged the formation of many more Korean game publishers. Many of these companies started experimenting with business models in the hope that breaking away from the fixed subscription charges used in *Lineage* could bring online gaming to an even bigger audience. Several hit on the idea of micropayments where, instead of charging people for the privilege of playing the game, players were charged for buying extra equipment for their virtual characters.

There were sound business reasons for taking such an approach. Few Koreans had credit cards but most had mobile phones, which could be used to make small payments. Nexon's hugely popular 2004 game *Crazyracing Kartrider*, a *Super Mario Kart*-inspired go-kart racing game with cheery kawaii characters, typified the micropayments approach. While people could play for free, the game offered a vast range of vehicles, accessories and customisation options to buy that ranged in price from new paint jobs for 10 cents to top-of-the-range go-karts for $10.

By the early 2000s South Korean game companies, cushioned by the huge profits they were making back home, began expanding around the world. As well as moving into Japan, North America and Europe, they also used the free-to-play business model to break into the world's most populous nation: China.

Like South Korea prior to *Jurassic Park*, China's game industry faced endemic levels of illegal copying that had severely undermined the sales of the nation's tiny collection of game companies such as Beijing Golden Disc, the publisher of the jingoistic 1996 Korean War-themed PC game *Chinese Airforce*. But for a population unable to afford their own computers and lacking credit cards, the South Korean free-to-play online games were ideal. Korean games such as *MapleStory*, *Lineage* and *Crazyracing Kartrider* became as popular with the Chinese as they were back in South Korea, encouraging countless PC game rooms to open in China.

China's communist government responded by adopting a dual personality when it came to video games. Its nationalistic side, uneasy with the number of Chinese people playing foreign games, decided to spend $1.8 billion over five years on supporting the development of 100 home-grown online games. By the latter half of the first decade of the 2000s China had a thriving game industry producing games geared towards the Chinese market such as *Learn from Lei Feng Online*, a 2006 game where players had to follow the example of Lei Feng, the People's Liberation Army soldier idolised in China for his selflessness and dedication to Maoism, in the hope of meeting Chairman Mao Zedong. Most popular of all was NetEase's *Fantasy Westward Journey*, a 2004 multiplayer game based on the famous Chinese novel *Journey to the West*, sometimes known as *Monkey*, which is set during the Warring States period of Chinese history (475 BC to 221 BC) that ended with China's unification. NetEase's online interpretation of the book became a major success, attracting more than 1.5 million active players at its height.

At the same time, the Chinese government viewed video games as a threat; a corrupting influence on the nation's youth. In 2004 China's Ministry of Culture began searching all foreign-made games for content that was "damaging the nation's glory", "disturbing the social order" or "threatening national unity". Swedish historical strategy game *Hearts of Iron* was banned because it treated Tibet, Xinjiang and Manchuria as separate from China[2].

Other bans were more puzzling, such as the outlawing of Electronic Arts' *FIFA*

2. *Hearts of Iron* is set in the years 1936-1948. During that time Tibet, Xinjiang and Manchuria were not part of China.

Soccer 2005. The government also banned video game adverts in the media. And, after an April 2007 report by the Beijing Reformatory for Juvenile Delinquents concluded that a third of inmates had committed crime because of video games and online porn, China's president Hu Jintao ordered a clean up of "internet culture" that resulted in a blanket ban on the opening of new internet cafés and game rooms from May to December 2007.

China's most invasive attempt to control the influence of video games was the anti-addiction system it introduced in 2007. The system monitors how long people play and if they exceed three hours in a day the game switches off or important features are disabled. This tracking software, which companies are required by law to include in their online games, is also linked to individuals' national ID numbers so the authorities can pinpoint those it regards as playing too much. Originally the system was intended to cover every player of online games but, after objections from adult gamers, the Chinese only applied it to under 18s. China got the idea from South Korea, which introduced a very similar system in 2004 amid growing public concern about young people's obsession with online games.

As well as bringing video games to prominence in China, the Korean free-to-play model also had a significant influence on Germany's game industry. German companies took the micropayments formula pioneered in Korea, but sought to deliver such games via web browsers rather than the Asian-style game rooms that did not exist in Europe.

One of the earliest German companies to pioneer the approach was Bigpoint. The Hamburg-based company began life in 2002, creating online sports team management games such as *Hockey-Manager*.

"Our key mission is very simple games you can just enter through the browser," said Nils-Holger Henning, Bigpoint's director of business development. "You don't need any download, installation or administrator rights, which means we can offer these games wherever you have internet access – the office, the university, your home, the internet café, wherever. We offer all our games for free and monetise the active users by selling virtual goods or items."

Although the initial games produced by Bigpoint and others were static, graphically unsophisticated titles such as *Hockey-Manager*, as broadband access and web browser capabilities increased Germany's browser game firms started offering more complex and visually impressive games that emphasised the community and social aspects of online gaming to keep players coming back, such as Bigpoint's 3D game *Pirate Galaxy*, which came out in 2010. These companies have also expanded outside of Germany, bringing free-to-play games to the rest of Europe and North America.

By late 2009, it was clear that the innovations of South Korean developers in the late 1990s and early 2000s had reconfigured the video game business on a global level. They had pioneered revolutionary business models, provided the foundation for the Chinese game industry, turned game playing into a national spectator sport and ended the hegemony of Japanese, American and European game developers.

Big brother: Will Wright inspects one of his sims
Courtesy of Electronic Arts

CHAPTER 25

Little Computer People

Maxis' executives looked on with an expression of uncomprehension. In front of them the company's co-founder Will Wright, whose *Sim City* had kept the game publisher in business for most of the 1990s, was pitching what sounded like a strong contender for worst game ever imagined.

It's called *Doll's House*, explained Wright, because it's a bit like one. The executives shifted uneasily in their seats. You have these little people in a house that you design and they watch TV, cook meals, go to work and sleep at night. And then something dramatic happens right, asked the executives hopefully, like someone coming to repossess the house? "No. It's simulation, there's not the save-the-world aspect," replied Wright. Oh.

Maxis declined to fund Wright's new pet project. "It sounded so mundane: going to the toilet, taking out the trash," said Wright. "At that point most games were about saving the world or flying a jet fighter. It didn't seem like an aspirational game. People were very conditioned to the idea that if you had a game about specific characters there had to be a story."

The idea for *Doll's House*, which would eventually be released as *The Sims* in 2000, grew out of Christopher Alexander's *A Pattern Language*, an architectural theory book Wright had been reading. "Alexander is a physics guy who went into architecture and was frustrated because architecture wasn't enough of a science for him," said Wright. "He wanted all the principles of architecture to be clearly reducible back to fundamental principles, which was what this book was." *A Pattern Language* consisted of 253 'patterns', short principles that Alexander and his co-writers believed architects and town planners should embrace when designing living spaces.[1] "I wanted *The Sims* to be an architecture game that assessed what you did from a human point of view," said Wright. "The people in *The Sims* were originally there to score the architecture."

To create the virtual people that would assess the player's building work, who were

---

1. Pattern 159, for example, argued that every room in a building should have natural light on at least two sides.

nicknamed sims after the residents of *Sim City*, Wright drew on his research for his 1991 game *Sim Ant*. *Sim Ant* was an ant colony simulation based on Bert Hölldobler and E.O. Wilson's Pulitzer Prize-winning science book *The Ants* – a comprehensive 752-page survey of ant behaviour, ecology and physiology. In *Sim Ant*, Wright had modelled the secretion of pheromones by individual ants that influenced the behaviour of other ants from the same nest to create what at, a macro-level, looked like intelligent co-ordinated behaviour across the whole nest.

"In most computer games you have a really good sense of the environment, very rarely does the player design the environment," said Wright. "We had to be able drop a sim into any possible situation and have them behave reasonably intelligently and ants seemed like a really good model. The sims follow pheromone trails in a weird sort of way. Everything in the game is advertising what needs they fulfil and, depending on the sim's need, they are attracted to those pheromones."

To Wright's surprise the emergent behaviour of the ant-like sims proved compelling to watch: "The humans worked a lot better than we thought they would and they became a lot more fascinating to watch and interact with." While the house-building roots of *The Sims* remained in place, Wright's game became more and more about watching and interacting with his virtual people. "We had to dumb them down a bit because they were so good at meeting their needs there was no reason for the player to ever interact with the game, so we actually made them stupid so the player had a role in the game."

For Wright, his evolving game idea tapped right into the kind of innate voyeurism that lay behind the appeal of the reality TV shows that became popular around the time of *The Sims*' release in 2000, such as *Big Brother* and *American Idol*. "I've always noticed, even felt myself, that people are inherently narcissistic – anything about them is going to be 10 times more fascinating than anything else, no matter how boring it is or exciting something else is," said Wright. "In *The Sims* one thing almost everybody does at some point, usually right off the bat, is put themselves in the game along with their family, house and neighbours. Now they were playing a game about their life, they become the superhero on screen – even though it's not so super. It surprised me that television went in the same direction. You wanted to see these glamorous people in exotic places, now it's these average Joes sitting around drinking beer and arguing with their wives. That's the voyeuristic aspect, *The Sims* feels very voyeuristic."

Wright was not the first person to investigate video gaming's potential for voyeurism and narcissism. In 1984 Activision bought the rights to Rich Gold's *Pet Person*, a software program inspired by the Pet Rock craze of the 1970s where people bought millions of named pebbles with eyes glued on. Users couldn't interact with *Pet Person*. Instead they got to watch an animated person wander around a virtual house. "*Pet Person* came to us virtually bankrupt," said Activision co-founder David Crane, who landed the job of turning Gold's creation into a game. "The idea of doing a computer-based Pet Rock was a grand one, but I came to the conclusion that as a non-interactive fishbowl it couldn't recoup its costs, so I added two-way interactivity. Now you could have an artificial life form that even communicated back to you."

After Crane's improvements, players were able to interact with their virtual person by buying them food, encouraging them to write letters to you and by playing card games with them. Released in 1985 as *Little Computer People*, the game became a cult favourite, but a loss-maker for Activision. "The game had a small, but very dedicated following. Those who got the product were fanatics," said Crane. "We had a letter from a grandmother who bought two Commodore 64s and two monitors so that her two grandchildren could each have a pet person of their own when they came to visit. The Commodore 64 also had an operating system bug that would damage the floppy disk where each little computer person's personality and status was saved one time in a thousand. People got so distraught at the death of their person that I had to design a piece of 'hospital' software for our consumer relations department. You could send in your floppy and your little computer person could be resuscitated, complete with intact personality in most cases."

Activision ventured into similar territory again the following year with the life simulation game *Alter Ego*. Peter Favaro, a recent psychology PhD graduate, wrote *Alter Ego* while trying to get his career as a psychologist off the ground. "Poverty prompted me to make a game," he said. "I was a starving new PhD in psychology who was only 26 years old and looked a good bit younger. I was so young that even crazy people were sane enough not to trust their mental health to such a baby. I lived in a small apartment in an upscale area of the North Shore of Long Island, outflanked by more mental health professionals than there are Starbucks in Seattle." In need of an income, Favaro decided to make a video game that drew on his knowledge of psychology. "I wanted people to experiment with choices and outcomes they would normally encounter in life, freely and without fear of real consequences. I wanted people to see what it would be like to be the villain, the scientist, the priest or the slut they've always dreamed of becoming but never had the nads to follow through on," he said.

To gather the information that would underpin the options and situations in *Alter Ego*, Favaro interviewed more than a thousand people to work out what made them tick and what life decisions, really affected their lives. "I did this for a year and I was quite a lunatic about it," he said. "I interviewed people to get a feel for what people thought were important events in their lives, then I embellished them to make them entertaining and emotionally evocative. One of the things that I found odd was that the vast majority of people were quite eager to talk to me – a complete stranger – about highly personal and sensitive topics. If I were a smarter man I would have been able to predict that some 25 years later people would clamour to publicise, without the promise of anonymity, the details of their personal lives on *YouTube* and *MySpace*."

Alter Ego put players in control of a virtual person, making decisions at key junctures in their life to see how their situation and personality evolved from birth to death.[2] The game pulled no punches with its choices, which sometimes delved into sex, drugs, violence and other controversial subjects. "Activision supported everything I wanted to put in the game and nothing was removed. When the marketing department expressed concern over the violence and sexuality, which in my mind are necessary to portray an accurate cross section of life experiences, the only thing I had to accept was a parental warning on the packaging," said Favaro.

While, like *Little Computer People*, Favaro's experimental game gained no more than a cult following it did touch many people's lives. "I have received hundreds of letters and emails from people saying that playing *Alter Ego* was a life-changing experience for them," said Favaro. "That's fine and flattering, but I never intended it to be that way. I am always shocked when people talk about how much they've learned about themselves from playing *Alter Ego*."

The poor sales of *Little Computer People* and *Alter Ego* largely killed off similar experiments and boded ill for *The Sims*. With no backing from Maxis' management and little funding, Wright continued to work on his doll's house game even though there seemed little prospect it would be released. In 1997, however, the prospects for Wright's project changed dramatically. The turning point was Electronic Arts' decision to buy Maxis. "Electronic Arts was doing due diligence, deciding whether to buy our company or not," said Wright. "They were originally thinking about buying Maxis for *Sim City*, but some of the Electronic Arts executives saw *The Sims* and went 'what's this?' because we didn't even tell them we were working on this project. They got very excited by it." Keen to see *The Sims* released, Electronic Arts bought Maxis and immediately handed Wright the money and larger team he needed to complete the game.

By the time Electronic Arts came to the rescue, Wright had developed the game's concept further in response to Id Software's 1996 first-person shooter *Quake*. *Quake* had marked the end of the Romero-Carmack partnership that had taken Id Software to the apex of the video game industry, but it also marked a significant step forward for the modding ideas the pair first pioneered in *Doom*. It came with its own programming language – *QuakeC* – that allowed players to not only create maps for players to fight in

---

2. *Alter Ego* was released in two separate editions: male and female. In keeping with the low proportion of female game players in the 1980s, the male version sold far better.

but offered enough flexibility for them to build whole new games on the back of the *Quake* engine. *QuakeC* accelerated the growth of the modding culture that Id had already encouraged with *Wolfenstein 3D* and *Doom*. By the end of 1996, within a few months of *Quake*'s launch, a team of three fans had produced one of the first mods to gain recognition as a great game in its own right: *Team Fortress*.

Designed as a multiplayer game *Team Fortress* divided players into competing teams composed of soldiers of different abilities who battled each other in a number of matches such as paintball-style 'capture the flag' missions and escort games where one team has to guide a VIP through the level while fending off assassination attempts from the opposing team. *Team Fortress* became one of the most popular mods ever made and by 1999 professional game developers Valve had teamed up with its creators to remake it as a commercial game. *Quake* also introduced another feature that, to Id's surprise, would spawn another outbreak of player creativity. Since death matches had proved to be a highlight for many fans of *Doom*, Id thought its players might enjoy the option to rewatch their games afterwards and added a game-record feature to *Quake*.[3]

It didn't take long for *Quake* fans to find an alternative use for the record feature. A group of players called The Rangers were first to tap into the feature's hidden potential. "We had an internet relay chat channel of our own and had many members and non-members alike sitting there chatting one night," said Heath Brown, a co-founder of The Rangers who played online as ColdSun. "At the time we were making videos of our skills to show the masses and a member named Sphinx jokingly said 'we should create a movie or something'. He laughed as if the thought was silly, but several of us went silent as the wheels began turning. I felt a chill go up my spine. I'm a very creative individual and I thought this would be a cool way to show some of my stories and ideas to people."

The Rangers set about trying to use *Quake*'s record function to create a brief test movie that they called *Diary of a Camper*, a 100-second short telling the story of the killing of a 'camper' – the term used in online games to refer to players who camp out in advantageous positions on a game map. They decided the camper in the film would be *Quake* designer John Romero. "Back then we actually played every day on the Id Software *Quake* servers doing beta tests and fragging other elite clans that hung out on those servers with the developers typing messages in the console.[4] John Romero left Id to make *Diakatana* and at the time there was some hard feeling from those of us close to Id," explained Brown.

---

3. The idea of games that could record and playback the action had been knocking around for several years by 1996. It had even formed the basis for Disney Interactive's 1992 PC flight sim *Stunt Island*, where players got to arrange props and cameras around the game world before taking to the skies to perform stunts that could be recorded and played back later.

4. Fragging is online gaming slang for killing rival players.

Diary of a Camper's story was paper-thin and The Rangers never intended it to be anything more than an experiment; a way to see if the idea of making a movie within *Quake* was possible. Despite its primitive nature, The Rangers put the recording online while they got to work on their first proper *Quake* movie: *Ranger Gone Bad*. To their shock, *Diary of a Camper* became a sensation among *Quake* players. "Our website went crazy," said Brown. "Suddenly there were a lot more *Quake* players wanting to join the clan. There was excitement to see what we would do next and even a little jealousy and mean comments from other clans who said we weren't *Quake* players, just movie makers."

Diary of a Camper acted as a proof of concept for thousands of *Quake* players who realised that with the recording feature and the power of *QuakeC* they had a miniature animation studio within their game that allowed them to harness of the infinitely flexible 3D world of *Quake* to create films. Soon hundreds of people were using *Quake* as a desktop movie studio. Eventually someone coined a name for the burgeoning game movie scene: machinima. "We called them *Quake* movies," said Brown. "Machinima must have been termed by someone more educated than a bunch of game junky *Quake*-addicts. Great name though."

Machinima blossomed into a global movement, aided in its growth by game developers who started adding features to their games that were designed to make the production of such movies easier. "The machinima community has really turned into something special," said Brown. "The work being done now is titanic in comparison to what we did. They even have an awards ceremony each year, just like the Oscars."

The explosion of modding and machinima within *Quake* and later first-person shooters had a direct influence on Wright's approach to *The Sims*, which, thanks to its house building, already had player creativity at its heart. "I was really impressed looking online at games where players were going in and getting more creative involvement," said Wright. "I came across the *Quake* community that was doing custom skins and machinima. We just loved the idea that players could modify almost any part of the game and so we made these tools to make it easier for them. What with *The Sims* being about a doll's house in some sense, we wanted players to feel that almost anything in there could be at some level modified."

While *Quake* encouraged players to redesign their gun-wielding space marine characters, Wright's team created tools that allowed players to create new clothes for their sims and design new wallpaper patterns. *The Sims* team became hooked on the idea of breaking down some of the barriers that used to exist between player and creator. As well as creating tools so players could make their own content for the game, the team – at the suggestion of Luc Barthelet, who Electronic Arts had put in charge of Maxis after the takeover – linked up with *Sim City* fans online to exchange ideas about the game and spread the word. "*The Sims* was our first effort to build a community and help build it rather than let it accidently happen," said Wright. "We did a lot of things with the community that really paid off and were kind of experimental."

Wright's team carried out live demos of the game over the web and asked viewers to suggest what should happen next in the hope of conveying what people would be able to do in the game while gaining some insight into how people would play it. "We had the game running and twice a second it would send a screenshot online. Players would ask can you make them do this? They were remotely playing the game through us to get a sense of how open ended the game was," said Wright. "Most of these fan sites were really starved of content, but because we were sending out these images, they were capturing them and using them to build up parts of their website. For example 'Here's the whole story about how *The Sims* go to the bathroom' based upon these screenshots they were capturing. It didn't seem like a big deal at the time, but we got the original few hundred people to very much be the evangelists for the next 1,000 people and the next 10,000."

By the time *The Sims* launched in February 2000, the internet buzz that began with those fan demos was at fever pitch. Some fans had even spent the weeks leading up to the launch designing clothes and wallpaper patterns for their sims-to-be, using tools released by Electronic Arts in advance of the game itself. "*The Sims* was a success from day one, mainly because of the community we built before," said Wright.

The Sims' mix of creativity, voyeurism and humanity tapped into an audience far beyond that of normal video games. It displaced *Myst* as the biggest-selling PC game of all time and estimates suggest more than half of the people who bought *The Sims* were women – a huge proportion for a traditionally male-dominated entertainment media.

The way players used *The Sims* was also unusual, but reflected the trend towards user-generated content seen in *Quake* and other first-person shooters. "People used the game to tell stories," said Wright. "It was something we anticipated, but it was even more successful than we thought. You could take screenshots and add text to tell stories. We put a feature in so by clicking on one button in the game it would upload to the web on our server, where anyone could read your story. That was far more popular than we thought it would be – before we knew it we had 100,000 stories."

For many of the players who created photo stories using *The Sims*, the game became a means of self-expression. "Inherently *The Sims* is a toy and people play with it and, at some point, they get wrapped up in the characters and start directing the story," said Wright. "There's a transition from open play to more directed narrative, for some people it goes from a form of entertainment to a form of self-expression. Most of the stories have some message to convey. There was one woman who wrote about how her sister was in an abusive relationship and how she managed to get out of it eventually. You feel she wanted to write that story because she wanted other people to realise they could get out of these relationships. This is the kind of person who would probably never write a book or short story or anything else, but she has, with *The Sims*, been able to convey this message that really had a strong resonance with her to a wider group."

Quake and *The Sims* took the concept of games that encouraged user-generated content into the mainstream of video gaming. The blurring of the boundary between player and creator would have a powerful influence on the development of video games during the 2000s. Every first-person shooter on the PC began to include modding features and by the late 2000s it had even spread onto consoles via shooters such as 2007's *Halo 3*. *The Sims 2*, released in 2004, sought to develop player storytelling by including machinima-influenced video making tools. There were mods that would turn into full-blown commercial games.[5] One, the *Autumn Tower Defense* map created for the 2002 strategy game *WarCraft III*, even spawned a new genre – the tower defense game where players had to stop enemies crossing the map by building defensive towers with archers that shot them as they ran past. Internet games *Flash Element TD* and *Desktop Tower Defense*, both launched in 2007, honed the concept and popularised the genre.

Some mods even caused game companies major embarrassments. In 2003, Dutch hacker Patrick Wildenberg was playing around with ideas for modding the PC version of Rockstar Games' 2004 crime game *Grand Theft Auto: San Andreas* when he discovered a hidden sex game. Rockstar had dropped the interactive sex scene from its game prior to its release, but the difficulty of removing it from the millions of lines of computer code that formed *Grand Theft Auto: San Andreas*, meant the company simply disabled that part of the game.

Wildenberg created a mod called *Hot Coffee* that switched the half-finished sex game back on and released it onto the internet so players could try it. The *Hot Coffee* mod appeared just as Rockstar was busy defending itself against claims that the *Grand Theft Auto* series had inspired a number of violent crimes in the US. Wildenberg's mod poured petrol onto the political firestorm already raging around *Grand Theft Auto*. Democrat Senator Hilary Clinton demanded a Federal Trade Commission investigation and accused Rockstar of stealing children's innocence. Rockstar's initial claim that the sex game was nothing to do with them quickly fell apart as other hackers uncovered it in the PlayStation 2 and Xbox versions of the game. For Rockstar the *Hot Coffee* mod was a disaster. It prompted the Australian authorities to ban *Grand Theft Auto: San Andreas* and Europe's

---

5. Canadian student Minh Le and American student Jess Cliffe's *Counter-Strike* mod for the first-person shooter *Half-Life* was probably the biggest of these. Created in 1999, *Counter-Strike* was an urban showdown between two teams of players – one playing terrorists, the other a counter-terrorist squad. It became a runaway success and by the early 2000s an estimated 1.7 million people were playing it online every month.

game age rating organisation to increase its age rating to adults only. In the US, Rockstar pulled the game from the shelves while it set about removing the sex scene entirely to try and calm down the outrage.

Despite the *Hot Coffee* incident, by the middle of the 2000s whole games were being designed around the idea of player content creation. British game designer Peter Molyneux's 2005 game *The Movies* embraced the machinima movement to create a business simulation where players ran a movie studio but also got to create machinima films, using virtual actors and sets, that could then be uploaded them to the web.

To help deal with the expected rush of online content, Molyneux's Lionhead Studios hired former press officer Sam van Tilburgh as a full-time community liaison officer who would vet the players' films and manage relations with *The Movies'* player community. "When you think of making movies, most players will want to create their favourite flick or even use stars they admire," said van Tilburgh. "In the first few months of *The Movies*, the number of *Batman*, *James Bond* and *Star Wars*-titled movies was astronomical. We had to remove thousands of movies every week due to copyright infringement."

Despite the rip offs, *The Movies* spawned one of the highest-profile machinima movies created during the first decade of the 2000s: *The French Democracy*, a 13-minute film created by Paris resident Alex Chan. *The French Democracy* was Chan's reaction to race riots that hit the French capital in November 2005. It told the story of a group of Parisians of African descent whose encounters with racism in their day-to-day lives reach boiling point when two black teenagers die while being chased by police. "It stood out because it made a point, it was a personal message about something that was relevant in the global world news coverage," said van Tilburgh. "Alex was living in Paris and experiencing what was going on and what loads of people were only seeing through their television screens. It grabbed people's imagination. For me that's when video games enter the mainstream and can become art. When they convey a message that makes people discuss certain issues in their or other people's lives."

The integration of game and player creativity tools that *The Movies* sought to achieve would be taken further by Media Molecule, a UK game studio formed in 2006 by a group of former Lionhead staff. Media Molecule had formed off the back of co-founder Mark Healey's *Rag Doll Kung Fu*, a crazed and amusing martial arts game featuring toy-like fighters controlled like puppets that Healey wrote in his spare time and released in 2005. *Rag Doll Kung Fu* was a surprise hit, but Media Molecule had a more ambitious title in mind for its debut: a game that let players make games. "Right from the start content creation was definitely one of the aspects for the game, but we didn't really know how much it was going to be," said Healey. "We had this slight internal struggle with ourselves. On one side it was 'we just need to make a really good platform game and get people to make stuff as well', on the other it was 'this should be a totally full-on game creation kit kind of thing'."

Sony Computer Entertainment solved the dilemma when Media Molecule went to its London offices to demo their prototype game. "We were slightly nervous about talking about the content creation too much and scaring them," said Healey. "A part of us thought they might go for something that's more traditional and safe, but Sony picked up on the content creation side of it. That was obviously a pivotal moment in realising this was an important part of what we were doing."

Encouraged by Sony, Media Molecule created the PlayStation 3 title *LittleBigPlanet*, an incredibly powerful game creation tool capable of producing a wide variety of 2D games that hid its complexity beneath an inviting and unthreatening scrapbook visual aesthetic. "Designing the content creation aspect was a process of iteration," said Media Molecule co-founder Alex Evans. "We started with very, very crude tools and they were basically completely physical – there was a hair dryer to melt things, a paint roller on the end of a long stick. You literally had to run around to paint. We got quite far into the project before we realised that it was not fun enough. So we had to do several about-faces, several rewrites. The physical thing was too much, so we ended up swinging too far the other way and made this incredibly complicated system. That was about a year into the three years of development; we were a third of the way through. We had already thrown it away once and now we had to throw away this very complex system. There was this stamping system, where you stamp things down to fix them in place, and that was the only fun bit of creating at the time. From that we extended the idea of stamping stuff down to the whole create mode. Everything that you create is essentially a process of stamping stuff down, smearing out shapes or cutting."

LittleBigPlanet's game creation system paved the way for a burst of player creativity rarely seen outside of the PC game scene. By the end of July 2009, eight months on from the game's November 2008 debut, players had uploaded more than a million levels to *LittleBigPlanet*'s social networking-inspired portal where others could download and play their work. "It's amazing, just totally amazing," said Healey. "It's far more prolific than I ever thought it would be. Probably the first thing I was really impressed with was the calculator thing. It's obviously not a fun level to play or anything, but just knowing that some mad guy spent time to work out how to make a functioning calculator in the game. I was in awe of his OCD'ness. And then there were the guys who clubbed together to redo *Contra*, the old arcade game. It's just unbelievable really."

The social networking websites that inspired Media Molecule's creation sharing system were also an important influence on Wright's *Spore*, an evolution-themed game that took the player-as-creator concepts of *The Sims* to a whole new level.

Spore put players in charge of the evolution of a lifeform from primordial zooplankton to space-travelling civilization. As well as evolutionary biology, Wright incorporated a library's worth of scientific ideas, proven and unproven, into the game. From the visual style of electron microscope imagery and panspermia theory – the belief that life on earth began when comets brought organic matter from outer space – to the assumptions used in radio astronomer Frank Drake's calculation that around 10,000 planets in our galaxy harbour intelligent life.

Wright hoped *Spore* would advance players' understanding of evolution, although the model of the final game was closer to the concept of intelligent design that was being used in the US to challenge attempts to teach evolutionary theory in schools. "I really just wanted to convey the idea that creatures evolved incrementally over a long period of time in response to their environment. Whether it was through selection or directed evolution was really a tactical question," said Wright. "It is kind of ironic that in some sense *Spore* is showing the intelligent design thing, which is not a theological or scientific philosophy but a political tactic. If you go around asking people, even religious conservative people, nobody believes in intelligent design as it's stated; they believe in creationism, which is different."

The core of *Spore* was its simple but powerful creature-creation tool that let players mould their lifeforms into new shapes and add or remove appendages such as antennae, jaws, tails and eyes. Once created, players could load their beasts onto *Sporepedia*, a *Flickr* and *Facebook*-influenced website that integrated with the game. *Sporepedia* not only let players share their creations, but also imported the lifeforms created by other players into the game, quietly populating each player's world with the work of others.

"With *Spore* one of our fundamental things was how do we make the content creation curve flatter so that everyone is participating, so that when you play *Spore*, whether you intend to be a content creator or not, that automatically populates on the web so almost everybody becomes a content creator," said Wright. "A big part of that was making the tools not just easy to use, but fun to use. With *The Sims* we were thinking about how to make the tools easy to use. That was ok, but with *Spore* we thought if we can make the tools fun to play then you get a lot more output." *Spore*'s fusion of content creation and social networking meant much of what players saw and interacted with in the game was created by their fellow players. Fittingly, given its subject matter, *Spore*'s embrace of player-generated content underlined just how much the video game medium had changed over the years. Having begun life as a medium defined wholly by developers, video games were rapidly evolving into one that turned consumers into artists. Games had become interactive not just in terms of the experience of playing, but interactive as a medium, subject to constant reshaping, modification and extension by those who used it.

Hey You, Pikachu!: A Nintendo fan comes face to face with the Pokémon star
Courtesy of Nintendo of America

CHAPTER 26

All-Access Gaming

When Japan's newspaper readers opened their morning papers on the 21st May 1998 they were confronted with a full-page picture of a battlefield littered with slaughtered samurai. The advert posed a bold question: "Has Sega been defeated for good?"

It certainly wasn't unthinkable. The Sony PlayStation and Nintendo 64 had crushed Sega's Saturn console, leaving the company's consumer division nursing losses of $242 million for the year to March 1998. But Sega already had the answer to its own question. The next day the newspapers once again showed the battlefield picture, but this time the samurai were rising to their feet to fight again. That November Sega returned to the fray with a brand new console: the Dreamcast. It would represent both a creative zenith and commercial nadir for the firm.

Sega's internal development studios pulled out all the stops for the Dreamcast. They created innovative music games from the intergalactic dance-off *Space Channel 5* to the Latin-flavoured maraca shaking of *Samba de Amigo*. They pioneered the use of cel-shaded graphics in the Japanese urban cool of *Jet Set Radio*.[1] Game designer Reiko Kodama delivered one of the best Japanese role-playing games of the early 2000s with her Jules Verne-influenced *Skies of Arcadia*. And Sega's AM-2 team created *Outtrigger*, a rare example of a Japanese-made first-person shooter, which used the Dreamcast's in-built modem to allow online play. Sega also used the Dreamcast's dial-up modem to bring the online role-playing game onto the home console with *Phantasy Star Online*. "The president of Sega at the time, Isao Okawa, was very vocal about the future of games being online," said Yuji Naka, the head of Sega's Sonic Team studio, which created *Phantasy Star Online*. "We chose to make an online *Phantasy Star* game because we thought the series was most suited to the online environment. Since it was my first

---

1. The cel-shading technique gives 3D games the look of a hand-drawn cartoon. The style became popular with developers during the early 2000s appearing in a diverse number of genres including car racing (Capcom's *Auto Modellista*) and first-person shooters (Ubisoft's *XIII*). Most famously Nintendo used cel-shading in 2002's *The Legend of Zelda: The Wind Waker*, which divided fans of the series who disagreed on whether its more child-like visuals were a welcome development.

attempt at such a big online game it was trial and error, we tested and re-tested and discarded a lot of things before coming up with the final product. It was a very different experience of game design and development." Set on the brightly coloured planet of Ragor, *Phantasy Star Online* was a far cry from the gloomy and dark fantasy worlds of *Ultima Online* and *EverQuest* both in terms of its visuals and in how it played. Instead of encouraging competition between players, Naka's game encouraged players to battle Ragor's alien monsters as a team. "I thought that in an online environment people would enjoy co-operating rather than competing and that a different approach would get players interested in the online version," said Naka. To help players from across the world play together, Naka developed a system that let players who spoke different languages communicate with each other using pre-defined lists of words or symbols that the game could translate instantly. As a result of Naka's communications aids and focus on co-operation, *Phantasy Star Online* boasted a sense of community between players that other more competitive online games from the time lacked.

Most ambitious of all, however, was Yu Suzuki's 1999 game *Shenmue*. At the time of its release, *Shenmue* was the most expensive game ever made. Suzuki spent five years and at least $20 million realising his vision with the aid of an army of artists, programmers and musicians.[2] Set in the Japanese city of Yokosuka in the mid-1980s, *Shenmue* cast players as Ryo Hazuki, a young man out to avenge the murder of his father. Its ambitious story hinged on not only on Hazuki's sense of loss and anger, but also on Suzuki's detailed recreation of 1980s Yokosuka, which provided a snapshot of a society at a cultural crossroads as the traditions of the past gave way to a new, modern vision of Japan.[3] Suzuki gave players the freedom to explore his virtual Yokosuka, taking in the sights and sounds of '80s Japan by taking on part-time jobs, talking with residents, eating at restaurants and visiting game centres to play Suzuki's arcade games from era.

But for all this ambition and creativity, Sega's efforts to convince the world to buy its console failed. When Sony released its PlayStation 2 console in March 2000 sales of the Dreamcast dried up entirely. In January 2001, a defeated Sega stopped production of the Dreamcast and reinvented itself as a publisher of games for the consoles made by its former rivals. Two months after pulling out of the console business, Sega quietly released the Japan-only Dreamcast game *Segagaga*, a bitter epitaph to its own fall from grace. *Segagaga* challenged players to do what Sega failed to do and make the Dreamcast a success. Its designer, Tetsu Okano, used the game to lampoon both the video game industry and his employer. At every step in their battle against the faceless Dogma

2. So many people were involved in making *Shenmue* that it took 10 minutes for the credits to roll.

3. *Shenmue's* story was to be told over a series of six games. Only two got released leaving the story unfinished.

Corporation, *Segagaga* lectured players on the reality of the video game industry. Game developers, the player is told, are "subhuman" but an unavoidable necessity. Elsewhere business advisors urge the player to sacrifice creativity, copy their rivals and remember, at all times, that video games are just product. It was a playable polemic from a designer who worried that the creativity of the video game industry was being sidelined by the pursuit of profit and that Japan was losing its position as the world's foremost producer of video games. On the surface Okano's fears about the future of the Japanese game industry seemed misplaced. Japan was still the world's biggest consumer of video games, Japanese consoles still dominated the international gaming landscape and, starting with *Final Fantasy VII*, the Japanese role-playing game – the most popular game genre in Japan – had become internationally successful. And then there was *Pokémon*.

Pokémon was an unstoppable video game phenomenon that had fanned out across the world like a tsunami following its Japanese debut in 1996. Images of Pikachu, the bright yellow tubby mouse-like creature with rosy-red cheeks that fronted the game, appeared everywhere converting children to its charms by the million. Although it started out as a video game, *Pokémon* had become a multimedia brand by the time it reached North America in September 1998. It was a game, an anime TV series, a trading card game and a manga comic. And, thanks to widespread merchandising, Pikachu could be found on everything from bed sheets and takeaway hot dog trays to airplanes and toy shop shelves. Within a month of arriving in the US, *Pokémon*'s anime show had become the country's most-watched children's show. Within seven months more than 2.5 million *Pokémon* video game cartridges and 850,000 sets of *Pokémon* trading cards had been sold. On top of that the hype was already building up for *Pokémon: The First Movie*, an animated feature that premiered in November 1999. It would earn more than $160 million at the box office and spawn numerous sequels. By the dawn of the year 2000, the custard-coloured Pikachu was everywhere, earning adoration from kids and befuddled glances from parents.[4]

Pokémon's journey to global domination began 10 years earlier, when Japanese game designer Satoshi Tajiri became interested in making a game based on linking together Nintendo's handheld Game Boy consoles. "Everyone was using the link feature to compete," he told *Time Asia*. "The idea I had was for information to go back and forth." Tajiri's information-sharing idea dovetailed nicely with his own interest in entomology. As a child, he had been fascinated with collecting creepy crawlies, so much so that he earned himself the nickname 'Dr Bug' for his obsession with beetles and other

---

4. The arrival of *Pokémon* was not always welcomed. In March 2001, the Supreme Council for Research in Saudi Arabia issued a fatwa, or religious ruling, banning *Pokémon* on the grounds that it promoted evolution, encouraged gambling and included a 'Zionist' symbol (a hexagram).

invertebrates. He envisaged *Pokémon*, a game where players set out to collect 'pocket monsters' that they could share with each other or use to battle creatures in other players' collections via the Game Boy's link cable.

Pokémon was also Tajiri's response to the changing attitudes to childhood in Japan. He worried that the pressure on Japanese children to study was isolating them from each other and that increased urbanisation was separating them from the natural world. He hoped *Pokémon* would help reconnect children with one another and offer them the chance to experience the joy he used to get from fishing creatures out of streams and searching for unusual beetles hiding under rocks. It took Tajiri six years to complete *Pokémon*. The result was nothing if not Japanese. Its childish 'kawaii' visuals owed a clear debt to Hello Kitty cuteness and manga comics, and the game itself fitted comfortably into the Japanese role-playing game mould. But with the original Game Boy in its twilight years, Tajiri's publisher Nintendo expected *Pokémon* would fare poorly at retail. Indeed, Tajiri was surprised that Nintendo even bothered to release his game.

The simultaneous 1996 release of the first two *Pokémon* games, *Pokémon Red* and *Pokémon Green*, however, struck a chord in Japan. Within three months, more than three million *Pokémon* game cartridges had been sold and its popularity only grew from there. *Pokémon*'s international success became a tipping point for the spread of Japanese pop culture, paving the way for the rapid spread of anime and manga into North America and Europe. *Pokémon*'s global appeal became a source of national pride in Japan, a confirmation that the country was a cultural, as well as economic, power capable of influencing culture on a global scale. Given the huge success of *Pokémon* and Japan's continued dominance of the home console market, Okano's view that the golden years of the Japanese gaming were over seemed out of step. But it turned out that Okano was right. Although the Japanese economy had sunk into a long drawn-out recession back at the start of the 1990s, the country's video game industry initially seemed immune. But after peaking in 1997, video game sales in Japan began to drop. In 2002 North America overtook Japan as the world's biggest consumer of video games, the following year European game sales edged above those in Japan.

And while sales in North America and Europe continued to grow, sales in Japan flat-lined. By 2009 the UK had displaced Japan as the world's second biggest consumer of games, after the US. Few had more to fear from this fall than Nintendo, the former king of the video game industry. While it was still enjoying spectacular profits from *Pokémon* and its Game Boy handhelds, Nintendo's share of the home console market was melting away. The PlayStation 2 had not only wiped out Sega, but also utterly eclipsed Nintendo's rival Gamecube console. To add to the pressure, Microsoft had just stormed into the console business with its Xbox.

Until the Xbox, Microsoft had sat on the fringes of the video game business. It had a few popular PC games to its name, most notably the *Microsoft Flight Simulator* series and the historical strategy game *Age of Empires*, but it had never really challenged the likes of Electronic Arts. It was a situation the Xbox set out to change for good. Backed by Microsoft's deep pockets and its flagship first-person shooter *Halo: Combat Evolved*, an elegant and dynamic burst of sci-fi action that liberated players from the confined corridors of most games of the genre, the Xbox became the first American-made games console to gain a significant share of the market since the Atari VCS 2600 in early 1980s. Microsoft managed to sell 24 million Xboxes, almost entirely in North America and Europe, just ahead of the 22 million Gamecubes sold by Nintendo.[5]

Nintendo's defeat led to speculation that the Kyoto giant may end up going the way of Sega, swapping console making for publishing. As a dedicated video game company, Nintendo lacked the sprawling business interests of Sony and Microsoft to fall back on and, while *Pokémon* was keeping the Kyoto firm profitable, it was hard to imagine Pikachu could carry the company indefinitely. The situation forced Nintendo's new president, Satoru Iwata, who replaced Hiroshi Yamauchi in 2002, to rethink the company's whole approach to console and game design. He concluded that the game industry had become enslaved by its pursuit of the latest technology, producing ever more complex games for ever more expensive consoles. It was a trend, he concluded, that had opened up a chasm between the dedicated game player and the broader mass market who wanted simpler, less involving entertainment. Iwata decided it was time for Nintendo to do something different, something surprising. As he put it: "You can't open up a new market of customers, if you can't surprise them."

In November 2004 Nintendo unveiled the first fruit of its soul-searching: the Nintendo DS, a handheld console that flipped open like a book to reveal two screens, one of which was a touch screen that players interacted with using a stylus. It resembled the multi-screen Game & Watch handhelds that Nintendo introduced back in 1982 with *Oil Panic*. In addition to its unusual twin-displays, the Nintendo DS had an in-built microphone and wireless connectivity, so people could play together without the need for cables.

Most game designers greeted the Nintendo DS with confusion. Many were at a loss as to what to do with the Japanese company's strange new handheld. Sony dismissed it as a gimmick, a "knee-jerk reaction" to the launch of its hi-tech PSP handheld console.

---

5. The Xbox struggled in Japan hampered by a lack of big name Japanese games and Japan's lack of interest in western games such as *Halo: Combat Evolved*. The Gamecube, meanwhile, sold better in Japan than in North America and Europe.

The initial scepticism quickly melted away, however, when Nintendo began serving up games such as *Nintendogs*, a virtual pet game where players brushed or stroked their digital hound using the touch screen and talked to it through the microphone. While *Nintendogs'* pixellated puppies won over younger players, *Dr Kawashima's Brain Training* (known as *Brain Age* in North America), a puzzle game based on the ideas of Japanese neuroscientist Dr Ryuta Kawashima, captured an older audience attracted by its collection of mental agility tests and Sudoku.

The Nintendo DS, the 'gimmick' handheld, appealed far beyond the standard game-player demographic and its sales far outstripped those of Sony's more traditional PSP. By 2009, more than 125 million Nintendo DS handhelds had been sold making it one of the most popular game consoles ever created. The DS experiment confirmed Nintendo's hunch that the expensive battle to provide players with the most technologically advanced games was a zero sum game. The rush for technical complexity and depth had narrowed the audience for video games. The next move was to apply those lessons to its next home console, which Nintendo had appropriately codenamed the Revolution.

"The industry places more value on the look and complexity of a game than it does on the amount of fun a person has while playing," Iwata wrote in Nintendo's 2005 annual report. "In today's world people are busy and the time and energy required to play games are seen as a burden. This is why more people are now saying 'video games are not for me' before they begin to play."

With this in mind, Nintendo made reinventing the game controller the focus of the Revolution project. The basic concept behind video game controllers had not changed since the launch of the Famicom back in 1983, but an ever-growing number of buttons, triggers and sticks had turned them from simple two-button pads into dauntingly complex control panels. For those unfamiliar with these devices, they were confusing and unnatural to use, making it near impossible to play any game. Nintendo's solution was to create a new type of controller, one that was instantly understandable.

Nintendo was not the first to try and find an answer to the controller problem. In 1984, just before it was broken in two, Atari was touting its Mindlink system, a headband controller for the VCS 2600 that let players control games by moving their forehead muscles. But, thanks to its tendency to induce headaches and Atari's financial troubles, it was never released.

More successful was the Mattel Power Glove, a glove controller for the NES released in 1989 that tapped into the hype about virtual reality. The $100 Power Glove measured the movement of wearers' fingers and the location of the players' hand in a room. It was a heavily pared-down version of the Data Glove, a $8,800 glove controller developed by VPL Research, the virtual reality research firm founded by former Atari Research employees Jaron Lanier and Thomas Zimmerman. "The first prototypes of the Power Glove were on the Amiga and we actually had some 3D glasses and had some really impressive, interesting games we'd done – 3D ball games and other wild surreal things," said Lanier. "But the commercial opportunity that came up was for a lower-end platform and a bigger market than the Amiga, which was a shame."

The Power Glove was promoted heavily. It got a star billing in the Nintendo-promoting feature film *The Wizard* and advertising that boasted about how it was bringing virtual reality to the shops "years ahead of schedule". It sounded incredibly exciting for the time, but the reality behind the hype disappointed many with its inaccuracy and underwhelming games. It was also uncomfortable. "When you closed your fist when wearing the Power Glove there was a little bit of plastic that would dig into your hand," said Lanier. Although it never lived up to the promise, the Power Glove was one of the first virtual reality devices to be made available to a mass audience. "Whatever's said that was the glove that got out there," said Lanier. "It got a lot of gloves out to people who could hack with it and play with it. There was a real Power Glove community for a while."

Virtual reality technology also inspired another attempt to escape the limitations of video game controllers around the same time when US singer-songwriter Dean Friedman formed his own virtual reality business: InVideo Systems. "I'd witnessed a demonstration of camera-based virtual reality at an Amiga computer convention," said Friedman, who is best known for his 1977 hit single *Ariel*. "People stood in front of a chroma key screen and were able to see themselves on a large TV, while interacting in real-time with animated objects.[6] No helmet, no gloves. Peripheral free. I tried it myself and it was like magic. They were selling developers' kits and I was so amazed that I bought one." Friedman teamed up with US children's TV channel Nickelodeon to create a game called *Eat a Bug* for its Sunday morning show *Total Panic*. "The idea was akin to *Honey, I Shrunk the Kids*," said Friedman. "The player stood against the screen and saw themselves shrunken in a garden environment – grass, dandelions – surrounded by animated insects. The object was to grab all the little bugs like flies and mosquitoes while avoiding the big bugs like the spider, bee and centipede."

While *Eat a Bug* was a popular addition to the show, the game's technology cost $15,000 – making it unviable as a consumer product. So instead of trying to reach the home market, Friedman's company sold its camera-based games to theme parks, museums and arcades around the world. "As soon as people walked into a game they

---

6. Chroma key screens are the blank green or blue screens used by filmmakers and TV studios to record actors or weather presenters and project them onto a computer-generated scene or weather map.

understood the game play intuitively and loved it," he said. "Millions enjoyed InVideo games, but it still took almost two decades for the technology to finally be embraced by the mainstream video game industry. Cost was a huge factor, but so was the job of acclimatising the industry to the idea that the public would so easily embrace a brand new mode of full-bodied game play."

Fourteen years after Friedman started experimenting with camera-based games, the home games business finally caught up with the pop star when Sony's London game studio produced the EyeToy. Released in 2003, the EyeToy was a camera that let players interact with games just by moving, thanks to its ability to recognise players' movements and gestures. Millions of PlayStation 2 owners bought the EyeToy after seeing the simple but fun games available in the accompanying *EyeToy: Play* collection that – just like *Eat a Bug* – fed their image directly onto the screen, such as *Wishi Washi*, where players had to vigorously wipe the air in front of them to clean a dirty screen, and *Kung Foo*, where players karate chopped miniature assailants.

Encouraged by the success of the EyeToy, Sony agreed to publish *Buzz! The Music Quiz*, a 2005 TV quiz show game that came with its own set of replica game show buzzers. While most quiz games available at the time suffered from low production values, *Buzz!'s* British creators Relentless Software focused on delivering a polished game that could justify the extra cost of its set of four controllers. "At the time a lot of developers would put their B or C team on such a game," said David Amor, the co-founder of Relentless. "We really bought into it as an idea and gave it our all. Just because it wasn't a first-person shooter didn't seem like a reason for the production values to be any lower."

For Relentless, the custom-designed buzzers were a vital part of the game. "When we announced the buzzers at the 2005 E3 trade show, some people's reaction was to point out that the same functionality could have been achieved with a regular controller," said Amor. "That's missing the point. We asked people to play our game show game using a game show buzzer, which really lowers the barriers to entry. There's no learning curve. Additionally we could ensure that four players could play at once, which is important for creating off-screen interaction."

Around the time that *Buzz! The Music Game* became a popular game on the PlayStation 2 in 2005, Nintendo's bid to reinvent the controller was beginning to take shape. While developing its 1996 console, the Nintendo 64, the company had briefly explored the idea of creating a wristwatch controller that tracked the movement of the player's wrist. It proved too confusing for players and the company shelved the idea, but with the company out to tear up the rulebook, the concept of motion-controlled games was revived. Nintendo built a simplified game controller designed to look like an unthreatening TV remote that allowed people to play simply by moving the remote around. And in keeping with Iwata's belief that the latest graphical breakthroughs were no longer impressing game players, the Revolution console itself was simply a souped-up Gamecube rather than a brand new machine. Nintendo called their new console the Wii.

Ahead of Wii's launch in November 2006, the video game industry greeted it in much the same way as the Nintendo DS: intrigued but wary. Microsoft had already launched its new Xbox 360 console, which boasted high-definition graphics, a larger hard drive and more support for online multiplayer games. And in the same month as the Wii reached the shops, Sony released the PlayStation 3 with boasts of even more high-definition visuals and an in-built Blu-Ray movie player. The Wii looked severely underpowered by comparison.

But Nintendo's instincts were spot on. The Wii's flagship launch game, *Wii Sports*, demonstrated the potential of the console's Wiimote controllers perfectly with its mix of tennis, baseball, golf, boxing and bowling. Millions of people, many of who had never owned a game console before, headed to the shops to get themselves a Wii. The Wii's appeal to a broader audience received an additional boost the following year with the release of *Wii Fit*, a collection of exercise games that producer Shigeru Miyamoto was inspired to make after going on a diet. *Wii Fit* came with the Balance Board, a controller similar to a bathroom scale that measured the position and weight distribution of players who stood on it.[7] The game acted as a virtual personal trainer, berating players for not exercising enough or being overweight and trying to get them fit with its selection of exercise-themed games ranging from yoga and push-ups to skiing and jogging.

By late 2009, Nintendo had sold close to 68 million Wiis. Microsoft's Xbox 360 and Sony's PlayStation 3 had both sold around half that amount. Nintendo's decision to break from the pack had taken them back to the top of the video game tree and prompted Sony and Microsoft to start developing motion controllers of their own. Nintendo had also taken games back to their coin-op roots, reviving the simple spirit of the arcades to provide lucid video games that were accessible, designed for short bursts of play and – above all – pure entertainment.

But while Nintendo won over millions of people previously alienated by the complexity of video games with its 'all-access gaming' mantra, other game designers were seeking to expand imaginative scale of video games by creating epic, big-budget titles that, like *Shenmue*, offered a vision of electronic entertainment that was more about art and ambition than concentrated fun.

---

7. The ideas that led to *Wii Fit* had been knocking around the video game business for some time. Prior to creating the Amiga computer, the Amiga Corporation developed 1982's *Mogul Maniac*, a skiing game for Atari's VCS 2600 that used the Joyboard, which let players control in-game movements by standing on it and leaning in different directions. Bandai's 1986 pressure-sensitive *Family Trainer* mat controller and the music game dance mats that became popular thanks to Konami's *Dance Dance Revolution* also foreshadowed the Balance Board.

Back to the '80s: Grand Theft Auto: Vice City resurrects the glam of Miami Vice
Rockstar Games

CHAPTER 27

The Grooviest Era Of Crime

Grand Theft Auto III opened with a bang. The player's nameless character is being transported to prison with two other convicts when the police convoy is attacked while crossing a bridge.[1] The attackers free one prisoner before fleeing, blowing up the bridge behind them in a bright flash of white. Seconds later the player is handed control. The bridge is broken in half and the burning wreckage of prisoner escort vehicles lay scattered. Only the player and the other prisoner have survived. One empty and undamaged car offers a chance to escape to freedom.

Directed by the other criminal, the player heads to a safe house in the city. The player drives past people milling around and passing traffic; up hills, under railway bridges and along darkened streets lit by the orange glow of streetlights. As Chinatown whizzes past, a squeeze of a button switches the car radio from Debbie Harry's *Rush Rush* to Double Clef FM, which is playing *La donna è mobile* from Giuseppe Verdi's 1851 opera *Rigoletto*. The escape car screeches around corners, leaving behind dark tyre tracks and startled pedestrians. Soon the run-down looking red light district comes into view and then the barely noticeable back alley leading to the safe house.

Dawn breaks, bleaching the sky pink and purple. Dirty clouds hover, threatening rain. A taxi hoots its horn, piercing the snatched chatter of passing pedestrians. A train rattles along the elevated tracks with a thundering noise. A passing van runs down a man in a Hawaiian shirt and knee-length shorts before speeding off in panic. Seconds later an ambulance races into view emergency lights flashing to treat the injured man whose prone body has attracted a crowd of shocked on-lookers. Just past the dual carriageway where the hit and run took place is a large river across which lies another part of the city with a Manhattan-esque skyline of skyscrapers pointing to the heavens. In the far distance a plane is descending down towards an airport that must lie beyond the skyscrapers leaving a vapour trial that cuts through the dawn sky in its wake.

1. The character would eventually be named as Claude during a brief appearance in 2004's *Grand Theft Auto: San Andreas.*

This is Liberty City, the world of *Grand Theft Auto III* and what happens next is anyone's guess. There's a story to follow where you could seek work as a thug-for-hire but it's not compulsory. You could steal a car and tour the city, see the sights and become a virtual tourist. Or you could grab that discarded baseball bat and prowl the streets beating passers-by to see how long it will be before the cops show up. Or drive off ramps to perform stunts with your car. Or take a ride on the train to get a better view of the city, visit a prostitute, or become a cab driver. Or maybe you will do all of this and more.

Released in October 2001 with little pre-release hype, *Grand Theft Auto III*'s three-dimensional city teemed with life and offered a sense of freedom, openness and possibility that no other game had achieved before. Many had, of course, tried before. Not least the first two *Grand Theft Auto* games.

The origins of the series date back to 1994 when Mike Dailly, a software engineer working for the research and development team at Scottish game studio DMA Design, developed a new graphics demo. The demo, *Rotator*, showed an isometric view city that could be rotated at will. "This was very fast and worked well, so the idea was to make a strategy game that could use it," said Dailly. "*Grand Theft Auto* actually started out as a gang versus gang game. However there were problems when the team tried to implement the rendering and Bullfrog's *Syndicate Wars* had just come out and done something very similar, so the idea was dropped."

Dailly, however, had an alternative to hand: "I had been working on a second engine called *Dino*. This one was based on an idea I got when watching a Sega Saturn game called *Clockwork Knight* being played." *Clockwork Knight* was a fairly unremarkable side-view platform game in the *Super Mario Bros* mould. Its visuals, however, were reminiscent of a fish-eye lens. Only the objects in the centre of the screen were seen as flat; those to the left and right of the screen's centre appeared to curve off to the sides. "It occurred to me that although I had a side-on engine, all I needed was to add a floor and it could be an above engine," said Dailly. "So with sad programmer graphics I set about using the previous prototype engine as a base."

It didn't take long for the DMA Design team working on the prototype *Grand Theft Auto*, then known as *Race and Chase*, to embrace Dailly's new engine. "I showed Keith Hamilton, David 'Oz' Ozbourn and Dave Jones," said Dailly.[2] "They decided to restart *Race and Chase* using the new engine since it allowed far more freedom. It was simply a case of me showing Dave, Oz and Keith the new demo and 30 minutes later everyone agreed to start again. It was a good meeting."

---

2. David Jones, like Dailly, was one of DMA Design's co-founders back in 1988. Keith Hamilton was the team leader and programmer on *Grand Theft Auto*. David Ozbourn was an artist at DMA.

Race and Chase was envisaged as a cops and robbers game. Players could be the police chasing the robbers or adopt the role of the fleeing criminal hoping to lose the pursuing squad cars. It didn't take long before the option to be the police was dropped. "Nobody wants to be the cop, they want to be bad and that evolved into *Grand Theft Auto*," said Gary Penn, who joined the Dundee-based studio as the game's producer halfway through its development.

The team envisaged the game's three cities as a playground where players could go wherever they liked.[3] It was an idea heavily influenced by *Elite*, the open-ended 1984 space sim created by Ian Bell and David Braben. "I had worked on *Frontier: Elite II* and there were other people on the team who had *Syndicate*, *Mercenary* and *Elite* very much in their minds as well," said Penn. "That combination definitely led to the more open structure that we condensed into, basically, *Elite* in a city. You take on jobs in a slightly different way but it's incredibly similar structurally, it's just a more acceptable real-world setting."

For months, however, the *Grand Theft Auto* project teetered on the edge of abandonment hampered by unstable code, boring action and overly complex controls. "When I joined DMA it was a mess," said Penn. "It was a mess for years, it never moved on, it never went anywhere. It was almost canned. The publisher, BMG Interactive, wanted to can it as it didn't seem to be going anywhere. It was no fun at all. The core of the play was fundamentally broke and it had a broken structure as well."

But the compelling idea of an open city to drive around was enough to keep the project alive. A renewed push to fix the game's rickety code proved to be the turning point, allowing the team to concentrate on improving the game experience itself. Piece by piece the game came together – sometimes more by accident, than design.

"The way the police originally worked was just rubbish and then one day – I think it was a bug – the police suddenly became mental and aggressive because they were trying to drive through you," said Penn. "That was an awesome moment because you got real drama where you went 'oh my god, the police are real psycho, they are trying to ram me off the road'."

But on its release in October 1997, *Grand Theft Auto*'s key feature - the offer of a city to roam in – went largely ignored. Instead people talked about the game's overhead viewpoint, which was seen as a 2D relic in an era when 3D was fast becoming the norm, and the subject matter.[4] Egged on by the game's publicist Max Clifford, *Grand Theft Auto* was greeted with howls of outrage thanks to its embrace of criminality. Most games portrayed players as heroes regardless of their in-game actions, or at the very least

---

3. The three cities were Liberty City, Vice City and San Andreas.

4. *Grand Theft Auto* did use 3D, but its fixed overhead view was more reminiscent of 2D racing games such as *Super Sprint*.

victims who have been forced into carrying out criminal acts. *Grand Theft Auto* offered no such veneer of respectability. The player was an amoral crook out to rob, kill and maim for personal gain and nothing more. Politicians and players alike were shocked by DMA Design's refusal to apply any morality to the player's actions. One reviewer called it the "most violent piece of gaming on the PlayStation" thanks to features such as the ability to mow down whole processions of Hare Krishna for extra money. In the UK Parliament, Conservative peer Lord Campbell of Croy accused the game of setting an "alarming precedent" and of glamourising crime.

But beyond the notoriety, *Grand Theft Auto* was a rare example of an action game taking up the challenge presented by *Elite* and Will Wright's *Sim City*: to create games that give players virtual 'sandboxes' to play in, where they created the narrative through their own actions and choices.

Prior to *Grand Theft Auto*, only a handful of games had managed to come anywhere close to offering players this level of freedom. Foremost was Bethesda Softworks' fantasy role-playing game series *The Elder Scrolls*. *The Elder Scrolls* games handed players a vast open-ended fantasy world where the main adventure was optional and covered just a fraction of what the game could offer. In effect Bethesda offered two narratives – the one created by the developers and the one defined by players within the game's world. "Giving players' freedom of choice was our main goal," said Todd Howard, the executive producer of the series. "To have the game react to you and remove as many boundaries as possible to what you can do. I think players often start a new game by trying things, asking the game 'can I do this?' and the more the game says 'yes' the better."

But by the start of the 2000s *Grand Theft Auto* was already looking outdated thanks to *Driver*, a 1999 driving game inspired by 1970s movies and TV shows such as *The Driver* and *Starsky and Hutch*. *Driver* hinged on its over-the-top car chases, but also gave players the freedom to drive around its cities as they pleased.[5] By comparison the same year's *Grand Theft Auto II* offered little beyond that contained the first game in the series and, thanks to the lack of sales-boosting public condemnation, it sold below expectations.

For its third attempt at *Grand Theft Auto*, DMA Design tried once again to deliver on its dream of a 3D city to walk, drive and fly around. "We had tried to do a 3D city with the first one but it was definitely beyond the team's capability at the time," said Penn. "With *Grand Theft Auto II* we tried other 3D elements, but the risk was too high. It made more sense to build on what we'd established with the first one. In all there were three or four attempts to do a 3D one before *Grand Theft Auto III*."

---

5. *Driver*, however, cast the players as an undercover cop out to win the trust of a group of gangsters by helping them evade the police rather than embracing the nihilism of *Grand Theft Auto*.

The breakthrough came from another group of DMA Design employees who had just completed work on *Space Station Silicon Valley*, a 3D platform game for the Nintendo 64.[6] "They were an incredibly capable team. They had just done *Space Station Silicon Valley* in 3D so they had the attitude and the ability to take the 2D game and put that in 3D," said Penn, who worked on *Grand Theft Auto III* for its first six months of development before leaving DMA Design. "The core team were so capable I didn't have a doubt that it would come out – it was really a case of how long it would take because it's such an involved thing. It's a really fucking hard game to make. It's a really hard game to make in 2D and really, really hard to make in 3D. So that the third one ever came out really impressed me."

Like *Doom* before it, *Grand Theft Auto III* reshaped the video game landscape. It proved that believable and open 3D game worlds could be created and sold millions. But while *Doom*'s success inspired a spate of copycats, *Grand Theft Auto III* faced relatively little competition largely because the challenge of creating a virtual world or city of comparable scope or vision was so difficult and expensive. Only Bethesda offered anything comparable in size and scope to DMA Design's epic with 2002's *The Elder Scrolls III: Morrowind* and, later, 2006's *The Elder Scrolls IV: Oblivion*.

But *Grand Theft Auto III* did encourage more games to try and maximise the freedom of choice available to players. From the open city of 2008 racing game *Burnout Paradise* to the uncharted solar systems of the 2007 space opera *Mass Effect*, many of which players could explore but most of which had no connection to the game's primary story. Having proved the concept with *Grand Theft Auto III*, Rockstar North – as DMA Design was renamed in 2002 – set about applying a stronger sense of time and place to their creation in the 2002 follow up *Grand Theft Auto: Vice City*. Inspired largely by the 1980s TV crime series *Miami Vice*, the city of *Grand Theft Auto: Vice City* transported players into a decade of pastel-coloured suits with rolled-up sleeves, hairspray, yuppie aspiration and cocaine, all accompanied with a soundtrack of '80s pop and rock hits. Sam Houser, the game's executive producer, described the 1980s as "the grooviest era of crime because it didn't even feel like crime". "You had Cuban hit men coming across and gunning people down in the street, but it was still celebrated in a sort of haze of cocaine and excess and Ferraris and Testarossas, and it was a totally topsy-turvy, back-to-front period of time," he said.

The game also fizzed with an acidic wit. Its lampooning took few prisoners. It took pop shots at video gaming's past with radio ads for the Degenatron console, where every game involved a red square battling green dots. It took aim at US gun culture with

---

6. Programmers Leslie Benzies, Adam Fowler and Obbe Vermeij plus artist Aaron Garbut, who had moved to DMA Design's new offices in Edinburgh that were opened after it was bought by Rockstar Games in 1999.

Ammu-Nation, a survivalist-staffed chain store that sold everything from handguns to rocket launchers, and satirised the poodle-haired rockers of the '80s with Love Fist, a fictional group of bare-chested, big-haired and vacuous Scottish rockers. While the 1980s retro chic gave *Grand Theft Auto: Vice City* a sense of time and place, the story remained straightforward tale of a thuggish criminal on the make. But with 2004's *Grand Theft Auto: San Andreas*, Rockstar North began to move beyond the nihilistic characters that had dominated the series up to then. Inspired by Los Angeles' gang culture of the early 1990s and set across the vast state of San Andreas, *Grand Theft Auto: San Andreas* cast the player as CJ, an African-American anti-hero who is dragged back into the gang culture he had been trying to escape after being framed for murder by two corrupt cops. With its more involved story and willingness to touch on the racial tension that existed in California in the early 1990s, *Grand Theft Auto: San Andreas* moved the series beyond the moral void of its earlier incarnations and embraced the growing trend among game developers to inject meaning and depth into the stories of the games they made during the 2000s.

The end of the 1990s had witnessed a significant shift in the way stories were conveyed in games. Ever since Will Crowther first created *Adventure* back in 1975, storytelling in video games had been dominated by two genres: the adventure game and, to a lesser extent, role-playing games. And as the year 2000 approached, narrative-driven games appeared to be entering a kind of golden age thanks to titles such as Jordan Mechner's 1997 murder mystery *The Last Express*, which combined distinctive Art Nouveau visuals with a cinematic storytelling, *Planescape: Torment*, a convention-defying role-playing game that cast players as an amnesic immortal on a philosophical pilgrimage to discover what kind of person they were, and the Norwegian adventure game *The Longest Journey*, where the female protagonist's quest was as much as a voyage of self-discovery as about saving the world.

Tim Schafer's comedy film noir *Grim Fandango* typified the maturation of video game storytelling. It told the story of Manny Calavera, a dead man stuck in a dead-end job as a travel agent in the Land of the Dead.[7] The adventure saw Calavera battle corruption to secure a high-speed train ticket to the afterlife for the newly deceased Mercedes Colomar within a world inspired by Mexico's Day of the Dead celebrations and Art Deco design.

Witty, inventive and chic, *Grim Fandango* in many ways marked the apex of adventure games. It won dozens of awards and gushing praise from critics the world over. But for its publisher LucasArts and others, it simply reinforced the belief that the adventure game, the crucible of video game storytelling, was past its sell-by date.

---

7. A place where, according to Aztec folklore, people went after death before embarking on a perilous four-year journey to the after world of Mictlan.

Despite being showered with accolades, *Grim Fandango* sold nowhere near enough to cover the cost of making it. And if a game as praised, inventive and memorable as *Grim Fandango* could not justify itself commercially, what hope was there for other, less distinguished adventure games, publishers asked themselves. As word of its commercial disappointment spread, video game companies began turning their backs on the genre.

Most surprising of all was Sierra Online's decision to ditch the genre it was founded on. "The people who had bought Sierra from Ken and Roberta Williams in 1996 didn't like adventure games and looked for information that built their case," said game designer Al Lowe, who had started work on *Leisure Suit Larry 8* for Sierra just before *Grim Fandango*'s release. "*Grim Fandango* didn't sell well and it was an adventure game. It got good reviews because it was a good game, but it was a game about death with really odd graphics. It was kind of off-putting. And then there was *Phantasmagoria: A Puzzle of Flesh* – that was odd. There was a lot of bondage and it was a horror film, so that didn't sell very well, so they used that as an excuse too." *Leisure Suit Larry 8* was cancelled and Lowe's involvement with Sierra came to an abrupt end.

Similar things were happening across the Atlantic where French game designer Philippe Ulrich was finding the tide had turned decisively against the adventure games that he had been making. "Publishing houses abandoned the genre," said Ulrich, who co-founded adventure game specialists Cryo Interactive after quitting Infogrames in the early 1990s. "It was too expensive, too complex; you needed authors, interactive brainstorming sessions, super-duper artificial intelligence. You really had to get into bed with movie industry people. I was fired from Cryo exactly for that: 'Phil, sorry, but you're too old and you come up with complex and expensive games. We want to emulate Kodak and offer disposable games'. So I decided to turn to music production and succeeded in selling two million albums within a few months. But I consider the whole thing as my big failure: I did not succeed in showing the world that adventure games were a major genre." Within three years of *Grim Fandango*'s release, adventure games had faded into obscurity, kept alive only by a small-but-loyal following in mainland Europe, particularly Germany.

Almost as surprising as the sudden death of adventure games was the emergence of first-person shooters as a new vehicle for video game storytelling around the same time. While first-person shooters had become hugely popular in North America and Europe after *Doom*, the genre was known for its focus on straightforward action. Most first-person shooters subscribed to Id co-founder John Carmack's view that video game stories were as unnecessary as plots in pornographic movies and concentrated on an adrenaline-pumping gun fights instead. The few first-person shooters that tried to add more substance to their back story, such as Epic Games' 1998 title *Unreal*, offered little that could be considered comparable to depth to the narratives of adventure games.

For Epic, the story was simply there to help distinguish *Unreal* from the kill 'em all action of Id's *Quake II*. "We wanted to make the anti-*Quake*," said Cliff Bleszinski, who co-designed *Unreal* with James Schmalz. "We were always for years in the shadow of Id Software. We looked at *Quake* and what it was doing and, as much as we were fans of it, we wanted to do a counter-program. They were doing dark, brown dungeons and we wanted to make beautiful, colourful sky cities and things like that."[8]

But just 20 days after *Grim Fandango*'s launch, Valve - a Washington state game studio - took the genre in a new direction with the release of *Half-Life*. *Half-Life* opened with 20 minutes of nothing much. Cast as scientist George Freeman, the player spent the opening minutes of Valve's first-person shooter waiting for a monorail to transport them to the research lab they worked in. Once there they had to put on their hazmat gear and pass through a number of security checkpoints on his way to the lab where Freeman worked. Such a sedate and uneventful introduction to an action game was unheard of. Most games sought to throw players headfirst into the action with minimal amounts of build-up or scene setting, preferring a 'here's a gun, let's go' approach. But *Half-Life*'s 20 minutes of near inaction allowed it to convey the background to its world, plot and lead character without the need to resort to text. And even once Freeman's experiment goes wrong, opening a portal through which deadly creatures flood onto Earth, the game continued to develop the story through visual cues rather than reams of text or dialogue.

One of the most effective was when the military arrived at the lab. Initially the player assumes they are a rescue team but when they begin executing survivors the understanding that they are there to eliminate the witnesses is conveyed instantly as part of the action. While *Half-Life*'s story lacked the depth of the stories offered in games such as *Grim Fandango* and *Planescape: Torment*, its 'show don't tell' approach to narrative made the case for a new way of thinking about how to explain and present stories within a video game. *Half-Life*'s within-the-game storytelling heralded a beginning of an augmentation process that would turn first-person shooters from straightforward trigger-happy adrenaline pumping games into experiences that combined action, interactive fiction, roleplaying and cinematic presentation. By the time *Half-Life* was released the next great leap forward for this ambitious fusion of genres was already in development at Ion Storm, the Dallas game studio formed by John Romero after his departure from Id in 1997.

---

8. *Unreal* did not unseat *Quake* from its dominance over the first-person shooter genre, but it did establish the reputation of Tim Sweeney as a creator of 3D graphics engines comparable to Id's Carmack. And when Epic turned *Unreal* from a single-player series to an multiplayer-focused game with 1999's *Unreal Tournament*, its work began to eclipse that of Id. "The first *Unreal* had a certain magic to it, but it had a completely broken multiplayer," said Bleszinksi, who designed levels for *Unreal Tournament*. "One of our programmers, Steve Polge, was really good at creating these kind of autonomous AIs who fight with a player and essentially we started making some multiplayer changes and improvements, basically fixing all that was broken. And it began to take on a life of its own. Epic co-founder Mark Rein looked at this and said this could be its own product and made the case for it. That's how *Unreal Tournament* was born."

When Romero founded Ion Storm he was one of world's premiere game designers. He had a track record for creating multi-million-selling games that sent players wild with excitement and a rock star image, complete with long waist-length black hair, to match. With such a reputation Romero quickly landed a huge publishing deal for Ion Storm. UK game publisher Eidos Interactive, empowered by the success of *Tomb Raider*, agreed a three-to-six game deal with Romero's new outfit that was worth at least $13 million. "It wasn't quite a blank cheque but it had a lot of zeroes on it," said Jeremy Heath-Smith, an executive on the Eidos board at the time.

With the money pouring in, Romero turned Ion Storm into his dream game studio. The company leased the glass-roofed penthouse offices of the 54-storey JPMorgan Chase Tower in Dallas and hired an expensive interior design agency to redecorate it to their tastes.[9] Romero also hired Warren Spector, a game producer who had worked on landmark titles such as *Wing Commander* and *Ultima Underworlds: The Stygian Abyss*, to head a second branch of the company in Austin, luring him to Ion Storm with the promise that he could make the game of his dreams.

Spector's dream game was *Deus Ex*, a fusion of first-person shooter and role-playing game that built on the ideas he had explored in 1994's sci-fi game *System Shock*. *Deus Ex* was a PC game about terrorism that explored conspiracy theories, government bureaucracy, the nature of capitalism, access to medical treatments and genetic engineering.[10] While Romero was a maker of primal and purist games, Spector used the opportunity Ion Storm's cash afforded him to create an ambitious, intelligent and literary game. Set in 2052, *Deus Ex* cast players as a United Nations counter-terrorist agent fighting the widespread terrorism of the era.[11]

As well as exploring current affairs, *Deus Ex* gave players the freedom to decide how to carry out their missions, instead of forcing them down pre-determined paths. Players could sneak into a building to steal documents or charge in all guns blazing or hack into computer systems and turn the building's security guns against any guards. Each mission could be completed in different ways and the choice was the players' alone. Some of choices made during missions also affected the game's narrative. A gun battle that led to the death of civilians could, for example, result in the player's character, JC Denton, being berated by his bosses. The whole approach felt revolutionary, said Sheldon Pacotti, one

9. The glass roof would become a big problem for Ion Storm as the glare from the sun made it hard for staff to see what was on their computer screens.

10. The game featured a genetically engineered plague called the Grey Death. The lack of access to the cure for the Grey Death among the poor of *Deus Ex's* world was a clear nod to the lack of access to anti-HIV treatments in Africa.

11. Although it pre-dated the 9/11 terrorist atrocities of 2001, the game's vision of a future where terrorism and fear of terrorism were part of everyday life was remarkably, if unintentionally, prescient.

of *Deus Ex*'s scriptwriters: "What I got excited about was trying to make something real in a video game: real characters, real environments."

The *Deus Ex* team aimed to reinforce the sense that players were part of a real world with the addition of everyday objects in the game's locations that ranged from basketballs that could be bounced to books containing readable extracts of John Milton's 17th century poem *Paradise Lost*. Pacotti also paid a lot of attention to ensuring the dialogue for the game was as realistic as he could make it. "Writers go through these evolutions where they discover the basics of writing," he said. "At the time I was discovering voice and was working hard on learning how to write a variety of different voices, so I put a lot of energy and thought into trying to make every type of character sound unique and authentic. I used the bus a lot then and I had a notepad and would write down phrases that I heard people say. There's no way to write voice without primary source. There was one book I had that was interviews with teenage gang members in Los Angeles – just transcripts of discussions."

Released in 2000, *Deus Ex* marked another major step forward for attempts to tell stories within the context of an action game. Its grey moral choices and the appearance that the player's decisions affected the game was a revelation at a time when linearity was the norm, even though *Deus Ex*'s choices were largely illusionary. "There were changes based on the decision you made but at the end of the day the players were all going to go through the same missions, so there weren't huge diversions of the plot," said Pacotti. "Some choices felt bigger than they might have been. There's a point where you get confronted by one of your co-workers and told to assassinate an informant. You can choose to follow the order or you can disobey or you can actually turn and kill your co-worker. And for about a mission after that you're a fugitive – you've basically committed a crime and you have to come back and report to your boss. All the conversations for that mission depend on that. It's a little bit smoke and mirrors though because you're going to go through the mission in the same way and get the same brief at the end but the mission itself is coloured differently because of this choice and hopefully that makes players feel the world is reacting to them and that they really have changed something."

By the time *Deus Ex* was published, however, Ion Storm was in serious trouble. The Dallas office had burned through all of the money Eidos had stumped up and Romero's own overhyped game *John Romero's Daiktana* had become the butt of internet jokes and a commercial disaster. "Ion Storm spun out of control and as much as we were monitoring that it took us a while to get our arms around exactly what we were dealing with and we were dealing with an unbelievably out of control beast," said Heath-Smith.

While Ion Storm would eventually close down, *Deus Ex*'s literary and moralistic aspirations would be an important influence on other story-driven action games most notably

Fallout 3, Bethesda's 2008 reboot of the 1997 role-playing game *Fallout* as an action-driven role-playing game set in a vast world comparable in scale to the studio's *The Elder Scrolls* games. "We knew *Fallout 3* was going to be primarily first-person but would also be a role-playing game at heart with lots of systems controlling things 'under the hood'," said Emil Pagliarulo, the lead designer and writer for *Fallout 3*. "It just so happens that that description also perfectly describes *Deus Ex*. It did what we wanted to do years earlier and did it brilliantly. It was a shooter, it was a role-playing game, it had great environments and it offered players choices that really mattered."

Fallout 3 took place in a world that experienced an alternative history after the end of the Second World War. Microprocessors were never invented, but laser guns were. Society, culture and fashion remained stuck in the 1950s and, most importantly of all, the Cold War did result in all-out nuclear war that turned most of the US into an inhospitable and irradiated wasteland where survivors and their descendents scratched out living under constant threat of attack from mutant beasts, bandits and slave masters. "In *Fallout 3* there are really two major stories being told at the same time," said Pagliarulo. "There's the story of the present with the Brotherhood of Steel and the Enclave and all that stuff. But there's also this story of a country that was annihilated by nuclear war. It was a very distinct world, a sort of Walt Disney-esque World of Tomorrow where you had the mores and fashions of the 1950s combined with the technology of the 25th century."

In line with the storytelling approach pioneered by *Half-Life*, *Fallout 3* told the story of the world that once was by showing rather than telling; from the ruined landmarks of Washington D.C. to the skeletal remains of the protestors, who had hoped to gain entry to the nuclear bunker the player's character grew up in after the bombs fell, that still gripped placards reading 'we're dying out here'. "The set dressing – the dead bodies and the empty cribs – that's the stuff that tells the story of the world that came before. The world that led to the Wasteland," said Pagliarulo. "You can't have one without the other and it's that juxtaposition that makes the *Fallout* world so special."

Fallout 3 also sought to challenge players with moral choices that, in contrast to the good versus evil choices of most video games, delved into sometimes uncomfortably grey decisions. "We do require the player to make some really difficult moral decisions sometimes," said Pagliarulo. "Using *The Pitt* downloadable content as an example, we force players to choose the fate of an innocent baby.[12] I think we really pushed the limits of players' comfort zones with that."

Another game to use its story to explore morality was 2007's *BioShock*, a *Deus*

---

12. The *Pitt* was an add-on to *Fallout 3* first made available as a paid-for download. It allowed the player to travel to the ruins of Pittsburgh, which is rebuilding itself off the back of slave labour.

Ex-esque fusion of first-person shooter and role-playing game developed by Irrational Games, the creators of *System Shock 2*. The game took place in Rapture, an underwater Art Deco city created by business tycoon Andrew Ryan in 1946 as a libertarian paradise where the world's greatest minds would be free from the dead hand of government that was inspired by novelist and libertarian philosopher Ayn Rand's 1957 novel *Atlas Shrugged*.[13] "I read the book many years ago. I was always attracted to her as a storyteller first, she was a great storyteller," said Ken Levine, the game's creative director who modelled Ryan on Rand and her beliefs. "She was absolutely certain of what she believed in, without a doubt, and that's what makes for great heroes and villains. I'm frankly sacred by certainty; I never really know what the right thing to do is. I think the people that do are often bullish – they come up with the most amazing things. The fact that Rand stood up in a time where all these elements - the church, the government – were saying we're just part of this larger thing, for her to say 'you know what, selfishness is not a bad word, it's ok to just be interested in your interest, why do I have a responsibility beyond that?' - that's so brave, that's so empowering in a lot of ways because you realise that you have all these other obligations because those people want you to do things for them. It was a freeing philosophy but it's not all right, it's not what I live my life by."

In *BioShock* the player ends up in Rapture 14 years after Ryan and his followers sealed the doors and left the restrictions of government behind them. In the intervening years, however, Ryan's dream of a laissez-faire society had crumbled to dust after breakthroughs in stem cell technology resulted in people giving themselves superhuman powers and turning on each other. Eventually, the widespread use of the technology caused the population to become increasingly mutated and insane.

BioShock's absorption of Rand's ideas provoked a lot of discussion about whether it was a critique of libertarianism, a defence of communism or a game that aimed to question the idea of political certainty. "Hopefully I got people, through a video game, to talk about that kind of thing," said Levine. "Games have been so self-limiting for so long about the topics they will take on. There's an issue of a lack of security and I'll put myself in here – I'll be working on *BioShock* and I'll be writing some of these pseudo-Jacobist streams and I'm thinking 'Jesus Christ who wants to listen to this?' I'm still surprised it sold as many units as it did – three million or whatever – because it's pretty out there topic wise. I think game developers underestimate the average gamer. Their tastes are broad; they are not a monolithic or monogamous group. They have a lot of tastes that are diverse and *BioShock* demonstrates that."

---

13. Rand's book, which promotes her libertarian political ideas, featured a town called Galt's Gulch, where the greatest minds in the US fled to escape the control of bureaucrats and a left-wing government.

The following year, *Grand Theft Auto IV*, sought to bridge the divide between the open-world freedom the series offered and the richer, more nuanced narratives and characters that games were increasingly embracing. The result was the anti-hero Niko Bellic, a veteran of the Yugoslavian wars of the early 1990s who moves to Liberty City in search of the American Dream. "The more we dug about and researched, the more fascinating the eastern European situation became," said Houser. "You've got people who came to the US 15 years ago who may have been involved in very intense, terrifying conflicts in eastern Europe and they've experienced the sort of post-communism meltdown that's taken place. Some of these people have been in wars and they all split – everyone went elsewhere." The complex and torn Bellic represented one of the most emotionally rounded lead characters ever seen in a game. A man haunted by the horrors of a past he tries in vain to escape, who is both naïve about the reality of life in the US and bemused by the cultural divide. And, as the game's story progresses, Bellic becomes a man ever more mournful of the loss of the hope he once had about his new life in America.

But while the writing had come on in leaps and bounds, *Grand Theft Auto* remained primarily a game about freedom offering a redesigned Liberty City that not only delivered the expected leap in visual detail but felt more alive and more real than ever before. To create the Liberty City of *Grand Theft Auto IV*, Rockstar North took thousands of photographs of the virtual metropolis's real-life inspiration, New York City, so it could rebuild the city piece by piece out of polygons. And more than any previous game in the series, *Grand Theft Auto IV* allowed players to enjoy just experiencing its digital city. From spending a day on the pier eating hot dogs and taking in the sights with a tourist telescope to surfing the web at one of the TW@ internet cafés dotted around the city and spending a night at a comedy club watching performances by real-life comedians such as Ricky Grevais and Katt Williams, Liberty City circa 2008 offered an unrivalled sense of freedom to explore and experience.

Grand Theft Auto IV, which had taken a team of around 150 people four years to make, brought together the two major trends of big-budget games in 2000s – richer storytelling and player freedom – to create the video game equivalent of a James Cameron blockbuster. It sold more than 13 million copies, making it one of the biggest hits of 2008, but more than that it summed up the scale of ambition and artistry of the big budget games that had emerged during that decade.

But as *Grand Theft Auto IV* and others delivered ever more expensive and bold visions of video gaming, a lo-fi counter-weight to these swaggering, globe-straddling, blockbusters was taking shape, driven by a desire to revive the artistic free-for-all that existed back in the early days of video games.

Coffee shop coders: Ron Carmel (left) and Kyle Gabler, aka 2D Boy
Courtesy of 2D Boy

CHAPTER 28

Magic Shooting Out Of People's Fingers

Kyle Gabler arrived at Electronic Arts' studios just outside San Francisco full of dreams about his new life as a game designer. He would get to work with game design legend Will Wright on titles such as *The Urbz: Sims in the City*, a 2004 console-only offshoot of *The Sims*, right in the Bay Area heartland of the video game industry. At the time Electronic Arts was the largest game publisher in the world, having emerged as one of the victors of the complex web of acquisitions, mergers and closures during the 1990s, which had restructured the business into an oligopoly dominated by a small number of multinational publishers.

Back in 1990 it was still possible for a team of less than five people to create a hit game with a budget of less than $200,000. But by the early 2000s, ever-rising technological complexity and consumer expectations had made the creation of a smash hit video game a multi-million-dollar endeavour that required teams containing of dozens of programmers, artists, designers, quality assurance testers and audio specialists.

Encouraged by dot.com boom investors, who were shovelling cash into every computing-related company going in the late 1990s, and the need to grow fast to cope with the rising costs of game development, publishers began buying each other out and merging in a desperate race to become the biggest. Such was the resulting concentration of power that in 2003 Electronic Arts' $2.48 billion annual turnover was 13 times that of Midway Games, the world's 20th largest game publisher.

The two to three dozen publishers who dominated the games business controlled the funding of almost all game development in the early 2000s. If these companies didn't give a game the go-ahead, that game would have little chance of ever being made, let alone marketed or distributed in the shops. And with millions at stake, publishers shied away from funding experimental and untested game ideas. Instead they bankrolled games that followed tried-and-tested styles of play and scenarios that market data suggested players were already comfortable with.

Countless experimental and off-the-wall games bit the dust as video game publishing underwent a transition similar to that experienced by Hollywood during the mid-1970s to late 1980s when studios refocused on blockbuster movies.

Among the games abandoned as a result of the changing business climate was Sensible Software's *Sex 'n' Drugs 'n' Rock 'n' Roll*, an adults-only comedy about a wannabe rock star's outrageous sex-and-drugs-fuelled rise to stardom. "The main problem was our publisher Warner Interactive sold up to GT Interactive," said Jon Hare, the co-founder of Sensible Software, the British game studio best known for its *Sensible Soccer* games. "GT were a Bible Belt company who couldn't be seen to be near a game where people snorted coke and shagged people in toilets, even if it was a comedy. It is depressing when I think about it, it could have been a really groundbreaking game."

Before joining Electronic Arts, however, Gabler had thought little about the industry's transformation into a multi-billion dollar business where failing to make a hit game could leave publishers out of pocket to the tune of tens of millions. The reality of life as a game maker in a large corporation came as a shock. "Before I worked at Electronic Arts, I imagined the halls would be filled with tricycles and rainbows, and everyone's office would be a ball pit and game development would be like magic shooting out of people's fingers," said Gabler. "It wasn't exactly like that. It turns out humans make games by typing on keyboards and having meetings with Post-It notes and telephone calls and spreadsheets."

Ron Carmel, a software engineer at Electronic Arts' online game service Pogo. com, shared Gabler's disillusionment.[1] They also felt, as salaried employees, detached from the success or failure of their work. The pair gravitated to each other. They would discuss their sense that game making was not the creative nirvana they always dreamed it would be. They talked about what they would do if they weren't constrained by the pressures of big-budget projects and corporate conservatism. Eventually they began to question whether it had to be this way. They concluded it didn't and decided it was time to act.

"Maybe it should have been obvious earlier that anybody can do those game development things out of their own bedroom or cafés and without millions of dollars, but it wasn't," said Gabler.

"Before taking the leap towards unemployment, we calculated how long we could survive without an income. We found that by living very lightly, we would be

---

1. Pogo.com was an advertising-funded website offering a wide variety of online games, including Bingo, Poker and board games such as *Monopoly*.

able to survive for a little over a year. If we ran out of time, we would have to crawl back to the world of employment."

The pair quit Electronic Arts in 2006 and became 2D Boy, a two-man indie game studio that used San Francisco coffee shops with wi-fi internet access as their offices. Five years earlier such a move would have been deemed suicidal. Without publisher support it seemed impossible to make, market and distribute a commercially viable game. But in the five years leading up to 2D Boy's formation, the adoption of broadband internet connections had begun to erode publishers' and retailers' grip on access to consumers.

By the time 2D Boy formed, the internet was already having a seismic affect on other publishing industries, most notably record labels who had been dragged – kicking and screaming for the most part – into a world of downloadable music. Online music stores such as Apple's iTunes had begun luring shoppers away from the shops, undermining the record labels' control of the retail space. The internet also allowed artists to release and market their music online for almost no cost. The rise of the British rock band Arctic Monkeys on the back of internet word-of-mouth personified the shifting balance of power in the music business. Promoted online by their fans, the Arctic Monkeys built up a loyal following who sent their 2005 debut song *I bet you look good on the dancefloor* straight to the top of the UK charts. Publishers in the film, newspaper and book industries soon found themselves facing a similar erosion of their control of what people consumed. And video games, the most digital of all mediums, were no exception.

Video games' shift towards online distribution began in the late 1990s with the emergence of websites where players could play games through their web browsers. Most were brief, throwaway diversions designed to drive traffic to sites that hoped to profit by selling advertising off the back of visitor numbers. By the end of the 1990s, some of these games started to attract legions of fans. The first big success was 1999's *Moorhuhn*, a game designed to promote Johnnie Walker whiskey to German drinkers.

Created by Dutch developers Witan for German publisher Phenomedia, *Moorhuhn* challenged players to shoot down as many grouses as possible in 90 seconds. "You might even guess which competitor they had in sight," said Thomas Daniels, the director of sales and marketing at Phenomedia. "The game was first used by promoters going into bars and letting people play on laptops. Achieving sufficient points got you a drinks voucher. You could also get a copy of the game on disc. Hardcore gamers put it online and because it was simple and fun it attracted interest and started spreading."

Moorhuhn became an internet sensation in German-speaking countries as millions of people logged onto websites to shoot grouses out of the sky. Within a decade of the game's creation more than 80 million copies of the original game and its 20 online sequels and spin-offs had been downloaded across the world, some 15 million offline *Moorhuhn* games had been bought, and the quirky bird that fronted the game had been immortalised in a German TV cartoon series. *Moorhuhn's* simple point-and-click bird-hunting was a world away from the ever more complex and involving games being created at a cost of millions by giant publishers such as Electronic Arts. And it was this simplicity that Phenomedia believed was vital to its massive success.

"One of the big differences when designing a successful game for a non-hardcore audience is complexity, especially in terms of accessibility," said Daniels. "A good casual game is designed so that you do not have to read a manual and it will ease you into the game, although they can get fairly complex or rather difficult over the course of the game. A hardcore game that gets enjoyed by hardcore gamers is usually complex in terms of things you have to take care of in the game right from the start – just think about those strategy and simulation titles."

The spread of the internet also encouraged small game companies to start thinking about reinventing the shareware business model used by Id Software for *Doom* for the dot.com era. Sexy Action Cool was one of the first companies to try selling shareware games via the internet. The company started life as a developer of online games for sites such as Pogo.com in 2000, just as the internet boom of the late 1990s turned into bust.

"It was pretty much the worst time to start an internet company, but we were pretty confident business wise," said co-founder Jason Kapalka, who like his fellow co-founders, Brian Fiete and John Vechey, worked at Pogo.com before forming Sexy Action Cool. "We thought it was improbable we could do this as publishers, rather than developers, so we didn't even bother to get a URL for ourselves. We called ourselves Sexy Action Cool, which was a weird inside joke from the Antonio Banderas' film *Desperado*. There was a poster for it with a critic quote that said 'Sexy, Action, Cool'. I thought it was the strangest thing that could ever be used in a sentence and it became an inside joke."

Sexy Action Cool's business plans and name quickly changed, however, when it was suggested they do a shareware version of *Diamond Mine*, a diamond-matching puzzle game they had created for Microsoft's online game site Zone.com. "At the time it seemed like an odd choice because we were trying to make a game that people would play for free online," said Kapalka. "The question was why would someone pay for this game when they could just play it on the Microsoft site for free? In 2000 one reason was that they couldn't be online all the time as they won't be able to use their phone and so maybe they'd want to play offline and the graphics would be better."

Figuring that the name Sexy Action Cool would do little to attract the audience they wanted, the company renamed itself PopCap Games and began selling *Diamond Mine Deluxe* as a downloadable game through their website. Selling the game online offered clear advantages over the mail-order shareware approach of old. "Things like *Doom* were pre-internet," said Kapalka. "If you wanted to buy *Doom*, you sent a cheque away and a couple of weeks later these disks came in the mail from Texas. It worked, but it was time consuming and a pain."

Diamond Mine Deluxe, which was later renamed *Bejeweled*, had no such delays. People could try the game online, pay for it and download it in a matter of minutes. Presented in the upbeat, colourful visuals that PopCap would make its trademark, *Bejeweled* challenged players to rearrange a grid of different diamonds by swapping adjacent diamonds to form horizontal or vertical lines of three or more of the same type. Much like *Tetris,* its simplicity belied its compulsive appeal. *Bejeweled* also challenged the status quo of game design thinking, which regarded its lack of a time limit and randomness as bad design. "The version of the game that became most popular was not timed and you were making moves until you ran out of them and then you lose," said Kapalka. "We showed it to a professional game designer who said it had terrible game design because the game just ended randomly. There was really no way to control that, you just keep swapping until eventually the configuration didn't give you any moves and then that was it. This guy was like 'this is bad game design, you don't have any control over whether you win or lose, it just randomly happens – it could happen on your second or third turn or it could go on for a long time'."

But like many developers who used the internet to sell their creations, PopCap discovered that such theories were often based on more on video game tradition than player preferences. "One of things that is unusual about *Bejeweled* is that it is largely luck-based, but there's nothing wrong with luck-based games they've just been severely underrepresented in computer games because of their birth in the arcades," said Kapalka. "Computer games have primarily been skills-based and a lot of that has been because of their arcade roots. They all test whether you can master the game and get better at it. That's fair enough, but the truth is in the realm of games that humans play and have played since the dawn of time games of luck have probably outnumbered games of skill."

Bejeweled's flouting of video game design theory did it no harm. A decade on from its 2000 debut as *Diamond Mine*, the game had sold more than 50 million copies across the world. PopCap's games had also tapped into an audience with markedly different backgrounds to the young male demographic targeted by most big-budget games with a 2006 survey of 2,191 players on PopCap's website reporting that 76 per cent were women and 47 per cent were aged 50 or older.

PopCap's confounding of industry thinking would be repeated over and over again as small, internet-based online game studios sprung up during the 2000s. *Alien Hominid*, a 2002 Flash game created by San Diego developers The Behemoth, also challenged industry assumptions. The game reinvented the run-and-gun style of shoot 'em up pioneered by the 1987 coin-op game *Contra*, where players world race through a 2D landscape blasting anything that got in the way. While the game style dated back years, it had largely died out by the time The Behemoth staged its revival of the genre.[2]

"A lot of Flash games weren't really testing the limits of what Flash could do, so the run 'n' gun action of *Alien Hominid* felt pretty unique for a web game in 2002," said Tom Fulp, co-founder of The Behemoth. John Baez, another co-founder of The Behemoth, felt that for many people the game style would feel fresh: "*Alien Hominid* has its roots in retro to be sure, but so many young kids have missed that whole part of the history of video games because they weren't born. For many kids in 2004, realistic 3D games were all they knew, so a side-scrolling 2D shooter was a novelty."

Alien Hominid won a huge online following, attracting 18 million players on the Flash games website Newgrounds alone - enough to inspire The Behemoth to convert the game to the leading consoles of the day: the PlayStation 2, Gamecube and Xbox. But while the web browser games of the early 2000s paved the way for the independent games movement, it would be the arrival of online game stores in the mid-2000s that really opened the creative floodgates. The first of these platforms piggybacked its way onto PCs across the world via *Half-Life 2*, Valve's 2004 sequel to its revolutionary first-person shooter *Half-Life*. In order to run *Half-Life 2*, PC owners were required to install Steam, an iTunes-esque application that managed their games and allowed them to buy new ones. As a result of its compulsory adoption by millions of *Half-Life 2* players, Steam became the dominant game download store on the PC.

---

2. SNK's cartoonish *Metal Slug* series had kept the flame alive for run and gun games both in the arcades and on home consoles, but beyond that the genre had faded away in the early 1990s.

Initially Steam only sold Valve's own games, but in 2005 the platform was opened up to other game makers. British game designer Mark Healey's comical marital arts title *Rag Doll Kung Fu* became the first non-Valve game released on Steam. Healey had started out in the games business as a programmer but ended up becoming a graphics artist for Peter Molyneux's Lionhead Studios. By 2004, however, he was growing restless: "I got to that stage where I was feeling the itch to code something so I went and learned C++, which is the programming language most people use to make games, and the best way to learn a language is to give yourself a project. I decided I was going to do a little fighting game."

Healey's training project snowballed into a more ambitious and visually unique game. "The game was getting to the stage where you could just punch and kick each other," said Healey. "I started putting platforms in and my colleague at Lionhead Alex Evans looked at it and suggested I put in rope swings and gave me a piece of code that was a simple rope simulation. I took it home and was just playing around with it really, and funnily enough one of the ropes fell onto the ground in the shape of almost-a-matchstick man. That sparked it off. I thought I could use this rope stuff to make mad characters. That was the Eureka moment."

Healey's fighting game turned into a chaotic multiplayer punch-up where players controlled floppy and amusing puppet-like characters that would wobble, flop and flail around the screen hoping to whack the living daylights out of each other. And after he put a link to his side project on Lionhead's website, *Rag Doll Kung Fu* became the talk of the youthful indie games movement. "It took on a life of its own on the internet," said Healey. "I got invited to the Experimental Games Workshop at the Game Developers Conference in San Francisco to show the game. So I trundled off there and in my mind it was going to be a small little room with maybe 10 people showing each other what they had done, but I turned up and it was a huge room with about 500 people in it. I totally panicked because I hadn't prepared a proper talk or anything."

The Experimental Games Workshop had established itself as the annual gathering of the indie games movement, a forum where the individuals and teams that were taking game creation back to the free-for-all experimentation of the early 1980s could unite. The workshop started out in 2002 as a low-key offshoot of the Game Developers Conference that showcased unusual lo-fi creations such as Jonathan Blow's *Air Guitar*, which used a web cam to allow players to strum imaginary six-strings to create music, and Eric Zimmerman's *Arcadia*, a collection of simple VCS 2600-style games that are displayed and played simultaneously.

Since that modest beginning it had grown in size and influence, serving up a mixture of the bizarre and groundbreaking to an audience of inspiration-seeking game developers and acting as an early champion of new ideas such as the tower defense mods created by players of *WarCraft III* that would later became a new game genre.

The year Healey arrived to show off *Rag Doll Kung Fu*, the indie games movement was reaching critical mass. Other demos shown that year included an early version of Blow's *Braid*, a brain-aching platform game coated in a narrative that subverted the traditional princess-rescuing storylines of *Super Mario Bros* and revolved around manipulating time to solve puzzles. Alongside that was *Attack of the Killer Swarm*, a game put together in a day by future 2D Boy Gabler. Set on a backdrop of a faded sepia photograph of a street scene, the player controls the 'killer swarm' - a mass of sketchy pencil lines that could be swished around the screen and used to hurl small people wandering on the streets into the air to the accompaniment of jaunty classical music and their screams of horror.

Healey's presentation, put together just an hour earlier, went down a storm. "I got a lot of laughs and it just so happened there were some people from Valve in the audience and they came up to me afterwards and said it's perfect for what we want to do with Steam – sell games. I flew to Seattle the next day to meet their founder Gabe Newell and did the deal there and then. It was all a bit of a mad rollercoaster. That was the first non-Valve game they ever released on Steam, so I was a bit of a guinea pig in some ways."

Rag Doll Kung Fu was quickly joined on Steam by more games. Among them was *Darwinia*, an unusual real-time strategy game dressed in distinctive 1980s retro-computer graphics. *Darwinia* was the creation of Introversion, a British game studio formed in 2001 by three university friends: Thomas Arundel, Chris Delay and Mark Morris.

From the day it formed, Introversion set out to distance itself from the market-driven creations of the mainstream games business, seeking instead to revive the freewheeling games culture of the early UK games industry. "I remember growing up playing so many different and varied video games, in my case on the Spectrum and then the Amiga," said Morris. "The games that you would have in your collection would all be quite different in form and layout – just crazy concepts. You would get games magazine cover disks and you really didn't know what you were going to put in and play next. But by the end of the '90s it had all become less about the innocent and child-like gaming experience and more about the attempt to have higher-resolution graphics, like the driving games that stopped being push left and you'll turn left, push right and you'll turn right, into these more complicated driver sims. What we were seeing was because of the publisher-developer model and the amount of money it cost to get a game in front of the consumer, anything that was creative and didn't fit into the publisher's master control spreadsheets was going to get cut."

So when Delay, the studio's creative driving force, came up with *Uplink*, a paranoia-inducing game where players were hired by corporations to damage their rivals, frame the innocent and crash the stock market by hacking into computers, the three figured they could use the internet to bypass the publishing world. "It was 2001, so the internet had been around but probably not mainstream at that point – my parents probably weren't wired in at that stage, but we knew that we could, as a company, handle people's transactions over the internet," said Morris. "We couldn't distribute electronically, but we could ship out physical units so the view was we were going to put *Uplink* in front of the masses and if they liked it they would be able to buy it."

Uplink struck a nerve with the video game press, which promoted it heavily. Soon Introversion was inundated with orders and approaches from video game publishers interested in signing their surprise hit. "We were getting publisher offers for *Uplink* but they were just atrocious," said Morris. "We were selling 30,000 units a month and publishers were saying we'll give you $10,000. That kind of made us realise that the industry, the whole model in terms of development, in terms of creativity was just fundamentally broken at its core back then. We went on the campaign trail to a certain degree really, yelling at publishers and being very angry and trying to convince developers – not big developers but smaller ones – that there was a different way."

Valve's opening up of Steam made this different way even easier. "Steam allowed us to reach so many more people," said Morris. "I actually argued not to go with Steam. I thought all we were going to do by giving Valve the opportunity to sell our games on Steam was to potentially lose some of the sales from our own website and give away some security and control. I was so wrong. The number of sales that Valve can deliver through Steam is just phenomenonal."

The ability for indie game makers to reach audiences without the need for a publisher expanded further in 2005 when Microsoft launched its second games console, the Xbox 360. With its hard drive and wireless broadband connection, the Xbox 360 allowed users to buy and download games via its Xbox Live Arcade service that stocked games ranging from remakes of early 1980s coin-op hits and indie games to full Xbox 360 games more usually sold through shops. The following year both the Nintendo Wii and Sony PlayStation 3 consoles launched boasting similar services of their own. The arrival of the touch-screen Apple iPhone and its accompanying AppStore in 2007 added another platform through which small developers could reach out to a mass audience via the internet. The establishment of these online game stores finally gave indie game developers the access and profile they needed to really get their work in front of potential buyers.[3] Gabler and Carmel's 2D Boy was one of the first indie developers to really reap the benefits of the new connections between player and creator that resulted from these platforms.

The pair's first creation was *World of Goo*, a spiritual successor to Gabler's earlier non-commercial indie game *Tower of Goo*. Like *Attack of the Killer Swarm*, Gabler created *Tower of Goo* for the Experimental Gameplay Project he helped form while a student at Carnegie Mellon University. In *Tower of Goo* players built wobbly towers by connecting blobs of slime that linked together like scaffolding. The challenge was to try and build the tallest tower possible without the structure toppling over. "It started with 100 little goo balls on a little green hill and the goal was to build a tower up to a cloud 25 metres high," said Gabler. "People on the internet had a lot of fun with it, competing to build the tallest towers, and even sculpting things like cats and pensises. It seemed like a good idea to expand on the basic goo-construction mechanic and turn *Tower of Goo* into a bigger, more detailed *World of Goo*."

Working on laptops while sat in San Francisco coffee shops, the pair evolved *Tower of Goo* into the more rounded *World of Goo*. Players now, in an echo of DMA Design's *Lemmings*, had to get their goo balls sucked out of each level through a pipe that they could only reach with their gooey construction projects. Each level presented a different construction challenge. It could be building a bridge across a gaping chasm or constructing a tower of goo inside a rotating washing machine drum that constantly knocked over their structures. Released on Nintendo's Wii-Ware platform, Steam and through 2D Boy's own website in late 2008, *World of Goo* became one of the biggest-selling indie games of the 2000s. "*World of Goo* has surpassed every one of our expectations," said Gabler. "It has sold a few hundred thousand. Nintendo told us it holds the record for being in the number one spot on WiiWare for the most weeks."

The year of *World of Goo*'s release marked a turning point for the commercial viability of indie games. As well as *World of Goo*, Blow's critically acclaimed *Braid* became the second-best selling game on Xbox Live Arcade in 2008. The Behemoth also scored a success with *Castle Crashers*, a million-selling scrolling fighting game

---

3. Japanese indie developers had more outlets for their work. Japan's thriving doujin scene – which is largely based around indie manga – is so popular that there are dedicated shops that will sell indie games. The internet has, however, helped doujin-soft developers get their work seen beyond Japan's shores.

sold through the Xbox Live Arcade that was reminiscent of Sega's 1989 coin-op *Golden Axe*. By the start of the 2010s indie game development had blossomed into a diverse and eclectic movement that was reinventing old genres, exploring new frontiers of game design and often outperforming major publishers with surprise hits such as *Braid*. As well as trying out new forms of games and rebooting faded genres, indie game teams also began to inject their work with more personal or provocative themes.

Introversion's *Defcon*, a nuclear war sim inspired by the fictional game *Global Thermonuclear War* featured in the 1983 movie *WarGames*, turned players into a bunker-dwelling general plotting nuclear strikes on a map of the world in the hope that once the nuclear war is over you emerge victorious because you killed more people than the other side.

"We didn't release it to make a political statement, but the game does that because you're sitting in your bunker and the data that is coming back to you is the city name and the number of people that you killed. That stark 6.2 million people dead figure," said Morris. "I don't think *Defcon* would ever have got commissioned by a publisher. Never in a million years. Publishers always want to be mass market, they look at things and ask what is going to be the turn off in this game. How many people are we going to offend by calling it a genocide 'em up? Why would you want to play a game that's miserable? My response is look at all the melancholy albums that are out there and sell in massive numbers. How many weepy films? There's absolutely no reason why games can't evoke that same kind of slightly depressed, contemplative emotional state in the player that other mediums do."

Every Day the Same Dream also served up a melancholy game experience although it focused on the plight of a lone office worker to deliver what its Italian creator Paolo Pedercini described as a comment on "alienation and refusal of labour". Set in a drab grey Art Deco cartoon world, *Every Day the Same Dream* presents players with their faceless office worker's monotonous routine of dress, commute, work, where only attempting to escape the tedium offers any reward. In stark contrast to most video games, Pedercini's game was a joyless experience where even the brief respites from the tedium ultimately failed to result in the kind of escapist catharsis most games would offer for solving their mysteries.

Others used their games to create more upbeat emotions. Thatgamecompany's *Flower*, a meditative game released via the PlayStation Network in 2009, gave players control of the breeze so that they could create swirling ribbons of flower petals that swoosh across the countryside bringing life to dead fields of grass and wheat.

"The idea for *Flower* grew organically from a number of inspirations," said Kellee Santiago, who co-founded Thatgamecompany with *Flower*'s designer Jenova Chen. "One was to try and capture the feeling of being in a large flower field. To capture both the sense of beauty when you see them all, but also the visceral feeling of being up close to an individual flower. If you do a Google image search for 'flower' you discover photographs from people all over the world, all fascinated with this aspect of nature. Technically, it was an exciting challenge. What would happen if we took this aspect of video games that is normally an afterthought on the edge of the world – the bushes and grasses – and put it right in front and make the entire game about it?"

By 2010 the creative boom of the indie games movement had begun to percolate upwards through the game industry, slowly but surely influencing the makers of the multi-million dollar games that helped prompt the movement's formation. "One of the reasons I love going to the Game Developers' Conference is to go to the independent games festival because it's absolutely amazing," said Cliff Bleszinski, designer of the 2006 action blockbuster *Gears of War*. "Look at how *Portal* came about.[4] It was a little independent game called *Narbacular Drop* and then Valve took it up, nursed it and worked with the creators and came up with one of the most amazing titles ever."

For Lionhead Studios' Molyneux the ideas flooding out of the indie scene were already having an influence on his thinking about games: "*Braid* was a great example, not only was there that time mechanic but when you launch *Braid* there's this little character on screen and straight away you're in the game. I've been inspired by that. *Castle Crashers* putting the weapon-changing stuff in the world rather than outside the world is, I think, really good. There's not been one example of freshness and newness, there's like 10 going: 'Hey you, old-codger game designers, why have you been doing it like that?'"

As 2009 drew to a close video games stood on the crest of a new era of creativity powered by both the grand visions of leading game designers and the fizzing experimental wildness of the indie movement.

Nearly 50 years had elapsed since the creators of *Spacewar!* became the first people to really experience what a video game was. In that half century the video game had evolved into an entertainment medium that encompassed experiences as diverse as *Tetris*, *Grand Theft Auto IV*, *Wii Fit*, *BioShock*, *Pac-Man*, *Mortal Kombat* and *Every Day the Same Dream*.

---

4. *Portal* was a 2007 puzzle game set in Valve's *Half-Life 2* world.

Yet the creativity that had taken the primitive chess games, shoot 'em ups, maze chases and ping-pong games of the 1960s and early 1970s and turned it not only into a huge international business, but a powerful and diverse artistic medium was showing no sign of slowing down. Far from settling into some kind of creative maturity, the video game remains an art form that still feels as if it has barely got started.

GAMEOGRAPHY

With the origins of video games dating as far back as the late 1940s, this overview is necessarily partial, selective and brief. Online game databases MobyGames (www.mobygames.com) and Arcade History (www.arcade-history.com) provide more comprehensive listings.

Playing games designed for obsolete platforms is fraught with difficulty, although re-releases on digital game stores such as Steam, PSN, Xbox Live Arcade and WiiWare together with retail compilations have improved matters greatly. Beyond the re-releases, there's the second-hand market and the murky world of emulation on PCs and – to a lesser extent – Macs. Emulator software mimics old systems, allowing digital copies of games to be played. Emulation often breaches copyright law.

The syntax for the game information below is: *Game title* (Year released, Publisher, Developer/Designer, Platform [Recommended platform if not original], Country of origin). The information is based on the first release of the game. Where the publisher and developer are the same only one is listed.

Welcome to the maze of twisty little passages…

Spacewar!

The first game built for entertainment, the two-player-only *Spacewar!* (1962, Tech Model Railroad Club, PDP-1, USA), is still good fun. Try it at: http://spacewar.oversigma.com

Spacewar! directly inspired the first two coin-op games:
- *Galaxy Game* (1971, Computer Recreations, Bill Pitts & Hugh Tuck, Coin-op, USA): *Spacewar!* Xeroxed.
- *Computer Space* (1971, Nutting Associates, Nolan Bushnell & Ted Dabney, Coin-op, USA): Great '70s sci-fi cabinet, primitive game. It in turn inspired the slight but fun *Tank* (1974, Atari, Steve Bristow & Lyle Rains, Coin-op, USA). *Tank* was later remade as *Combat* (1977, Atari, VCS 2600, USA).

Spacewar! also inspired the first vector graphics arcade game: *Space Wars* (1977, Cinematronics, Larry Rosenthal, Coin-op, USA). After that retina-searing vector graphics became a common sight in the arcades until they were cast aside around 1984. Many great moments:
- *Tailgunner* (1979, Cinematronics, Tim Skelly, Coin-op, USA): Wireframe 3D space fighting.
- *Warrior* (1979, Cinematronics, Tim Skelly, Coin-op, USA): Ghostly overhead view sword duels.
- *Lunar Lander* (1979, Atari, Rich Moore & Howard Delman, Coin-op, USA): A tense battle against gravity. Based on *Lunar Lander* (1973, DEC, Jack Burness, DEC GT40, USA), the vector update of the text-only *Lunar* (1969, Jim Storer, PDP-8, USA).
- *Asteroids* (1979, Atari, Ed Logg & Lyle Rains, Coin-op, USA): Vector gaming's finest moment

and Atari's biggest-selling coin-op. Enduring rock blasting.
- *Battlezone* (1980, Atari, Ed Rotberg, Coin-op, USA): Groundbreaking wireframe 3D tank sim.
- *Gravitar* (1982, Atari, Rich Adams & Mike Hally, Coin-op, USA): *Lunar Lander* with galactic exploration. Led to the far superior: *Thrust* (1986, Superior Software, Jeremy Smith, BBC Micro, UK) and *Oids* (1987, FTL, Dan Hewitt, Atari ST, USA). *Thrust* creator Jeremy Smith went on to make the glorious *Gravitar*-influenced arcade adventure *Exile* (1988, Superior Software, Peter Irvin & Jeremy Smith, BBC Micro [Amiga], UK).
- *Star Wars* (1983, Atari, Mike Hally, Coin-op, USA): The original on-rails shooter, where movement is taken out of the players' hands so they can concentrate on the shooting. Relive the rebels' raid on the Death Star in colour wireframes.
- *Major Havoc* (1983, Atari, Owen Rubin & Mark Cerny, Coin-op, USA): Overambitious but interesting *Tailgunner*, *Lunar Lander*, maze game mash-up.

Many other on-rail blasters followed *Star Wars*.
- *Operation Wolf* (1987, Taito, Eigo Okajima, Coin-op, Japan): A bloody hostage rescue mission powered by an Uzi controller.
- *Star Wars: Rebel Assault* (1993, LucasArts, Vince Lee, PC: MS-DOS, USA): Movie footage enhanced high-speed space battles.
- *Time Crisis* (1995, Namco, Takashi Sano, Coin-op, Japan): Trigger-happy race against the clock and a foot pedal for taking cover.
- *The House of the Dead* (1996, Sega, Rikiya Nakagawa, Coin-op, Japan): Deal death to the undead.
- *Rez* (2001, Sega, United Game Artists, Dreamcast, Japan): Abstract shoot 'em up possessed by the pulsing heartbeat of the nightclub. Now in high definition thanks to *Rez HD* (2008, Microsoft Game Studios, Q Entertainment, Xbox 360, Japan).
- *Killer7* (2005, Capcom, Grasshopper Manufacture, PlayStation 2, Japan): Otaku developer Suda51 welds puzzle solving and a choice of routes to the on-rails shooter.

While the vector monitor was left behind, the phosphorous burn look continues to inspire:
- *Geometry Wars: Retro Evolved* (2005, Microsoft Game Studios, Bizarre Creations, Xbox 360, UK): Frantic blasting throwback.
- *Groov* (2009, Funkmasonry Industries, Julian Kantor, Xbox 360, USA): *Asteroids* turned jazz odyssey.

Shoot 'em ups
After *Spacewar!* the next big evolutionary leap forward for the shoot 'em up was the coin-gobbling megahit *Space Invaders* (1978, Taito, Tomohiro Nishikado, Coin-op, Japan). Its ominous soundtrack and the relentless march of the alien horde remains an iconic moment in video game history and, while slow compared to later games, its quality is undeniable. Namco then upped the pressure on player with *Galaxian* (1979, Namco, Kazunori Sawano, Coin-op, Japan), which ripped away the security blankets of *Space Invaders* by erasing the shields and having aliens break from the pack to dive bomb players with a shower of missiles. The sequel, *Galaga* (1981, Namco, Coin-op, Japan), further stacked the odds in the aliens favour by arming them with tractor beams capable of capturing player's craft.

Around the same time Atari's Dave Theurer delivered a trio of landmark blasters:
- *Missile Command* (1980, Atari, Dave Theurer, Coin-op, USA): Born out of nuclear nightmares.
- *Tempest* (1981, Atari, Dave Theurer, Coin-op, USA): Dizzying abstract shooting fury.
- *I, Robot* (1983, Atari, Dave Theurer, Coin-op, USA): The first game with 3D polygon graphics.

Williams' lead designer Eugene Jarvis, meanwhile, served up some of the rawest and most exhilarating shoot 'em up moments ever made:

- *Defender* (1981, Williams, Eugene Jarvis & Larry DeMar, Coin-op, USA): Ferocious save 'em up action. Inspired the excellent hostage rescue helicopter action of *Choplifter* (1982, Brøderbund, Dan Gorlin, Apple II, USA). *Elite* co-creator David Braben revived *Defender* in 3D with the swirling and soaring *Zarch* (1987, Acornsoft, David Braben, Archimedes, UK).

- *Robotron: 2084* (1982, Williams, Eugene Jarvis & Larry DeMar, Coin-op, USA): Claustrophobic battles against swarms of killer robots.

After a brief spell away from the industry, Jarvis returned with the anti-drugs shooter *Narc* (1988, Williams, Eugene Jarvis, Coin-op, USA): Brutal zero tolerance policing that makes RoboCop look liberal.

- *Smash TV* (1990, Williams, Eugene Jarvis, Coin-op, USA): TV game show parody with over-the-top explosions and gory destruction all in the name of winning a toaster.

- *Target: Terror* (2004, Raw Thrills, Eugene Jarvis, Coin-op, USA): Jarvis' rapid-fire response to the 9/11 terrorist attacks.

The left-to-right space battle of *Scramble* (1981, Konami, Coin-op, Japan) and the eye-catching *Xevious* (1982, Namco, Masanobu Endo, Coin-op, Japan) heralded a new era for the shoot 'em up that neatly divided into the horizontally and the vertically scrolling.

The vertical scrollers:

- *1942* (1984, Capcom, Yoshiki Okamoto, Coin-op, Japan): Second World War aerial battle against the Japanese air force.

- *SWIV* (1991, Storm, Random Access, Amiga, UK): Thundering shoot 'em up where players controlled either a helicopter or an armoured jeep.

- *Batsugun* (1993, Toaplan, Coin-op, Japan) and *DonPachi* (1995, Atlus, Cave, Coin-op, Japan) took the gaming public into 'bullet hell' for the first time.

- *Radiant Silvergun* (1998, Treasure, Hiroshi Iuchi, Coin-op [Saturn], Japan): Awesome and much sought after shoot 'em up with a unique RPG-style levelling-up approach to its weaponry.

- *Ikaruga* (2001, Treasure, Hiroshi Iuchi & Atsutomo Nakagawa, Coin-op [Xbox 360], Japan): A good introduction to the bullet hell genre where your defensive and offensive capabilities are based around the ability to change your ship's colour from black to white.

- *Perfect Cherry Blossom* (2003, Team Shanghai Alice, PC: Windows, Japan): Pyrotechnic firestorm where bullets fizz out at you like sparks from Catherine wheel fireworks.

- *Warning Forever* (2004, Hikware, Hikoza Ohkubo, PC: Windows, Japan): Stylish (and free) indie game where you battle a succession of giant 'boss' spaceships that evolve to counter your method of attack.

The horizontal scrollers:

- *Uridium* (1986, Hewson, Graftgold, Commodore 64, UK): Daring battles above giant galactic dreadnoughts.

- *R-Type* (1987, Irem, Coin-op, Japan): The defining horizontal shooter of the late 1980s with its biomechanical graphical style, giant bosses and multilayered power-ups.

- *Contra* (1987, Konami, Koji Hiroshita, Coin-op, Japan): Helped establish another shoot 'em up sub-genre, the run and gun game, where progress is made by running and jumping through levels on foot.

- *Parodius* (1988, Konami, MSX, Japan): Konami mocks its own horizontally scrolling shooter *Gradius* (1985, Konami, Coin-op, Japan). Also see the risqué send-up of *Sexy Parodius* (1996, Konami, Coin-op, Japan).

- *Thunderforce IV* (1992, Technosoft, Megadrive, Japan): High-pressure shooter that demands lightning reactions.
- *Cho Aniki Bakuretsu Ranto Hen* (1995, Nippon Computer Systems, Masaya, Super NES, Japan): Bizarre homoerotic shooter with a cult following.
- *Einhänder* (1997, Square, PlayStation, Japan): Square broke off from making RPGs to create this first-rate and highly polished shooter.

Running and gunning highlights:
- *Commando* (1985, Capcom, Tokuro Fujiwara, Coin-op, Japan): Vertically scrolling run-and-gun game that inspired the superior *Ikari Warriors* (1986, SNK, Coin-op, Japan), which offered a two-player mode and tanks to drive.
- *The Killing Game Show / Fatal Rewind* (1990, Psygnosis, Raising Hell Software, Amiga, UK): Metallic platforming shoot 'em up where the action can be rewound or fast forwarded.
- *Turrican* (1990, Rainbow Arts, Factor 5, Amiga, West Germany): *Metroid*-esque in its desire to let players roam. Boasted some great weapons, not the least the 360°-rotation lightning gun that burns through everything in its way.
- *Gunstar Heroes* (1993, Sega, Treasure, Megadrive, Japan): Raw fast-paced action. One of the greatest run-and-gun games along with:
- *Metal Slug 3* (2000, SNK, Neo Geo, Japan): Punchy action and great cartoony visuals.
- *Alien Hominid* (2002, The Behemoth, Online: Flash, USA): An internet sensation that introduced a new generation to run-and-gun games.

Zaxxon (1982, Sega, Coin-op, Japan) applied isometric visuals to the shooter, but few followed its lead. The most notable are the impressive but tough *Viewpoint* (1992, SNK, Aicom, Coin-op, Japan) and the Gulf War-inspired *Desert Strike: Return to the Gulf* (1992, Electronic Arts, Mike Posehn, Megadrive, USA).

Others also fell outside the convenient vertical-horizontal divide:
- *Raid on Bungeling Bay* (1984, Brøderbund, Will Wright, Commodore 64, USA): A multi-directional helicopter-based race to destroy the enemy's ever-growing defences and factories. Will Wright's first game.
- *Gauntlet* (1985, Atari Games, Ed Logg, Coin-op, USA): Four-player fantasy fun. Also see the Bitmap Brothers' steam punk reworking: *The Chaos Engine* (1993, Renegade, Bitmap Brothers, Amiga, UK)
- *Space Harrier* (1985, Sega, Yu Suzuki, Coin-op, Japan): Lurid colours, one-eyed mammoths, swooping Chinese dragons and giant mushrooms make this into-the-screen shooter a freaky ride.
- *After Burner* (1987, Sega, Yu Suzuki, Coin-op, Japan): Jet fighter air battles. Evolved into Suzuki's head-spinning 360°-motion game: *R-360 G-Loc Air Battle* (1990, Sega, Yu Suzuki, Coin-op, Japan).
- *Worms* (1994, Ocean Software, Team 17, PC: MS-DOS [Xbox 360], UK): A wickedly funny update of the artillery games of the early 1980s that is as vital now as it was back in 1994.
- *Max Payne* (2001, Gathering of Developers, Remedy Entertainment, PC: Windows, Finland): Gritty noir third-person shooter that stood out for its 'bullet time' effect where players could slow time during the intense gun fights. Its equally good sequel: *Max Payne 2: The Fall of Max Payne* (2003, Rockstar Games, Remedy Entertainment, PC: Windows, Finland).

Ping-Pong & Pong

Ping-Pong (1972, Magnavox, Ralph Baer & Bill Rusch, Magnavox Odyssey, USA): The first bat

'n' ball video game. Originally made in 1967 but had to wait five years to escape the workshop. Within a few months of its release there was:
- *Pong* (1972, Atari, Al Alcorn, Coin-op, USA): The moment the world woke up to video games. *Pong* clones dominated the 1970s:
- *TV Pingame* (1973, Chicago Coin, Coin-op, USA): The bastard child of *Pong* and pinball.
- *Rebound* (1974, Atari, Coin-op, USA): *Pong* reimagined as volleyball.
- *Quadrapong* (1974, Atari, Coin-op, USA): Four-player *Pong*.

Breakout (1976, Atari, Coin-op, USA) revived the whole bat 'n' ball genre with its block-smashing action. Followers:
- *Circus* (1977, Exidy, Howell Ivy & Edward Valeau, Coin-op, USA): Balloon-popping take on *Breakout* that replaces the ball with circus acrobats on a seesaw.
- *Arkanoid* (1986, Taito, Akira Fujita, Coin-op, Japan): *Breakout* with power-ups. The best *Breakout* variant.
- *Plump Pop* (1987, Taito, Yoshihisa Nagata, Coin-op, Japan): Tooth-decayingly cute.
- *Super Glove Ball* (1990, Mattel, Rare & William Novak, NES, UK & USA): Power Glove-controlled 3D *Breakout*.
- *Cosmic Smash* (2001, Sega, Sega Rosso, Coin-op [Dreamcast], Japan): Retro sci-fi visuals plus drum and bass soundtrack bring *Breakout* into the 21st century.

Pinball & pachinko
Both have rich histories of their own, but in video game form:
- *Pinball Construction Set* (1983, BudgeCo, Bill Budget, Apple II, USA): D.I.Y. pinball.
- *Devil's Crush* (1990, NEC, Compile, PC Engine, Japan) and *Pinball Dreams* (1992, 21st Century Entertainment, Digital Illusions, Amiga, Sweden): Video game pinball's elite.
- *Microsoft Pinball Arcade* (1998, Microsoft, PC: Windows, USA): A playable tour through pinball history. Includes *Spirit of '76*, the microprocessor-enhanced pinball table made by Dave Nutting Associates.
- *Peggle* (2007, PopCap Games, Brian Rothstein & Sukhbir Sidhu, PC: Windows, USA): A joyous fusion of pinball and pachinko that oozes puppyish charm.
- *Sho Chiku Bai Pachinko* (2009, Mission One, iPhone, Japan): Pachinko in your pocket.

Sports
American football
- *Atari Football* (1978, Atari, Coin-op, USA): The trackball-based original US football game.
- *John Madden Football* (1990, Electronic Arts, Park Place Productions, Megadrive, USA): The first great American football game.
- *Madden NFL 09* (2008, EA Sports, EA Tiburon, Xbox 360, USA)

Athletics
- *Track & Field* (1983, Konami, Coin-op, Japan)

Australian Rules Football
- *AFL Premiership 2006* (2006, Sony Computer Entertainment, IR Gurus, PlayStation 2, Australia)

Baseball
- *World Series Major League Baseball* (1983, Mattel, Don Daglow & Eddie Dombrower, Intellivision, USA): Started the embrace of TV presentation.

- *Earl Weaver Baseball* (1987, Electronic Arts, Don Daglow and Eddie Dombrower, Amiga, USA): Still one of the most exacting baseball sims.
- *MLB 06: The Show* (2005, Sony Computer Entertainment, Sony San Diego, PlayStation 2, USA)
- *Baseball Mogul 2007* (2006, Enlight Software, Sports Mogul, PC: Windows, USA): For would-be baseball team executives.

Basketball
- *NBA 2K10* (2009, 2K Sports, Visual Concepts, PlayStation 3, USA): All the trappings of professional basketball.
- *NBA Street Homecourt* (2007, EA Sports Big, EA Canada, Xbox 360, Canada): Street-level basketball.

BMX
- *Dave Mirra Freestyle BMX 2* (2001, Acclaim, Z-Axis, PlayStation 2, USA)

Bowling
- *Wii Sports* (2006, Nintendo, Wii, Japan)

Boxing
- *Punch Out!!* (1984, Nintendo, Coin-op, Japan): Arcade boxing classic.
- *Ready 2 Rumble Boxing* (1999, Midway, Dreamcast, USA)
- *Rocky* (2002, Ubisoft, Rage, PlayStation 2, UK)

Cheerleading
- *We Cheer* (2008, Namco Bandai Games, Machatin & Land Ho, Wii, Japan): Pom-pom exercise.

Cricket
- *Brian Lara International Cricket 2005* (2005, Codemasters, Swordfish Studios, PlayStation 2, UK)

Cycling
- *Tour de France: Centenary Edition* (2003, Konami, DC Studios, PlayStation 2, UK): Cycle your way to victory in the Tour de France.
- *Pro Cycling Manager: Season 2008* (2008, Focus Home Interactive, Cyanide Studios, PC: Windows, France): Tour de France cycling team management game.

Darts
- *PDC World Championship Darts 2009* (2009, Oxygen Interactive, Rebellion, Wii, UK)

Diving
- *Endless Ocean* (2007, Nintendo, Arika, Wii, Japan)

Future and fantasy sports
- *Speedball 2: Brutal Deluxe* (1990, Image Works, Bitmap Brothers, Amiga, UK): Still unrivalled. The sub-title says it all.
- *Chaos League* (2004, Focus Home Interactive, Cyanide, PC: Windows, France): Fantasy RPG version of American football.
- Also see the same team's remake of the tabletop sports RPG *Blood Bowl* (2009, Focus Home Interactive, Cyanide, PC: Windows, France).

Fishing
- *Sega Bass Fishing* (1998, Sega, Sega AM1, Coin-op, Japan)

Golf
- *Will Harvey's Zany Golf* (1988, Electronic Arts, Sandcastle Productions, Apple II, USA): Miniature golf delights.
- *Golden Tee Fore! 2004* (2003, Incredible Technologies, Coin-op, USA): Trackball enhanced.
- *Tiger Woods' PGA Tour 09* (2008, EA Sports, EA Tiburon, Xbox 360, USA)

Horse racing
- *G1 Jockey 4* (2005, Koei, PlayStation 2, Japan)

Hunting
- *Duck Hunt* (1984, Nintendo, NES, Japan): Zapper gun blasting.
- *Deer Hunter Tournament* (2008, Atari Interactive, Southlogic Studios, PC: Windows, Brazil)
- *Afrika* (2008, Sony Computer Entertainment, Rhino Studios, PlayStation 3, Japan): For those who prefer shooting animals with cameras.

Ice hockey
- *NHL 09* (2008, EA Sports, EA Canada, Xbox 360, Canada)

Mountaineering and climbing
- *Crazy Climber* (1980, Nichibutsu, Coin-op, Japan): Skyscraper-scaling arcade fun.
- *Bivouac* (1987, Infogrames, Amstrad CPC, France): Attempt to simulate the mountaineering challenge of the French Alps.

Poker
- *Texas Hold 'em* (2006, Microsoft Game Studios, TikGames, Xbox 360, USA)

Rugby
- *Rugby '08* (2007, EA Sports, HB Studios, PlayStation 2, Canada)

Skateboarding
- *Tony Hawk's Pro Skater 4* (2002, Activision O_2, Neversoft, PlayStation 2, USA): Before the series lost its edge.
- *Skate.* (2007, Electronic Arts, EA Black Box, Xbox 360, Canada): More restrained stunts than the Tony Hawk games but more rewarding for that, especially when coupled with a city to skate and grind your way through.

Skiing and snowboarding
- *SSX Tricky* (2001, EA Sports Big, EA Canada, PlayStation 2, Canada): Joyous snowboard racing fun.
- *Amped 3* (2005, 2K Sports, Indie Built, Xbox 360, USA): Snowboarder chic alternative to the hyperrealism of the *SSX* series.
- *Family Ski* (2008, Namco Bandai, Wii, Japan): Balance Board-enhanced gentle ski fun.

Snooker and pool
- *Jimmy White's 'Whirlwind' Snooker* (1991, Virgin Games, Archer MacLean, Amiga, UK)
- *Pool Paradise* (2004, Ignition Entertainment, Awesome Studios, Gamecube, UK)

Soccer
- *Football Manager* (1981, Addictive Games, Kevin Toms, ZX81, UK): The start of the soccer management genre.
- *Sensible World of Soccer* (1994, Renegade, Sensible Software, Amiga, UK): 2D footy's pinnacle
- *FIFA 10* (2009, EA Sports, EA Canada, PlayStation 3, Canada)
- *Football Manager 2010* (2009, Sega, Sports Interactive, PC: Windows, UK)

Surfing
- *Kelly Slater's Pro Surfing* (2002, Activision O$_2$, Treyarch, Xbox, USA)

Table tennis
- *Rockstar Games Presents Table Tennis* (2006, Rockstar Games, Rockstar San Diego, Xbox 360, USA)

Tennis
- *Virtua Tennis 3* (2006, Sega, Sega AM3, Coin-op, Japan)
- *Wii Sports* (2006, Nintendo, Wii, Japan)

Volleyball
- *Dead or Alive Xtreme Beach Volleyball* (2003, Tecmo, Team Ninja, Xbox, Japan): Almost forgotten beneath the pixel titillation is a decent enough volleyball game.

Wrestling
- *Super Fire Pro Wrestling Special* (1994, Human Entertainment, Super NES, Japan)

Others
- *World Games* (1986, Epyx, Commodore 64, USA): Ranges from the traditional, such as skiing, to the unusual, such as caber tossing and bull riding.
- *California Games* (1987, Epyx, Commodore 64, USA): Sunny sporting delights.
- *Wii Sports Resort* (2009, Nintendo, Wii, Japan): 11 sports wrapped in an island resort theme. Also see the previously mentioned *Wii Sports* (2006, Nintendo, Wii, Japan).

Fighting

The boxing game *Heavyweight Champ* (1976, Sega, Coin-op, Japan) was the earliest example. Then came the genre-defining double whammy of *Kung-Fu Master* (1984, Irem, Coin-op, Japan) and *Karate Champ* (1984, Data East, Technos Japan, Coin-op, Japan).

Kung-Fu Master's followers:
- *Final Fight* (1989, Capcom, Coin-op, Japan): Over-the-top '80s US urban grit and three-player mode.
- *Golden Axe* (1989, Sega, Makoto Uchida, Coin-op, Japan): Swords and sorcery battling that laid the template for the chirpy indie smash *Castle Crashers* (2008, The Behemoth, Xbox 360, USA).
- *River City Ransom / Street Gangs* (1989, Technos Japan, Mitsuhiro Yoshida & Hiroyuki Sekimoto, NES, Japan): Beat up gangs, steal their money, buy a spa treatment. A unique, cute and humorous marriage of RPG and beat 'em up.
- *Dynasty Warriors 4* (2003, Koei, Omega Force, PlayStation 2, Japan): Exhilarating lone warrior versus an army action.
- *Viewtiful Joe* (2003, Capcom, Clover Studio, Gamecube, Japan): Riotous punch-up action inspired by Japan's tokusatsu ('live action') TV shows. Think *Mighty Morphin' Power Rangers*.

- *Yazuka* (2005, Sega, Amusement Vision, PlayStation 2, Japan) and *The Warriors* (2005, Rockstar Games, Rockstar Toronto, PlayStation 2, Canada): Both reinvented the scrolling fighting for the 3D era.

Karate Champ's disciples:
- *International Karate + / Chop 'n Drop* (1987, System 3, Archer MacLean, Commodore 64, UK): Three-player cheat-fuelled fun.
- *Street Fighter II* (1991, Capcom, Yoshiki Okamoto, Coin-op, Japan): One of the most influential and enduring fighting games ever made. Its secret moves were a revolution. *Street Fighter IV* (2008, Capcom, Capcom & Dimps, Coin-op, Japan) proved the series hasn't lost its appeal.
- *Mortal Kombat* (1992, Midway Games, Ed Boon & John Tobias, Coin-op, USA): Now best remembered for the controversy than the entertainment value. SNK's *Street Fighter II* followers – *Fatal Fury: King of Fighters* (1991, SNK, Coin-op, Japan) and *Samurai Shodown II* (1994, SNK, Neo Geo, Japan) – were superior.
- *Virtua Fighter* (1993, Sega, Yu Suzuki, Coin-op, Japan): Took fighting into the third dimension.
- *Bushido Blade* (1997, Squaresoft, Light Weight, PlayStation, Japan): A daring jettison of fighting game tradition. No energy bars, instant kills and limb-disabling damage juxtaposed against beautiful backdrops. Also see the instant death decapitations of *Barbarian: The Ultimate Warrior / Death Sword* (1987, Palace Software, Steve Brown, Commodore 64, UK).
- *Soul Calibur* (1998, Namco, Coin-op [Dreamcast], Japan): Astonishing weapon-based combat.
- *Super Smash Bros Melee* (2001, Nintendo, HAL Laboratory, Gamecube, Japan): Nintendo's characters gathered together for deliriously fun cartoon punch-ups.
- *Dead or Alive 3* (2001, Tecmo, Team Ninja, Xbox, Japan) and *Tekken 6* (2007, Namco Bandai, Coin-op, Japan): Equally matched rival fighting game series.
- *Rag Doll Kung Fu* (2005, Valve, Mark Healey, PC: Windows, UK): Elastic puppetry underpins this crazed and irreverent martial arts game.

Driving
Restricted by technology, the earliest driving games opted for a bird's eye view of the road, such as in *Gran Trak 10* (1974, Atari, Coin-op, USA).

The best of the overhead tarmac burners:
- *Ivan 'Ironman' Stewart's Super Off-Road Racer* (1989, Leland, John Morgan, Coin-op, USA): Rough and tumble off-road racing on suitably bumpy tracks. As with most overhead racers best played with friends.
- *Super Cars II* (1991, Gremlin Graphics, Magnetic Fields, Amiga, UK): Tongue-in-cheek take on the overhead racer with weapons to destroy your rivals and between-race run-ins with bureaucratic safety officials.
- *Micro Machines V3* (1997, Codemasters, PlayStation, UK): Based on the toy cars and set on tracks set up on kitchen tables and in household gardens. The overhead racer's finest moment.

A rare few tried an isometric viewpoint:
- *Racing Destruction Set* (1985, Electronic Arts, Rick Koenig, Commodore 64, USA): Build-your-own-tracks racing. A precursor to the racecourse-building game *Trackmania* (2003, Focus Multimedia, Nadeo, PC: Windows, France).
- *R.C. Pro-Am* (1988, Nintendo, Rare, NES, UK): Zippy isometric races.

Nürburgring/1 (1976, Dr.-Ing. Reiner Foerst, Coin-op, West Germany) and *Night Driver* (1976, Atari, Dave Shepperd, Coin-op, USA) introduced the driver's perspective viewpoint, while *Turbo*

(1981, Sega, Coin-op, Japan) was first with the behind-the-car view. These three games formed the basis of most of what follows and the divide between the simulation-emphasis of *Nürburgring/1* and the arcade thrills of *Night Driver* is still very much visible.

First, the simulation leaning:
- *Revs* (1984, Acornsoft, Geoff Crammond, BBC Micro, UK): The start of physics degree holder Crammond's efforts to simulate professional racing, which continued with *Formula 1 Grand Prix* (1992, Microprose, Geoff Crammond, PC: MS-DOS, UK) and *Grand Prix 4* (2002, Microprose, Geoff Crammond, PC: Windows, UK). Also see the fantastical rollercoaster drag racing of *Stunt Car Racer / Stunt Track Racer* (1989, Microprose, Geoff Crammond, Amiga, UK).
- *Hard Drivin'* (1988, Atari Games, Coin-op, USA): 3D driving game pioneer.
- *Indianapolis 500: The Simulation* (1989, Electronic Arts, Papyrus Design Group, PC: MS-DOS, USA): Landmark use of 3D to simulate Indy 500 racing complete with instant replays and multiple camera angles.
- *Gran Turismo* (1997, Sony Computer Entertainment, Polyphony Digital, PlayStation, Japan): Petrolhead wish fulfilment. Fill your virtual garage with every flash car you ever wanted. Try *Gran Turismo 3: A-Spec* (2001, Sony Computer Entertainment, Polyphony Digital, PlayStation 2, Japan).
- *Grand Prix Legends* (1998, Sierra, Papyrus Design Group, PC: Windows, USA): Recreates the 1967 Formula 1 season in glorious detail.
- *F1 2002* (2002, EA Sports, Image Space, PC: Windows, USA): Stole Geoff Crammond's thunder with its choice between full-on simulation and arcade action.
- *rFactor* (2005, Image Space, PC: Windows, USA): Incredibly detailed. A true simulation.
- *X-Motor Racing* (2007, Exotypos, PC: Windows, USA): So accurate the car industry uses its technology for research and development.

On the arcade side:
- *Out Run* (1986, Sega, Yu Suzuki, Coin-op, Japan): All the glamour of the 1980s.
- *Super Hang-On* (1987, Sega, Yu Suzuki, Coin-op, Japan): Motorbike racing.
- *Cisco Heat* (1990, Jaleco, Coin-op, Japan): San Francisco-based racer that makes the most of the city's famous hills.
- *Lotus Esprit Turbo Challenge* (1990, Gremlin Graphics, Magnetic Fields, Amiga, UK): Slick racer that stands the test of time.
- *Super Mario Kart* (1992, Nintendo, Super NES, Japan): Madcap tyre-screeching fun that kick started the whole kart-racing genre.
- *Daytona USA* (1993, Sega, Toshihiro Nagoshi, Coin-op, Japan): One of the most enduring coin-op games of all-time and with good reason.
- *Burnout* (2001, Acclaim, Criterion Games, PlayStation 2, UK): Daredevil racer that rewards dangerous driving. *Burnout 3: Takedown* (2004, EA Games, Criterion Games, PlayStation 2, UK) had the most panache. *Burnout Paradise* (2008, Electronic Arts, Criterion Games, PlayStation 3, UK) opened up the city streets.
- *Crazyracing Kartrider* (2004, Nexon, PC: Windows, South Korea): Kart racing goes online. The biggest development in the sub-genre since *Super Mario Kart*.
- *Need for Speed: Underground* (2004, EA Games, EA Black Box, PlayStation 2, Canada): Custom car racing inspired by the 2001 film *The Fast and the Furious*.

Other driving games of note:
- *Excitebike* (1984, Nintendo, Shigeru Miyamoto, NES, Japan): Side-view motocross racing.
- *Kikstart 2* (1987, Mastertronic, Mr Chip, Commodore 64, UK): Evel Knievel daredevil stunt

biking. Modern equivalent: *Trials HD* (2009, Microsoft Game Studios, RedLynx, Xbox 360, Finland).
- *Wave Race: Blue Storm* (2001, Nintendo, Nintendo Software Technology, Gamecube, USA): Jet ski racing
- *F-Zero GX* (2003, Nintendo, Amusement Vision (Sega), Gamecube, Japan): High-speed futuristic racing with excellent-looking tracks.
- *Project Gotham Racing 3* (2005, Microsoft Game Studios, Bizarre Creations, Xbox 360, UK): Racing game series distinguished by its Kudos time bonuses for stylish driving.
- *Wipeout HD* (2008, Sony Computer Entertainment, Studio Liverpool, PlayStation 3, UK): High-definition, breakneck, futuristic racing.
- *Colin McRae: Dirt 2 / Dirt 2* (2009, Codemasters, Xbox 360, UK): Off-road racing.
- *18 Wheels of Steel: Extreme Trucker* (2009, ValuSoft, SCS Software, PC: Windows, Czech Republic): Truck driving simulation game. For more racing-orientated trucking: *18 Wheeler* (1999, Sega, AM2, Coin-op, Japan).

Grand Theft Auto
Although the original came across as a driving game, it was and is so much more.

The spiritual ancestor is the emancipating space sim *Elite* (1984, Acornsoft, Ian Bell & David Braben, BBC Micro, UK), which gave players a galaxy to play in. *Elite* has its own, more obvious, followers in *Privateer* (1993, Electronic Arts, Origin Systems, PC: MS-DOS, USA) and the massively multiplayer online world of *EVE Online* (2003, Simon & Schuster Interactive, CCP, PC: Windows, Iceland). Also see the official sequels: *Frontier: Elite II* (1993, Gametek, David Braben, Amiga, UK) and *Frontier: First Encounters* (1995, Gametek, Frontier Developments, PC: MS-DOS, UK).

The simulated cities of Sean Cooper's *Syndicate* (1993, Electronic Arts, Bullfrog, Amiga, UK), a gritty sci-fi game of corporate conquest, and its sequel *Syndicate Wars* (1996, Electronic Arts, Bullfrog, PC: MS-DOS, UK) are another influence. As were DMA Design's own early dabbles in 3D: the non-linear *Body Harvest* (1998, Gremlin Interactive, DMA Design, Nintendo 64, UK) and the 3D platforming of *Space Station Silicon Valley* (1998, Take-Two Interactive, DMA Design, Nintendo 64, UK). Incidentally, the latter's possess-the-enemy action owes a significant debt to the cerebral shoot 'em up *Paradroid* (1985, Hewson, Graftgold, Commodore 64, UK). See its prettier remake: *Paradroid '90* (1990, Hewson, Graftgold, Amiga, UK).

The first *Grand Theft Auto* (1997, BMG Interactive, DMA Design, PC: MS-DOS, UK) contains most of the core features of the series, but the move to 3D was a revelation:
- *Grand Theft Auto III* (2001, Rockstar Games, DMA Design, PlayStation 2, UK): The stunning 3D breakthrough that made level-based games look old-fashioned overnight.
- *Grand Theft Auto: Vice City* (2002, Rockstar Games, Rockstar North, PlayStation 2, UK): Eighties retro cool with a wicked sense of humour.
- *Grand Theft Auto: San Andreas* (2004, Rockstar Games, Rockstar North, PlayStation 2, UK): Three cities, one state. The most ambitious of series.
- *Grand Theft Auto IV* (2008, Rockstar Games, Rockstar North, Xbox 360, UK): The most detailed of Rockstar's city sims yet and the best written story of the series.

Few of the attempts to copy *Grand Theft Auto*'s virtual city action have delivered. One of the few that did:
- *Crackdown* (2007, Microsoft Game Studios, Realtime Worlds, Xbox 360, UK).

Single-player role-playing games

TSR's 1974 pen-and-paper game *Dungeons & Dragons* started it all, but the leap to the digital world was near instant, starting with the long-lost pioneer *Pedit5* (1974, Rusty Rutherford, PLATO, USA). Then:
- *Dungeon* (1975, Don Daglow, PDP-10, USA): Letters and punctuation in lieu of graphics.
- *Oubliette* (1977, Jim Schwaiger, PLATO, USA) and *Moria* (1978, Kevet Duncombe & Jim Battin, PLATO, USA): First-person view dungeons.

Most enduring of all the mainframe RPGs was *Rogue* (1980, Michael Toy & Glenn Wichman, Unix, USA): Beneath its ugly alphanumeric character visuals lurks a game of constant variety thanks to its randomly generated dungeons. *Rogue*'s influence and fanatical following spawned a huge number of so-called 'roguelikes', among them:
- *Moria* (1983, Robert Koeneke & Jimmey Todd, VAX-11/780, USA): No relation to the PLATO game. It was an important influence on the excellent action RPG *Diablo* (1996, Blizzard Entertainment, Blizzard North, PC: Windows, USA). *Diablo*'s followers: *Dungeon Siege* (2002, Microsoft Game Studios, Gas Powered Games, PC: Windows, USA), the spandex-clad superhero RPG *Freedom Force* (2002, Crave Entertainment, Irrational Games, PC: Windows, USA) and, of course, *Diablo II* (2000, Blizzard, Blizzard North, PC: Windows, USA).
- *NetHack* (1987, The NetHack DevTeam, Unix, USA): The most notable roguelike. Created by open-source development.

Highlights of the first wave of home computer role-playing games:
- *Space II* (1979, Edu-Ware, David Mullich, Apple II, USA): Recreational drug taking and religious missionary adventures in space. An expansion to the original *Space* (1979, Edu-Ware, Steven Pederson & Sherwin Steffin, Apple II, USA).
- *Wizardry: Proving Grounds of the Mad Overlord* (1981, Sir-Tech, Andrew Greenberg & Robert Woodhead, Apple II, USA): The biggest RPG of the early 1980s but the series evolved too slowly to maintain its early lead.
- *Tunnels of Doom* (1982, Texas Instruments, Kevin Kenney, TI-99/4a, USA): Replaced line-drawing first-person view dungeons with solid walls.

Alakabeth: World of Doom (1979, Richard Garriott, Apple II, USA) became the foundation stone of the legendary *Ultima* series:
- *Ultima: The First Age of Darkness* (1981, California Pacific Computer, Richard Garriott & Ken Arnold, Apple II, USA): The start of the genre-defining series.
- *Ultima IV: Quest of the Avatar* (1985, Origin Systems, Richard Garriott, Apple II, USA): The addition of a moral backbone raised the bar for all RPGs and it's still regarded by many as the best in the series. Marked the start of *Ultima*'s 'enlightenment' trilogy that also includes the brilliant *Ultima V: Warriors of Destiny* (1988, Origin Systems, Richard Garriott, PC: MS-DOS, USA) and *Ultima VI: The False Prophet* (1990, Origin Systems, Richard Garriott, PC: MS-DOS, USA).
- *Ultima VII: The Black Gate* (1992, Origin Systems, Richard Garriott, PC: MS-DOS, USA) and *Ultima VII Part Two: The Serpent's Isle* (1993, Origin Systems, Richard Garriott, PC: MS-DOS, USA): The last great moments for the single-player *Ultima*s.
Also the spin-offs:
- *Ultima Worlds of Adventure 2: Martian Dreams* (1991, Origin Systems, PC: MS-DOS, USA): Jules Verne steam punk, where the celebrities of the Victorian era travel to Mars in a steam-powered rocket ship.
- *Ultima Underworlds: The Stygian Abyss* (1992, Origin Systems, Blue Sky Productions, PC: MS-DOS, USA): Started Id Software on the path to *Doom*, but a great RPG in its own right.

Mid-1980s to mid-1990s:
- *Mandragore* (1984, Infogrames, Marc Cecchi, Thomson MO5, France): *Ultima*-style RPG with an eco-message and references to the ancient Greek poems *The Iliad* and *Odyssey*.
- *Wasteland* (1988, Electronic Arts, Interplay, Commodore 64, USA): Post-nuclear war epic that paved the way for the *Fallout* series. One of the finest RPGs of the 1980s.
- *B.A.T.* (1990, Ubisoft, Computer's Dream, Atari ST, France): *Blade Runner* goes Gallic. Features a programmable computer that is attached to the forearm of the player's character. Also its sequel: *The Koshan Conspiracy* (1992, Ubisoft, Computer's Dream, Amiga, France).
- *Darklands* (1992, Microprose, Arnold Hendrick, PC: MS-DOS, USA): Staggeringly detailed and open-ended quest for celebrity and wealth in medieval Germany.
- *Legend / The Four Crystals Of Trazere* (1992, Mindscape, Anthony Tagilone & Pete James, Amiga, PC, UK): A proto-*Diablo* with an impressively flexible design-a-spell magic system.

The trendsetting *Dungeon Master* (1987, FTL, Atari ST, USA) took the genre into real-time and won over many who had previously been put off by the lack of action in RPGs. Provided the blueprint for:
- *Eye of the Beholder* (1990, SSI, Westwood Associates, PC: MS-DOS, USA): Official *Advanced Dungeons & Dragons* tie-in that outshined its inspiration.
- *Captive* (1990, Mindscape, Anthony Crowther, Amiga, UK): Sci-fi *Dungeon Master* with thousands of levels. Its 3D sequel, set in a large virtual city, was even better: *Liberation: Captive 2* (1994, Mindscape, Anthony Crowther, CD32, UK).
- *Ishar: Legend of the Fortress* (1992, Silmarils, Atari ST, France): Made managing the personalities and relationships of your band of adventurers central to the game. Also attempted to dissuade players from saving too often by deducting gold for each save.

From the late 1990s onwards, three companies dominated western RPGs: Interplay, BioWare and Bethesda Softworks.
Interplay's late 1990s output marked a high point for RPG storytelling:
- *Baldur's Gate* (1998, Interplay, BioWare, PC: Windows, Canada): Fantasy RPG writing at its very best.
- *Planescape: Torment* (1999, Interplay, Black Isle Studios, PC: Windows, USA): A quest of self-discovery that ripped up the RPG rulebook. Deep, unique and often funny.
- *Fallout* (1997, Interplay, Black Isle Studios, PC: Windows, USA) and *Fallout 2* (1998, Interplay, Black Isle Studios, PC: Windows, USA): An uncompromisingly bleak journey into the horrors of a post-nuclear war world. Its mix of retro-technology and 1950s Americana was sheer genius.

After *Baldur's Gate*, BioWare went on to make:
- *Neverwinter Nights* (2002, Atari Interactive, BioWare, PC: Windows, Canada): The inclusion of the Aurora Toolset turned it into a RPG construction kit.
- *Star Wars: Knights of the Old Republic* (2003, LucasArts, BioWare, Xbox, Canada): BioWare begins its investigation of branching plotlines, moral grey zones and action that it continued in:
- *Mass Effect* (2007, Microsoft Game Studios, BioWare, Xbox 360, Canada): Ambitious space opera.
- *Mass Effect 2* (2010, Electronic Arts, BioWare, Xbox 360, Canada): Ironed out the flaws of the original to deliver an arresting sci-fi adventure that was as intimate and subtle as it was action-packed and epic.

While BioWare focused on player-driven stories, Bethesda sought to create RPGs that were as open as possible:

- *The Elder Scrolls IV: Oblivion* (2006, 2K Games, Bethesda Softworks, Xbox 360, USA): An astonishingly big and diverse fantasy world.
- *Fallout 3* (2008, Bethesda Softworks, Xbox 360, USA): The scene-setting intro within the confines of the Vault 101 nuclear bunker gives little indication of the vast and harrowing journey amid the ruins of Washington D.C. that follows. Also try the ethical quandaries of the expansion: *Fallout 3: The Pitt* (2009, Bethesda Softworks, Xbox 360, USA).

Finally, *Fable II* (2008, Microsoft Game Studios, Lionhead Studios, Xbox 360, UK). While the first game in the series, *Fable* (2004, Microsoft Game Studios, Lionhead Studios & Big Blue Box, Xbox, UK), was an audacious statement of intent, the sequel's balletic combat, willingness to reject RPG cliché and commitment to letting players be the hero they want to be is a very strong contender for being the best game to have Peter Molyneux's name attached to it.

Japanese role-playing games

After being introduced to the concept by *Ultima*, *Wizardry* and *The Black Onyx* (1984, Bullet-Proof, Henk Rogers, NEC PC-8801, Japan), Japan reinterpreted the whole genre, starting with:
- *Dragon Quest / Dragon Warrior* (1986, Enix, Chunsoft, NES, Japan): Its manga-influenced art, scene-setting soundtrack, random encounters with enemies, emphasis on narrative and simple controls provided the blueprint for almost every subsequent role-playing game made in Japan. *Dragon Quest VIII: Journey of the Cursed King* (2006, Square Enix, Level-5, PlayStation 2, Japan) is as good an introduction to the series as any.

The first wave of Japanese role-playing games (JRPGs from now on):
- *Phantasy Star* (1987, Sega, Yuji Naka, Master System, Japan): The start of Sega's fantasy sci-fi fusion series. Best starting point: *Phantasy Star II* (1989, Sega, Yuji Naka, Megadrive, Japan).
- *Final Fantasy* (1987, Square, Hironobu Sakaguchi, NES, Japan): Darker and more sombre than *Dragon Quest*. Series highlights: *Final Fantasy VI* (1994, Square, Super NES, Japan): The series 2D peak, which took place within an eye-catching world inspired by the Industrial Revolution; *Final Fantasy VII* (1997, Square, PlayStation, Japan) and *Final Fantasy VIII* (1999, Square, PlayStation, Japan): Took the series to international adoration with delightful cinematography, melodramatic plots and memorable characters; *Final Fantasy XII* (2006, Square Enix, PlayStation 2, Japan): Five years in the making and worth the wait.
- *Mother* (1989, Nintendo, Shigesato Itoi, NES, Japan): World-saving adventure set in 1980s small-town America that mixes humour with an underlying theme of childhood abandonment. The excellent sequel *Earthbound* (1994, Nintendo, Ape Inc, Super NES, Japan) made it out of Japan.

Shigeru Miyamoto's *Zelda* series skilfully captured the essence of the RPG within the template of an action game:
- *The Legend of Zelda* (1986, Nintendo, NES, Japan): Mind-blowingly open-ended for its time.
- *The Legend of Zelda: A Link to the Past* (1991, Nintendo, Super NES, Japan): For many the best of the series. Timeless.
- *The Legend of Zelda: Link's Awakening DX* (1998, Nintendo, Game Boy Color, Japan): Portable *Zelda* and no weaker for that.
- *The Legend of Zelda: Ocarina of Time* (1998, Nintendo, Nintendo 64, Japan): The series' makes the jump into 3D and loses none of its magic.
- *The Legend of Zelda: Majora's Mask* (2000, Nintendo, Nintendo 64, Japan): Eiji Aonuma takes the helm from Miyamoto and delivers the darkest game in the series: a 72-hour scramble to save the world from impending destruction.
- *The Legend of Zelda: The Wind Waker* (2002, Nintendo, Gamecube, Japan): Refreshingly bright

cartoon looks. Same *Zelda* brilliance.

Post-first wave JRPG highlights:
- *Secret of Mana* (1993, Square, Super NES, Japan): Sterling *Zelda*-esque action game that evolved out of the *Final Fantasy* series
- *Chrono Trigger* (1995, Square, Super NES, Japan): A masterpiece created by a 'super group' team that combined the talents of the then-separate Square and Enix.
- *Super Mario RPG: Legend of the Seven Stars* (1996, Nintendo, Square, Super NES, Japan): Nintendo's mascot gets the JRPG treatment
- *Kingdom Hearts* (2002, Square, PlayStation 2, Japan): Strange but compelling collision between the characters of Walt Disney and *Final Fantasy*.
- *Xenogears* (1998, Square, PlayStation, Japan): Philosophical and intricate JRPG that dwells on the nature of religion. Following in its footsteps: *Xenosaga Episode I: Der Wille zur Macht* (2002, Namco, Monolith Soft, PlayStation 2, Japan).
- *Skies of Arcadia* (2000, Sega, Overworks, Dreamcast, Japan): Dreamy airship pirate adventure.
- *Monster Hunter Freedom* (2005, Capcom, PSP, Japan): Monster safari RPG that comes to life when played with other PSP owners.

The most popular JRPG series of all by a long, long way is the cute character cockfighting of *Pokémon*.
- *Pokémon Green* and *Pokémon Red* (1996, Nintendo, Game Freak, Game Boy, Japan): Where the collect 'em all craze began.
- *Pokémon Gold* and *Pokémon Silver* (1999, Nintendo, Game Freak, Game Boy Color, Japan): Added pokémon breeding to the already beguiling mix.
- *Hey You, Pikachu!* (1998, Nintendo, Ambrella, Nintendo 64, Japan): Surreal voice recognition game for kids, where you befriend Pikachu.
- *My Pokémon Ranch* (2008, Nintendo, Ambrella, Wii, Japan): Import pokémon from the Nintendo DS editions of the game into a virtual ranch run by your Mii.

JRPGs also led to the rise of strategy role-playing games: turn-based games that mix the traits of the genre with tactical planning. Kicking off the sub-genre was *Fire Emblem: Ankoku Ryu to Hikari no Tsurugi* (1990, Nintendo, Intelligent Systems, NES, Japan), which was later remade as *Fire Emblem: Shadow Dragon* (2008, Nintendo, Intelligent Systems, Nintendo DS, Japan). Strategy RPG highlights: *Tactics Ogre: Let us Cling Together* (1995, Quest, Super NES, Japan); *Final Fantasy Tactics* (1997, Square, PlayStation, Japan); the musical theatre-influenced *Rhapsody: A Musical Adventure* (1998, Nippon Ichi Software, PlayStation, Japan); and the deep-but-light-hearted *Disgaea: Afternoon of Darkness* (2003, Nippon Ichi Software, Otwo, PSP, Japan).

Massively multiplayer online games

Revisiting the history of massively multiplayer online (MMO) games is inherently problematic; not just because of changing technologies, but also because they are as much defined by the people playing them at a single point in time as the design of the game itself. Consequently the heyday of many of the titles that follow here are long over.

The MUDs:
- *MUD* (1980, Richard Bartle & Roy Trubshaw, PDP-10, UK): The starting point for almost everything that followed.
- *AberMUD* (1989, Alan Cox, Richard Acott, Jim Finnis & Leon Thrane, Unix, UK): Spread MUDs far and wide with its combat emphasis.

- *TinyMUD* (1989, James Aspnes, Unix, USA): Pushed socialising to the fore.
- *LambdaMOO* (1990, Pavel Curtis, Unix, USA): Divorced from combat and all about player interaction and creativity.
- *DikuMUD* (1991, Sebastian Hammer, Michael Seifert, Hans Henrik Staerfeldt, Tom Madsen & Katja Nyboe, Unix, Denmark): The basis of most big name commercial massively multiplayer role-playing games.
- *Jurassic Park* (1994, Samjung Data Service, PC: MS-DOS, South Korea): The seed that grew into South Korea's influential game industry.

The early graphics-based multiplayer games:
- *Habitat* (1986, Quantum Link, Lucasfilm Games, Commodore 64, USA): The most ambitious massively multiplayer game of the 1980s.
- *Air Warrior* (1986, GEnie, Kesmai, Macintosh, USA): Online Second World War dog fights.
- *Neverwinter Nights* (1991, AOL/SSI, Stormfront Studios, PC: MS-DOS, USA): Based on SSI's single-player role-playing game engine as used in the likes of *Secret of the Silver Blades* (1990, SSI, PC: MS-DOS, USA).

The commercial pioneers:
- *Meridian 59* (1996, The 3DO Company, Archetype Interactive, PC: Windows, USA): *DikuMUD* meets *Doom*.
- *Ultima Online* (1997, Electronic Arts, Origin Systems, PC: Windows, USA): Expansive and intricate game that defined the future of the genre through both its failings and its successes.
- *EverQuest* (1999, Verant Interactive, 989 Studios, PC: Windows, USA): A more co-operative experience that learned the lessons of *Ultima Online*'s chaotic early days.
- *Phantasy Star Online* (2000, Sega, Sonic Team, Dreamcast, Japan): Co-operative monster slaying that introduced the genre to the games console.

South Korean games:
- *The Kingdom of the Winds* (1996, Nexon, Jake Song, PC: Windows, South Korea): Online game set in ancient Korea.
- *Lineage* (1998, NCSoft, Jake Song, PC: Windows, South Korea): Multiplayer castle raiding that brought South Korean video games to global attention.
- *Ragnarök Online* (2002, Gravity Co., PC: Windows, South Korea): Norse mythology and Korean 'manhwa' comic visuals.
- *MapleStory* (2003, Nexon, Wizet, PC: Windows, South Korea): Free-to-play multiplayer role-playing game that racked up more than 100 million registered players.

Others:
- *Argentum Online* (2000, Pablo Marquez, Matías Pequeño & Fernando Testa, PC: Windows, Argentina): Argentina's *Ultima Online*. The protests that followed the country's economic collapse in 2001 seeped into the open-source game when players barricaded the virtual streets.
- *Disney's Toontown Online* (2001, Walt Disney Company, Walt Disney Internet Group, PC: Windows, USA): Massively multiplayer action for kids.
- *Final Fantasy XI Online* (2002, Square, PlayStation 2, Japan): Japanese role-playing reinvented for online.
- *Second Life* (2003, Linden Lab, PC: Windows, USA): A vast, social and artistic experiment whose origins can be traced back to *TinyMUD*.
- *A Tale in the Desert* (2003, eGenesis, PC: Windows, USA): Civilization and culture building in Ancient Egypt. Governed by a representative democracy that gives players the power to change

the rules of the game at the (virtual) ballot box.
- *World of WarCraft* (2004, Blizzard Entertainment, PC: Windows, USA): Polished online role-playing designed for the mass appeal it achieved.
- *Fantasy Westward Journey* (2004, NetEase, PC: Windows, China): China's first big online gaming success.

Adventure
Will Crowther's *Adventure* (1976, Will Crowther, PDP-10, USA) started it all, but the game really took off in the wake of Don Woods' remix, *Adventure* (1977, Will Crowther & Don Woods, PDP-10, USA).

The roots of Crowther's game can be seen in:
- *Eliza / Doctor* (1965, Joseph Weizenbaum, IBM 7094, USA): Virtual psychotherapy. Try it at: www.chayden.net/eliza/Eliza.html
- *Highnoon* (1970, Christopher Gaylo, unknown mainframe, USA): Wild West show down in text. Relive it and whole Teletype gaming experience at www.mybitbox.com/highnoon
- *Hunt the Wumpus* (1972, Gregory Yob, unknown mainframe, USA): Often mislabelled as an adventure game, but it's a monster-hunting puzzle game. Texas Instruments' graphical version, *Hunt the Wumpus* (1982, Texas Instruments, Kevin Kenney, TI-99/4a, USA), is more inviting than the text original.

Adventureland (1978, Adventure International, Scott Adams, TRS-80, USA) took text adventures onto home computers, but Adams' best was his *Treasure Island*-inspired follow-up *Pirate Adventure* (1978, Adventure International, Scott Adams, TRS-80, USA).

Infocom, however, were the masters of the text adventure. The company's output spanned all genres of fiction:
- *Zork! The Great Underground Empire – Part 1* (1980, Infocom, Apple II, USA): A fantasy adventure landmark, based on the original mainframe *Zork!* (1979, Tim Anderson, Marc Blank, Bruce Daniels & Dave Lebling, PDP-10, USA).
- *Deadline* (1982, Infocom, Marc Blank, Apple II, USA): Detective novel and the first Infocom game with feelies.
- *Planetfall* (1983, Infocom, Steve Meretzky, Apple II, USA): Superb sci-fi comedy quest, as is *The Hitchhiker's Guide to the Galaxy* (1984, Infocom, Douglas Adams & Steve Meretzky, Apple II, USA) and *Leather Goddesses of Phobos* (1986, Infocom, Steve Meretzky, Apple II, USA)
- *Plundered Hearts* (1987, Infocom, Amy Briggs, Apple II, USA): Mills & Boon romance.
- *Lurking Horror* (1987, Infocom, Dave Lebling, Atari ST, USA): Horror set in the grounds of MIT, the birthplace of Infocom.
Infocom's literary peaks were the anti-Reagan *A Mind Forever Voyaging* (1985, Infocom, Steve Meretzky, Apple II, USA) and the anti-nuclear *Trinity* (1986, Infocom, Brian Moriarty, Apple II, USA).

Infocom's big rival was Sierra. Roberta Williams delivered most of the company's adventuring high points:
- *Mystery House* (1980, On-Line Systems, Ken Williams & Roberta Williams, Apple II, USA): The first illustrated text adventure.
- *Time Zone* (1982, On-Line Systems, Roberta Williams, Apple II, USA): Roberta's bold visions for adventure gaming led to this vast work with its 1,500 locations, an enormous number for its time.

- *King's Quest* (1984, Sierra On-Line, Robert Williams, PCjr, USA): A fairy tale adventure that introduced animation into the text adventure.
- *King's Quest IV: The Perils of Rosella* (1988, Sierra On-Line, Roberta Williams, PC: MS-DOS, USA): Its female hero predated Lara Croft by nearly a decade.
- *King's Quest VI: Heir Today, Gone Tomorrow* (1992, Sierra, Roberta Williams & Jane Jensen, PC: MS-DOS, USA): The series' high point.
- *Roberta Williams' Phantasmagoria* (1995, Sierra, Roberta Williams, PC: Windows, USA): Williams' most ambitious work since *Time Zone*: a video-based interactive horror movie spread over 12 CDs. Encapsulated all the hopes and all the flaws of full-motion-video games.

Sierra's other notable adventure writer during the 1980s was Al Lowe:
- *Leisure Suit Larry in the Land of the Lounge Lizards* (1987, Sierra, Al Lowe, PC: MS-DOS, USA): A sarky lampooning of wannabe womanisers that sold in huge numbers to wannabe womanisers. Its precursor, *Softporn Adventure* (1981, On-Line Systems, Chuck Benton, Apple II, USA), is just crude.
- *Torin's Passage* (1995, Sierra, Al Lowe, PC: Windows, USA): A rare example of an adventure designed to captivate kids and their parents.

Beyond the dominant Sierra-Infocom axis of the 1980s:
- *The Prisoner* (1980, Edu-Ware, David Mullich, Apple II, USA): Strange, experimental adventuring inspired by the cult 1960s TV series. Edu-Ware's unusual output continued with the two-player terrorism sim *Terrorist* (1980, Edu-Ware, Steven Pederson, Apple II, USA).
- *The Hobbit* (1982, Melbourne House, Beam Software, ZX Spectrum, Australia): Tolkien's book brought to life.
- *Portopia Murder Serial Case* (1983, Enix, Yuji Horii, NEC PC-6001, Japan): Japanese murder mystery adventure that inspired *Metal Gear Solid* creator Hideo Kojima.
- *The Pawn* (1985, Rainbird, Magnetic Scrolls, Sinclair QL, UK): The debut release of Magnetic Scrolls, the UK's answer to Infocom.

France had a thriving adventure scene of its own. Highlights:
- *Paranoïak* (1984, Froggy Software, Jean-Louis Le Breton & Fabrice Gille, Apple II, France): A battle against a nervous breakdown.
- *Le Crime du Parking* (1984, Froggy Software, Jean-Louis Le Breton & Fabrice Gille, Apple II, France): Adult murder mystery.
- *Même les Pommes de Terre Ont des Yeux* (1985, Froggy Software, Clotilde Marion, Apple II, France): Latin American revolution comedy.
- *L'Affaire Vera Cruz* (1985, Infogrames, CPC, France): Classy detective game.
- *Méwilo* (1987, Coktel Vision, Muriel Tramis & Patrick Chamoiseau, CPC, France): An exploration of Martinique's culture and the history of slavery. The same team followed it with the anti-slavery 'war game' *Freedom: Rebels in the Darkness* (1988, Coktel Vision, Muriel Tramis & Patrick Chamoiseau, CPC, France).
- *Captain Blood* (1988, Infogrames, Didier Bouchon & Philippe Ulrich, Atari ST, France): Bold, different and bizarre space adventure.

Déjà Vu: A Nightmare Comes True (1985, Mindscape, ICOM Simulations, Macintosh, USA) marked the start of the point-and-click era of adventure games, which Lucasfilm dominated during its decade-long focus on the genre. Picks? Almost all of them: *Maniac Mansion* (1987, Lucasfilm Games, Ron Gilbert & Gary Winnick, Commodore 64, USA), *Indiana Jones & The Last Crusade* (1989, Lucasfilm Games, Noah Falstein, PC: MS-DOS, USA), *Loom* (1990, Lucasfilm Games, Brian Moriarty, Amiga, USA), *The Secret of Monkey Island* (1990, Lucasfilm Games, Ron Gilbert,

PC: MS-DOS, USA), *Monkey Island 2: Le Chuck's Revenge* (1991, Lucasfilm Games, Ron Gilbert, PC: MS-DOS, USA), *Indiana Jones & The Fate of Atlantis* (1992, LucasArts, Hal Barwood & Noah Falstein, PC: MS-DOS, USA), *Sam & Max Hit the Road* (1993, LucasArts, Steve Purcell, PC: MS-DOS, USA), *Maniac Mansion: Day of the Tentacle* (1993, LucasArts, Dave Grossman & Tim Schafer, PC: MS-DOS, USA), *Full Throttle* (1995, LucasArts, Tim Schafer, PC: MS-DOS, USA), *Grim Fandango* (1998, LucasArts, Tim Schafer, PC: Windows, USA).

Other point-and-click adventures of note:
- *Cruise for a Corpse* (1991, Delphine, Atari ST, France): As finely crafted as any of Lucasfilm's work.
- *Darkseed* (1992, Cyberdreams, Amiga, USA): H.R. Giger did the artwork for this creepy horror adventure.
- *I Have No Mouth and I Must Scream* (1996, Cyberdreams, PC: MS-DOS, USA): Disturbing sci-fi tale packed with ethical conundrums.
- *Broken Sword: The Shadow of the Templars / Circle of Blood* (1996, Virgin Interactive, Revolution Software, PC: Windows, UK): Mined Knights Templar conspiracy theories long before Darran Brown's *Da Vinci Code*.
- *The Last Express* (1997, Brøderbund, Smoking Car Productions, Macintosh, USA): Jordan Mechner's unique and beautiful tale of murder on the Orient Express. Cruelly overlooked.
- *The Longest Journey* (1999, Funcom, Ragnar Tørnquist, PC: Windows, Norway): One of the last great adventure games.

The embrace of full-motion video clips dominated mid-1990s adventure games, but few stand the test of time:
- *Myst* (1993, Brøderbund, Cyan Worlds, Macintosh, USA): Polished puzzle game set on a mysterious island that became one of the most popular games of all time. Its roots can be found in the kooky surrealism of *The Manhole* (1988, Cyan Worlds, Rand & Robyn Miller, Macintosh, USA) and *Cosmic Osmo and the Worlds of the Mackerel* (1989, Activision, Cyan Worlds, Macintosh, USA).
Jane Jensen's southern gothic *Gabriel Knight* trilogy is both a fossil record of CD-ROM's impact on video games and the highlight of the days of full-motion video:
- *Gabriel Knight: Sins of the Fathers* (1993, Sierra, Jane Jensen, PC: MS-DOS, USA).
- *The Beast Within: A Gabriel Knight Mystery* (1995, Sierra, Jane Jensen, PC: Windows, USA).
- *Gabriel Knight 3: Blood of the Sacred, Blood of the Damned* (1999, Sierra, Jane Jensen, PC: Windows, USA).

Cinemaware's interactive movies, which predated the CD-inspired marriage of Hollywood and Silicon Valley, were better than almost all the full-motion video games that came after the pioneering company's demise. *Defender of the Crown* (1986, Cinemaware, Kellyn Beck, Amiga, USA) defined the Cinemaware style, but the company's later work was superior. See: *The King of Chicago* (1986, Cinemaware, Doug Sharp, Macintosh, USA), *Rocket Ranger* (1988, Cinemaware, Bob Jacob & Kellyn Beck, Amiga, USA), *Lords of the Rising Sun* (1989, Cinemaware, Doug Barnett, Amiga, USA) and *Wings* (1990, Cinemaware, John Cutter, Amiga, USA). Best of all: *It Came from the Desert* (1989, Cinemaware, David Riordan, Amiga, USA) and its expansion disk *Antheads: It Came from the Desert II* (1990, Cinemaware, David Riordan, Amiga, USA).

By 2000, the adventure game had faded away. Keeping the flame alive:
- *Phoenix Wright: Ace Attorney* (2001, Capcom, Game Boy Advance, Japan): Legal eagle visual novel.
- *Masq* (2002, AlterAction, Javier Maldonado, PC: Windows, USA): Comic strip visuals, real-time

decisions and available for free. Go to: www.alteraction.com
- *Another Code: Two Memories* (2005, Nintendo, Cing, Nintendo DS, Japan): Touch screen and microphone enhanced adventuring.
- *Hotel Dusk: Room 215* (2007, Nintendo, Cing, Nintendo DS, Japan): Excellent writing and pencil sketch artwork that echoes the video for A-Ha's 1985 hit single *Take On Me*.
- *Heavy Rain* (2010, Sony Computer Entertainment, Quantic Dream, PC: Windows, France): A gripping serial-killer thriller that finally realised French game visionary David Cage's efforts to reinvent the adventure game. Cage's earlier attempts are fascinating if not as successful:
- *Omikron: The Nomad Soul* (1999, Eidos Interactive, Quantic Dream, PC: Windows, France): Sci-fi confusion starring David Bowie.
- *Fahrenheit / Indigo Prophecy* (2005, Atari Interactive, Quantic Dream, Xbox, France): Frustratingly flawed tale of occult conspiracies.

Arcade adventures

Possibly the broadest genre of all. Encompasses anything the mixes the story and puzzle-solving elements of adventure games with the action of the arcades. The earliest:
- *Adventure* (1979, Atari, Warren Robinett, VCS 2600, USA): Remoulded the text adventure into a text-free action experience.

Other notable arcade adventures not featured elsewhere in this guide:
- *Dragon's Lair* (1983, Cinematronics, Advanced Microcomputer Systems & Sullivan Bluth Studios, Coin-op, USA): Figurehead of the short-lived laserdisc game craze. An interactive cartoon, but mainly a cartoon.
- *Ghostbusters* (1984, Activision, David Crane, Commodore 64, USA): Based on the hit comedy film. Manage the Ghostbusters firm and then go bust some ghosts.
- *Another World / Out of This World* (1991, Delphine, Eric Chahi, Amiga, France): Uncluttered and refined action adventure superbly told without the need for speech or text.
- *Tomb Raider II* (1997, Eidos Interactive, Core Design, PlayStation, UK): Lara Croft's seminal globetrotting second coming. After that the series lost its way until *Lara Croft Tomb Raider: Anniversary* (2007, Eidos Interactive, Crystal Dynamics, Xbox 360, USA) recaptured the magic of the first two games.
- *Ico* (2001, Sony Computer Entertainment, Team Ico, PlayStation 2, Japan): Awe-inspiring journey into a sun-bleached world. Truly magical.
- *Okami* (2006, Capcom, Clover Studios, PlayStation 2, Japan): Beautiful action game that replicates the look of traditional Japanese artwork.
- *Shadow of the Colossus* (2005, Sony Computer Entertainment, Team Ico, PlayStation 2, Japan): An emotional rollercoaster where journeys through desolate wastes are punctuated with epic battles against enormous Colossi. Ends with a brutal twist.
- *Uncharted: Drake's Fortune* (2007, Sony Computer Entertainment, Naughty Dog, PlayStation 3, USA): Bombastic action game that echoes the Indiana Jones films. Even more spectacular was the sequel: *Uncharted 2: Among Thieves* (2009, Sony Computer Entertainment, Naughty Dog, PlayStation 3, USA).
- *No More Heroes* (2007, Marvelous Entertainment, Grasshopper Manufacture, Wii, Japan): A punky and scrappy otaku take on *Grand Theft Auto*.

Puzzles

Tetris (1984, Alexey Pajitnov, Elektronika 60 [Game Boy], USSR) remains the king of puzzle games: A hypnotic time-eater that is as compelling 25 years on as it was first time around. Those interested in what Alexey Pajitnov did after *Tetris* should try *Hexic* (2003, MSN Games, Alexey

Pajitnov, Online: Browser [Xbox 360], USA). Two other Russian puzzle games that appeared soon after *Tetris* reached the west are also worth tracking down:
- *7 Colors* (1991, Infogrames, Gamos, PC: MS-DOS, USSR): Goal is to take control of half the play area before your opponent by absorbing diamonds of a particular colour. Originally called *Filler*.
- *Colour Lines* (1992, Gamos, PC: MS-DOS, Russia): *Tetris* reimagined as a board game but with a dash of *Reversi*.

And while we're at it, another Eastern European puzzle gem is the tomb-exploration challenge of *Quadrax* (1994, Ultrasoft, David Durcak and Marian Ferko, ZX Spectrum [PC: Windows], Slovakia).

Tetris inspired many other games about lining up patterns or objects. The best:
- *Klax* (1989, Atari Games, Dave Akers and Mark Stephen Pierce, Coin-op, USA): Tile piling conveyor belt fun.
- *Bejeweled* (2000, PopCap Games, Online: Browser, USA): Pressure-free mixing and matching pleasure. Originally called *Diamond Mine*.
- *Puzzle Bobble* / *Bust-a-Move* (1994, Taito, Coin-op, Japan): Jaunty bubble matching spin-off from the *Bubble Bobble* series.

Saving things is another big theme in puzzle games:
- *Lemmings* (1991, Psygnosis, DMA Design, Amiga, UK): Save the charmingly cute and oblivious critters from 1,001 horrible deaths.
- *Pipe Mania* / *Pipe Dream* (1989, Lucasfilm Games, The Assembly Line, Amiga, UK): Frantic pipe-building to stop the 'ooze' escaping.
- *LocoRoco* (2006, Sony Computer Entertainment, Tsutomu Kouno, PSP, Japan): Wonderfully bright and cheery game where you roll wobbly jellies to safety. *Rolando* (2008, ngmoco, HandCircus, iPhone, London) ventured into similar territory, but with the added bonus of the iPhone's tilt controls.

Sticking with gooey blobs, there's the utterly satisfying construction challenge *World of Goo* (2008, 2D Boy, Wii, USA). Its forerunner: *Tower of Goo* (2005, Kyle Gabler, PC: Windows, USA). Both owe a debt to the challenging construction puzzle game *Bridge Builder* (2000, Alex Austin, PC: Windows, USA).

Managing air or rail traffic is another recurring puzzle game theme, although these are more about planning ahead than logic: Air traffic control sim *Final Approach* (1982, Apollo, Dan Oliver, VCS 2600, USA) and *The Kennedy Approach* (1985, Microprose, Andy Hollis, Commodore 64, USA) are two of the earliest, but *Flight Control* (2009, Firemint, iPhone, Australia) is much more fun, not least because of its pleasing finger-tracing mechanic. Also see the rail controller challenge of *The Train Game* (1983, Microsphere, ZX Spectrum, UK) and *Trains!* (2009, Armor Games, TigerTail Studios, Online: Flash, India).

Few of the above are, however, serious brainteasers. The real migraine-inducers:
- *Sokoban* (1982, Thinking Rabbit, Hiroyuki Imabayashi, NEC PC-8801, Japan): Maddening block-pushing puzzle. Its challenge – some of the toughest levels of the long-running series require hundreds of moves to complete – has made the game a focus of artificial intelligence researchers.
- *Spacestation Pheta* (1985, T&T Software, TI-99/4A [Macintosh], USA): Don't be fooled by the

Lode Runner looks.

- *The Sentinel* (1986, Firebird, Geoff Crammond, BBC Micro, UK): Unique game that demands Chess-like levels of concentration.
- *Minesweeper* (1990, Microsoft, Robert Donner & Curt Johnson, PC: Windows, USA): Alongside *Solitaire* the mine-spotting logic puzzle game is one of the world's most played games thanks, primarily, to its inclusion in the Windows operating system.
- *The Incredible Machine* (1992, Sierra, Dynamix & Jeff Tunnell Productions, PC: MS-DOS, USA): Takes the crazy contraption building idea of the board game *Mouse Trap* and turns it into a fiendish test of ingenuity.
- *Dr Kawashima's Brain Training / Brain Age* (2005, Nintendo, Nintendo DS, Japan): A daily dose of Sudoku, math tests and word games to boost your 'brain age' (aka score).
- *Professor Layton and the Curious Village* (2007, Level-5, Nintendo DS, Japan): 135 brainteasers held together with a point-and-click adventure game about a strange village. *The Fool's Errand* (1987, Miles Computing, Cliff Johnson, Macintosh, USA) also mixed challenging conundrums with a story.
- The mind-bending perspective blurring of the M.C. Escher-inspired *Echochrome* (2008, Sony Computer Entertainment, PlayStation 3, Japan) is also a must for those after a real brain wracking experience. Along similar lines is *Illusions* (1984, Coleco Electronics, Nice Ideas, Colecovision, France).
- *Braid* (2008, Microsoft Game Studios, Jonathan Blow, Xbox 360, USA): Devious puzzle-based platform game about turning back time. Ingenious and subversive.

Best of the rest:
- *Marble Madness* (1984, Atari Games, Mark Cerny, Coin-op, USA): Trackball-enhanced zany marble steering with a uniquely odd soundtrack.
- *Boulder Dash* (1984, First Star Software, Peter Liepa and Chris Gray, Atari 800, USA): Excellent arcade puzzle game where players gather diamonds while figuring out how to avoid monsters and falling rocks. Superior Software delivered a just-as-good-copy with its *Repton* games, of which *Repton 2* (1985, Superior Software, Tim Tyler, BBC Micro, UK) is the best.
- *Shanghai* (1986, Activision, Brodie Lockard, Macintosh, USA): Based on the tile-based board game Mah-Jong Solitaire. Other Mah-Jong influenced titles include *Ishido: The Way of Stones* (1990, Accolade, Publishing International & Michael Feinberg, Macintosh, USA), which incorporated elements of the *I Ching* into its new age mix.
- *Gobliins 2: The Prince Buffoon* (1992, Coktel Vision, Pierre Gilhodes & Muriel Tramis, Amiga, France): Use two goblins with different abilities to solve various challenges.
- *Zen Bound* (2009, Chilingo, Secret Exit, iPhone, Finland): Tranquil puzzle where you wind string around wooden carvings.
- *Scribblenauts* (2009, Warner Bros Interactive, 5th Cell, Nintendo DS, USA): Incredibly inventive puzzle game where you solve challenges by summoning objects simply by typing in any word you feel like.

Platform games
Although it lacked the jumping that became a hallmark of the genre, the ladder climbing, platform action of *Space Panic* (1980, Universal, Coin-op, Japan) was the first game that could truly be described as a platform game. The genre's forefathers:
- *Maneater* (1975, Project Support Engineering, Coin-op, USA): *Jaws* cash-in where you dive for treasure while avoiding sharks. The fibreglass cabinet set the TV within the jaws of a great white shark's mouth.
- *Frogs* (1978, Gremlin, Coin-op, USA): Jumping game where you control a frog that must jump

up to catch insects.

- *Heiankyo Alien* (1980, Denki Onkyo, The University of Tokyo's Theoretical Science Group, Coin-op, Japan): A maze game where players dig holes to trap aliens that can then be knocked out. The hole digging used in *Space Panic* to trap monsters is similar.

Donkey Kong (1981, Nintendo, Shigeru Miyamoto, Coin-op, Japan) ushered in the platform era. Its success was followed by numerous single-screen platform games, including:
- *BurgerTime* (1982, Data East, Coin-op, Japan): Fast-food themed platform run-around.
- *Donkey Kong Jr.* (1982, Nintendo, Shigeru Miyamoto, Coin-op, Japan): A role-reversal from the first game. Mario is now the villain and the caged Donkey Kong is the victim. You play Donkey Kong's son. Donkey Kong reappeared in many more games, including the over-hyped eye candy of *Donkey Kong Country* (1994, Nintendo, Rare, Super NES, UK).
- *Chuckie Egg* (1983, A&F Software, Nigel Alderton, ZX Spectrum, UK): Simple but oh-so-perfect egg-collecting bliss.
- *Congo Bongo* (1983, Sega, Sega, Coin-op, Japan): Sega's isometric retort to *Donkey Kong*.
- *Lode Runner* (1983, Brøderbund, Douglas E. Smith, Apple II, USA): Excellent ladder-based action. One of the first games to feature a level editor.
- *Bubble Bobble* (1986, Taito, Fukio Mitsuji, Coin-op, Japan): The bubble gum visuals and soundtrack compliment one of the finest platform games ever devised. Near perfect in two-player mode.
- *Rodland* (1990, Jaleco, Coin-op, Japan): *Bubble Bobble*-apeing platfomer but no less great for that.

Donkey Kong also introduced Mario. Nintendo's extensive back catalogue of Mario games is a treasure trove of great gaming moments. Highlights:
- *Mario Bros* (1983, Nintendo, Shigeru Miyamoto & Gunpei Yokoi, Coin-op, Japan): Mario became a plumber and Luigi made his debut in this bouncy slice of turtle stomping. Echoed the winged ostrich-riding game *Joust* (1982, Williams, John Newcomer, Coin-op, USA).
- *Super Mario Bros* (1985, Nintendo, Shigeru Miyamoto & Takashi Tezuka, NES, Japan): The game that sold the NES and made Mario a star. Simply dazzling. After that came three sequels in the same mould that packed more inventiveness into every one of their levels than some game design teams produce during their whole career. All three are magnificent: *Super Mario Bros 3* (1988, Nintendo, Shigeru Miyamoto & Takashi Tezuka, NES, Japan); *Super Mario World* (1990, Nintendo, Shigeru Miyamoto & Takashi Tezuka, Super NES, Japan); *Super Mario World 2: Yoshi's Island* (1995, Nintendo, Shigeru Miyamoto & Takashi Tezuka, Super NES, Japan).
- *Super Mario 64* (1996, Nintendo, Shigeru Miyamoto, Nintendo 64, Japan): A complete reinvention of the platform game and series for the 3D era but still as gloriously spellbinding as ever.
- *Super Mario Galaxy* (2007, Nintendo, Shigeru Miyamoto & Yoshiaki Koizumi, Wii, Japan): Gravity-defying platforming that throws up a constant stream of surprises and marvels.
- *New Super Mario Bros Wii* (2009, Nintendo, Shigeru Miyamoto & Shigeyuki Asuke, Wii, Japan): Proof the original 2D formula is just as compelling as it was 24 years earlier.

Super Mario Bros' scrolling platform action inspired many:
- *The New Zealand Story* (1988, Taito, Coin-op, Japan): Bright and bubbly old-school platforming.
- *Rainbow Islands: The Story of Bubble Bobble 2* (1987, Taito, Fukio Mitsuji, Coin-op, Japan): An excellent sequel to *Bubble Bobble* despite having almost nothing in common with its pre-equal.
- *Ghouls 'n Ghosts* (1988, Capcom, Coin-op, Japan): Rock-hard platform game. Only for the most determined.
- *A Boy and His Blob: Trouble on Blobolonia* (1989, Imagineering, David Crane, NES, USA): Puzzle-

inflected twist on the genre.

- *Sonic the Hedgehog* (1991, Sega, Yuji Naka, Megadrive, Japan): Platforming meets pinball. The best of the many sequels is *Sonic Adventure* (1998, Sega, Sonic Team, Dreamcast, Japan), the speedy hedgehog's 3D debut.

- *Earthworm Jim* (1994, Playmates Interactive, Shiny Entertainment, Megadrive, USA) and *Earthworm Jim 2* (1995, Playmates Interactive, Shiny Entertainment, Megadrive, USA): Great lead character, distinctive levels.

- *Rayman* (1995, Ubisoft, Michel Ancel, Atari Jaguar [PlayStation], France): Traditional platforming in a vibrant world. Also see its crazed spin-off, *Rabbids Go Home* (2009, Ubisoft, Michel Ancel & Jacques Exertier, Wii, France).

Not every 2D platform game, however, was an exercise in the cute or cartoony:

- *Infernal Runner* (1985, Loriciels, Eric Chahi, Commodore 64, France): Gory take on the genre packed with lethal contraptions.

- *Castlevania* (1986, Konami, Famicom Disk System, Japan): The start of the long-running vampire hunting series that peaked with the inventive *Castlevania: Symphony of the Night* (1997, Konami, Toru Hagihara, PlayStation, Japan).

- *Metroid* (1986, Nintendo, Gunpei Yokoi & Yoshio Sakamoto, Japan): Non-linear sci-fi platform adventure. The outstanding *Super Metroid* (1994, Nintendo, Yoshio Sakamoto & Makoto Kanoh, Super NES, Japan) and *Metroid Fusion* (2002, Nintendo, Yoshio Sakamoto, Game Boy Advance, Japan) both stuck to the original's format. *Metroid Prime* (2002, Nintendo, Retro Studios, Gamecube, USA) expertly reinvented the series giving it a first-person viewpoint without reducing it into a first-person shooter.

- *Shinobi* (1987, Sega, Noriyoshi Ohba, Coin-op, Japan): Ninja action that delivered a harmonious blend of fighting and platforming.

- *Prince of Persia* (1989, Brøderbund, Jordan Mechner, Apple II, USA): A excellently animated and designed platform game with a distinctive silent movie feel. Also see: *Prince of Persia: The Sands of Time* (2003, Ubisoft, Ubisoft Montreal, PlayStation 2, Canada), a graceful parkour-influenced return.

- *Oddworld: Abe's Oddysee* (1997, GT Interactive, Oddworld Inhabitants, PlayStation, USA): Wonderfully original.

Other 3D platformers of note:

- *Alpha Waves* (1990, Infogrames, Christophe de Dinechin, Atari ST, France): Platforming's first, and rather odd, move into 3D.

- *Crash Bandicoot* (1996, Sony Computer Entertainment, Naughty Dog, PlayStation, USA): Narrow paths typical of 3D platforming prior to *Super Mario 64*, but the game's larger-than-life character carries it through.

- *NiGHTS: Into Dreams* (1996, Sega, Sonic Team, Saturn, Japan): Sega's dreamy answer to *Crash Bandicoot* and *Super Mario 64*.

- *Sly Cooper and the Thievius Raccoonus* (2002, Sony Computer Entertainment, Sucker Punch Productions, PlayStation 2, USA): Stealth-based platforming.

- *Ratchet & Clank: Up Your Arsenal* (2004, Sony Computer Entertainment, Insomniac Games, PlayStation 2, USA): Wacky and idiosyncratic weaponry in this trigger-happy platformer.

- *Psychonauts* (2005, Majesco, Double Fine Productions, Xbox, USA): Peyote-tripping platform strangeness.

- *LittleBigPlanet* (2008, Sony Computer Entertainment, Media Molecule, PlayStation 3, UK): The loveable scrapbook looks signal just how different a platformer this is with its game-making tools and a vast swarm of user creations to try. A source of endless interest and novelties.

British surrealism

The early days of the UK games industry were filled with games that were bizarre and surreal, yet often hugely popular. Agit-prop game designer Mel Croucher was first to serve up slices of the weird and wonderful:

- *Can of Worms* (1982, Automata, Mel Croucher & Christian Penfold, ZX81, UK): Tabloid newspaper enraging bad taste mini-games.
- *Pimania* (1982, Automata, Mel Croucher & Christian Penfold, ZX81, UK): Abstract puzzle adventure featuring the creepy PiMan character that became Automata's mascot.
- *Deus Ex Machina* (1984, Automata, Mel Croucher, ZX Spectrum, UK): Croucher's finest moment. A truly unique video game experience that grinds together Shakespeare, Aldous Huxley and well-known British TV personalities.

During the early 1980s, however, the teenage Matthew Smith was the poster-child for British surrealism thanks to:

- *Manic Miner* (1983, Bug-Byte, Matthew Smith, ZX Spectrum, UK): A nutty remix of the US platform hit *Miner 2049'er* (1982, Big Five Software, Bill Hogue, Atari 800, USA).
- *Jet Set Willy* (1984, Software Projects, Matthew Smith, ZX Spectrum, UK): The surreal platform game that defined Britain's taste for the odd. Smith then became a legendary figure by quitting the industry before completing another game. His 'disappearance' resulted in a spate of 'where's Matthew Smith?' websites during the late 1990s that equated him to a video gaming Syd Barrett much to his bemusement.

The furry ruminant loving Jeff Minter has made a career out of keeping the flame alive for British surrealism and early '80s shoot 'em ups:

- *Gridrunner* (1982, Llamasoft, Jeff Minter, VIC-20, UK): Minter's breakthrough slice of hi-octane blasting. It landed him a US publishing deal, which he promptly lost thanks to his subsequent output of sheep, llama, goat and camel themed shooters.
- *Attack of the Mutant Camels / Advance of the Megacamel* (1983, Llamasoft, Jeff Minter, Commodore 64, UK): *Defender* reimagined as a battle against giant camels.
- *Trip-a-Tron* (1988, Llamasoft, Jeff Minter, Atari ST, UK): When not making sweaty palmed shooters, Minter spent his time exploring his vision of 'light synthesizers', interactive light show software. *Trip-a-Tron* was the first to come close to realising his vision. His experiments eventually resulted in the Xbox 360's interactive musical visualiser *Neon* (2005, Microsoft Game Studios, Llamasoft, Xbox 360, UK).
- *Llamatron: 2112* (1991, Llamasoft, Jeff Minter, Atari ST, UK): Minter reinvents *Robotron: 2084* with screaming fractals, bouncing cans of killer cola, sheep noises and non-stop blasting. Even better than its inspiration.
- *Tempest 2000* (1994, Atari Corporation, Llamasoft, Atari Jaguar, UK): The influence of acid house seeps into Minter's work in this spectacular remake of the Atari coin-op.
- *Space Giraffe* (2007, Llamasoft, Jeff Minter & Ivan Zorzin, Xbox 360, UK): Storming shooter with hazy, trippy psychedelic visuals.
- *Gridrunner Revolution* (2009, Llamasoft, Jeff Minter & Ivan Zorzin, PC: Windows, UK): A remake of Minter's breakthrough game.

Other moments of British madness:
- *Wanted: Monty Mole* (1984, Gremlin Graphics, Peter Harrap, ZX Spectrum, UK): Confused social commentary about a coal-stealing mole.
- *A Day in the Life* (1985, Micromega, ZX Spectrum, UK): A digital hymn to Clive Sinclair where you steer his disembodied head to Buckingham Palace while dodging killer microchips

and other oddities.

- *Skool Daze* (1985, Microsphere, David Reidy, ZX Spectrum, UK): Relatively sane and rather brilliant schoolboy sim: avoid getting lines while trying to steal your terrible school report. Its equally good sequel: *Back to Skool* (1985, Microsphere, David Reidy, ZX Spectrum, UK).

- *Wizball* (1987, Ocean Software, Sensible Software, Commodore 64, UK): Bullet-shooting bouncing ball battles abstract objects in a quest to bring colour to a grey world. Odd but utterly sane compared to its mad-as-a-box-of-frogs sequel *Wizkid* (1992, Ocean Software, Sensible Software, Amiga, UK). Sensible Software also made the wonderful *Mega Lo Mania / Tyrants: Fight Through Time* (1991, Virgin Interactive, Sensible Software, Amiga, UK), a humorous strategy game in the *Populous* mould where technological advances could leave cavemen fighting biplanes and tanks. As the game says it's "ergonomically terrific".

- *Head Over Heels* (1987, Ocean Software, John Ritman & Bernie Drummond, ZX Spectrum, UK): Wonderfully strange. One of the best arcade adventures ever devised.

- *Rock Star Ate My Hamster* (1988, Codemasters, Colin Jones, Amiga, UK): Low-budget rock manager parody. One of the better budget games to fill Britain's shelves during the late 1980s.

Other budget game picks:

- *Dizzy: The Ultimate Cartoon Adventure* (1986, Codemasters, The Oliver Twins, ZX Spectrum, UK): This hard-boiled hero was the nearest Britain got to producing a Mario of its own. The puzzle-focused arcade adventure was followed by numerous sequels including *Treasure Island Dizzy* (1987, Codemasters, The Oliver Twins, Amiga, UK) and air-bubble hopping of *Bubble Dizzy* (1990, Codemasters, The Oliver Twins, Atari ST, UK).

- *Rescue* (1987, Mastertronic, Icon Design, ZX Spectrum, UK): Action-packed race-against-time on board a space station to save the "ultimate discovery".

- *Werewolves of London* (1988, Mastertronic, Viz Design, CPC, UK): Free-roaming action adventure where you trawl London to munch on a group of occultists who have turned you into a werewolf.

Head Over Heels was directly inspired by *Knight Lore* (1984, Ultimate Play the Game, Tim Stamper & Chris Stamper, ZX Spectrum, UK): A groundbreaking game that made isometric-view action adventures the staple diet of British computer owners in the 1980s. Other isometric delights:

- *Ant Attack* (1983, Quicksilva, Sandy White, ZX Spectrum, UK): Got there before *Knight Lore* with the distinctive architecture of Antchester.

- *The Great Escape* (1986, Ocean Software, Denton Designs, ZX Spectrum, UK): Escape a Nazi prisoner of war camp. How is up to you, just don't caught. The moment when your character's morale reaches rock bottom and he reverts to being a submissive model prisoner is one of gaming's most depressing game overs.

- *Spindizzy* (1986, Electric Dreams, Paul Shirley, CPC, UK): A fiendish puzzle game reminiscent of Atari Games' *Marble Madness*.

- *The Last Ninja* (1987, System 3, Mark Cale & Tim Best, Commodore 64, UK): Elegant ninja action adventure.

- *La Abadía de Crimen* (1987, Opera Soft, Paco Menéndez & Juan Delcán, CPC, Spain): An involving and sophisticated murder mystery game that shows that the UK wasn't the only source of great isometric arcade adventures. Other 1980s highlights from Spain: *Army Moves* (1986, Dinamic, ZX Spectrum, Spain): Super hard drive-and-gun arcade action; *Goody* (1987, Opera Soft, Gonzalo Suárez, ZX Spectrum, Spain): Crafty bank robber platforming; *Rescate Atlantida* (1989, Dinamic, ZX Spectrum, Spain): Vast underwater exploration game but – like almost every Spanish game from the time – very, very hard.

- *Get Dexter* (1986, Ere Informatique, Rémi Herbulot, CPC, France): Marvellously bizarre prob-

lem solving game where the player is accompanied by a strange creature that consists of a head on a large foot.
- *D/Generation* (1991, Mindscape, Robert Cook & James Brown, Amiga, USA): Colourful isometric adventure enhanced by the use of lasers that bounce off walls.
- *Little Big Adventure / Relentless: Twinsen's Adventure* (1994, Electronic Arts, Adeline, PC: MS-DOS, France): The last great isometric arcade adventure. Also see: *Little Big Adventure 2 / Twinsen's Odyssey* (1997, Electronic Arts, Adeline, PC: Windows, France).

Maze
Pac-Man (1980, Namco, Toru Iwatani, Coin-op, Japan) remains the genre's finest moment. *Pac-Man: Championship Edition* (2009, Namco Bandai, Toru Iwatani, Xbox 360, Japan) is a fine update, but the original's just as good.

Other maze-like games of note:
- *Q*bert* (1982, Gottlieb, Warren Davis & Jeff Lee, Coin-op, USA): If M.C. Escher made *Pac-Man*.
- *Dig Dug* (1982, Namco, Coin-op, Japan): Hectic underground tunnelling where the goal is to inflate cute monsters until they burst.
- *Bomber Man* (1983, Hudson Soft, NEC PC-6001, Japan): The start of Hudson Soft's long running bomb 'em up maze game. Comes alive with friends, try *Bomberman Ultra* (2009, Hudson Soft, PlayStation 3, Japan).
- *Super Monkey Ball* (2001, Sega, Amusement Vision, Gamecube, Japan): A zany test of skill where you roll a monkey in a Perspex ball through mazes suspended in the air.
- *The Last Guy* (2008, Sony Computer Entertainment, PlayStation 3, Japan): Monster movie-inspired save 'em up that turns satellite images of real-life cities into mazes.

First-person shooter
The genre's origins stretch all the way back to *Maze* (1974, Steve Colley, Greg Thompson & Howard Palmer, Imlac PDS-1, USA), but the world had to wait for Id Software's releases before it became a permanent feature of the gaming landscape:
- *Catacomb 3-D* (1991, Softdisk, Id Software, PC: MS-DOS, USA): Introduced most of the basic concepts of the first-person shooter, but lacked the intensity of Id's subsequent work
- *Wolfenstein 3D* (1992, Apogee, Id Software, PC: MS-DOS, USA): Nazi-slaying brutality that ushered in the Id revolution
- *Doom* (1993, Id Software, PC: MS-DOS, USA): Id's primal masterpiece. One of the most significant video games ever made.
- *Quake* (1996, GT Interactive, Id Software, PC: Windows, USA): The Romero-Carmack partnership's parting shot took user-generated content to new heights and spawned the machinima movement. The multiplayer-focused *Quake III: Arena* (1999, Activision, Id Software, PC: Windows, USA) has more lasting appeal.

After Id's revolution:
- *Duke Nukem 3D* (1996, Apogee, 3D Realms, PC: MS-DOS, USA): Rip-roaring destruction fronted by the semi-ironic, politically incorrect, alpha-male knucklehead that is Duke Nukem.
- *Half-Life* (1998, Sierra, Valve, PC: Windows, USA): Landmark integration of storytelling and first-person shooter. The exceptional *Half-Life 2* (2004, Sierra, Valve, PC: Windows, USA) is a must.
- *Unreal Tournament* (1999, GT Interactive, Epic Games & Digital Extremes, PC: Windows, USA & Canada): Pure multiplayer magic.

- *The Operative: No-one Lives Forever* (2000, Fox Interactive, Monolith Productions, PC: Windows, USA): *Austin Powers*-style James Bond parody with a heroine that echoes actress Joanna Lumley's days as Purdy in *The New Avengers*.
- *Deus Ex* (2000, Eidos Interactive, Ion Storm, PC: Windows, USA): RPG character development, powerful storytelling and expert game design. Owes a big debt to *System Shock* (1994, Origin Systems, Looking Glass Studios, PC: MS-DOS, USA).
- *Serious Sam: The First Encounter* (2001, Gathering of Developers, Croteam, PC: Windows, Croatia): An anything but serious high-action rampage.
- *Halo: Combat Evolved* (2001, Microsoft Game Studios, Bungie, Xbox, USA): Taut, iconic and thrilling sci-fi blockbuster that stamped all over the competition. Two equally vital sequels: *Halo 2* (2004, Microsoft Game Studios, Bungie, Xbox, USA) and *Halo 3* (2007, Microsoft Game Studios, Bungie, Xbox 360, USA)
- *Far Cry* (2004, Ubisoft, Crytek, PC: Windows, Germany): Majestic open-ended tactical shooting in lush jungle environments. Also see its Africa-set sequel *Far Cry 2* (2008, Ubisoft, Ubisoft Montreal, Xbox 360, Canada) and the original developers' post-*Far Cry* effort *Crysis* (Electronic Arts, Crytek, PC: Windows, Germany): The video game equivalent of a supercar.
- *Tom Clancy's Rainbow Six: Vegas* (2006, Ubisoft, Ubisoft Montreal, Xbox 360, Canada): Counterterrorism operations set amid the glitz of Las Vegas.
- *BioShock* (2007, 2K Games, Irrational Games, Xbox 360, USA): A tour de force that mixed vigorous blasting with one of gaming's most effective political statements. The decaying grandeur of the underwater city of Rapture and the Big Daddies, with their whale-like moans, confirmed the game's iconic status.

Second World War first-person shooters are almost a genre in themselves. Picks: *Medal of Honor: Allied Assault* (2002, EA Games, 2015 Inc., PC: Windows, USA), *Battlefield 1942* (2002, EA Games, Digital Illusions, PC: Windows, Sweden) and *Call of Duty: World at War* (2008, Activision, Treyarch, Xbox 360, USA). *Call of Duty*'s massively popular present day offshoot: *Call of Duty: Modern Warfare 2* (2009, Activision, Infinity Ward, Xbox 360, USA).

The modding of first-person shooters has also produced several great games. The prime cuts all got Valve makeovers:
- *Team Fortress* (1996, John Cook, Robin Walker & Ian Caughley, PC: Windows, USA): Carefully balanced team-versus-team multiplayer mod.
- *Counter-Strike* (1999, Minh Le & Jess Cliffe, PC: Windows, Canada & USA): A multiplayer phenomenon with a terrorism theme.
- *Portal* (2007, Valve, Xbox 360, USA): First-person brainteaser that grew out of the free student project *Narbacular Drop* (2005, Nuclear Monkey Software, PC: Windows, USA).

Stealth
- *Castle Wolfenstein* (1981, Muse Software, Silas Warner, Apple II, USA): The inspiration for *Wolfenstein 3D* but unlike Id's creation it was a game of covert infiltration not gung-ho slaughter.
- *Metal Gear* (1987, Konami, Hideo Kojima, MSX2, Japan): The starting point for Hideo Kojima's genre-defining series. The cinematic flair may have been years away, but the hide and seek play was already evident.
- *Metal Gear Solid* (1998, Konami, Hideo Kojima, PlayStation, Japan): A stunning return for the series. A tense test of player ingenuity and patience wrapped in the cinematic presentation Kojima first explored with the *Blade Runner*-isms of *Snatcher* (1988, Konami, Hideo Kojima, NEC PC-8801, Japan) and the point-and-click adventure *Policenauts* (1994, Konami, Hideo Kojima, NEC PC-9821, Japan).

- *Metal Gear Solid 2: Sons of Liberty* (2001, Konami, Hideo Kojima, PlayStation 2, Japan): Kojima's movie director aspirations take centre-stage in the series' most divisive game. For those who loved Kojima's world its convoluted and intricate story was a triumph, for everyone else it was an indulgence.
- *Metal Gear Solid 3: Snake Eater* (2004, Konami, Kojima Productions, PlayStation 2, Japan): Set during the Cold War, this revisits series hero Solid Snake's youth. The jungle environments felt expansive and liberating after the confined industrial complexes of the previous two games.
- *Metal Gear Solid 4: Guns of the Patriots* (2008, Konami, Kojima Productions, PlayStation 3, Japan): Reflective entry in the series where an old and weary Solid Snake returns to the battlefield. Kojima's anti-war message was never clearer.

As well as the first *Metal Gear Solid*, 1998 saw the release of two other landmark stealth games:
- *Tenchu: Stealth Assassins* (1998, Sony Computer Entertainment, Acquire, PlayStation, Japan): Ninja assassinations. The grappling hook-aided rooftop-to-rooftop movement is great fun.
- *Thief: The Dark Project* (1998, Eidos Interactive, Looking Glass Studios, PC: Windows, USA): Medieval thievery where audio clues were as important as visual ones for success. Try the third game in the series *Thief: Deadly Shadows* (2004, Eidos Interactive, Ion Storm, PC: Windows, USA), which opened up the thieving opportunities.

Other great moments in stealth gaming:
- *Hitman 2: Silent Assassin* (2002, Eidos Interactive, IO Interactive, PC: Windows, Denmark): Cold, calculated and precise assassinations are the order of the day.
- *Tom Clancy's Splinter Cell: Chaos Theory* (2005, Ubisoft, Ubisoft Montreal & Ubisoft Annecy, Xbox, Canada & France): Ubisoft's *Metal Gear Solid* rival finally moved out of Kojima's shadow with this taut test of patience and skill.
- *Assassin's Creed II* (2009, Ubisoft, Ubisoft Montreal, Xbox 360, France): Parkour-inspired rooftop escapes and daring assassins in Renaissance Italy.
- *Batman: Arkham Asylum* (2009, Eidos Interactive, Rocksteady Studios, PlayStation 3, UK): The first Batman game to really capture the essence of DC Comics' dark knight. The trippy psychological battles with the Scarecrow owe a debt to the fourth-wall-breaking encounter with Psycho Mantis in *Metal Gear Solid*. Of the earlier Batman titles two stand out:
- *Batman the Caped Crusader* (1988, Ocean Software, Special FX, Atari ST, UK): The game is plodding but the comic book panel visual approach was inspired. Sega's beat 'em up *Comix Zone* (1995, Sega, Sega Technical Institute, Megadrive, USA) repeated the approach.
- *Batman the Movie* (1989, Ocean Software, Amiga, UK): Ocean's best movie tie-in, mixing platform-based, bat-rope action with high-speed Batmobile driving.

Horror
Much groping in the dark during the 1980s, but few scares:
- *Alien* (1984, Argus Press, Concept Software, Commodore 64, UK): Based on the 1979 sci-fi film. The elusive alien keeps players on edge.
- *The Rats!* (1985, Hodder & Stoughton, Five Ways Software, ZX Spectrum, UK): Panic-inducing real-time text adventure.
- *Project Firestart* (1989, Electronic Arts, Dynamix, Commodore 64, USA): Had all the ingredients for what followed – dramatic close ups, careful use of audio, player vulnerability – but lacked the frights.

Horror games came of age with *Alone in the Dark* (1992, Infogrames, Frédérick Raynal, PC: MS-DOS, France) and the zombie-terror of *Resident Evil* (1996, Capcom, Shinji Mikami, PlayStation

[Gamecube], Japan) sealed the deal. After *Resident Evil*, the world fell in love with video game nasties:

- *Silent Hill* (1999, Konami, Team Silent, PlayStation, Japan): A more psychological take on the horror game where running away was often the best option. *Silent Hill 2* (2001, Konami, Team Silent, PlayStation 2, Japan) was the fog-shrouded horror series' high point.
- *Fatal Frame / Project Zero* (2001, Tecmo, Keisuke Kikuchi, PlayStation 2, Japan): Player vulnerability taken to the max. Your only defence: a camera obscura.
- *The Thing* (2002, Vivendi Universal, Computer Artworks, PlayStation 2, UK): The nagging fear that one of your party could be an alien makes this an exercise in looking over your shoulder.
- *Eternal Darkness: Sanity's Requiem* (2002, Nintendo, Silicon Knights, Gamecube, Canada): Insanity as the source of horror.
- *Manhunt* (2003, Rockstar Games, Rockstar North, PlayStation 2, UK): Sadistic murder sim. Equally unsettling is the *Lord of the Flies*-esque horror of *Rule of Rose* (2006, Sony Computer Entertainment, Punchline, PlayStation 2, Japan).
- *The Path* (2009, Tale of Tales, PC: Windows, Belgium): A creepy and avant-garde take on *Little Red Riding Hood*.

Resident Evil had numerous sequels and offshoots, including the acrobatic action of *Devil May Cry* (2001, Capcom, Hideki Kamiya, PlayStation 2, Japan), but the most significant was *Resident Evil 4* (2005, Capcom, Shinji Mikami, Gamecube, Japan), which realigned the horror genre by dumping shuffling zombies of old in favour of fast-moving horrors. Followed by:

- *Gears of War* (2006, Microsoft Game Studios, Epic Games, Xbox 360, USA): Steroid-enhanced action with tense gun battles from behind cover and a lead character who conveys a feeling of brute force. Not horror, but *Resident Evil 4* was an important influence.
- *Dead Space* (2008, Electronic Arts, EA Redwood Shores, PlayStation 3, USA): Terrifying sci-fi horror with exceptional audio.
- *Left 4 Dead* (2008, Valve, Certain Affinity, Xbox 360, USA): Desperate 'flee the zombies' co-operative action.

Finally, the natural disaster 'horror' games:

- *Zettai Zetsumei Toshi / Disaster Report / S.O.S.: The Final Escape* (2002, Irem, PlayStation 2, Japan): Earthquake survival adventure in the mould of '70s disaster movies.
- *Disaster: Day of Crisis* (2008, Nintendo, Monolith Soft, Wii, Japan): Multiple disasters and terrorists to boot, lots of side games divert from the main adventure.

Music games

A protracted birth with isolated dabbles until the mid-1990s:

- *Moondust* (1983, Creative Software, Jaron Lanier, Commodore 64, USA): Experimental sonic adventure. Also see Jaron Lanier's previous video game oddity: *Alien Garden* (1982, Epyx, Bernie DeKoven & Jaron Lanier, Atari 800, USA).
- *Synthétia* (1984, Vifi-Nathan, Michel Galvin, Thomson TO7, France): A music creation toy where players use the TO7's in-built light pen to draw sound waves.
- *Dance Aerobics* (1987, Bandai, Human Entertainment, NES, Japan): Early dancing game that used Bandai's Power Pad controller.
- *Hostages: Rescue Mission* (1988, Infogrames, Atari ST, France): Anti-terrorist action featuring early example of music that responds to player actions.
- *First Samurai* (1991, Image Works, Vivid Image, Amiga, UK): Each swoop and hit of the players' sword adds to the musical accompaniment.
- *ToeJam & Earl in Panic on Funkotron* (1993, Sega, Johnson Voorsanger Productions, Megadrive,

USA): Hip-hop loving aliens and a mini-game where players must dance to the rhythm.

The vibrant *PaRappa the Rapper* (1996, Sony Computer Entertainment, NanaOn-Sha, PlayStation, Japan) and DJing cool of *Beatmania* (1997, Konami, Yuichiro Sagawa, Coin-op, Japan) established music games as a genre.

Japan dominated initially:
- *Dance Dance Revolution* (1998, Konami, Coin-op, Japan): Booty-shaking megahit.
- *Guitar Freaks* (1999, Konami, Coin-op, Japan): Introduced the guitar controller.
- *Samba de Amigo* (1999, Sega, Sonic Team, Coin-op [Dreamcast], Japan): Latin-flavoured maraca shaking.
- *Space Channel 5* (1999, Sega, United Game Artists, Dreamcast, Japan): Camp sci-fi chic starring Michael Jackson.
- *Vib-Ribbon* (1999, Sony Computer Entertainment, NanaOn-Sha, PlayStation, Japan): A platform game that generated levels based on the content of audio CDs inserted into the PlayStation. Followed by the Japanese calligraphy rapping game *Mojib-Ribbon* (2003, Sony Computer Entertainment, NanaOn-Sha, PlayStation 2, Japan).
- *Mad Maestro* (2001, Sony Computer Entertainment, Desert Planning, PlayStation 2, Japan): Classical music madness.
- *Gitaroo Man* (2001, Koei, iNiS, PlayStation 2, Japan): Ebullient guitar solo battles with your foes.

The US finally caught up thanks to Harmonix's work:
- *Frequency* (2001, Sony Computer Entertainment, Harmonix Music Systems, PlayStation 2, USA): Too arty for mass appeal but well worth tracking down.
- *Karaoke Revolution* (2003, Konami, Harmonix Music Systems, PlayStation 2, USA): Added vocals to the music game mix, paving the way for the sing-a-long party extravaganza that is *SingStar* (2004, Sony Computer Entertainment, Sony Studio London, PlayStation 2 [PlayStation 3], UK).
- *Guitar Hero* (2006, Red Octane, Harmonix Music Systems, PlayStation 2, USA): Wedded the music game to the air guitar dreaming of western rock fans.
- *Guitar Hero III: Legends of Rock* (2007, Activision, Neversoft, PlayStation 3, USA): Improved on the original with the addition of downloadable tracks to buy and cameos from rock guitar gods such as Slash from Guns 'n' Roses and Rage Against the Machine's Tom Morello.
- *Rock Band* (2007, MTV Games, Harmonix Music Systems, Xbox 360, USA): The singing, guitar-playing and drum-beating strands of music games united at last.

Not to be forgotten:
- *Digital Praise* (2008, Digital Praise, PC: Windows, USA): Christian rock joins the guitar-controller party
- *DJ Hero* (2009, Activision, FreeStyleGames, PlayStation 3, UK): *Guitar Hero* for the turntable set. The more freeform play of *DJ: Decks & FX* (2004, Sony Computer Entertainment, Relentless Software, PlayStation 2, UK) got there first.
- *Wii Music* (2008, Nintendo, Shigeru Miyamoto & Kazumi Totaka, Wii, Japan): Miyamoto's attempt to make playing music simple. Uses the Wii controller to mimic dozens of instruments.

Simulations
Non-combat flight simulations:
- *Flight Simulator* (1980, SubLogic, Bruce Artwick & Stu Moment, Apple II, USA): Brought virtual aviation to the home computer. The ever-improving series, later rebranded as *Microsoft Flight*

Simulator, concentrated on civil aviation. *Microsoft Flight Simulator X* (2006, Microsoft Game Studios, PC: Windows, USA) offers incredible detail and very good tutorials making it ideal for those unfamiliar with the genre.
- *Pilotwings* (1990, Nintendo, Shigeru Miyamoto & Tadashi Sugiyama, Super NES, Japan): Not a sim, but great aeronautical fun all the same.
- *Stunt Island* (1992, Disney Interactive, The Assembly Line, PC: MS-DOS, UK): Perform and film aerial stunts. The video recording options foreshadowed the rise of machinima.
- *Microsoft Space Simulator* (1994, Microsoft, Bruce Artwick Organization, PC: MS-DOS, USA): Getting on a bit now but its realism carries it.
- *Google Earth Flight Simulator* (2005, Google, PC: Windows, USA): Not the most realistic but it does integrate the maps of the Google Earth allowing you to fly anywhere in the world.
- *Orbiter* (2006, Martin Schweiger, PC: Windows, UK): Free, super-accurate space travel simulation.
- *X-Plane 9* (2008, Laminar Research, PC: Windows, USA): *Microsoft Flight Simulator*'s biggest rival. The professional version is certified by the US Federal Aviation Administration as a pilot training tool.

Combat flight simulations:
- *Red Baron II* (1997, Sierra, Dynamix, PC: Windows, USA): Sterling First World War fighter plane battles that developed a strong modder community.
- *European Air War* (1998, Microsoft, PC: Windows, USA): Microsoft take a break from civilian flight sims to serve up this Western Front air combat classic.
- *IL-2 Sturmovik* (2001, 1C Company, 1C: Maddox Games, PC: Windows, Russia): The often neglected Eastern Front of the Second World War gets the attention it deserves with this thrilling flight sim.
- *Comanche 4* (2001, NovaLogic, PC: Windows, USA): The first choice for those seeking helicopter gunship action.
- *Falcon 4.0: Allied Force* (2005, Graphsim Entertainment, Lead Pursuit, PC: Windows, USA): Update of the 1998 modern fighter jet sim that added many of the mods created by the original's fans as well as improved physics.
- *Ace Combat 6: Fires of Liberation* (2006, Namco Bandai, Namco Project Aces, Xbox 360, Japan): Fast-paced aerial action, more shoot 'em up than sim.

Train simulations:
- *Southern Belle* (1985, Hewson, Mike Male & Bob Hillyer, BBC Micro, UK): Early 1900s steam engine driving sim. Take the Southern Belle from London Victoria to Brighton.
- *Densha de Go!* (1996, Taito, Coin-op, Japan): Modern day train driving that replicates the commuter lines of Japan, starting with the quiet rural lines before heading to the urban crush of the rush hour.
- *Microsoft Train Simulator* (2001, Microsoft, Kuju Entertainment, PC: Windows, UK): Does for trains what the *Microsoft Flight Simulator* series did for flight sims. Microsoft later abandoned the train sim genre leaving a gap that *Trainz Simulator 2009* (2008, Auran, PC: Windows, Australia) has sought to fill.

Other simulations:
- *Silent Service II* (1990, Microprose, PC: MS-DOS, USA): Highly accurate submarine simulator.
- *Search & Rescue 4: Coastal Heroes* (2003, Just Flight, InterActive Vision, PC: Windows, Denmark): Coast guard heroism.
- *Rigs of Rod* (2003, Pierre-Michel Ricordel, PC: Windows, France): Open-source vehicle simula-

tion software that began as a off-road truck sim but has evolved into one of the most accurate and detailed simulators around.
- *Silent Hunter 4: Wolves of the Pacific* (2007, Ubisoft, Ubisoft Romania, PC: Windows, Romania): Submarine game that emphasises 'action' over simulation.
- *Ship Simulator 2008* (2007, Lighthouse Interactive, VSTEP, PC: Windows, Netherlands): Seasickness inducing journeys across the waves.

Strategy and management
The earliest strategy games stuck rigidly to the format of tabletop war games. Chris Crawford, however, broke the mould:
- *Tanktics: Computer Game of Armored Combat on the Eastern Front* (1978, Chris Crawford, PET, USA): Primitive text-only strategy but introduced the use of fog of war.
- *Eastern Front 1941* (1981, Atari Program Exchange, Chris Crawford, Atari 800, USA): Turn-based decisions with real-time following of orders. The real-time/turn-based hybrid approach resurfaced in the exceptional *Combat Mission II: Barbarossa to Berlin* (2002, CDV, Battlefront.com, PC: Windows, USA)
- *Excalibur* (1983, Atari Program Exchange, Chris Crawford, Larry Summers & Valerie Atkinson, Atari 800, USA): Arthurian war game designed to encourage diplomacy over conflict, an approach Crawford expanded on with:
- *Balance of Power* (1985, Mindscape, Chris Crawford, Macintosh, USA): Cold War geopolitics where diplomatic manoeuvring rather than military action is the name of the game.

Other political sims:
- *President Elect* (1981, SSI, Nelson Hernandez, Apple II, USA): US presidential election campaign simulator.
- *Hidden Agenda* (1988, Springboard, Trans Fiction Systems, Macintosh, USA): Text-based simulation of running a post-revolution Central American state.
- *Conflict: Middle East Political Simulator* (1990, Virgin Interactive, David Eastman, Amiga, UK): Diplomacy focused game where you play the new Israeli prime minister.
- *Floor 13* (1992, Virgin Games, David Eastman, PC: MS-DOS, UK): Take charge of a sinister secret police force in early 1990s Britain. Your job: to torture, assassinate and smear opponents of Her Majesty's Government.
- *Republic: The Revolution* (2003, Eidos Interactive, Elixir Studios, PC: Windows, UK): Post-communism revolutionary politics sim, set in a fictional ex-Soviet state. The designers' grand vision got the better of them, but the result was still a fascinating, if flawed, attempt.

Germany's business and kingdom management sims eschewed combat for planning, creating a sub-genre that owes a significant debt to German board games.
- *Kaiser* (1984, Ariolasoft, Dirk Beyelstein, Atari 800, West Germany): Simple kingdom management game.
- *Hanse* (1986, Ariolasoft, Ralf Glau & Bernd Westphal, Commodore 64, West Germany): Middle Ages trading sim set around the ports of the Baltic Sea.
- *Ports of Call* (1987, Aegis International, International Software Development, Amiga [iPhone], West Germany): Much-loved maritime trading business sim.
- *Pizza Tycoon / Pizza Connection* (1994, Software 2000, Cybernetic Corporation, Amiga, Germany): Run your own pizza parlour managing everything from overheads to pizza topping choices.
- *Die Fugger II* (1996, Sunflowers, PC: MS-DOS, Germany): Political and business intrigue in medieval Germany.
- *The Guild 2* (2006, JoWooD, 4Head Studios, PC: Windows, Germany): *Die Fugger*'s spiritual

successor.

- *Catan* (2007, Big Huge Games, Xbox 360, USA): Video game version of classic German board game *The Settlers of Catan*.

Non-German management sims:

- *The Oregon Trail* (1971, Don Rawitsch, Paul Dillenberger & Bill Heinemann, unknown mainframe [iPhone], USA): Classic educational game that expertly tells the story of the Wild West pioneers. Another educational game worth seeking out is the logic puzzle title *Rocky's Boots* (1982, The Learning Company, Warren Robinett & Leslie Grimm, Apple II, USA)

- *Railroad Tycoon* (1990, Microprose, Sid Meier, PC: MS-DOS, USA): Railroad empire building in the early days of steam trains. Sid Meier revisited the series with *Sid Meier's Railroads!* (2006, 2K Games, Firaxis, PC: Windows, USA).

- *Transport Tycoon* (1994, Microprose, Chris Sawyer, PC: MS-DOS, UK): The artificial intelligence is showing its age, but this transport empire-building game is still one of the finest business games ever made.

- *Theme Park* (1994, Electronic Arts, Bullfrog, PC: MS-DOS, UK): Peter Molyneux's slapstick take on running a funfair. *Rollercoaster Tycoon 3* (2004, Atari Interactive, Frontier Developments, PC: Windows, UK) is its spiritual successor.

- *Dungeon Keeper* (1997, Electronic Arts, Bullfrog, PC: Windows, UK): Molyneux turns the RPG on its head. Build dungeons to stop marauding 'heroes' from stealing your treasure.

- *The Movies* (2005, Activision, Lionhead Studios, PC: Windows, UK): Machinima-inspired business sim where you double as Hollywood studio mogul and film director.

- *Chocolatier* (2007, PlayFirst, Big Splash Games, PC: Windows [iPhone], USA): Chocolate tycoon.

- *Monopoly* (2008, Electronic Arts, EA Bright Light Studio, Wii, UK): An excellent video game version of the iconic property tycoon board game.

Dani Bunten Berry's *M.U.L.E.* (1983, Electronic Arts, Ozark Softscape, Atari 800, USA) was a delicate balancing act of player competition and inter-dependence that is justifiably regarded as one of the all-time best-designed games. Desperately needs a modern remake. Dani Bunten Berry's premature death from lung cancer in 1997 robbed video gaming of one of its brightest talents just as the online technology her work aspired to use came of age. Her sterling legacy:

- *Wheeler Dealers* (1978, Speakeasy Software, Dan Bunten, Apple II, USA): Pioneered the auction battles of *M.U.L.E.*

- *Cartel$ & Cutthroat$* (1981, SSI, Dan Bunten, Apple II, USA): Early business sim for up to eight players.

- *The Seven Cities of Gold* (1984, Electronic Arts, Ozark Softscape, Commodore 64, USA): Single-player game of exploration in the New World. Inspired the high seas adventuring of *Sid Meier's Pirates!* (1987, Microprose, Sid Meier, Commodore 64 [Amiga], USA).

- *Heart of Africa* (1985, Electronic Arts, Ozark Softscape, Commodore 64, USA): Beneath the now primitive visuals lurks an expansive and deep game of exploring the wilderness. The sun blindness that blurs out the screen if you spend too long in the desert and the mixed-up joystick controls that result from delirium due to a lack of water foreshadowed the insanity effects of *Eternal Darkness: Sanity's Requiem*.

- *Modem Wars* (1988, Electronic Arts, Ozark Softscape, Commodore 64, USA): Early online real-time strategy.

The action-oriented, real-time strategy model pioneered in *Modem Wars* and the likes of *Herzog Zwei* (1989, TechnoSoft, Megadrive, Japan) came of age in *Dune II: The Building of a Dynasty / Dune*

II: Battle for Arrakis (1992, Virgin Interactive, Westwood Associates, PC: MS-DOS, USA). After that real-time strategy ruled the roost:
- *Cannon Fodder* (1993, Virgin Interactive, Sensible Software, Amiga, UK): Humorous point-and-click strategy action.
- *StarCraft* (1998, Blizzard Entertainment, Chris Metzen & James Phinney, PC: Windows, USA): Refined real-time strategy that conquered South Korea.
- *Commandos: Behind Enemy Lines* (1998, Eidos Interactive, Pyro Studios, PC: Windows, Spain): Real-time strategy game focused on managing a small group of commandos.
- *Shogun: Total War* (2000, Electronic Arts, Creative Assembly, PC: Windows, UK): The start of Creative Assembly's bold strategy series that is based around epic real-time battles involving hundreds of troops. The series has continued to improve with each addition, see *Empire: Total War* (2009, Sega, Creative Assembly, PC: Windows, UK).
- *Conflict Zone: Modern War Strategy* (2001, Ubisoft, MASA Group, PC: Windows, France): So-so real-time strategy game, but the need to win the media propaganda battle as well as the actual conflict made it stand out.
- *Pikmin* (2001, Nintendo, Shigeru Miyamoto & Masamichi Abe, Gamecube, Japan): Shigeru Miyamoto's bonsai real-time strategy game.
- *The Settlers IV* (2001, Blue Byte, PC: Windows, Germany): Accessible real-time strategy game that avoids micromanagement of your troops. Also offers a conflict-free mode for those just wanting to build their kingdoms in peace.
- *Cossacks: European Wars* (2001, CDV, GSC Game World, PC: Windows, Ukraine): Intricate strategy based on Eastern European history.
- *WarCraft III: Reign of Chaos* (2002, Blizzard Entertainment, Rob Pardo, PC: Windows, USA): Excellent fantasy themed strategy action. Its mods led to the tower defense genre, where the goal is to defend your bases, be they towers or something else, from marauding hordes. *Ramparts* (1990, Atari Games, John Salwitz, Coin-op, USA) could be described as the first tower defense game but it took two free Flash games to establish it as a distinct branch of gaming, namely: *Desktop Tower Defense* (2007, Kongregate, Paul Preece, Online: Flash, UK) and *Flash Element TD* (2007, David Scott, Online: Flash, UK). Other tower defense highlights: *Plants vs. Zombies* (2009, PopCap Games, George Fan, PC: Windows, USA): Vibrant crossover hit; *Bailout Wars* (2009, Gameloft, iPhone, France): A tower defense reaction to the 2008 credit crunch. Stop the bankers from raiding the coffers of the White House.
- *Perimeter* (2004, 1C Company, K-D Lab, PC: Windows, Russia): Terraforming and unit evolution mark this out as an innovator.
- *Command & Conquer 3: Tiberium Wars* (2007, Electronic Arts, EA Los Angeles, PC: Windows, USA): Sci-fi edged strategy.

In terms of popularity turn-based strategy has lost out to its action-oriented cousin, but its more cerebral pace has still produced many great moments:
- *Nobunaga's Ambition* (1987, Koei, MSX, Japan): The start of Koei's long running historical strategy game series.
- *UFO: Enemy Unknown / X-COM: UFO Defense* (1994, Microprose, Mythos Games, PC: MS-DOS, UK): Manage the global effort to repel alien invaders from Earth in this sublime mix of battlefield tactics, research management and defensive base building. Its creator Julian Gollop has served up a succession of first-class squad-based strategy games over the years including: *Laser Squad* (1988, Target Games, Mythos Games, ZX Spectrum, UK): The 2D blueprint for *UFO: Enemy Unknown*; *X-COM: Apocalypse* (1997, Microprose, Mythos Games, PC: MS-DOS, UK): Turn-based and real-time hybrid follow-up to *UFO: Enemy Unknown*; *Rebelstar: Tactical Command* (2005, Namco, Codo Technologies, Game Boy Advance, UK): Portable tactical strategy; Also see

his earlier RPG strategy gem *Chaos* (1985, Games Workshop, Julian Gollop, ZX Spectrum, UK) and its multiplayer wizard battles.
- *Advance Wars* (2001, Nintendo, Intelligent Systems, Game Boy Advance, Japan): Expertly crafted and user-friendly game that even the strategy adverse should love.
- *Dynasty Tactics* (2002, Koei, PlayStation 2, Japan): An invigorating strategy spin-off from Koei's action-packed *Dynasty Warriors* series.
- *Hearts of Iron* (2002, Strategy First, Paradox Interactive, PC: Windows, Sweden): Complex and historically accurate strategy that is not for the faint hearted, but is rewarding for those willing to delve deep. Also see: *Europa Universalis III* (2007, Paradox Interactive, PC: Windows, Sweden).

The king of turn-based strategy games has, however, been the long-running and utterly compulsive *Civilization* series. *Sid Meier's Civilization* (1991, Microprose, Sid Meier, PC: MS-DOS, USA) marked the start but *Sid Meier's Civilization IV* (2005, 2K Games, Firaxis Games, PC: Windows, USA), which introduced religion into mix, is the one to go for. The cut-down *Civilization Revolution* (2008, 2K Games, Firaxis, Xbox 360, USA) provides a gentler introduction, but at the expense of much the series' subtle beauty.

God games
Will Wright's *Sim City* (1989, Maxis, Will Wright, Macintosh, USA) opened the floodgates, rejecting stories and goals for open-ended play. While the original's raw appeal still lingers, a more accessible introduction to the series that made town planning fun is *Sim City 4* (2003, EA Games, Maxis, PC: Windows, USA).

The society-building rivalry of *Utopia* (1982, Mattel, Don Daglow, Intellivision, USA) ventured into similar territory before *Sim City* although it pales against Wright's creation. The emergent dynamics of *Life* (1970, John Conway, PDP-7, UK) was an important influence on Wright's game, try it at: www.bitstorm.org/gameoflife

Wright's non-*Sim City* work during the 1990s was more miss than hit, but his virtual ant farm *Sim Ant* (1991, Maxis, Will Wright, PC: Windows, USA) laid the foundation for his next major creation: the voyeuristic doll's house that is *The Sims* (2000, EA Games, Maxis, PC: Windows, USA). The increased emphasis on the lives of your virtual people in *The Sims 2* (2004, EA Games, Maxis, PC: Windows, USA) only enhanced the appeal. Again others had dabbled in similar territory before:
- *Little Computer People* (1985, Activision, David Crane, Commodore 64, USA): *The Sims* 15 years too early.
- *Alter Ego: Male Version / Alter Ego: Female Version* (1986, Activision, Peter Favaro, Apple II [iPhone], USA): It may be text only but the colourful and witty writing brings it to life. The iPhone version includes both the male and female versions of this live-another-life game.

Released a few months after *Sim City* was Peter Molyneux's deity sim *Populous* (1989, Electronic Arts, Bullfrog, Amiga, UK). *Populous II: Trials of the Olympian Gods* (1991, Electronic Arts, Bullfrog, Amiga, UK) was better though. Molyneux returned to playing god with *Black & White* (2001, EA Games, Lionhead Studios, PC: Windows, UK) and *Black & White 2* (2005, Electronic Arts, Lionhead Studios, PC: Windows, UK), but with stronger ties between you and your followers.

Life & dating simulations
- *Tokmeki Memorial: Forever With You* (1995, Konami, Koji Igarashi, PlayStation, Japan): Highschool romance game that was incredibly popular in Japan.

- *Harvest Moon* (1996, Natsume, Pack-in-Video, Super NES, Japan): Live an idyllic rural existence in this farm sim. *Harvest Moon: A Wonderful Life* (2003, Natsume, Victor Interactive, Gamecube, Japan) captures the series' country living delights best.
- *Animal Crossing* (2001, Nintendo, Gamecube, Japan): Vastly entertaining life simulation where you begin a new life in a village filled with colourful characters and a seemingly never-ending supply of constant surprises. The game uses the Gamecube's internal clock to introduce seasonal surprises on days such as Halloween.

Virtual pets
- *Dogz* (1995, PF Magic, Rob Fulop, PC: Windows, USA): Loveable computer puppies. And for those who prefer cats: *Catz* (1995, PF Magic, Rob Fulop, PC: Windows, USA). *Nintendogs* (2005, Nintendo, Kiyoshi Mizuki, Nintendo DS, Japan) took the dog rearing to the next level.
- *Tamagotchi* (1996, Bandai, Aki Maita, Electronic toy, Japan): The egg-shaped toy captivated millions and stretched the patience of everyone else with its constant demands for attention.
- *Seaman* (1999, Sega, Vivarium, Dreamcast, Japan): Look after and chat with a frankly freaky fish-man-parasite thing.
- *Alien Fish Exchange* (2001, nGame, mobile phones, UK): Fish breeding and sharing (and if you are feeling particularly mean, cooking) that foreshadowed many of the ideas that would later become the basis of social networking games.

Social networking
The explosive growth of social networks after 2005 has been accompanied by a rush of games tied to websites such as Facebook and MySpace that revolve around co-operation between friends.
- *Habbo* (2000, Sulake Corporation, Sampo Karjalainen & Aapo Kyrölä, Finland): A social network for teenagers presented as a hotel, which has a video game look as well as games integrated into it.
- *Lexulous* (2007, Lexulous, Rajat Agarwalla & Jayant Agarwalla, Facebook, India): Essentially a remake of the board game *Scrabble*. Originally called *Scrabulous* until Hasbro's lawyers objected.
- *Parking Wars* (2007, A&E Television Network, area/code, Facebook, USA): Park on other people's streets and fine those who park on yours. Like the moment in *Monopoly* where you hope your opponent won't notice you landed on their property but on a global scale.
- *Restaurant City* (2007, Playfish, Facebook, UK): Build and manage restaurants for people to visit and enlist your friends as employees.
- *Farmville* (2009, Zynga, Facebook, USA): Insanely popular crop growing.
- *Mob Wars* (2009, PsychoMonkey, David Maestri, Facebook, USA): Team up with your friends to build a mafia empire by beating up rival players.

Indie & Doujin-soft
Both the western indie game scene and Japan's equivalent doujin-soft movement are treasure troves of surprising, confusing, delightful and appalling games that are growing in number at a bewildering rate. This list can only scratch the surface:
- *Pencil Whipped* (2000, ChiselBrain, Lorne Flickinger, PC: Windows, USA): First-person shooter set in a pencil sketch world that looks like a disturbed child's nightmare.
- *Uplink* (2001, Introversion Software, Chris Delay, PC: Windows, UK): Knife-edge hacking game. Introversion went on to make: *Darwinia* (2005, Introversion Software, Chris Delay, PC: Windows, UK): Real-time strategy reinvention; *Defcon* (2006, Introversion Software, Chris Delay, PC: Windows, UK): An unsettling game of global thermonuclear war.
- *Cave Story* (2004, Studio Pixel, Daisuke Amaya, PC: Windows, Japan): A glorious Japanese indie platformer with a RPG heart.

- *Torus Trooper* (2004, ABA Games, Kenta Cho, PC: Windows, Japan): Gut-wrenchingly fast fusion of *Wipeout* and *Tempest*.
- *Tumiki Fighters* (2004, ABA Games, Kenta Cho, PC: Windows, Japan): Fly a toy plane through a chunky polygon world and use the debris of those you destroy as a shield. Formed the basis for *Blast Works: Build, Trade, Destroy* (2008, Majesco, Budcat Creations, Wii, USA)
- *Dan! Da! Dan!* (2005, Omega, PC: Windows, Japan): Vertically scrolling shooter where you blast your way through pastel-coloured blocks to avoid crashing.
- *I'm O.K. – A Murder Simulator* (2005, Thompsonsoft, PC: Windows, USA): Side-scrolling shooter retort to the bombastic anti-video game campaigner Jack Thompson, who said he'll donate $10,000 to charity if *Grand Theft Auto* publisher Take-Two released a game about a father of a child killed by a computer gamer who takes revenge by murdering people in the industry.
- *Passage* (2007, Jason Rohrer, PC: Windows, USA): A five-minute slice of video game social commentary exploring life, death and marriage.
- *AaaaaAAaaaAAAaaAAAAaAAAAA!!!: A Reckless Disregard for Gravity* (2009, Dejobaan Games, PC: Windows, USA): A game about hurtling to earth after leaping off a skyscraper. An exhilarating rush of out-of-control panic.
- *Blueberry Garden* (2009, Erik Svedäng, PC: Windows, Sweden): Eccentric platform game.
- *Bonsai Barber* (2009, Zoonami, Wii, UK): A quirky game where you trim the foliage of vegetable customers at your barber's shop.
- *Canabalt* (2009, Adam Atomic, Daniel Baranowsky, Online: Flash, USA): Many indie games champion the ideal of games controlled by one button alone. This is one of the best – a panicked and frantic race over rooftops to escape the monsters that are destroying the city.
- *Everyday the Same Dream* (2009, Molleindustria, Paolo Pedercini, Online: Flash, Italy): Stylish agit-prop with an addictive soundtrack.
- *Machinarium* (2009, Amanita Design, Jakub Dvorsky, PC: Windows, Czech Republic): Beautifully illustrated indie adventure game.
- *Zombie Pub Crawl* (2009, Orange Crane Games, iPhone, USA): Marauding zombies are getting in the way of you and a pint of lager. Time to get out the gun.
- *Enviro-Bear 2010: Operation: Hibernation* (2010, Justin Smith, iPhone, Canada): Bear tries to drive car. Cue clumsy driving chaos.

Miscellaneous

The square pegs that won't fit the round holes:
- *Scram* (1980, Atari Program Exchange, Chris Crawford, Atari 800, USA): Nuclear power plant management.
- *Frogger* (1981, Sega, Konami, Coin-op, Japan): Traffic-dodging gem that defies categorisation.
- *Midwinter* (1989, Rainbird, Maelstrom Games, Atari ST, UK): Organise a guerrilla war to overthrow an evil dictator. Its giant polygon 3D island was a technical marvel at the time although the game's an unforgiving experience. A few years later the excellent *Hunter* (1991, Activision, Paul Holmes, Amiga, UK) offered a similar experience – minus the character recruitment elements – across a large 3D archipelago.
- *North & South* (1989, Infogrames, Stéphane Baudet, Atari ST, France): Zany strategy board game with excursions into real-time strategy and platforming.
- *Storm Master* (1991, Silmarils, André Rocques & Louis-Marie Rocques, Atari ST, France): Design strange flying machines to wage war on the enemy in action sequences.
- *Transarctica* (1993, Silmarils, André Rocques, Amiga, France): Giant war trains battle across the frozen wastes in this unusual fusion of action, strategy and adventure.
- *Chop Suey* (1995, 20th Century Fox, Magnet Interactive Studio, Theresa Duncan & Monica Gesue, PC: Windows, USA): Two girls go on a giddy trip through a Midwest town. One of the

opening shots of the games for girls movement of the late 1990s.

- *Barbie Fashion Designer* (1996, Mattel Media, PC: Windows, USA): Design clothes for Barbie dolls and print them out on special fabric printer paper. The first game aimed at girls to become a big success and ahead of the curve of user-generated content to boot. Create-and-print play was also explored by another girl game: *Anime Land* (1995, Casio, Casio Loopy, Japan): Create and decorate manga images and then turn them into stickers by printing them out on the Loopy's thermal printer.

- *Snake* (1997, Nokia, Nokia 6610, Finland): The first mobile phone game to grab people's attention. Eat dots to grow your tail while avoiding the walls or your tail. Its roots can be found in *Blockade* (1976, Gremlin, Coin-op, USA).

- *Power Shovel / Power Diggerz* (1999, Taito, Coin-op, Japan): Very silly fun with diggers.

- *Magic Pengel: The Quest for Colour* (2003, Taito, Garakuda-Studio, PlayStation 2, Japan): *Pokémon*-style creature battling where you draw beasts that the game then brings to life.

- *Segagaga* (2001, Sega, Tetsu Okano, Dreamcast, Japan): A tongue-in-cheek state of the industry address.

- *Jet Set Radio Future* (2002, Sega, Smilebit, Xbox, Japan): Icily cool inline skating graffiti 'em up where you race around a city-turned-playground while decorating buildings with your spray can.

- *Wario Ware Inc.: Mega Microgame$! / Wario Ware Inc.: Minigame Mania* (2003, Nintendo, Game Boy Advance, Japan): Rapid fire bursts of bright, simple and bizarre mini-games that veer from the nostalgic to the wildly inventive.

- *Katamari Damacy* (2004, Namco, Keita Takahashi, PlayStation 2, Japan): Roll a sticky ball around a crude-but-charismatic 3D world until you become a messy mass of people, animals, pianos and other everyday objects. Silly but fun. Keita Takahashi followed it up with the how-far-can-you-stretch madness of *Noby Noby Boy* (2009, Namco Bandai, Keita Takahashi, PlayStation 3, Japan).

- *Cooking Mama* (2006, Taito, Cooking Mama, Nintendo DS, Japan): Cookery themed mini-games. Prompted a protest game from People for the Ethical Treatment of Animals, who objected to the meat-based recipes: *Cooking Mama – Mama Kills Animals* (2008, People for the Ethical Treatment of Animals, Online: Flash, USA)

- *The Simpsons Game* (2007, Electronic Arts, EA Redwood Shores, PlayStation 3, USA): The game's simple beat 'em up action is nothing special but its wicked lampooning of the video game industry and game design conventions demands attention, not least Will Wright's power-crazed cameo.

- *Fl0w* (2007, Sony Computer Entertainment, Thatgamecompany, PlayStation 3, USA): Relaxing zooplankton adventures. Drift through the ocean nibbling on other wee beasties to evolve.

- *Flower* (2009, Sony Computer Entertainment, Thatgamecompany, PlayStation 3, USA): A dreamy, beautiful, enchanting and, above all, joyous game where you control the wind to create streams of swirling petals that cavort around the country fields bringing them back to life.

HARDWARE GLOSSARY

A brief overview of the computers and consoles mentioned in this book. If you want to learn more the Old Computers website is an excellent starting point: www.old-computers.com

#123

3DO Interactive Multiplayer
Manufacturers: Creative Labs, Goldstar, Panasonic, Sanyo
Year released: 1993
Origin: USA
Type: Home console

Electronic Arts founder Trip Hawkins' bid to create a common standard for video games. Several companies produced versions of the 3DO, starting with the Panasonic FZ-1. Creative Labs' Creative 3DO Blaster was an expansion card that turned PCs into 3DO systems.

A

Acorn Archimedes
Manufacturer: Acorn Computers
Year released: 1987
Origin: UK
Type: Personal computer

The first computer to use Acorn's ARM microprocessor technology.

Altair 8800
Manufacturer: MITS
Year released: 1975
Origin: USA
Type: Kit computer

One of the first computers available for home use. Inspired the formation of Microsoft.

Amiga
Manufacturers: Commodore International, Amiga Technologies
Year released: 1985

Origin: USA
Type: Personal computer

After Commodore went bust in 1994, a German company called Amiga Technologies bought the rights and continued making the Amiga line of computers until it too shut down in 1997. The Amiga now lives on as the AmigaOS operating system.

Amstrad CPC
Manufacturer: Amstrad
Year released: 1984
Origin: UK
Type: Personal computer

A popular competitor to the Commodore 64 and ZX Spectrum in Europe. Especially big in France.

Amstrad GX4000
Manufacturer: Amstrad
Year released: 1990
Origin: UK
Type: Home console

Amstrad's bid to create a European rival to the NES. Based on the technology of the Amstrad CPC line of computers.

Apple I
Manufacturer: Apple Computer
Year released: 1976
Origin: USA
Type: Personal computer

One of the first fully assembled home computers. Only around 200 were made.

Apple II
Manufacturer: Apple Computer
Year released: 1977
Origin: USA
Type: Personal computer

Part of the first wave of mass-produced home computers.

Astrocade
See Bally Professional Arcade

Atari 400 / 800
Manufacturer: Atari Inc.
Year released: 1979
Origin: USA
Type: Personal computer

Originally the 400 had 4Kb of RAM and the 800 had 8Kb. Later this was upped to 48Kb.

Atari 5200 SuperSystem
Manufacturer: Atari Inc.
Year released: 1982
Origin: USA
Type: Home console

Based on the hardware of the Atari 400 and 800 computers.

Atari 7800 ProSystem
Manufacturer: Atari Inc., Atari Corporation
Year released: 1984 and 1986
Origin: USA
Type: Home console

Originally released in 1984 but when Jack Tramiel bought Atari's consumer division soon after it was pulled from sale. Re-released in 1986 to compete with the NES.

Atari Jaguar
Manufacturer: Atari Corporation
Year released: 1993
Origin: UK
Type: Home console

Designed by Flare II, a company founded by a team that had previously worked for Sinclair Research and on the unreleased Konix Multisystem.

Atari Lynx
Manufacturer: Atari Corporation
Year released: 1989
Origin: USA
Type: Handheld console

Full colour handheld rival to the Game Boy. Designed to cater for left-handed players too. Created by some of the same team that developed the Amiga computer.

Atari Pong
See Sears Tele-Games Pong

Atari ST
Manufacturer: Atari Corporation
Year released: 1985
Origin: USA
Type: Personal computer

The first home computer to support the MIDI electronic music standard and became widely used by professional electronic music acts such as 808 State. The Amiga's main rival in Europe.

Atari VCS 2600
Manufacturer: Atari Inc.
Year released: 1977
Origin: USA
Type: Home console

The system that popularised the cartridge-based console. Released in Japan as the Atari 2800 in 1983. *Combat*, a remake of Atari's 1974 coin-op *Tank*, was built in.

Auto Race
Manufacturer: Mattel Electronics
Year released: 1976
Origin: USA
Type: Handheld game

The first handheld video game. Used LED lights for its visuals.

B

Ball
See Game & Watch

Bally Professional Arcade
Manufacturer: Bally, Astrocade
Year released: 1978
Origin: USA
Type: Home console

Designed by Dave Nutting Associates. In 1982, Bally sold the rights to Astrocade, who relaunched it as the Astrocade.

Bandai Intellivision
See Intellivision

Bandai SuperVision 8000
See SuperVision 8000

Bandai TV Jack 1000
Manufacturer: Bandai
Year released: 1977
Origin: Japan
Type: Home console

Part of Japan's home Pong boom of 1977. Bandai's first step into the video game business.

BBC Micro
Manufacturer: Acorn Computers
Year released: 1981
Origin: UK

Type: Personal computer

Line of home computers commissioned by the BBC.

Brown Box
See Magnavox Odyssey

Casio Loopy
Manufacturer: Casio
Year released: 1995
Origin: Japan
Type: Home console

Short-lived console aimed at female players. Included a built-in thermal printer that allowed users to print colour stickers. It could also capture images from videocassette recorders.

C

Cassette Vision
Manufacturer: Epoch
Year released: 1981
Origin: Japan
Type: Home console

Until Nintendo released its Famicom, this was Japan's leading console.

CD32
Manufacturer: Commodore
Year released: 1993
Origin: USA
Type: Home console

A CD-based console reincarnation of Commodore's Amiga computer.

CD-i
Manufacturer: Magnavox, Philips
Year released: 1991
Origin: Netherlands
Type: Home console

Released under the Magnavox brand in North America.

CDTV
Manufacturer: Commodore International
Year released: 1991
Origin: USA
Type: Home console/Personal computer

Sold as a console but could be turned into an Amiga computer.

ColecoVision

Manufacturer: Coleco
Year released: 1982
Origin: USA
Type: Home console

Its initial success was cut brutally short by the US video game crash.

Color TV Game 6

Manufacturer: Nintendo
Year released: 1977
Origin: Japan
Type: Home console

Nintendo's first games console containing six *Pong*-style games.

Color TV Game 15

Manufacturer: Nintendo
Year released: 1978
Origin: Japan
Type: Home console

Yet another *Pong* console with 15 variations on offer.

Color TV Game Block Breaker

Manufacturer: Nintendo
Year released: 1979
Origin: Japan
Type: Home console

Home version of Nintendo's 1978 coin-op game *Block Breaker*.

Color TV Game Racing 112

Manufacturer: Nintendo
Year released: 1978
Origin: Japan
Type: Home console

Console offering a selection of driving games. The console's casing was designed by Shigeru Miyamoto.

Commodore 64

Manufacturer: Commodore International
Year released: 1982
Origin: USA
Type: Personal computer

The best-selling home computer model of all-time with around 17 million sold worldwide. Included the SID sound chip that was one of the most advanced for its time.

Commodore PET
Manufacturer: Commodore International
Year released: 1977
Origin: USA
Type: Personal computer

One of the first mass-produced personal computers. PET stood for Personal Electronic Transactor. Came encased in 1970s futuristic plastic that fused keyboard, monitor and computer together. The PET lacked the graphical capabilities of its key rival the Apple II, being monochrome and only capable of displaying ASCII characters.

D

Data General Nova
Manufacturer: Data General
Year released: 1969
Origin: USA
Type: Minicomputer

Created by a team of former Digital Equipment Corporation staff.

DEC GT40
Manufacturer: Digital Equipment Corporation
Year released: 1972
Origin: USA
Type: Terminal

Vector graphics terminal that connected to the PDP-11 minicomputer.

Dendy
Manufacturer: Steepler Company
Year released: 1993
Origin: Russia
Type: Home console

Pirated copy of Nintendo's NES console. Became very popular in the former USSR. The games on sale were usually illegal copies of games available in Japan and the US.

Donkey Kong
See Game & Watch

Dreamcast
Manufacturer: Sega
Year released: 1998
Origin: Japan
Type: Home console

Sega's final console. Featured an in-built 56k dial-up modem. A keyboard and mouse was also available.

E

EDSAC
Manufacturer: University of Cambridge
Year released: 1949
Origin: UK
Type: Vacuum tube computer

The first computer with memory.

Elektronika 60
Manufacturer: Elektronika
Year released: Unknown, probably late 1970s
Origin: USSR
Type: Minicomputer

Soviet clone of the Digital Equipment Corporation's PDP-11 minicomputer.

ENIAC
Manufacturer: University of Pennsylvania
Year released: 1946
Origin: USA
Type: Vacuum tube computer

The first programmable multi-purpose computer.

Epoch TV Block
Manufacturer: Epoch
Year released: 1979
Origin: Japan
Type: Home console

The Japanese version of Atari's Video Pinball console

Epoch TV Tennis
Manufacturer: Epoch
Year released: 1975
Origin: Japan
Type: Home console

Japan's first games console – a version of the Magnavox Odyssey.

Exelvision EXL100
Manufacturer: Exelvision
Year released: 1984
Origin: France
Type: Personal computer

Created by a group of former Texas Instruments employees.

F

Fairchild Channel F
Manufacturer: Fairchild Semiconductor, Zircon International
Year released: 1976
Origin: USA
Type: Home console

The first console to use cartridges. Zircon International bought the rights in 1979 and launched a redesigned version, the Channel F System II.

Famicom
See NES

Famicom Disk System
Manufacturer: Nintendo, Sharp
Year released: 1986
Origin: Japan
Type: Home console add-on

A disk storage device for the Famicom, the Japanese version of the NES. Offered more memory for games than cartridges and the possibility to save games. Sharp released the Twin Famicom, an all-in-one Famicom console and Famicom Disk System, in the same year as the Disk System was released.

FM-7
Manufacturer: Fujitsu
Year released: 1982
Origin: Japan
Type: Personal computer

One of Japan's most popular computers during the 1980s.

FM Towns
Manufacturer: Fujitsu
Year released: 1989
Origin: Japan
Type: Personal computer

A multimedia-based PC for the Japanese market. A home console version of the computer, the FM Towns Marty, was launched in 1991.

Football
Manufacturer: Mattel Electronics
Year released: 1977
Origin: USA
Type: Handheld game

Portable LED game.

G

Game & Watch
Manufacturer: Nintendo
Year released: 1980
Origin: Japan
Type: Handheld game

Nintendo released 60 Game & Watch games between 1980 and 1991.

Game Boy
Manufacturer: Nintendo
Year released: 1989
Origin: Japan
Type: Handheld console

The original was monochrome. 1998's Game Boy Color added basic colour visuals. The Game Boy Advance, launched in 2001, was a full upgrade of the handheld.

Game Boy Advance
See Game Boy

Game Boy Color
See Game Boy

Game Gear
Manufacturer: Sega
Year released: 1990
Origin: Japan
Type: Handheld console

A portable Master System.

Gamecube
Manufacturer: Nintendo, Panasonic
Year released: 2001
Origin: Japan
Type: Home console

Panasonic released a DVD-playing version, the Panasonic Q, in Japan. The Wii can play Gamecube games.

Genesis
See Megadrive

H

Halcyon
Manufacturer: RDI Video Systems

Year released: 1985
Origin: USA
Type: Home console / Personal computer

The Halcyon was a Laserdisc player, games console and home computer. It cost $2,500 and just two games were released before RDI Video Systems closed down.

Hector 1
Manufacturer: Micronique
Year released: 1983
Origin: France
Type: Personal computer

A souped-up version of Micronique's 1981 computer the Victor Lambda, which in turn was based on the US-made Interact Computer that was launched in 1979.

Hitachi TV-Game VG-104
Manufacturer: Hitachi
Year released: 1977
Origin: Japan
Type: Home console

Part of Japan's 1977 home *Pong* boom. Monochrome visuals and four built-in games.

I

IBM 1130 Computing System
Manufacturer: IBM
Year released: 1965
Origin: USA
Type: Minicomputer

IBM minicomputer that made computing affordable to smaller businesses.

IBM 701
Manufacturer: IBM
Year released: 1952
Origin: USA
Type: Vacuum tube computer

One of the first computers produced in quantity rather than being custom-made.

IBM 7094
Manufacturer: IBM
Year released: 1962
Origin: USA
Type: Mainframe computer

Transistor-based update of the IBM 709 mainframe, which used vacuum tubes.

IBM PC
See PC

Imlac PDS-1
Manufacturer: Imlac Corporation
Year released: 1970
Origin: USA
Type: Minicomputer

One of the first, if not the first, computers with a graphical user interface. Used a light pen instead of a mouse to point and a foot pedal to click.

Intellivision
Manufacturer: Mattel Electronics, Bandai
Year released: 1979
Origin: USA
Type: Home console

Bandai released it in Japan. Also home to PlayCable, the world's first downloadable games service, which was the result of a partnership between General Instrument and Mattel. The service, launched in 1981, supplied games via cable TV although the games could not be stored on the Intellivision.

Interton VC-4000
Manufacturer: Interton
Year released: 1978
Origin: West Germany
Type: Home console

One of many European consoles released in the late 1970s and early 1980s that were based on the 1292 Advanced Programmable Video System, which was created by another West German firm: Radofin.

iPhone
Manufacturer: Apple Inc.
Year released: 2007
Origin: USA
Type: Smartphone

Touchscreen smartphone that became an outlet for thousands of downloadable games. Its phone-less sister product, the iPod Touch, can also play most of these games.

K

KIM-1
Manufacturer: MOS Technology
Year released: 1976
Origin: USA
Type: Kit computer

Created by microprocessor firm MOS Technology prior to its purchase by Commodore International.

Konix Multisystem
Manufacturer: Konix
Year released: Never released
Origin: UK
Type: Home console

Was to be a UK rival to the Megadrive and Super NES. Its peripherals were set to include a light gun with recoil and the Power Chair, a motorised seat designed to mimic the hydraulics used in arcade games such as *Out Run*. Also known as the Slipstream. Its release was cancelled in August 1989.

L

Lisa
Manufacturer: Apple Computer
Year released: 1983
Origin: USA
Type: Personal computer

The first personal computer with a graphical user interface.

M

Macintosh
Manufacturer: Apple Computer
Year released: 1984
Origin: USA
Type: Personal computer

The first popular personal computer with a graphical user interface.

Magnavox Odyssey
Manufacturer: Magnavox
Year released: 1972
Origin: USA
Type: Home console

The first video game console, developed by Sanders Associates as the Brown Box. Sold with plastic overlays to put over the TV screen to enhance its primitive graphics. It was also home to the first light gun controller, *Shooting Gallery*.

Magnavox Odyssey^2
Manufacturer: Magnavox, Philips
Year released: 1978
Origin: USA
Type: Home console

Released in Europe as the Philips Videopac G7000 (Philips C52 in France). Also released in Brazil as the Philips Odyssey.

Magnavox Odyssey3

Manufacturer: Magnavox, Philips
Year released: 1983
Origin: USA
Type: Home console

Never released in North America but did get a brief release in Europe as the Philips Videopac G7400.

Magnavox Odyssey 100

Manufacturer: Magnavox
Year released: 1975
Origin: USA
Type: Home console

The first of several home *Pong* consoles released by Magnavox.

Master System

Manufacturer: Sega
Year released: 1985
Origin: Japan
Type: Home console

Outsold the NES in Europe. Also popular in Brazil, where it was distributed by Tec Toy. Originally released in Japan as the Sega Mark III.

MBX Expansion System

Manufacturer: Milton Bradley
Year released: 1983
Origin: USA
Type: Personal computer add-on

Originally planned as a stand-alone console, Milton Bradley eventually turned the MBX into an add-on for the TI-99/4a home computer. Featured speech synthesis, voice recognition and an analogue joystick that also allowed 360° rotation. Discontinued almost as soon as released after Texas Instruments quit the home computer market. Versions for the Atari VCS 2600 and Atari 5200 consoles were also mooted.

Megadrive

Manufacturer: Sega
Year released: 1988
Origin: Japan
Type: Home console

Called the Genesis in North America. The Sega CD add-on, also known as the Mega CD, allowed users to play CD-ROM games.

Merlin

Manufacturer: Parker Brothers
Year released: 1978
Origin: USA
Type: Handheld game

Multi-game handheld that doubled as a musical instrument. Created by former NASA scientist Bob Doyle.

MicroVision

Manufacturer: Milton Bradley
Year released: 1979
Origin: USA
Type: Handheld console

The first handheld console. Created by Jay Smith who later designed the Vectrex console.

MK14

Manufacturer: Science of Cambridge
Year released: 1977
Origin: UK
Type: Kit computer

Clive Sinclair's first venture into the computer business.

MSX / MSX2

Manufacturer: Various (designed by ASCII Corporation and Microsoft Japan)
Year released: 1983
Origin: Japan
Type: Personal computer

Touted as a home computer standard and produced by a bewildering number of manufacturers across the world including Casio, Daewoo, GoldStar, Hitachi, Panasonic, Philips, Sony, Spectravideo and Yamaha (to name just a few). The MSX2, the second generation version of the computer, was introduced in 1986. Eventually the line was discontinued in 1995.

N

Nascom 1

Manufacturers: Nascom Microcomputers
Year released: 1977
Origin: UK
Type: Kit computer

An early British computer.

NEC PC-6001

Manufacturer: NEC
Year released: 1981

Origin: Japan
Type: Personal computer

Released in North America as the NEC TREK.

NEC PC-8001
Manufacturer: NEC
Year released: 1979
Origin: Japan
Type: Personal computer

One of the earliest home computers made in Japan.

NEC PC-8801
Manufacturer: NEC
Year released: 1981
Origin: Japan
Type: Personal computer

Popular Japanese computer throughout the 1980s, also known as the PC88.

NEC PC-9801 / NEC PC-9821
Manufacturer: NEC
Year released: 1982
Origin: Japan
Type: Personal computer

Japanese rival to the PC that was popular well into the late 1990s before losing out to the PC. The PC-9821Ra43 model, the last in the line, was released in 2000.

NEMO
Manufacturer: Hasbro
Year released: Never released
Origin: USA
Type: Home console

Used VHS video cassettes rather than cartridges for its games. NEMO (Never Ever Mention Outside) was the console's working name. Hasbro abandoned the January 1989 launch of the console as the Control-Vision in late 1988.

Neo Geo
Manufacturer: SNK
Year released: 1990
Origin: Japan
Type: Home console

Very expensive home version of SNK's coin-op video game technology. Game cartridges cost upwards of $200 at the time. The rarest now sell to collectors for more than $1,000. In 1994 SNK released a CD-ROM version of the console.

NES

Manufacturer: Nintendo, Sharp
Year released: 1983
Origin: Japan
Type: Home console

Called the Famicom in Japan. The Famicom Disk System, released in Japan in 1986, allowed owners to play and save games on floppy disks. Sharp released the Twin Famicom, a combined Famicom and Famicom Disk System, in the same year.

Nintendo 64

Manufacturer: Nintendo
Year released: 1996
Origin: Japan
Type: Home console

Introduced analogue joysticks and vibration features to console joypad controllers.

Nintendo DS

Manufacturer: Nintendo
Year released: 2004
Origin: Japan
Type: Handheld console

Dual-screen reinvention of the handheld console with an in-built microphone, wireless multi-player gaming and a stylus for interacting with its touch screen.

Nokia 6610

Manufacturer: Nokia
Year released: 2002
Origin: Finland
Type: Mobile phone

One of the first mobile phones that could connect to the internet. Nokia went on to launch the N-Gage mobile phone game console in 2003. In 2005 Nokia turned the N-Gage brand into a game download service for its smartphones.

O

Oil Panic

See Game & Watch

Oric-1

Manufacturer: Tangerine Computer Systems
Year released: 1983
Origin: UK
Type: Personal computer

Successful in France but not in the UK. Quickly succeeded by the Oric Atmos in 1984.

P

Panasonic FZ-1
See 3DO Interactive Multiplayer

PC
Manufacturer: Various
Year released: 1981
Origin: USA
Type: Personal computer

Created by IBM using off-the-shelf technology, which allowed other companies to produce copies without the fear of legal action. By the end of the 1980s IBM PC compatibles made by rival firms were outselling IBM's own PCs. Microsoft's MS-DOS and Windows operating systems became the common software standard for PCs and, after initial success in the business market, the PC spread into homes as people adopted multimedia and, later, internet technology.

PC Engine
Manufacturer: NEC
Year released: 1987
Origin: Japan
Type: Home console

Created in collaboration with Japanese games publisher Hudson Soft. Called the TurboGrafx-16 in North America. The TurboGrafx-CD briefly became the most widespread CD drive format in Japan.

PCjr
Manufacturer: IBM
Year released: 1984
Origin: USA
Type: Personal computer

A slimmed-down version of the IBM PC targeted at the educational and home computer markets.

PDP-1
Manufacturer: Digital Equipment Corporation
Year released: 1960
Origin: USA
Type: Minicomputer

The first of DEC's long-running PDP series of minicomputers. The birthplace of *Spacewar!*.

PDP-6
Manufacturer: Digital Equipment Corporation
Year released: 1963
Origin: USA

Type: Minicomputer

Just 26 were sold worldwide.

PDP-7
Manufacturer: Digital Equipment Corporation
Year released: 1965
Origin: USA
Type: Minicomputer

The first version of the Unix operating system was made on this computer.

PDP-8
Manufacturer: Digital Equipment Corporation
Year released: 1965
Origin: USA
Type: Minicomputer

With its $18,000 price tag it was the cheapest of the PDP minicomputers.

PDP-10
Manufacturer: Digital Equipment Corporation
Year released: 1967
Origin: USA
Type: Minicomputer

Its success with universities in the 1970s made it a hotbed of early computer games.

PDP-11
Manufacturer: Digital Equipment Corporation
Year released: 1970
Origin: USA
Type: Minicomputer

New models were still being made as late as 1990.

Ping-O-Tronic
Manufacturer: Zanussi
Year released: 1974
Origin: Italy
Type: Home console

Based on similar technology to the Magnavox Odyssey.

PLATO
Manufacturer: n/a
Year released: 1960
Origin: USA
Type: Computer network

A computer network designed to serve terminals in schools and created by the University of Illinois. PLATO stands for Programmed Logic for Automated Teaching Operations. Most of the groundbreaking games designed for it ran on PLATO IV terminals, which were introduced in 1972.

PlayStation
Manufacturer: Sony Computer Entertainment
Year released: 1994
Origin: Japan
Type: Home console

Born out of Sony and Nintendo's falling out over the creation of a CD version of the Super NES. In 1997 Sony launched Net Yaroze, a development kit aimed at hobby programmers who wanted to write PlayStation games.

PlayStation 2
Manufacturer: Sony Computer Entertainment
Year released: 2000
Origin: Japan
Type: Home console

Its built-in DVD player helped encouraged the shift from VHS cassettes to DVDs. Also plays PlayStation games.

PlayStation 3
Manufacturer: Sony Computer Entertainment
Year released: 2006
Origin: Japan
Type: Home console

Includes an in-built Blu-Ray player. Early models could play PlayStation 2 and PlayStation games. Later ones only played PlayStation games.

PSP
Manufacturer: Sony Computer Entertainment
Year released: 2004
Origin: Japan
Type: Handheld console

Short for the PlayStation Portable. Used Sony's Universal Media Disc (UMD) storage format.

R

RCA Studio II
Manufacturer: RCA
Year released: 1977
Origin: USA
Type: Home console

Black and white competitor to the Fairchild Channel F.

S

Saturn
Manufacturer: Sega
Year released: 1994
Origin: Japan
Type: Home console

Sega's most successful console in Japan.

Sears Tele-Games Pong
Manufacturer: Atari
Year released: 1975
Origin: USA
Type: Home console

Atari's first home version of *Pong*. Atari released its own version, Atari Pong, in early 1976.

Sega CD
See Megadrive

Sharp MZ-80K
Manufacturer: Sharp
Year released: 1978
Origin: Japan
Type: Personal computer

Early home computer with monochrome visuals.

Simon
Manufacturer: Milton Bradley
Year released: 1978
Origin: USA
Type: Handheld game

With its flying saucer design and its *Close Encounters of the Third Kind* musical motif, Simon became a pop culture icon. Given the status it went on to achieve, it was appropriately launched at New York City's legendary Studio 54 nightclub.

Sinclair QL
Manufacturer: Sinclair Research
Year released: 1984
Origin: UK
Type: Personal computer

Sinclair Research's successor to the ZX Spectrum, marketed as a computer for small business-es rather than home users. Discontinued in 1986 after Amstrad took over Sinclair Research.

Slipstream
See Konix Multisystem

Speak & Math
Manufacturer: Texas Instruments
Year released: 1980
Origin: USA
Type: Educational toy

Used in the Pet Shop Boys' song *Two Divided by Zero*.

Speak & Read
Manufacturer: Texas Instruments
Year released: 1980
Origin: USA
Type: Educational toy

One of Texas Instruments' three speech synthesis-enhanced portable educational aids.

Speak & Spell
Manufacturer: Texas Instruments
Year released: 1978
Origin: USA
Type: Educational toy

Its speech synthesis was the height of technology at the time and its educational benefits helped it sell by the truckload.

Super Cassette Vision
Manufacturer: Epoch
Year released: 1984
Origin: Japan
Type: Home console

Epoch's last home console. Released in response to the Famicom.

Super NES
Manufacturer: Nintendo
Year released: 1990
Origin: Japan
Type: Home console

In 1995, Nintendo launched a Japan-only satellite modem add-on for the Super NES called Satellaview. Users could download games and online magazines through the service at a set hour of the day when the satellite service, run by St.GIGA, broadcast the data.

SuperVision 8000
Manufacturer: Bandai
Year released: 1979

Origin: Japan
Type: Home console

The first Japanese-designed home console to use game cartridges.

T

Tamagotchi
Manufacturer: Bandai
Year released: 1996
Origin: Japan
Type: Handheld game

Portable virtual pet. More than 40 versions have been released and tens of millions sold across the world.

Telstar
Manufacturer: Coleco
Year released: 1976
Origin: USA
Type: Home console

The first console to use General Instruments' AY-3-8500 chip.

Thomson MO5
Manufacturer: Thomson
Year released: 1984
Origin: France
Type: Personal computer

Designed to compete with the ZX Spectrum and the Commodore 64. Widely used in French schools.

Thomson TO7
Manufacturer: Thomson
Year released: 1982
Origin: France
Type: Personal computer

Came with a built-in light pen.

TI-99/4
Manufacturer: Texas Instruments
Year released: 1979
Origin: USA
Type: Personal computer

Sold with a 13-inch colour Zenith monitor as TI lacked an approved modulator for connecting to televisions.

TI-99/4A
Manufacturer: Texas Instruments
Year released: 1981
Origin: USA
Type: Personal computer

TI's mass market computer that helped spark the home computer price war of the early 1980s, which in turn helped cause the video game console crash.

TK-80
Manufacturer: NEC
Year released: 1976
Origin: Japan
Type: Kit computer

NEC's first venture into home computing.

TR-DOS
See ZX Spectrum

TRS-80
Manufacturer: Tandy / Radio Shack
Year released: 1977
Origin: USA
Type: Personal computer

One of the first mass-produced home computers.

TX-0
Manufacturer: Massachusetts Institute of Technology (MIT)
Year released: 1955
Origin: USA
Type: Minicomputer

Experimental computer built at MIT's Lincoln Laboratory.

TurboGrafx-16
See PC Engine

TurboGrafx-CD
See PC Engine

U

UFO A500 II
Manufacturer: Selection
Year released: Unknown (late 1980s or early 1990s)
Origin: China
Type: Home console

Clone of Nintendo's Famicom. This and other clones were widespread in mainland Asia and Russia at the time as official home consoles were either unavailable or too expensive for most people to buy.

Unix
Manufacturer: n/a
Year released: 1969
Origin: USA
Type: Operating system

A widely used operating system. Forms the basis of the Macintosh's MacOS and Linux.

V

VAX-11/780
Manufacturer: Digital Equipment Corporation
Year released: 1977
Origin: USA
Type: Minicomputer

Could emulate the PDP-11.

Vectrex
Manufacturers: General Consumer Electric, Milton Bradley and Bandai
Year released: 1982
Origin: USA
Type: Home console

Vector graphics console. Created by Smith Engineering but manufactured by General Consumer Electric and later Milton Bradley. Bandai released it in Japan.

VIC-20
Manufacturer: Commodore International
Year released: 1980
Origin: USA
Type: Personal computer

Commodore's first mass-market home computer.

Video Pinball
Manufacturer: Atari
Year released: 1977
Origin: USA
Type: Home console

Included a version of Atari's coin-op hit *Breakout*.

Videomaster Home T.V.
Manufacturer: The Sales Team

Year released: 1974
Origin: UK
Type: Home console

Another of the European-made analogue circuit *Pong* consoles that were based on the Magnavox Odyssey and predated Atari's home *Pong* game.

Videopac G7000
See Magnavox Odyssey[2]

Video Sport MK2
Manufacturer: Henry's
Year released: 1974
Origin: UK
Type: Home console

Henry's were a British hi-fi and TV retailer. The console used circuits similar to those in the Magnavox Odyssey. Wood casing look.

W

Wii
Manufacturer: Nintendo
Year released: 2006
Origin: Japan
Type: Home console

Its motion-based Wiimote controllers gave it broad appeal. Also runs Gamecube games. By the start of 2010 more than 60 million had been sold worldwide.

X

Xbox
Manufacturer: Microsoft
Year released: 2001
Origin: USA
Type: Home console

The first American-made games console to sell in large quantities since the Atari VCS 2600 and the first console to include a hard drive.

Xbox 360
Manufacturer: Microsoft
Year released: 2005
Origin: USA
Type: Home console

Microsoft's second games console. Placed a strong emphasis on its online multiplayer games service Xbox Live.

Z

ZX80

Manufacturer: Science of Cambridge
Year released: 1980
Origin: UK
Type: Personal computer

Clive Sinclair's super-cheap home computer. Inspired Commodore's VIC-20.

ZX81

Manufacturer: Sinclair Research
Year released: 1981
Origin: UK
Type: Personal computer

Released in North America as the Timex Sinclair 1000.

ZX Spectrum

Manufacturer: Sinclair Research
Year released: 1982
Origin: UK
Type: Personal computer

The UK's leading home computer during the mid-1980s. Also became a popular computer format in Eastern Europe and Russia during the 1990s thanks to low-cost versions of the computer that ran TR-DOS, an operating system for the Spectrum first developed by the British company Technology Research in 1985.

REFERENCES

1 | HEY! LET'S PLAY GAMES!

Allan, Roy A. (2001) *A History of the Personal Computer: The People and the Technology*. London, Ontario, Canada: Allan Publishing
A history of early computers

Baer, Ralph H. (2005) *Videogames: In the Beginning*. Springfield, New Jersey: Rolenta Press
Ralph Baer's meticulously detailed account of the creation of the Brown Box

Bennett, J.M. (1994) 'Autobiographical snippets'. In: Bennett, J.M. et al (1994) *Computing in Australia – The Development of a Profession*. Sydney, Australia: Hale and Iremonger. p55 – as cited at www.goodeveca.net/nimrod/bennett.html [Last accessed: 2 March 2010]
The reaction to and creation of the Nimrod

Bernstein, Alex and de V. Roberts, Michael (1958) 'Computer v. chess-player'. *Scientific American*, June 1958
Article on the value of computer chess research and figure on the time it would take a computer to calculate all possible chess moves

Computer History Museum (no date) *Mastering the Game: A History of Computer Chess*. [Online] www.computerhistory.org/chess [Last accessed: 27 February 2010]
The history of computer chess

Crowley, David and Pavitt, Jane (editors) (2008) *Cold War Modern: Design 1945-1970*. London, UK: V&A Publishing
Background on Cold War, 2001: A Space Odyssey and early computers

Ferguson, Niall (2006) *The War of the World: History's Age of Hatred*. London, UK: Allen Lane
Second World War and Cold War history

Gaddis, John Lewis (2005) *The Cold War*. 2007 edition. London, UK: Penguin Books
Cold War history

Goldsmith Jr., Thomas T. and Mann, Estle Ray (1948) *Cathode-Ray Tube Amusement Device*. [Patent] US Patent No. 2,455,992
Details of the Cathode-Ray Tube Amusement Device

Jansen, Marius B. (2000) *The Making of Modern Japan*. Cambridge, Massachusetts: Belknap Press, Harvard University Press
Japan's surrender in the Second World War

Kotok, Alan (1962) 'A Chess playing program for the IBM 7090 computer'. BSc paper, Massachusetts Institute of Technology
Background on Alan Kotok's computer work

Levy, Steven (1984) *Hackers: Heroes of the computer revolution*. 2001 edition. London, UK: Penguin Books
Spacewar's creation

Plant, Sadie (1997) *Zeros + Ones*. 1998 paperback edition. London, UK: Fourth Estate
Feminist re-examination of computer history with background on Alan Turing and early computing

Podfather (2009) [TV broadcast] London: BBC Four. 12 October 2009. GMT: 21.00

History of the development of integrated circuits

Polkinghorn, Frank (1973) *Thomas Goldsmith, Electrical Engineer, an Oral History*. New Brunswick, New Jersey: IEEE History Center [Online] www.ieee.org/portal/cms_docs_iportals/iportals/aboutus/history_center/oral_history/pdfs/Goldsmith008. pdf [Last accessed: 14 September 2008]
Background on Thomas Goldsmith's work

Shannon, Claude E. (1950) 'Programming a computer for playing chess'. *Philosophical Magazine*, 41 (7), No. 314, March 1950
Explains why computer chess is useful for research

'The Making of Spacewar!' (2003) *Edge Presents Retro: The Making of Special*. Bath, UK: Future Publishing. pp6-9
Spacewar!'s creation and source of quotes from Steve Russell and Martin Graetz, used with the kind permission of Future Publishing

2 I AVOID MISSING BALL FOR HIGH SCORE

Arcade History (no date) [Online] www.arcade-history.com [Last accessed: 2 March 2010]
Information on coin-operated games

Atomic Planet Entertainment (2005) *Taito Legends* [PlayStation 2] Empire Interactive
Information on Taito's early years

Baer, Ralph H. (2005) *Videogames: In the Beginning*. Springfield, New Jersey: Rolenta Press
Detailed information on the launch of the Odyssey and the creation of Pong

Barth, Linda (2004) 'Pinball: It took most of a century to evolve, and its best days may be past'. *American Heritage*. 20 April 2004. [Online] www.americanheritage.com/articles/magazine/it/2004/4/2004_4_20.shtml [Last accessed: 2 April 2010]
Details about New York City's pinball ban

Brand, Stewart (1972) 'Spacewar: Fanatic life and symbolic death among the computer bums'. *Rolling Stone*. 7 December 1972. [Online] www.wheels.org/spacewar/stone/rolling_stone.html [Last accessed: 2 March 2010]
Spacewar and early computer game culture

Bushnell, Nolan K. (1972) *Computer Space Instructions*. Mountain View, California: Nutting Associates Inc.
How Computer Space was presented to arcade operators

Digital Eclipse (2004) *Atari Anthology* [PlayStation 2] Atari Interactive
Interview with Nolan Bushnell

Kent, Steven L. (2001) *The Ultimate History of Video Games*. New York, New York: Three Rivers Press
History of Atari, Pong and New York City's pinball ban

'Oldies but Goodies: Celebrate Atari's 6th Birthday' (1978) *St. Pong*, Volume 4, July/August 1978. Sunnyvale, California: Atari Inc.
Details of when staff joined Atari Inc.

Penner, Jeremy (2005) 'Important failures in videogame history'. *The Gamer's Quarter*. Issue 2, 3rd quarter 2005. St Louis, Missouri: The Gamer's Quarter. pp56-57
Timings of Galaxy Game and Computer Space releases

Range, Peter Ross (1974) 'Space age pinball machines'. *The Ledger*, Lakeland, Florida. Sunday 15 September 1974. Section D, p7
Overview of the video game's impact on the arcade industry and the Odyssey

Russell, Bruce (1976) 'Penny arcade games – It's a multi-million dollar world'. *Daily News*, Kingsport, Tennessee. Wednesday 13 October 1976. p8
How video games changed attitudes to arcades

'The untold Atari story' (2009) *Edge*, Issue 200, April 2009, pp94-99
Interview with Ted Dabney, co-founder of Atari

3 | A GOOD HOME RECREATION THING

Albright Jr., Ronald G. (1985) *The Orphan Chronicles*. San Dimas, California: Millers Graphics
Texas Instruments' involvement in the microprocessor business

Amis, Martin (1982) *Invasion of the Space Invaders*. London, UK: Hutchinson & Co.
Still uses term TV Games

Arcade History (no date) [Online] www.arcade-history.com [Last accessed: 2 March 2010]
Information on coin-operated games

'Atari Corporation' (2004) *International Directory of Company Histories*, Volume 66. Farmington Hills, Michigan: St James Press.
[Online] www.fundinguniverse.com/company-histories/Atari-Corporation-Company-History.html [Last accessed: 2 March 2010]
History of Atari Inc.

Atomic Planet Entertainment (2005) *Taito Legends* [PlayStation 2] Empire Interactive
Information on Taito's early years

'Computer game tycoon runs up score of profits' (1978) *The Spokesman Review*. Spokane, Washington. Thursday 25 May 1978.
p44
Interview with Nolan Bushnell

DeMaria, Rusel and Wilson, Johnny L. (2002) *High Score! The Illustrated History of Electronic Games*. Berkeley, California: McGraw-Hill/Osborne. p17
Atari Inc's manifesto

Digital Eclipse (2004) *Atari Anthology* [PlayStation 2] Atari Interactive
Interview with Nolan Bushnell

Kent, Steven L. (2001) *The Ultimate History of Video Games*. New York, New York: Three Rivers Press
History of Atari

'King Pong explains home-game craze' (1978) *Wilmington Morning Star*. Wilmington, North Carolina. Monday 24 April 1978.
p4-D.
Interview with Nolan Bushnell

Lewis, Jesse (1977) 'Americans play space age games'. *Ocala Star-Banner*. Ocala, Florida. Thursday 28 April 1977. p8A
The impact of video games on the arcades

Old Computers (no date) [Online] http://old-computers.com [Last accessed: 2 March 2010]
Information on 1970s consoles

Provenzo Jr, Eugene (1991) *Video Kids: Making Sense of Nintendo*. Cambridge, Massachusetts: Harvard University Press. p48
The significance of video games in changing people's relationship with TV

'The untold Atari story' (2009) *Edge*, Issue 200, April 2009, pp94-99
Interview with Ted Dabney, co-founder of Atari

4 | CHEWING GUM, BAILING WIRE AND SPIT

Arcade History (no date) [Online] www.arcade-history.com [Last accessed: 2 March 2010]
Information on coin-operated games

'Coin-operated game outrages safety council' (1977) *The Blade*. Toledo, Ohio. Saturday 1 January 1977. p4
Controversy about Death Race

'Computer game for morbid fans' (1976). *The Kingman Daily Miner*. Kingman, Arizona. Friday 1 July 1976, p11
Controversy about Death Race

'Death Race popularity grows' (1977) *Spokane Daily Chronicle*. Spokane, Washington. Tuesday 9 August 1977, p6
Controversy increases Death Race sales

Greene, Bob (1976) 'Death Race: You pay quarter and try to kill pedestrian'. *The Free Lance-Star.* Fredericksburg, Virginia. Wednesday 27 October 1976, p2
Controversy about Death Race

Bally Flicker (1975) [Promotional Flyer] Bally Manufacturing Co.
Information about the Flicker pinball table

Hardwick, M. Jeffrey (2004) *Mall Maker: Victor Gruen, Architect of an American Dream.* Philadelphia, Pennsylvania: University of Pennsylvania Press
Biography of Victor Gruen, the inventor of the modern shopping mall

Nürburgring/1 (1976) [Promotional Flyer] Dr.-Ing. Reiner Foerst GmbH
Information on Nürburgring/1

Nürburgring/1 and Nürburgring/2 (1976) [Promotional flyer] Dr.-Ing. Reiner Foerst GmbH
Information on Nürburgring/1

Old Computers (no date) [Online] http://old-computers.com [Last accessed: 2 March 2010]
Information on 1970s computers and consoles

Podfather (2009) [TV broadcast] London: BBC Four, 12 October 2009. GMT: 21.00
The history of the microprocessor

'Steve Jobs profile: Apple's hard core' (2009) *Scotland on Sunday.* 11 January 2009. [Online] http://scotlandonsunday.scotsman.com/comment/Steve-Jobs-profile-Apple39s-hard.4863847.jp [Last accessed: 6 March 2010]
Profile of Steve Jobs

5 I THE BIGGEST EUREKA MOMENT EVER

Ahl, David H. (editor) (1978) *BASIC Computer Games: Microcomputer Edition.* New York, New York: Workman Publishing.
Pre-home computer games and background about their creation

Bagnall, Brian (2007) *On the Edge: The Spectacular Rise and Fall of Commodore.* Winnipeg, Manitoba, Canada: Variant Press
History of Commodore

Barton, Matt (2008) *Dungeons & Desktops.* Wellesley, Massachusetts: A K Peters
Background on pre-home computer role-playing games, including Pedit5

Briggs, Andrew (1983) 'Meet the men behind Infocom's mask'. *Micro Adventurer,* November 1983, pp10-11
Interview with Infocom

'Commodore International Ltd' (1993) *International Directory of Company Histories,* Volume 7, St. James Press, 1993. [Online] www.fundinguniverse.com/company-histories/Commodore-International-Ltd-Company-History.html [Last accessed: 11 March 2010]
History of Commodore

Computer History Museum (no date) *Mastering the Game: A History of Computer Chess* [Online] www.computerhistory.org/chess [Last accessed: 27 February 2010]
History of computer chess research

Crawford, Chris (1982) 'Eastern Front (1941)'. In: Salen, Katie and Zimmerman, Eric (editors) (2006) *The Game Design Reader: A Rules of Play Anthology.* Cambridge, Massachusetts: The MIT Press. pp714-724
Background on early computer strategy games

DigiBarn Computer Museum (2004) 'Maze War 30 Year Retrospective'. 7 November 2004. [Online] www.digibarn.com/history/04-VCF7-MazeWar/index.html [Last accessed: 6 March 2010]
History of Maze / Maze War

Dingler, David A. (2006) 'Where flight simulation really began'. *FlightSim.com.* 10 January 2006. [Online] www.flightsim.com/main/feature/link.htm [Last accessed: 6 March 2010]
History of flight simulations

'Gaming Our Way Through History: The Oregon Trail: Education, Nostalgia, and Memory' (no date) American Studies Pro-

gram, The College of William & Mary, Williamsburg, Virginia. [Online] http://web.wm.edu/amst/370/2005F/sp1/home.htm [Last accessed: 6 March 2010]
Background on The Oregon Trail game

Hafner, Katie (1994) 'Will Crowther Interview'. *The Internet Archive.* [Online] http://ia301514.us.archive.org/0/items/Will-CrowtherInterview/WC111.txt [Last accessed: 6 March 2010]
Interview with Will Crowther

Jerz, Dennis G. (2007) 'Somewhere nearby is Colossal Cave: Examining Will Crowther's original *Adventure* in code and in Kentucky'. *Digital Humanities Quarterly.* Volume 1, Number 2, Summer 2007. [Online] http://digitalhumanities.org/dhq/vol/001/2/000009/000009.html [Last accessed: 6 March 2010]
History of Will Crowther's Adventure

Library Information Network (no date) 'A bibliography of the Edwin A. Link Collection of the Evans Library at Florida Institute of Technology'. [Online] www.lib.fit.edu/pubs/linkbib [Last accessed: 6 March 2010]
Background on Edwin A. Link

Microchess (no date) [Mail order form] Toronto, Ontario, Canada: Micro-Ware Ltd.
Information of Micro-Ware

O'Neill, Judy (1990) 'An interview with William Crowther'. 12 March 1990. Charles Babbage Institute, Centre for the History of Information Processing, University of Minnesota, Minneapolis. [Online] www.cbi.umn.edu/oh/pdf.phtml?id=97 [Last accessed: 6 March 2010]
Interview with Will Crowther about his non-game work

Poundstone, William (2006) 'Game theory'. In: Salen, Katie and Zimmerman, Eric (editors) (2006) *The Game Design Reader: A Rules of Play Anthology.* Cambridge, Massachusetts: The MIT Press. pp382-408
Background on Kriegsspiel

Rheingold, Howard (1991) *Virtual Reality.* 1992 edition. London: Mandarin Paperbacks.
Background on flight simulation history

'Sierra On-Line Inc.' (1996) *International Directory of Company Histories*, Volume 15, St. James Press (1996). [Online] www.fundinguniverse.com/company-histories/Sierra-OnLine-Inc-Company-History.html [Last accessed: 11 March 2010]
History of Sierra On-Line

Smith, Steve (1994) *PC Pilot: The complete guide to computer aviation.* New York, New York: Avon Books
Background on early flight simulations

'The Joy of Computer Gaming' (1983) In: *The Best of 99'er Volume 1* (1983). Eugene, Oregon: Emerald Valley Publishing Co., pp223-225
Notes confusion about what to do with home computers

Thompson, Greg (2004) *The amazing history of Maze – it's a small world after-all.* [Presentation] Vintage Computer Festival, Mountain View, California. 7 November 2004. [Online] www.digibarn.com/collections/presentations/maze-war/The-aMazing-History-of-Maze.ppt [Last accessed: 6 March 2010]
History of Maze / Maze War

Wells, H.G. (1913) *Little Wars.* 2004 edition. Whitefish, Montana: Kessinger Publishing
The sci-fi author's Kriegsspiel rule book

Wozniak, Steve with Smith, Gina (2006) *iWoz.* 2007 paperback edition. London, UK: Headline Review
Steve Wozniak's account of Apple's formation and the creation of the Apple II computer

6 | HIGH-STRUNG PRIMA DONNAS

Amis, Martin (1982) *Invasion of the Space Invaders.* London, UK: Hutchinson & Co.
The success of Space Invaders. And yes, it is that Martin Amis

'Ask Hal: Frequently asked questions to the Blue Sky Rangers' (no date) *IntellivisionLives.com* [Online] www.intellivisionlives.com/bluesky/people/askhal/askhal.html#A1 [Last accessed: 6 March 2010]
Details on Mattel's move into home consoles

Atlas, Terry (1978) 'Simon toy an elusive bestseller'. *Evening Independent*. St Petersburg, Florida. Thursday 21 December 1978. p5-C
Simon sells out

Baer, Ralph H. (2005) *Videogames: In the Beginning*. Springfield, New Jersey: Rolenta Press
The creation of Simon

Bloom, Steve (1983) 'Parker Brothers strikes back'. *Video Games*, Volume 1, Number 4, January 1983. pp50-53 & 56
Parker Brothers' involvement in the handheld business

Cook, Wanda (1979) 'Electronic games a real challenge'. *The Blade*. Toledo, Ohio. Thursday 6 December 1979. p41
The popularity of handheld games in the late 1970s

DeMaria, Rusel and Wilson, Johnny L. (2002) *High Score! The Illustrated History of Electronic Games*. Berkeley, California: McGraw-Hill/Osborne
Background on Space Invaders and early consoles

Edwards, Benj (2009) 'Jerry Lawson, Black Video Game Pioneer'. *Vintage Computing and Gaming*. 24 February 2009. [Online] www.vintagecomputing.com/index.php/archives/545 [Last accessed: 5 March 2010]
Interview with the Fairchild Channel F designer

Gielens, Jaro (2000) *Electronic Plastic*. Berlin, Germany: Die Gestatlen Verlag
Information on early handhelds

Kemp, Leslie (1979) 'Electronic toys top list of Polk buyers'. *The Ledger*. Lakeland, Florida. Monday 3 December 1979. pp1A & 6A.
Report on sales of handhelds

Kent, Steven L. (2001) *The Ultimate History of Video Games*. New York, New York: Three Rivers Press
Background on early handhelds and consoles, also Space Invaders

Mattel Dimension 78 (1978) [Toy fair catalog] Mattel
Information on Mattel's handheld games

Old Computers (no date) [Online] http://old-computers.com [Last accessed: 2 March 2010]
Information on late 1970s consoles

Skow, John (1979) 'Those beeping, thinking toys'. *Time*. Monday 10 December 1979. [Online] www.time.com/time/magazine/article/0,9171,912568-1,00.html [Last accessed: 6 March 2010]
Report on the handheld game boom

Sullivan, George (1983) 'Looking to the future'. *Blip*, Volume 1, No. 4, May 1983, pp20-21
Profile of Coleco

Williams, Dmitri (2006) 'A (brief) social history of video games'. In: Vorderer, Peter and Bryant, Jennings (editors) (2006) *Playing Computer Games: Motives, Responses and Consequences*. Mahwah, New Jersey: Lawrence Erlbaum [Online] http://dmitriwilliams.com/research.html [Last accessed: 6 March 2010]
Sales data for the US video game industry

7 | PAC-MAN FEVER

Allison, Anne (2004) 'Cuteness as Japan's Millennial product'. In: Tobin, Joseph (editor) (2004) *Pikachu's Global Adventure: The Rise and Fall of Pokémon*. Durham, North Carolina: Duke University Press. pp34-49
The kawaii art style defined and explained

Amis, Martin (1982) *Invasion of the Space Invaders*. London, UK: Hutchinson & Co.
UK-centric view on the video game boom

Atari Inc. (1980) *The Book: A Guide to Electronic Game Operation and Servicing*. Sunnyvale, California: Atari Inc. pp3-3 to 3-6
Technical information on vector and raster scan displays

'Beating the game game' (1982) *Time*. Monday 18 January 1982. [Online] www.time.com/time/magazine/article/0,9171,949467-1,00.html [Last accessed: 6 March 2010]

Bally president Robert Mullane on his initial reaction to Pac-Man
Bloom, Steve (1983) '*Video Games* interview: Bill Grubb and Dennis Koble'. *Video Games*, Volume 1, Number 4, January 1983. pp22-29 & 81
Interview with the founders of Imagic

Bloom, Steve (1983) 'Parker Brothers strikes back'. *Video Games.* Volume 1, Number 4, January 1983. pp50-53 & 56
Parker Brothers enters the home video game business

Buckner & Garcia (1982) *Pac-Man Fever.* K-Tel
The Pac-Man Fever album

Burnham, Van (editor) (2001) *Supercade: A Visual History of the Videogame Age 1971-1984.* Cambridge, Massachusetts: The MIT Press
Information on early 1980s arcade games

Crawford, Chris (1984) *The Art of Computer Game Design.* 1997 electronic version. Berkeley, California: McGraw-Hill/Osborne Media. [Online] www.vancouver.wsu.edu/fac/peabody/game-book/Coverpage.html#ACKNOWLEDGMENT [Last accessed: 28 February 2010]
The first book on the theory of video game design, includes insights into game developer and industry thinking at the height of the boom

Digital Eclipse (2003) *Midway Arcade Treasures* [PlayStation 2] Midway Games
The creation of Defender

'Dona and the Candy Factory' (1983) *Blip*, Volume 1, No. 4, May 1983. pp6-8
Interview with Dona Bailey, co-designer of Centipede

Fisher, Anne B. (1986) 'Glamour: Getting it or getting it back'. *Fortune.* 12 May 1986. [Online] http://money.cnn.com/magazines/fortune/fortune_archive/1986/05/12/67557/index.htm [Last accessed: 6 March 2010]
Video game cartridge margins and Warner stock price growth

'Food & games' (1982) *Electronic Fun with Computers & Games.* Volume 1, No. 1, November 1982. p13
Quaker Oats buys U.S. Games

Goldhaber, Michael (1986) *Reinventing Technology: Policies for Democratic Values.* New York, New York: Routledge & Kegan Paul. p3
Background on the politics of the Atari Democrats

Gutman, Dan (1982) 'Gamemakers: Boy wonder'. *Electronic Fun with Computers & Games.* Volume 1, No. 1, November 1982. pp30-33 & 97
Interview with Mark Turnell of Sirius Software about his success

'Here comes...Pac-Man-imation!' (1982) *Electronic Fun with Computers & Games.* Volume 1, No. 1, November 1982. pp44-46
Feature on the Pac-Man cartoon series

Hulse, Ed (1982) 'Games go Hollywood'. *Electronic Fun with Computers & Games.* Volume 1, No. 1, November 1982. pp22-27 & 97
How movie studios began moving into video games publishing

Hutcheon, Stephen (1983) 'The video games boom has yet to come'. *The Age.* Sydney, Australia. 7 June 1983. p31
Australian report on popularity of video games in USA
Jacobson, Mark (1983) 'Zen and the art of Donkey Kong'. *Video Games.* Volume 1, Number 4, January 1983. pp30-33
Eugene Jarvis interview setting out his vision of 'sperm games'

Kent, Steven L. (2001) *The Ultimate History of Video Games.* New York, New York: Three Rivers Press
Background on the early 1980s video game boom

Kohler, Chris (2005) *Power-Up: How Japanese Video Games Gave the World an Extra Life.* Indianapolis, Indiana: BradyGames
Background on Donkey Kong

Koster, Raph (2005) *A Theory of Fun for Game Design.* Scottsdale, Arizona: Paraglyph Press
Evolution of shoot 'em ups in the late 1970s and early 1980s

Kraft, Joseph (1982) 'Atari Democrats may provide a base'. *Nashua Telegraph.* Nashua, New Hampshire. Tuesday 19 October

1982. p14.
Report on the Atari Democrats

Lekachman, Robert (1982) 'The Atari Democrats: Neo-liberal agenda has its minor virtues'. *Gainesville Sun*. Gainesville, Florida. Saturday 30 October 1982. p5A.
Analysis of the Atari Democrats and their beliefs

Logg, Ed (1980) 'Game description: Shoot the centipede'. [Company memo] Atari Inc. inter-office memo to Dan Van Elderen. 17 July 1980.
Original proposal for Centipede

Missile Command (1980) [Promotional Flyer] Sunnyvale, California: Atari Inc.
Details of official Missile Command scenario

'Missile Command' (1981) In: *Atari Video Computer System Catalog: 49 Game Program Cartridges* (1981) Sunnyvale, California: Atari Inc.
Details of official Missile Command scenario

'Pac-Man Fever! An Interview with Buckner & Garcia' (2006) *Jawbone Radio*, podcast 117. Wednesday 28 June 2006. [Online] http://jawboneradio.blogspot.com/2006/06/jawbone-117-pac-man-fever-interview.html [Last accessed: 6 March 2010]
Interview with Buckner & Garcia

Rheingold, Howard (1991) *Virtual Reality*. 1992 edition. London: Mandarin Paperbacks.
Information on Rocky's Boots

Robinett, Warren (2003) 'Foreword'. In: Wolf, M. and Perron, B. (editors) (2003) *Video Game Theory Reader*.
Reflections of the development of the Atari VCS 2600 game Adventure

Rubin, Michael (2006) *Droidmaker: George Lucas and the Digital Revolution*. Gainesville, Florida: Triad Publishing
Background on George Lucas' entry into video games

Sloane, Martin (1982) 'Video games come to supermarket!'. *Gainesville Sun*. Gainesville, Florida. Saturday 6 June 1982. p9E.
Report on supermarket plans to install arcade video games

Sullivan, George (1983) 'Video hall of fame'. *Blip*, Volume 1, No. 3, April 1983. pp8-9
Interview with high-scoring players

'The Making of Warrior' (2006) *Edge*, Issue 169, December 2006. pp100-103.
The creation of Cinematronics' Warrior

Thomas, E., Stacks, J.F. and Kelly, J. (1982) 'Basking in Reagan's troubles'. *Time*. Monday 12 July 1982. [Online] www.time.com/time/magazine/article/0,9171,925536-1,00.html [Last accessed: 6 March 2010]
Report on the Atari Democrats

Tsongas, Paul E. (1983) 'Nothing wrong with Atari, Mr President'. *Times-News*. Hendersonville, North Carolina. Saturday 12 March 1983. p4
Report on the Atari Democrats

'United Airlines Officials Toy With Airborne Computer Games' (1984) *Ocala Star-Banner*. Ocala, Florida. 8 June 1984. p9B.
United Airlines plans to offer in-flight games to passengers

'Video games – are they killing rock and roll?' (1982) *Electronic Fun with Computers & Games*. Volume 1, No. 1, November 1982. p17
Decline in record sales as video games take off

'Video game gift guide' (1983) *Video Games*. Volume 1, Number 4, January 1983. p44
Pretzel Invaders and other video game merchandise

Williams, Dmitri (2006) 'A (brief) social history of video games'. In: Vorderer, Peter and Bryant, Jennings (editors) (2006) *Playing Computer Games: Motives, Responses and Consequences*. Mahwah, New Jersey: Lawrence Erlbaum. [Online] http://dmitriwilliams.com/research.html [Last accessed: 6 March 2010]
US video game sales data

Woodbury, R., Alexander, C. P. and Towle, L. (1983) 'Video games go crunch!'. *Time*. Monday 17 October 1983. [Online] www.time.com/time/magazine/article/0,9171,952210,00.html [last accessed: 11 March 2010]
Post-crash view on the boom years

'Z-Z-Zap! Atari, Bally shooting it out for the arcade game business' (1980) *The Milwaukee Journal*. Tuesday 26 August 1980. Part 2, p10
Report on the arcade industry in 1980

8 | DEVILISH CONTRAPTIONS

'2 supervisors oppose video games' (1982) *Milwaukee Sentinel*. Tuesday 11 May 1982. Part 1, p5
Politicians react to public fears about video games

Albright Jr., Ronald G. (1985) *The Orphan Chronicles*. San Dimas, California: Millers Graphics
Account of the battle between Texas Instruments and Commodore

'Atari Corporation' (2004) *International Directory of Company Histories*, Volume 66. Farmington Hills, Michigan: St James Press. [Online] www.fundinguniverse.com/company-histories/Atari-Corporation-Company-History.html [Last accessed: 2 March 2010]
History of Atari Inc.

Atari Inc. (1982) 'Operator of the '80s': Stewart Burch of Cotati, CA'. *Atari Coin Connection*, Volume 6, Number 1, February 1982
Profile of an arcade that has restricted under 18s access to counter parental fears

Atari Inc. (1982) 'Atari launches community awareness program for video games'. *Atari Coin Connection*, Volume 6, Number 3, April/May 1982. p1
Atari's Community Awareness Program

'Atari parts are dumped' (1983) *New York Times*, 28 September 1983, pD-4
Report on the Atari landfill incident

'Atari refuses to let the video game fad die' (1984) *Business Week*, 21 May 1984, p46
James Morgan on taking over Atari

'Atari suing to halt X-rated video games' (1982) *The Tuscaloosa News*. Tuscaloosa-Northport, Alabama. Sunday 17 October 1982. p12A
Atari reacts to Custer's Revenge

Bagnall, Brian (2007) *On the Edge: The Spectacular Rise and Fall of Commodore*. Winnipeg, Manitoba, Canada: Variant Press
Details of Commodore's VIC-20 and Commodore 64 home computers

Bailey, Adam J. (2007) *The Atari Landfill Revealed*. [Online] Archived at: http://web.archive.org/web/20071213195747/http://atari.digital-madman.com [Last accessed: 6 March 2010]
Investigation examining the Atari landfill incident

'Bally Manufacturing Corporation' (1991) *International Directory of Company Histories*, Volume 3, St. James Press, 1991. [Online] www.fundinguniverse.com/company-histories/BALLY-MANUFACTURING-CORPORATION-Company-History.html [Last accessed: 11 March 2010]
History of Bally

Bloom, Steve (1983) 'Parker Brothers strikes back' *Video Games*, Volume 1, Number 4, January 1983. p50-53 & 56
Interview about Parker Brothers' entry into the home video game business

Bloom, Steve (1983) '*Video Games* interview: Bill Grubb and Dennis Koble'. *Video Games*. Volume 1, Number 4, January 1983. pp22-29 & 81
Interview with Imagic's founders

Brown, Z., Greenberg, D. and Rubin, J. (1982) 'Community restricts video games'. *Pittsburgh Post-Gazette*. 22 July 1982. p8
Rules imposed on arcades

Callistien, Debbie (1981) 'Belleair Bluffs receives static over video games'. *The Evening Independent*, St Petersburg, Florida. 21 October 1981, p12-A

Council debates control of arcades
Collins, Glenn (1981) 'Are video games harmless?'. *The Spokesman-Review*. Spokane, Washington. Tuesday 8 September 1981. p18
Public fears about video games

'Commodore International Ltd' (1993) *International Directory of Company Histories*, Volume 7, St. James Press, 1993. [Online] www.fundinguniverse.com/company-histories/Commodore-International-Ltd-Company-History.html [Last accessed: 11 March 2010]
History of Commodore

'Computer or video games' (1983) *The New York Times*. 28 April 1983. [Online] www.nytimes.com/1983/04/28/business/computer-or-video-games.html?&pagewanted=all [Last accessed: 7 March 2010]
Home computers challenge home consoles

'Control of Space Invaders and Other Electronic Games' (1981) *Hansard*. House of Commons debate, 20 May 1981, Vol. 5, cc287-91 [Online] http://hansard.millbanksystems.com/commons/1981/may/20/control-of-space-invaders-and-other#S6CV0005P0_19810520_HOC_167 [Last accessed: 6 March 2010]
UK Parliament debate on restricting Space Invaders and video game arcades

Del Giudice, Vincent and Keene, Thomas R. (2009) 'U.S. Joblessness may reach 13 percent, Rosenberg says'. *Bloomberg.com*. 9 November 2009. [Online] www.bloomberg.com/apps/news?pid=20601087&sid=aHmxIMR1DFq0 [Last accessed: 6 March 2010]
Report comparing unemployment levels in the early 1980s and late 2000s recessions

Fisher, Anne B. (1986) 'Glamour: Getting it or getting it back'. *Fortune*. 12 May 1986. [Online] http://money.cnn.com/magazines/fortune/fortune_archive/1986/05/12/67557/index.htm [Last accessed: 6 March 2010]
The impact of the crash on Atari and Warner Communications

Heard, Alex (1981) 'Court to rule on age-limit law for video-game parlors'. *Education Week*. 16 November 1981. [Online] www.edweek.org/login.html?source=http://www.edweek.org/ew/articles/1981/11/16/01110065.h01.html&destination=http://www.edweek.org/ew/articles/1981/11/16/01110065.h01.html&levelId=2100 [Last accessed: 6 March 2010]
Public fears about video games

'Here comes…Pac-Man-imation!' (1982) *Electronic Fun with Computers & Games*. Volume 1, No. 1, November 1982. pp44-46
Feature on the Pac-Man cartoon series

Hulse, Ed (1982) 'Games go Hollywood'. *Electronic Fun with Computers & Games*. Volume 1, No. 1, November 1982. pp22-27 & 97
Report on movie studios moving into video game business

'Indo ban on video games begin' (1981) *New Straits Times*. Saturday 19 December 1981, p16
Indonesia bans video games

Jacobson, Mark (1983) 'Zen and the art of Donkey Kong'. *Video Games*. Volume 1, Number 4, January 1983. pp30-33
Eugene Jarvis interview

Johnson, Tim (1981) 'Video games are targets of two ordinances in Belleair Bluffs'. *St. Petersburg Times*, Florida. Wednesday 18 November 1981. p9
Video games face restrictions

Kennedy, David (1981) 'Video games turn youths into junkies'. *Boca Raton News*. 16 August 1981. p8B
Fears about video games

Knox, Andrea (1984) 'Amusements: Pinball gains, video games lose ground in some areas'. *The Evening Independent*, St Petersburg, Florida. Tuesday 20 March 1984. p4-B
How the crash is reducing arcades' interest in video games

Koop, C. Everett (1982) 'Family violence: A chronic public health issue'. [Speech] Western Psychiatric Institute, Pittsburgh, Pennsylvania. 9 November 1982.
The lecture given by the US Surgeon General prior to his comments about video games

Koop, C. Everett (1982) 'Statement by Dr. E. Koop, Surgeon General'. [Press release] 10 November 1982.
The Surgeon General clarifies his comments on video games

Kraft, Scott (1982) 'Computer game invasion at full tilt'. *The Free Lance-Star*. Fredericksburg, Virginia. Saturday 2 January 1982. p19
Space Invaders wrist and public fears about arcades

Lee, Jeff (1998) 'The History of Q*Bert'. Unpublished
*The creation of Q*Bert*

Leibowitz, David (1983) 'Year of the home computer'. *Video Games*. Volume 1, Number 4, January 1983. pp75 & 80
Report on how home computers may effect on video game consoles

'Letters to the editor' (1982) *Electronic Fun with Computers & Games*. Volume 1, No. 1, November 1982. p8
Thoughts on how home computers might replace home consoles

Mandel, Howard (1983) 'They say it ain't porno'. *Video Games*. Volume 1, Number 4, January 1983. pp13-14
Custer's Revenge protests

McQuiddy, M.E. (1983) 'Dump here utilized'. *Alamogordo Daily News*, New Mexico. Sunday 25 September 1983, p1
Report on the Atari landfill dumping

McQuiddy, M.E. (1983) 'City to Atari: E.T. trash go home'. *Alamogordo Daily News*, New Mexico. Tuesday 27 September 1983
Alamogordo reacts to the influx of Atari waste

'Medicine: New Maladies' (1981) *Time*. Monday 29 June 1981. [Online] www.time.com/time/magazine/article/0,9171,951740,00.html [Last accessed: 6 March 2010]
Space Invaders wrist makes its debut

Miller, Jacquie (1983) 'Customs bans another pornographic video game'. *The Citizen*. Ottawa, Canada. Tuesday 11 January 1983. p5
Canada's ban on Custer's Revenge and another pornographic video game from the same company

Miller, Scott and Broussard, Greg (1982) *Shootout: Zap the Video Games*. Waxahachie, Texas: Grove Creek Publishing.
Tips book on beating arcade video games

'New software to give life to video games systems' (1984) *The Age*. Sydney, Australia. Tuesday 17 January 1984, p29
Report on dumping of Atari products in Alamogordo

Powell, Dwane (1982) 'Koop-Man'. [Cartoon] *The News and Observer*. Raleigh, North Carolina. 11 November 1982. p4A
Newspaper cartoon reacting to US Surgeon General Dr Everett Koop's criticism of video games

Rapoport, Roger (1983) 'A very brief video salvage game'. *The Philadelphia Inquirer*, 29 September 1983, pA-1
Report on the dumping of Atari equipment in Alamogordo

Schrage, M. (1985) 'A video star is zapped into oblivion'. *The Age*. Sydney, Australia. 31 December 1985. p9
Interview with Howard Warshaw, E.T.'s creator, about the crash

Sharpe, Rochelle (1980) 'Councilmen line up behind computer games'. *Concord Monitor*. New Hampshire. Tuesday 9 December 1980. p1
Public concerns about video games

Stout, J. (1979) 'Pets Corner'. *Liverpool Software Gazette*. No. 1, November 1979. p12
The Commodore PET's performance in Europe

Sullivan, George (1983) 'Looking to the future'. *Blip*, Volume 1, No. 4, May 1983. pp20-21
Profile of Coleco

Sunshine, Laura (1982) 'Colecovision'. Atari Inc inter-office memo to Riley Rowe. 25 October 1982.
Internal Atari Inc. memo on the findings of a focus group comparing the Colecovision and Atari 5200

Teets, John (1983) 'Plenty of video games at consumer electronics show'. *Pittsburgh Post-Gazette*. 2 June 1983. p24
Growth in VCR sales

Texas Instruments (1983) 'TI reports third quarter results'. [Press release] 28 October 1983.
Texas Instruments quits the home computer business

'Video games – are they killing rock and roll?' (1982) *Electronic Fun with Computers & Games*. Volume 1, No. 1, November 1982. p17
Decline in music sales due to video games

'Video games banned by Suharto' (1981) *New Straits Times*. Wednesday 16 December 1981, p13
Indonesia bans video games

'Video games turned off' (1981) *Spokane Daily Chronicle*, Spokane, Washington. Friday 20 November 1981. p8
Philippines shuts down arcades

Williams, Dmitri (2006) 'A (brief) social history of video games'. In: Vorderer, Peter & Bryant, Jennings (editor) (2006) *Playing Computer Games: Motives, Responses and Consequences*. Mahwah, New Jersey: Lawrence Erlbaum. [Online] http://dmitriwilliams. com/research.html [Last accessed: 6 March 2010]
US video game sales data

Woodbury, R., Alexander, C. P. and Towle, L. (1983) 'Video games go crunch!'. *Time* Monday 17 October 1983. [Online] www. time.com/time/magazine/article/0,9171,952210,00.html [last accessed: 11 March 2010]
Report on the bursting of the video game bubble

9 | UNCLE CLIVE

'$99! Computer' (1982) *Electronic Fun with Computers & Games*. November 1982, Vol. 1, No. 1. pp82-86
Report on the release of the Timex Sinclair 1000, the North American version of the ZX81

'Adventure File' (1983) *Micro Adventurer*, Issue 1, November 1983. pp45-48
Pimania details

Amis, Martin (1982) *Invasion of the Space Invaders*. London, UK: Hutchinson & Co.
Information on early European game consoles

Baer, Ralph H. (2005) *Videogames: In the beginning*. Springfield, New Jersey: Rolenta Press
Information on early European game consoles

'Clive Sinclair' (1982) *Practical Computing*. Volume 5, Issue 7, July 1982. [Online] www.worldofspectrum.org/CliveSinclairInterview1982 [Last accessed: 7 March 2010]
Clive Sinclair profile including his comments on the BBC computer

Everiss, Bruce (1980) 'Editor: 0.0'. *Liverpool Software Gazette*. No.3, March 1980, p4
Thoughts on the growth of home computing

Everiss, Bruce (1984) 'Imagine: What was the name of the game?'. *Your Computer*. November 1984. pp76-77
Influence of Microdigital in Liverpool

'Focus: Spectrum' (2000) *Edge Essential Hardware Guide 2000*, pp108-113
The story of the ZX Spectrum's creation

'Global underground' (2002) *Edge*, Issue 118, Christmas 2002, pp72-82.
Influence of Melbourne House on Australia's game business

Liverpool Software Gazette. November 1979. No.1
Microdigital's location and stock

Page, Barnaby (1988) 'Yes we have no mañanas'. *The Games Machine*, Issue 12, November 1988, pp13-14
Report on the Spanish game industry

Rodríguez, Fernando (2003) 'Historia del software español de entretenimiento'. *MSDOX*. 21 February 2003. [In Spanish] [Online] www.msdox.com/reportajes/historiasoftesp210203.php [Last accessed: 20 March 2010]
The history of the Spanish video game industry

Savage, Jon (1991) *England's Dreaming: Sex Pistols and Punk Rock*. London, UK: Faber & Faber
History of the punk rock movement in the UK
Spufford, Francis (2003) *Backroom Boys: The Secret Return of the British Boffin*. 2004 paperback edition. London, UK:

Faber & Faber
The creation of Elite

Stuckey, Helen and Harsel, Noé (2007) 'Hits of the 80s: Aussie games that rocked the world'. Australian Centre for the Moving Image. November 2007. [Online] www.acmi.net.au/hits_80s_essay.htm [Last accessed: 7 March 2010]
History of Australia's game industry

'The making of Football Manager' (2009) *Edge*, Issue 203, July 2009, pp106-109
Interview with Kevin Toms

'The making of Jet Set Willy' (2006) *Edge*, Issue 159, February 2006, pp102-105
Interview with Matthew Smith

'The making of Manic Miner' (2001) *Edge*, Issue 103, November 2001, pp92-95
Interview with Matthew Smith

'The Sinclair story – part one' (1985) *Crash*. No.15, April 1985. pp86-87
Background on Clive Sinclair

'The Sinclair story – part two' (1985) *Crash*. No.16, May 1985. pp126-127
Background on Clive Sinclair

'The Sinclair story – part three' (1985) *Crash*. No. 17, June 1985. pp56-57
Background on Clive Sinclair

'The Sinclair story – part four: The Televisionary' (1985) *Crash*, No. 18, July 1985. pp68-69
Background on Clive Sinclair

'The wisdom of Matthew Smith' (2004) *Retro Gamer*. Issue 7, August 2004, pp26-29
Interview with Matthew Smith

Thumb Candy (2000) [TV broadcast] London, UK: Channel 4. 31 March 2001. GMT: 23.35
Interview with Matthew Smith

Veronis, Dr Andrew M. (1980) 'Etcetera'. *Liverpool Software Gazette*. No.3, March 1980, pp79-80
Reaction to the launch of the ZX80

World of Spectrum (no date) [Online] www.worldofspectrum.org [Last accessed: 7 March 2010]
Source of raw data on the growth of the UK game publishing industry. Analysis by author.

10 I THE FRENCH TOUCH

'24 Ordinateurs' (1983) [In French] *Tilt*, Issue 8, November/December 1983, pp142-165
Home computers available in France in 1983

Amstrad '2690F' (1985) [Advert] [In French] *Tilt*, Issue 25, October 1985, p17
Sales of the CPC in France

Bagnall, Brian (2007) *On the Edge: The Spectacular Rise and Fall of Commodore*. Winnipeg, Manitoba, Canada: Variant Press
The history of Commodore

Chevènement, Jean-Pierre (1985) 'I.P.T.: y'a un micro dans le potache' [In French] *Tilt*, Issue 24, September 1985, pp8-10
Interview with the French education minister about the national school computer scheme

Commercial Breaks: The Battle for Santa's Software (1984) [TV broadcast] London, UK: BBC 2. 13 December 1982. GMT: 20.00
Documentary on UK publishers Imagine Software and Ocean Software, charting the implosion of the former

Curry, Andrew (2009) 'Monopoly killer: Perfect German board game redefines genre'. *Wired*. 23 March 2009. [Online] www.wired.com/gaming/gamingreviews/magazine/17-04/mf_settlers?currentPage=all [Last accessed: 7 March 2010]
Overview of Germany's board game culture

Davies, Jonathan (1991) 'Just why are French games so weird?'. *Amiga Power*, Issue 6, October 1991, pp74-77.
UK perspective on French games

Debord, Guy (1967) *The Society of the Spectacle*. Translated by Donald Nicholson-Smith (1994). New York, New York: Zone Books
The manifesto of the Situationists

Delalandre, Jean-Philippe (1986) 'Descente chez le créateur du mois' [In French] *Tilt*, Issue 35, October 1986, p21
Jean-Louis Le Breton profile

Delcourt, Guy (1984) 'La puce aux oeufs d'or' [In French] *Tilt*, Issue 14, July/August 1984, pp16-22 & 82-83
Compares the status of the French video game industry to that of the UK and US

'Eurovision' (1988) *Crash*, Issue 59, December 1988, pp120-122
Report on European game publishers outside the UK

Everiss, Bruce (2008) 'Piracy, Imagine Software and the Megagames'. *Bruce on Games* [Blog]. 25 March 2008. [Online] www.bruceongames.com/2008/03/25/piracy-imagine-software-and-the-megagames [Last accessed: 20 March 2010]
Bruce Everiss reflects on the effect of video game piracy and the demise of Imagine

Gros Pixels (no date) [In French] [Online] www.grospixels.com [Last accessed: 12 March 2010]
Background on French computers and consoles

Harbonn, Jacques (1985) 'Malédiction!' [In French] *Tilt*, Issue 25, October 1985, pp119-120
Overview of MIme les Pommes de Terre ont des Yeux

'Hector, L'ordinateur qui a la peche' (1983) [Advert] [In French] *Tilt*, Issue 8, November/December 1983, pp42-43
Advert for the French home computer the Hector

Hussey, Andrew (2001) *The Game of War: The Life and Death of Guy Debord*. 2002 paperback edition. London, UK: Pimlico
Biography of Guy Debord and details of the May 1968 riots in France

'Infogrames Entertainment S.A.' (2001) *International Directory of Company Histories*, Vol. 35, St James Press. [Online] www.fundinguniverse.com/company-histories/Infogrames-Entertainment-SA-Company-History.html [Last accessed: 12 March 2010]
History of Infogrames

Jean-Louis Le Breton (no date) [Personal website] [In French] www.jeanlouislebreton.com [Last accessed: 7 March 2010]
Biographical information

Kean, Roger (1984) 'The biggest commercial break of them all'. *Crash*, No. 12, Christmas Special. pp60-64
The collapse of Imagine

'L'Aigle d'Or: Ça plane pour lui' (1984) [In French] *Tilt*, Issue 12, May 1984, p54
Mentions the reaction to L'Aigle d'Or

Labaille, Louise (1983) 'Aux Commandes du G7200'. [In French] *Tilt*, Issue 3, January/February 1983, pp38-39
Information on the Videopac G7200 console and the rise of home computers

Lange, Andreas (2002) 'Report from the PAL zone'. In: King, Lucien (editor) (2002) *Game On: The History and Culture of Videogames*. London, UK: Laurence King Publishing, pp46-55
Overview of video game history in Europe

'Le Crime du Parking' (1985) [In French] *Tilt*, Issue 24, September 1985, p41
Game review

'Loriciels: L'année de toutes les ouvertures' (1985) [In French] *Tilt*, Issue 25, October 1985, p25
Loriciels sets out its European expansion plans

Old Computers (no date) [Online] http://old-computers.com [Last accessed: 2 March 2010]
Information on European home computer and console systems

Page, Barnaby (1987) 'The budget boom'. *Crash*, Issue 45, October 1987. pp45-46.
The rise of budget games in the UK

'The history of Future Crew' (1997) [Online] www.defacto2.net/apollo-x/fc.htm [Last accessed: 7 March 2010]
The demo crew's history

'Rush sur run' (1983) [In French] *Tilt*, Issue 6, July/August 1983, p14
News report noting the use of printed French instructions for English language games rather than translating the words in the game

Savage, Jon (1991) *England's Dreaming: Sex Pistols and Punk Rock*. London, UK: Faber & Faber
History of the punk rock movement in the UK that also covers the May 1968 riots

Siegler, Joe (1998) 'Remedy Entertainment & Future Crew'. 26 August 1998 [Online] www.3drealms.com/news/1998/08/remedy_entertai_1.html [Last accessed: 7 March 2010]
Explanation of the links between Future Crew and Remedy Entertainment

'Standard Europeen' (1983) [In French] *Tilt*, Issue 6, July/August 1983, p14
Report on the French-made Jopac console

Stilphen, Scott (2004) 'DP Interviews...Anthony Weber'. *Digital Press*. [Online] www.digitpress.com/library/interviews/interview_anthony_weber.html [Last accessed: 7 March 2010]
Interview with the creator of Galactic Chase

'Ubi Soft Entertainment S.A.' (2001) *International Directory of Company Histories*, Vol. 41, St James Press. [Online] www.fundinguniverse.com/company-histories/Ubi-Soft-Entertainment-SA-Company-History.html [Last accessed: 12 March 2010]
History of Ubisoft

11 I MACINTOSHIZATION

Adlum, Eddie (1986) 'A trip to the 1986 JAMMA arcade game show, part 1'. *RePlay* (unspecified late 1986 issue). As reproduced by *GameSetWatch* (8 January 2009) [Online] www.gamesetwatch.com/2009/01/replay_mag_article_contact_the.php [Last accessed: 7 March 2010]
Report on the industry's excitement about Out Run

Arcade History (no date) [Online] www.arcade-history.com [Last accessed: 2 March 2010]
Information on coin-op arcade games

Ardai, Charles (1986) 'Year in review'. *Computer Gaming World*, Issue 33, December 1986, pp20-21 & 24-26
The influence of the Macintosh on video games

'Atari talks Gauntlet, Paperboy' (1986). *RePlay*, January 1986. As reproduced by *GameSetWatch* (26 June 2007) [Online] www.gamesetwatch.com/2007/06/column_replay_atari_talks_gaun.php#more [Last accessed: 7 March 2010]
Interview with Atari Games covering the success of Gauntlet

Barton, Matt and Loguidice, Bill (2009) 'The history of the Pinball Construction Set: Launching millions of creative possibilities'. *Gamasutra*. 6 February 2009. [Online] www.gamasutra.com/view/feature/3923/the_history_of_the_pinball_.php [Last accessed: 12 March 2010]
Pinball Construction Set and its influence

Boctok Inc and Fukuda, Miki (editors) (2000) *Bit Generation 2000*. Kobe, Japan: Kobe Fashion Museum
Yu Suzuki profile

Boosman, Frank (1986) 'Designer profile: Chris Crawford'. *Computer Gaming World*, Issue 33, December 1986, p49
Interview with the Balance of Power designer

Boosman, Frank (1987) 'Designer profile: Chris Crawford (Part 2)'. *Computer Gaming World*, Issue 34, January/February 1987, pp56-59
Interview with the Balance of Power designer

Bunten, Dan (1984) 'Dispatches: Insights from the strategy game design front - M.U.L.E. Designer notes'. *Computer Gaming World*, Volume 4, No. 2, April 1984, pp17 & 42
Thoughts on the development of M.U.L.E.

Cignarella, Patricia (1989) '*Marketing Computer*'s cover story on Roberta Williams'. *Sierra News Magazine*. Volume 2, No. 2, Autumn 1989. pp22-25
Interview with Roberta Williams

Cousins, Mark (2004) *The Story of Film*. 2008 paperback edition. London, UK: Pavilion Books
The Hollywood studio system

Croft, Martin (1985) 'So long, and thanks for the adventure'. *Micro Adventurer,* Issue 16, February 1985. pp9-10
Interview with Douglas Adams about Infocom's Hitchhiker's Guide to the Galaxy

Davies, Jonti (2008) 'The making of: OutRun'. *Retro Gamer,* Issue 54, August 2008, pp26-31
Interview with Yu Suzuki

Ferrell, Keith (1987) 'The future of computer games'. *Compute!,* Issue 90, November 1987, p14
Industry thinking post-crash

Gaddis, John Lewis (2005) *The Cold War.* 2007 edition. London, UK: Penguin Books
Cold War history

Hague, James (editor) (1997) *Halcyon Days: Interviews with Classic Computer and Video Game Programmers.* Free web version 2002. Dadgum Games. [Online] www.dadgum.com/halcyon [Last accessed: 28 February 2010]
Interview with Dani Bunten Berry

Hang-On (1985) [Promotional Flyer] Sega
Background information of Hang-On sit-down machine

Jong, Philip (2006) 'Roberta Williams'. *Adventure Classic Gaming.* 12 March 2010. [Online] www.adventureclassicgaming.com/index.php/site/interviews/198 [Last accessed: 12 March 2010]
Interview with Roberta Williams

'Masters of the arts' (2009) *Edge,* Issue 200, April 2009. pp66-73
Interview with Trip Hawkins

'Missing in action' (2000) *Edge,* Issue 92, Christmas 2000. pp76-83
Douglas Engelbart profile

Moriarty, Brian (1998) 'Dani Bunten: Lifetime Achievement Award'. [Presentation] Computer Game Developers Association Awards, Long Beach, California, 7 May 1998. [Online] www.anticlockwise.com/dani [Last accessed: 12 March 2010]
Thoughts on Dani Bunten Berry's games

'Out Run' (1986) [Promotional flyer] Sega Europe
Highlights the game's Alpine and southern France inspired courses

Rheingold, Howard (1991) *Virtual Reality.* 1992 edition. London: Mandarin Paperbacks.
History of the graphical user interface

Smithe, Nancy (1989) 'Roberta Williams: The storyteller who started it all'. *Sierra News Magazine.* Volume 2, No. 2, Autumn 1989. pp3 & 35-37
Interview with Roberta Williams

Szczepaniak, John (2006) 'Mechanical Donkeys'. *The Gamer's Quarter.* Issue 6, 3rd Quarter 2006, pp66-81
Interviews with those involved in the creation of M.U.L.E.

'Time-Zone: An interview with Roberta Williams' (1982) *Computer Gaming World,* Volume 2, No. 3, May/June 1982. pp14-15
Interview with Roberta Williams

Williams, Dmitri (2006) 'A (brief) social history of video games'. In: Vorderer, Peter & Bryant, Jennings (editors) (2006) *Playing Computer Games: Motives, Responses and Consequences.* Mahwah, New Jersey: Lawrence Erlbaum. [Online] http://dmitriwilliams.com/research.html [Last accessed: 6 March 2010]
Figures on the US video game industry

12 I A TOOL TO SELL SOFTWARE

'Agog over new computer game' (1988) *New Straits Times.* Malaysia. Tuesday 23 February 1988, p12
Japanese reaction to the release of Dragon Quest III

Allison, Anne (2004) 'Cuteness as Japan's Millennial product'. In: Tobin, Joseph (editor) (2004) *Pikachu's Global Adventure: The Rise and Fall of Pokémon.* Durham, North Carolina: Duke University Press, pp34-49
The influence of 'cuteness' in Japan

Anderson Jr, Larry and Amos, Ryan (2003) *Intellivision FAQ v6.0*. [Online] www.gamefaqs.com/console/intellivision/file/916427/2894 [Last accessed: 13 March 2010]
Information on the Bandai Intellivision

Ashcraft, Brian (2009) 'Rape games officially banned in Japan'. *Kotaku*. 2 June 2009. [Online] http://kotaku.com/5275409/rape-games-officially-banned-in-japan [Last accessed: 13 March 2010]
The RapeLay controversy

'Behind the scenes: Bomberman' (2008) *Games*™, Issue 74, September 2008, pp134-137
History of the Bomberman series

'Bishojo game' (no date) *Economicexpert.com*. [Online] www.economicexpert.com/a/Bishojo:game.html [Last accessed: 13 March 2010]
Information about bishojo games

Boctok Inc and Fukuda, Miki (editors) (2000) *Bit Generation 2000*. Kobe, Japan: Kobe Fashion Museum
Video game history from a Japanese perspective

Brown, Damon (2008) *Porn & Pong: How Grand Theft Auto, Tomb Raider and Other Sexy Games Changed Our Culture*. Port Townsend, Washington: Feral House
Background on bishojo games

Chaplin, Heather and Ruby, Aaron (2005) *Smartbomb: The Quest for Art, Entertainment and Big Bucks in the Videogame Revolution*. 2006 paperback edition. Chapel Hill, North Carolina: Algonquin Books
Profile of Shigeru Miyamoto

'Consoles: Release Timeline' (no date) *Japan-Games.com*. [Online] www.japan-games.com/wiki/pmwiki.php?n=Consoles.ReleaseTimeline [Last accessed: 29 March 2010]
Japanese console release dates

Crigger, Lara (2007) 'Searching for Gunpei Yokoi'. *The Escapist*, 6 March 2007. [Online] www.escapistmagazine.com/articles/view/issues/issue_87/490-Searching-for-Gunpei-Yokoi [Last accessed: 13 March 2010]
Gunpei Yokoi's design philosophy explained

'East is Eden: A journey into the heart of Japanese game culture' (2002) *Edge*, Issue 108, March 2002, pp52-63
Examination of Japanese gaming tastes

Gameheadz: History of Video Games (2003) Discovery Channel [TV broadcast]
Interview with Shigeru Miyamoto

Gielens, Jaro (2000) *Electronic Plastic*. Berlin, Germany: Die Gestatlen Verlag
Information on Game & Watch and other Japanese handhelds

Goto, Hiroshige (2006) 'Interview with Nintendo president Satoru Iwata'. [In Japanese] *PC Watch*. 6 December 2006. [Online] http://pc.watch.impress.co.jp/docs/2006/1206/kaigai324.htm [Last accessed: 13 March 2010]
Nintendo's belief in Gunepi Yokoi's design theories

Hatano, Yoshiro and Shimazaki, Tsuguo (1997) 'Japan (Nippon)'. In: Francoeur, Robert T. (editor) (1997) *The International Encyclopedia of Sexuality*, Volume I-IV. New York, New York: The Continnum Publishing Company. [Online] www2.hu-berlin.de/sexology/IES/japan.html#0 [Last accessed: 13 March 2010]
Gender and sexuality in Japan

Iwabuchi, Koichi (2002) *Recentering Globalization: Popular Culture and Japanese Transnationalism*. Durham, North Carolina: Duke University Press.
Japan's cultural relationship with Asia and the West post-war

Iwabuchi, Koichi (2004) 'How 'Japanese' is Pokémon?'. In: Tobin, Joseph (editor) (2004) *Pikachu's Global Adventure: The Rise and Fall of Pokémon*. Durham, North Carolina: Duke University Press, pp53-79
Japanese popular culture

'Iwata asks – Punch-Out!!' (2009) *Wii.com*. 7th August 2009 [Online] http://uk.wii.com/wii/en_GB/software/iwata_asks_-_punch-out_2251.html [Last accessed: 27 March 2010]
Details about the game EVR Race

Jackson, David S. (1996) 'The Spielberg of video games'. *Time*. 20 May 1996. [Online] www.time.com/time/magazine/article/0,9171,984568,00.html [Last accessed: 13 March 2010]
Interview with Shigeru Miyamoto

'Japan bans rape-simulation video games' (2009) *Fox News*. 5 June 2009. [Online] www.foxnews.com/story/0,2933,525228,00.html [Last accessed: 13 March 2010]
The RapeLay controversy

Kelts, Roland (2006) *Japanamerica: How Japanese Pop Culture has Invaded the U.S.*. 2007 paperback edition. New York, New York: Palgrave Macmillan
Influence of Japanese culture on North America

Kent, Steven L. (2001) *The Ultimate History of Video Games*. New York, New York: Three Rivers Press
History of Nintendo

Kohler, Chris (2005) *Power-Up: How Japanese Video Games Gave the World an Extra Life*. Indianapolis, Indiana: BradyGames
Overview of Japanese video game culture

Masuyama (2002) 'Pokémon as Japanese culture?'. In: King, Lucien (editor) (2002) *Game On: The History and Culture of Videogames*. London, UK: Laurence King Publishing, pp34-43
Japanese popular culture and video games

Lieu, Tina (1997) 'Where have all the PC games gone?'. *Japan Inc*. August 1997. [Online] www.japaninc.com/cpj/magazine/issues/1997/aug97/0897pcgames.html [Last accessed: 13 March 2010]
PC gaming in Japan

Lunsing, Wim (2006) 'Yaoi ronso: Discussing depictions of male homosexuality in Japanese girls' comics, gay comics and gay pornography'. *Intersections: Gender, History and Culture in the Asian Context*. Issue 12, January 2006. [Online] http://intersections.anu.edu.au/issue12/lunsing.html [Last accessed: 13 March 2010]
Japanese homosexual manga explained

McCloud, Scott (1994) *Understanding Comics: The Invisible Art*. New York, New York: HarperPerennial
Differences in comic art styles in the US and Japan

McCurry, Justin (2009) 'Japan under pressure to clamp down on child pornography'. *Guardian.co.uk*. 11 May 2009. [Online] www.guardian.co.uk/world/2009/may/11/japan-child-pornography [Last accessed: 13 March 2010]
Controversy about RapeLay

McLelland, Mark (2005) *Queer Japan from the Pacific War to the Internet Age*. Lanham, MD: Rowman & Littlefield Publishers
Homosexuality in post-Second World War Japan

McLelland, Mark (2006) 'A short history of hentai'. *Intersections: Gender, History and Culture in the Asian Context*, Issue 12, January 2006. [Online] http://intersections.anu.edu.au/issue12/mclelland.html [Last accessed: 13 March 2010]
History of bishojo publications

McRoy, Jay (editor) (2005) *Japanese Horror Cinema*. Honolulu, Hawaii: University of Hawai'i Press
Background on the otaku and Japanese attitudes to technology

Moosa, Eugene (1983) 'Japanese video junkies turning to crime to support their habit'. *The Ledger*. Lakeland, Florida. Friday 11 February 1983, p8D
Figures on the number of arcades operating in Japan

Nintendo of America (2008) 'Biography of Shigeru Miyamoto'. [Press release] Nintendo of America, July 2008.
Shigeru Miyamoto's career history

'Nintendon't' (2007) *Games*[TM], Issue 63, October 2007, pp156-161
The history of Nintendo

Old Computers (no date) [Online] http://old-computers.com [Last accessed: 2 March 2010]
Information on Japanese computers and consoles

Outrageous Fortunes: Nintendo [TV broadcast] London, UK: BBC Three. Monday 19th April 2004, 9pm.
Background on the history of Nintendo

Palmer, Edwina (editor) (2005) *Asian Futures, Asian Traditions*. Folkestone, Kent: Global Oriental
Cultural thinking in Japan during the 1980s

Perry, David and DeMaria, Rusel (2009) *David Perry on Game Design: A Brainstorming Toolbox*. Boston, Massachusetts: Course Technology
Thoughs on Japanese role-playing games

Pollack, Andrew (1984) 'The keyboard stymies Japan'. *The New York Times*. 7 June 1984. pD-17
Language problems undermine computing in Japan

Porter, Michael E. et al (2009) *The Video Games Cluster in Japan*. [Online] www.isc.hbs.edu/pdf/Student_Projects/Japan_Video_Games_2009.pdf [Last accessed: 14 March 2010]
Figures on the Japanese game industry including the popularity of role-playing games

Richards, Evelyn (1983) 'Cracking Japan computer market surely no fast-food undertaking'. *Sunday Deseret News*. Salt Lake City, Utah. Sunday 6 November 1983. p6M
Home computing in Japan in the early 1980s

'Rising Sun: How Japan stole the videogame industry' (2002) *Edge*, Issue 108, March 2002, pp68-77
History of the Japanese video game industry

Sheff, David (1994) *Game Over: Nintendo's Battle to Dominate an Industry*. 1999 *Arcade* edition. London, UK: Hodder and Stoughton
The history of Nintendo

'The history of Epoch Co.' (2010) Epoch. [In Japanese] http://epoch.jp/info/ep02.html [Last accessed: 13 March 2010]
Overview of Epoch's history

'The making of The Black Onyx' (2008) *Edge*, Issue 185, February 2008, pp102-105
Interview with Henk Rogers

'World News' (1988) *Nintendo Power*, No. 1, July/August 1988, p92
Dragon Quest III's popularity in Japan

13 I I COULD HAVE SWORN IT WAS 1983

'Absolutely brilliant!' (1992) *Amiga Power*, Issue 9, January 1992. pp86-90
Interview with Codemasters

'An audience with Richard Darling' (1999) *Edge*, Issue 75, September 1999, pp118-123
Interview with Codemasters' Richard Darling

Blomquist, Cord and Lehrer, Eli (2007) *Politically Determined Entertainment Ratings and How to Avoid Them*. Issue Analysis No. 12, December 2007. Washington, DC: Competitive Enterprise Institute
Politics of the Comics Code and Hays Code

Brown, Damon (2008) *Porn & Pong: How Grand Theft Auto, Tomb Raider and Other Sexy Games Changed Our Culture*. Port Townsend, Washington: Feral House, p66
Recounts Nintendo's content rules

Burstein, Daniel (1989) 'A yen for New York: What the Japanese own – what they're after' *New York*. 16 January 1989, pp27-36
Feature on Japanese influence in New York

Code of the Comics Magazine Association of America Inc., adopted October 26, 1954 (1954) In: Senate Committee on the Judiciary (1955) *Comic Books and Juvenile Delinquency. Interim Report*. Washington, DC: United States Government Printing Office
The Comics Code

'Company lookback: Codemasters' (2003) *Games^TM*, Issue 7, June 2003, pp144-147
The history of Codemasters

'Console yourself' (2008) *Games^TM*, Issue 66, January 2008, pp158-163
Information on the Amstrad GX4000 console

Cousins, Mark (2004) *The Story of Film*. 2008 paperback edition. London, UK: Pavilion Books
The Hays Code and its effects

Crudele, John (1988) 'Video-game comeback'. *New York*. 25 January 1988, p20
Report on the return of the video game console

De Gale, Luther (1988) 'Whatever happened to the Nintendo?'. *The Games Machine*, Issue 11, October 1988. pp27-28
The failure of Nintendo to conquer Europe

DeMaria, Rusel and Wilson, Johnny L. (2002) *High Score! The Illustrated History of Electronic Games*. Berkeley, California: McGraw-Hill/Osborne
The performance of the Sega Master System in the US

'Dust to dust, attics to attics?' (1987) *The Games Machine*, Issue 1, October/November 1987, pp29-31
European perspective on the return of the games console

Farnham, Alan et al (1987) 'Video games make a Christmas comeback'. *Fortune*. 7 December 1987. [Online] http://money.cnn.com/magazines/fortune/fortune_archive/1987/12/07/69967/index.htm [Last accessed: 13 March 2010]
The return of the video game console

Get into the World of Nintendo (1990) [Catalog] Nintendo of America
Nintendo Power supplement showing the vast range of Nintendo-related merchandise available in the US in the late 1980s

Glasser, Ray and Vvolo, Art (1987) 'Inside the 1987 Consumer Electronics Show'. [Online video] www.youtube.com/watch?v=Jm-485Khm6E [Last accessed: 13 March 2010]
The technology on show at the 1987 Consumer Electronics Show

Gros Pixels (no date) [In French] [Online] www.grospixels.com [Last accessed: 12 March 2010]
Information on the Amstrad GX4000 console

Held, Tom (1988) 'TV video games are elusive target'. *Milwaukee Sentinel*. Tuesday 20 December 1988. p1 & 9
Limited supplies of Nintendo products to US shops

Herz, J.C. (1997) *Joystick Nation*. London, UK: Abacus
The scale of Nintendo's operations in the US

Jackson, David S. (1996) 'The Spielberg of video games'. *Time*. 20 May 1996. [Online] www.time.com/time/magazine/article/0,9171,984568,00.html [Last accessed: 13 March 2010]
Profile of Shigeru Miyamoto

Jansen, Marius B. (2000) *The Making of Modern Japan*. Cambridge, Massachusetts: Belknap Press, Harvard University Press
The history of Japan after the Second World War

'Japanese will play at bashing: Computer game will offer 400 ways to slam their own country' (1992) *Morning Star*. Wilmington, North Carolina. Friday 13 March 1992. p3A
Report on the launch of Japan Bashing

Kean, Roger and Wild, Nik (1988) 'Ultimaely playing a Rare game' *The Games Machine*, Issue 4, March 1988, pp27-34
Interview with Rare

Kent, Steven L. (2001) *The Ultimate History of Video Games*. New York, New York: Three Rivers Press
Nintendo's rise to dominance in the US and the lawsuits that followed

Kim, Min-kyu and Park, Tae-soon (2006) *Between Censorship and Rating: State of Global Screening Systems*. Seoul, South Korea: Korea Game Development Institute
Examination of ethical codes, such as the Hays Code, and their relationship with video games

Kohler, Chris (2005) *Power-Up: How Japanese Video Games Gave the World an Extra Life*. Indianapolis, Indiana: BradyGames
Biography of Shigeru Miyamoto and overview of the Super Mario Bros series

Koster, Raph (2005) *A Theory of Fun for Game Design*. Scottsdale, Arizona: Paraglyph Press
The Comics Code's effects on comic books

Lindstron, Bob (1991) 'The gaming globe: different cultures play different games. Or do they?'. *Compute!*, Issue 125, January 1991. p74
US perspective on European gaming in the late 1980s and early 1990s

Matthew, Eric (1991) 'If I'd known then…'. *Amiga Power*, Issue 2, June 1991. pp72-3
The Bitmap Brother reflects on his games

McGill, Douglas C. (1988) 'Nintendo scores big'. *The New York Times*. 4 December 1988. p31
Nintendo's success in reviving the games console

McGill, Douglas C. (1989) 'A Nintendo labyrinth filled with lawyers, not dragons'. *The New York Times*. 9 March 1989. pA1
The legal battle between Atari Games and Nintendo

Meigs, James B. (1987) 'Home video: Games bounce back'. *Popular Mechanics*. Volume 164, No. 10, October 1987. p28
Report on the return of the video game console

Moore, Lucy (2008) *Anything Goes: A Biography of the Roaring Twenties*. London, UK: Atlantic Books, pp112-3
Hollywood's Sin City image and the creation of the Hays Code

'Mr Sugar, it's triplets' (1990) *The Games Machine*. Issue 34, September 1990, pp14-5
Report on the Amstrad GX4000 console

'Nintendo is sued by Atari' (1989) *The New York Times*. 2 February 1989. pD13
Nintendo being sued by Atari Corporation and Atari Games

Nintendo of America (no date) 'Nintendo's History'. [Public relations information]
History of Nintendo

Page, Barnaby (1988) 'The Konix console – it's brilliant!'. *The Games Machine*. Issue 12, November 1988. pp8-9
Preview of the never-released Konix console

'Playing dirty in video games?' (1989) *St Petersburg Times*. St Petersburg, Flordia. 8 December 1989. Section E, page 1
Democrat senator questions Nintendo's dominance of the US video game industry

Provenzo Jr, Eugene (1991) *Video Kids: Making Sense of Nintendo*. Cambridge, Massachusetts: Harvard University Press
Provenzo's examination of the success of Nintendo in the US

Ramirez, Anthony (1990) 'Court backs Nintendo on video-game suits'. *The New York Times*. 15 March 1990. pD5
US Court of Appeals backs Nintendo in the Atari Games lawsuit

Rosenthal, Marshal M. (1989) 'A tale of two Ataris: Lawsuit update and exclusive interview'. *The Games Machine*, Issue 17, April 1989, p13
Report on Atari lawsuits against Nintendo in the US

Sanger, David E. (1992) 'He may buy the team and not see the game'. *Pittsburgh Post-Gazette*. 10 February 1992, p2
Report on Hiroshi Yamauchi's buy out of the Seattle Mariners baseball team

Sheff, David (1994) *Game Over: Nintendo's Battle to Dominate an Industry*. 1999 *Arcade* edition. London, UK: Hodder and Stoughton
The history of Nintendo

Smith, Steve (1986) 'Atari, Sega and Nintendo plan comeback for video games'. *HFD – The Weekly Home Furnishings Newspaper*. 23 June 1986. [Online] As reproduced at: www.escalonimaginario.com/elafountain/comeback.html [Last accessed: 13 March 2010]
News report on moves to bring the video game back to US homes

'The motion picture production code of 1930 (Hays Code)' (1930) Association of Motion Picture Producers and the Motion Picture Producers and Distributors of America. [Online] As reproduced online at: www.artsreformation.com/a001/hays-code.html [Last accessed: 13 March 2010]
The original version of the Hays Code

'The Nintendo Entertainment System' [Advert by Nintendo of America] *New York*. 25 November 1989. pp12-13
R.O.B. leads Nintendo's advertising campaign

'The original good egg' (2006) *Retro Volume 2: The Ultimate Retro Companion from Games™*. Bournemouth, UK: Imagine Publishing. pp12-17
The history of Dizzy

'The power generation' (1989) *The Games Machine*, Issue 16, March 1989. pp16-19
Feature on the Konix Multi-System console

The Wizard (1989) [Film] Directed by: Todd Holland. Universal Pictures.
The Nintendo-promoting movie

Tobin, Joseph (editor) (2004) *Pikachu's Global Adventure: The Rise and Fall of Pokémon*. Durham, North Carolina: Duke University Press
Information on US fears about Japan in the late 1980s

'Urge parents to ban death games' (1988) *The Bryan Times*. Bryan, Ohio. 19 November 1988. Entertainment section, p4
National Coalition on Television Violence takes aim at Nintendo

'Video game industry eyes comeback' (1986) *Ocala Star-Banner*. Ocala, Flordia. Sunday 28 September 1986. p8B
The comeback of the video game console

'Video games come back from brink' (1987) *Milwaukee Sentinel*. Wednesday 3 December 1987. Part 4, p10
Report on the comeback of the game console

Watson, Tom (1991) 'Renegade and The Bitmap Brothers'. *Amiga Power*, Issue 1, May 1991. pp70-74
Interview with The Bitmap Brothers

Williams, Dmitri (2002) 'Structure and competition in the US home video game industry'. *JMM – The International Journal on Media Management*. Volume 4, No. 1, pp41-54
Reflections on the history of the US video game industry

14 I INTERACTIVE MOVIES

'Agreement regarding the confidentiality of information' (1983) Amiga Corporation and Atari Inc. [Internal company document] 21 November 1983
Internal Atari Inc. document on the company's deal with the Amiga Corporation

Atari Inc & Lucasfilm (1984) 'Sound effects add new excitement to galactic gameplay'. [Press release] Issued: 8 May 1984. [Online] http://peterlangston.com/LFGames/PRSound.html [Last accessed: 13 March 2010]
The launch of Lucasfilm Games and its movie-inspired approach to games

Bagnall, Brian (2007) *On the Edge: The Spectacular Rise and Fall of Commodore*. Winnipeg, Manitoba, Canada: Variant Press
The history of the Commodore Amiga

'Behind the scenes: Maniac Mansion + Day of the Tentacle' (2007) *Games™*, Issue 63, October 2007, pp142-147
The creation of Lucasfilm's SCUMM adventure game system

Bielby, Matt (1991) 'The force: The Lucasfilm interview'. *Amiga Power*, Issue 7, November 1991, pp70-74
Interview with Lucasfilm's Doug Glen

Defender of the Crown (1986) [Video game manual] [Amiga] Cinemaware
Sets out Cinemaware's vision

DeMaria, Rusel and Wilson, Johnny L. (2002) *High Score! The Illustrated History of Electronic Games*. Berkeley, California: McGraw-Hill/Osborne
Infocom's 'We unleash the world's Ö' advert and background on Lucasfilm and Cinemaware

Greer, Jonathan (1984) 'Star Wars creator teams up with Atari'. *San Jose Mercury News*, 9 May 1984. [Online] http://peterlangston.com/LFGames/SJMN19840508.html [Last accessed: 13 March 2010]
Lucasfilm Games' links with the movie business

Grimes, William (1993) 'When the film audience controls the plot'. *The New York Times*, 13 January 1993. pC15
Background on Hasbro's NEMO

Jacob, Bob (1991) 'If I'd known then…'. *Amiga Power*, Issue 7, November 1991, pp76-77
Cinemaware co-founder reflects on his company's games

'Jay Miner interview, Pasadena, September 1992' (1992) *Craig's Retro Computing Page* [Online] www.craigsretrocomputingpage.com/jaystale/jaystale.html [Last accessed: 13 March 2010]
Interview with Jay Miner, the creator of the Commodore Amiga computer

Katz, Arnie (1984) 'Lucasfilm premieres first two games: Can it become a force in electronic gaming?'. *Electronic Games*, September 1984. [Online] www.atarihq.com/othersec/library/lucasflm.html [Last accessed: 13 March 2010]
Report on the launch of Lucasfilm Games

Katz, Arnie (1988) 'The Cinemaware story'. *ST-Log*, Issue 18, April 1988, p81
The rise of Cinemaware

Mechner, Jordan (1985-1993) *Old Journals.* [Online] http://jordanmechner.com/old-journals [Last accessed: 13 March 2010]
Details the development of The Prince of Persia

Reimer, Jeremy (2007) 'A history of the Amiga, part 1: Genesis'. *Ars Technica.* 1 August 2007. [Online] http://arstechnica.com/hardware/news/2007/07/a-history-of-the-amiga-part-1.ars [Last accessed: 13 March 2010]
History of the Commodore Amiga

Rinzler, J.W. (2007) *The Making of Star Wars.* 2008 Edition. London, UK: Ebury Press
Background on the special effects work in the movie Star Wars

Rubin, Michael (2006) *Droidmaker: George Lucas and the Digital Revolution.* Gainesville, Florida: Triad Publishing
The history of Lucasfilm and Lucasfilm Games

Smith, Rob (2008) *Rogue Leaders: The Story of LucasArts.* London, UK: Titan Publishing Group
The history of Lucasfilm Games

'Ten tips from the programming pros: Secrets from Lucasfilm's Game Group' (1984) *Atari Connection*, Spring 1984, pp34-35
[Online] http://peterlangston.com/LFGames/TenTips.html [Last accessed: 13 March 2010]
The Lucasfilm Games' team set out their beliefs on game design

'The making of Maniac Mansion' (2005) *Edge*, Issue 151, July 2005, pp104-109
Feature on the creation of Lucasfilm's first SCUMM-based adventure game

Vinciguerra, Rev. Robert A. (2008) 'Where are they now? Nolan Bushnell's Axlon'. *The Rev. Rob Times*, 17 January 2008 [Online] www.revrob.com/content/view/112/52 [Last accessed: 13 March 2010]
A look back at Nolan Bushnell's Axlon

15 | AH! YOU MUST BE A GOD

Carless, Simon (2006) 'Being Peter Molyneux'. *Gamasutra*, 31 October 2006. [Online] www.gamasutra.com/view/feature/1791/being_peter_molyneux.php [Last accessed: 31 October 2006]
Interview with Peter Molyneux

Chaplin, Heather and Ruby, Aaron (2005) *Smartbomb: The Quest for Art, Entertainment and Big Bucks in the Videogame Revolution.* 2006 paperback edition. Chapel Hill, North Carolina: Algonquin Books
Profile of Will Wright

Bielby, Matt (1991) 'Peter Molyneux, what have you started?'. *Amiga Power*, Issue 5, September 1991, pp56-59
Report on the rise of 'god games' following the release of Populous

Bunten Berry, Dani (no date) *Game Design Memoir* [Online] www.anticlockwise.com/dani/personal/biz/memoir.htm [Last accessed: 13 March 2010]
Dani Bunten Berry's views on her game Seven Cities of Gold

Edwards, Benj (2007) 'The history of Civilization'. *Gamasutra*, 18 March 2003. [Online] www.gamasutra.com/view/feature/1523/the_history_of_civilization.php [Last accessed: 13 March 2010]
The history of Sid Meier's Civilization

Herz, J.C. (1997) *Joystick Nation.* London, UK: Abacus

Interview with Will Wright

Gardner, Martin (1970) 'Mathematical Games: The fantastic combinations of John Conway's new solitaire game Life'. *Scientific American*, Issue 223, October 1970, pp120-123
Information on the original version of Life

Gorenfeld, John (2003) 'Get behind the M.U.L.E.'. *Salon.com*, 18 March 2003. [Online] http://dir.salon.com/story/tech/feature/2003/03/18/bunten/index.html [Last accessed: 13 March 2010]
A profile of Dani Bunten Berry

Hague, James (editor) (1997) *Halcyon Days: Interviews with Classic Computer and Video Game Programmers*. Free web version 2002. Dadgum Games. [Online] www.dadgum.com/halcyon [Last accessed: 28 February 2010]
Interview with Dani Bunten Berry

Hunt, Stuart (2008) 'In the chair with...Will Wright'. *Retro Gamer*. Issue 47, January 2008, pp56-59.
Interview with Will Wright

Molyneux, Peter (1991) 'If I'd known then...'. *Amiga Power*, Issue 1 May 1991, p112
Peter Molyneux reflects on his earliest games, including Populous

'PC Retroview: Dune II' (2000) *IGN: UK Edition*. 13 July 2000 [Online] http://uk.pc.ign.com/articles/082/082093p1.html [Last accessed: 13 March 2010]
The creation of Dune II

Pullin, Keith (2002) 'Games that changed the world: Sid Meier's Civilization'. *PC Zone*, Issue 121, November 2002. pp150-153
The history of Civilization

Seabrook, John (2006) 'Game master: Will Wright changed the concept of video games with The Sims. Can he do it again with Spore?'. *The New Yorker*, 6 November 2006. [Online] www.newyorker.com/archive/2006/11/06/061106fa_fact [Last accessed: 13 March 2010]
Interview with Will Wright

'Sid Meier's conversation' (2006) *Edge*, Issue 159, February 2006, pp62-67
Interview with Sid Meier

'Sid Meier's Pirates!' (2005) *UGO Strategy Gaming Online*, 31 January 2005. [Online] www.strategy-gaming.com/interviews/sid_meiers_pirates [Last accessed: 13 March 2010]
Interview with Sid Meier

'The book of god games' (2009) *Games^{TM}*, Issue 79, January 2009, pp80-87
God game retrospective

'The making of Dune II: Battle for Arrakis' (2008) *Edge*, Issue 196, Christmas 2008, pp106-111
The creation of Dune II

'The making of Populous' (2003) In: *Edge Presents Retro: The Making of Special* (2003). Bath, UK: Future Publishing. pp108-111
The creation of Populous

'The making of X-COM: Enemy Unknown' (2003) *Edge*, Issue 131, Christmas 2003, pp129-131
The influence of Civilization on X-COM: Enemy Unknown

'The railroad tycoon' (2006) *Games^{TM}*, Issue 51, November 2006, pp88-91
Interview with Sid Meier

Winstanley, Mark (1993) 'Dune II'. *Amiga Power*, Issue 22, February 1993, pp16-7
Preview of Dune II

16 | A PLANE TO MOSCOW

Arcade History (no date) [Online] www.arcade-history.com [Last accessed: 2 March 2010]
Background on Eastern European and USSR arcade games

'Behind the scenes of Tetris' (2006) *Games^{TM}*, Issue 51, November 2006, pp146-149

The making of Tetris

Bridge, Tony (1985) 'Eureka!'. *Micro Adventurer*, Issue 15, January 1985, pp34-5
Report on the launch of the Hungarian-made game Eureka!

Crowley, David (2008) 'Thaw modern: Design in Eastern Europe after 1956'. In: Crowley, David and Pavitt, Jane (editors) (2008) *Cold War Modern: Design 1945-1970*. London, UK: V&A Publishing
Goulash communism explained

DeMaria, Rusel and Wilson, Johnny L. (2002) *High Score! The Illustrated History of Electronic Games*. Berkeley, California: McGraw-Hill/Osborne
The making of Tetris

Edwards, Tim (2003) 'From Russia with Love'. *PC Gamer*, October 2003. pp70-75
The making of Tetris

'Eureka! £25,000' (1984) *Personal Computer Games*, Issue 11, October 1984, p13
News report on the launch of Eureka!

Extrema Ukraine (no date) 'The history of our company'. [In Russian] [Online] http://extrema-ua.com/history.php [Last accessed: 13 March 2010]
The history of the Ukrainian arcade game maker

'Global underground' (2002) *Edge*, Issue 118, Christmas 2002. pp72-82
The state of Russian and Eastern European game development circa 2002

Handy, Dominic (1989) 'Licence to sell'. *The Games Machine*, Issue 23, October 1989. pp61-62
The history of Domark, publishers of Eureka!

Kent, Steven L. (2001) *The Ultimate History of Video Games*. New York, New York: Three Rivers Press
The legal battle over Tetris

Lewis, Peter H. (1988) 'New software game: It comes from Soviet'. *The New York Times*, 29 January 1988, pD1
News report on the arrival of Tetris in the US

'Mr Tetris' (1999) *Edge*, Issue 74, August 1999, pp69-75
Interview with Henk Rogers

Museum of Soviet Arcade Machines (no date) [Online] [In Russian] www.15kop.ru [Last accessed: 13 March 2010]
Information on Soviet-era arcade games

'Russia: Borrowing from the capitalists' (1965) *Time*, Feburary 12 1965 [Online] www.time.com/time/magazine/article/0,9171,840532-1,00.html [Last accessed: 13 March 2010]
Compares the USSR's technology that of the US

Sheff, David (1994) *Game Over: Nintendo's Battle to Dominate an Industry*. 1999 *Arcade* edition. London, UK: Hodder and Stoughton
The battle for Tetris recounted in detail

Shoemaker, Richie (2004) 'Games that changed the world: Operation Flashpoint: Cold War Crisis'. *PC Zone*, Issue 140, April 2004. pp138-141
The making of Operation Flashpoint and background about Bohemia Interactive

'Soviet Union: Computer games' (1977) *Time*, 1 August 1977 [Online] www.time.com/time/magazine/article/0,9171,915171,00.html [Last accessed: 13 March 2010]
The computer technology gap between the US and USSR

Szczepaniak, John (2005) 'Global gaming'. *Retro Gamer*, Volume 2, Issue 3, April 2005. pp78-80
Video gaming in Russia

Szczepaniak, John (2005) 'It's a gamer's world out there'. *The Gamer's Quarter*, Issue 2, 3rd quarter 2005, pp82-5
Gaming in Russia with brief mention of the TR-DOS operating system

Tetris: From Russia with love (2004) [TV broadcast] London, UK: BBC Four. 6 July 2004. GMT: 20.30
The battle for the rights to Tetris

'The making of Tetris' (2001) *Edge*, Issue 100, August 2001, pp134-137
The making of Tetris

'This is not life, just business: Rényi Gábor' (2006) [In Hungarian] *Manager Magazin*, October 2006 [Online] www.managerma-gazin.hu/magazin.php?page=article&id=620 [Last accessed: 13 March 2010]
Information on Rényi Gábor and his father

U.S.S.R. Online (no date) [Online] http://soviet-empire.com [Last accessed: 13 March 2010]
Forum comments on Soviet-era arcade gaming

Xikluna, Nicky (1985) 'Behind the curtain: Big K goes to Hungary!'. *Big K*, Issue 12, March 1985, pp28-30
Profile of Hungarian game studios during the communist era

Zaitchik, Alexander (2007) 'Soviet-era arcade games crawl out of their Cold War graves'. *Wired*, 7 June 2007 [Online] www.wired.com/gaming/hardware/news/2007/06/soviet_games [Last accessed: 13 March 2010]
Report on Soviet Union arcade machines

Zaitchik, Alexander (2007) 'The lost arcade games of the Soviet Union'. *Wired*, 7 June 2007 [Online] www.wired.com/gaming/hardware/multimedia/2007/06/gallery_soviet_games [Last accessed: 13 March 2010]
Details of Soviet Union arcade machines

17 | SEGA DOES WHAT NINTENDON'T

Ashcraft, Brian with Snow, Jean (2008) *Arcade Mania! The Turbo-Charged World of Japan's Game Centers*. Tokyo, Japan: Kodansha International
Chapter 5 examines the popularity of fighting games in Japanese arcades

Cousins, Mark (2004) *The Story of Film*. 2008 paperback edition. London, UK: Pavilion Books
Background on Hong Kong's kung-fu movie industry

DeMaria, Rusel and Wilson, Johnny L. (2002) *High Score! The Illustrated History of Electronic Games*. Berkeley, California: McGraw-Hill/Osborne
Background on Electronic Arts relationship with Sega and Sega's worries about Altered Beast

Kent, Steven L. (2001) *The Ultimate History of Video Games*. New York, New York: Three Rivers Press
Covers the battle between Sega and Nintendo in the early 1990s, figures from NPD Group on Super NES and Genesis sales

Leeds, Matthew (1987) 'Earl Weaver Baseball'. *Commodore Magazine*, Volume 8, No. 5, pp48-49
Overview of Earl Weaver Baseball

Manly, Lorne (2006) 'Strat-O-Matic, the throwback, endures the era of the X-Boxes'. *The New York Times*, 13 January 2006 [Online] www.nytimes.com/2006/01/13/sports/13stratomatic.html?_r=2 [Last accessed: 13 March 2010]
The history of the Strat-O-Matic sports games

'Nintendo, Sega & Sony under one roof' (1996) *Next Generation*, Issue 2, August 1996, pp6-12
Console maker attitudes to each other in the US

Sheff, David (1994) *Game Over: Nintendo's Battle to Dominate an Industry*. 1999 *Arcade* edition. London, UK: Hodder and Stoughton
Recounts Sega's challenge to Nintendo

SportsNight Oklahoma (2008) 'Strat-O-Matic Baseball'. [TV news report] Broadcast: 2 July 2008. Cox Channel. [Online video] www.youtube.com/watch?v=_vMGRmWuBOQ [Last accessed: 13 March 2010]
Report on Strat-O-Matic's enduring popularity

Takoushi, Tony (1989) 'Sega's sweet sixteen'. *Advanced Computer Entertainment*, No. 17, February 1989, pp9-11
Preview of the Sega Megadrive

'The making of John Madden Football' (2003) *Edge*, Issue 119, January 2003, pp104-107
The creation of the Megadrive version of John Madden Football

'The making of Street Fighter II' (2002) *Edge*, Issue 108, March 2002, pp100-105
Covers the creation of Street Fighter II

18 | MORTAL KOMBAT

Brandt, Richard (1993) 'Video games: Is all that gore really child's play?'. *BusinessWeek*, 14 June 1993. [Online] www.business-week.com/archives/1993/b332324.arc.htm [Last accessed: 13 March 2010]
Report on the controversy about violent games and the Senate hearings

Bruch, Debra (2004) 'The prejudice against theatre'. *The Journal of Religion and Theatre*. Volume 3, No. 1, Summer 2004. [Online] www.rtjournal.org/vol_3/no_1/bruch.html [Last accessed: 4 April 2010]
Plato's views on theatre

Bryan, Kevin (2005) *PSX: The Guide to the Sony Playstation*. 2007 edition. Clifton, New Jersey: Digital Press. p93 [Online] www.digitpress.com/products/psxbook.htm [Last accessed: 10 March 2010]
Reflections on impact of the age rating system

Bryon Review (2008) *Safer Children in a Digital World: The Report of the Bryon Review*. London, UK: Department for Children, Schools and Families.
Research review of the effects of video games on children

Diamond, John (1993) 'Senators urge warning labels on violent video games'. *Ludington Daily News*, Ludington, Michigan. 2 December 1993, p14
Report on the controversy about violent games and the Senate hearings

Diamond, John (1993) 'Warning labels urged on games'. *Pittsburgh Post-Gazette*. 2 December 1993, pA-10
Report on the controversy about violent games and the Senate hearings

Elmer-Dewitt, Philip et al (1993) 'The amazing video game boom'. *Time*, 27 September 1993. [Online] www.time.com/time/magazine/article/0,9171,979289,00.html [Last accessed: 13 March 2010]
The expansion of the games industry in the early 1990s

Elmer-Dewitt, Philip and Dickerson, John F. (1993) 'Too violent for kids?'. *Time*, 27 September 1993. [Online] www.time.com/time/magazine/article/0,9171,979298,00.html [Last accessed: 13 March 2010]
Report on the controversy about violent games and the Senate hearings

'Gremlin: Nuking US Gold' (1985) *Sinclair User*, Issue 39, June 1985. p13
News in brief on CND's reaction to Raid Over Moscow

Grimes, William (1993) 'When the film audience controls the plot'. *The New York Times*, 13 January 1993. pC15
Report on CD-based games that mentions Night Trap

'Kay-Bee Toys stops selling Night Trap' (1993) *The News*, Boca Raton, Florida. 18 December 1993, p7B
Retailers react to the violent games controversy

Kent, Steven L. (1994) 'Video battles – The hottest games hitting the markets all share the same weapon: violence'. *The Seattle Times*, 13 October 1994 [Online] http://community.seattletimes.nwsource.com/archive/?date=19941013&slug=1935654 [Last accessed: 13 March 2010]
The impact of the age ratings system in the US

Kent, Steven L. (2001) *The Ultimate History of Video Games*. New York, New York: Three Rivers Press
Account of the Senate hearings and the formation of an age rating system

Kline, Michael (1994) 'Politicians aim firepower at video games'. *The Daily Collegian Online*, 15 February 1994. [Online] www.collegian.psu.edu/archive/1994/02/02-15-94tdc/02-15-94darts-2.asp [Last accessed: 13 March 2010]
Report on the controversy about violent games and the Senate hearings

Kohler, Chris (2005) *Power-Up: How Japanese Video Games Gave the World an Extra Life*. Indianapolis, Indiana: BradyGames
Attitudes to race in Japan and how they are reflected in video games

'Konception' (2002) *Mortal Kombat: Deadly Alliance – The Komeback* [*Edge* supplement] Bath, UK: Future Publishing. pp4-7
The creation of Mortal Kombat

Langberg, Mike (1994) 'Sega to withdraw the controversial game Night Trap from stores'. *Knight-Ridder/Tribune News Service.* 10 January 1994
Sega takes Night Trap off the shelves

'List of games generally banned in Germany' (2009) Bundesprüfstelle für Jugendgefährdende Medien
List of games banned in Germany, supplied direct to the author

'Minority Report' (2002) *Edge,* Issue 110, May 2002, pp72-79
Discrimination and prejudice in video games

'Nazi video games glorify Holocaust' (1991) *Gainesville Sun.* Gainesville, Florida. 4 May 1991, p5D
Report on neo-Nazi video games appearing in Germany and Austria

Nelson, Robert T. (1993) 'Video-game makers agree to rating system'. *The Seattle Times,* 9 December 1993 [Online] http://community.seattletimes.nwsource.com/archive/?date=19931209&slug=1736106 [Last accessed: 13 March 2010]
US video game industry proposes age rating system

Nichols, Peter M. (1993) 'Home video: Bills in many states'. *The New York Times,* 18 February 1993, pC22
Video game controversy goes nationwide in the US

Nichols, Peter M. (1993) 'Home video: Movies on a 5-inch CD'. *The New York Times,* 3 June 1993, pC18
Pre-hearing coverage of Night Trap

Nichols, Peter M. (1995) 'Home video'. *The New York Times,* 27 October 1995, pD22
The relaunch of Night Trap

'Philip Morris wants its logo off video games' (1991) *Toledo Blade.* Toledo, Ohio. 13 February 1991, p10
Tobacco manufacturer rows with Sega about in-game cigarette adverts

Puga, Ana (1993) 'Vile video shocks Senate'. *Pittsburgh Post-Gazette,* 10 December 1993, pA-17
Report on the controversy about violent games and the Senate hearings

'Ratings for video games' (1994) *The New York Times,* 4 January 1994, pD11
Covers the introduction of age rating system

Reeder, Sara (1992) 'Computer game ethics'. *Compute!,* Issue 137, January 1992, p100
Early 1990s attempts to restrict the content of video games in the US

Redburn, Tom (1993) 'Toys 'R' Us stops selling a violent video game'. *The New York Times,* 17 December 1993, pB1
Retailer reaction to the video game violence controversy

Sarrikoski, Petri (1999) 'Tietokonepelit Osana Audiovisuaalisen Kulttuurin Moraalipaniikka'. *Wider Screen,* Issue 1-2, 1999. [In Finnish] [Online] www.widerscreen.fi/1999/1-2/tietokonepelit_osana_audiovisuaalisen_kulttuurin_moraalipaniikkia.htm [Last accessed: 3 April 2010]
Details about the reaction to Raid Over Moscow in Finland

'The making of Lemmings' (2001) *Edge,* Issue 94, February 2001, pp134-137
Controversy about Lemmings

'The making of Moonstone: A hard days knight' (2007) *Edge,* Issue 172, February 2007, pp92-95
How attitudes to violent games differed before and after the hearings

'Video-game violence' (1994) *The Seattle Times.* 13 January 1994. [Online] http://community.seattletimes.nwsource.com/archive/?date=19940113&slug=1889415 [Last accessed: 13 March 2010]
The development of the age-rating system

19 | A LIBRARY IN A FISH'S MOUTH

'Art and science' (2006) *Edge,* Issue 162, May 2006, pp74-81
A retrospective look at Psygnosis

Asakura, Reiji (2000) *Revolutionaries at Sony.* New York, New York: McGraw-Hill
Details of Sony and Nintendo's joint work on a CD console

Atkin, Denny (1994) 'Silicon Hollywood'. *Compute!*, Issue 161, February 1994, p98
Report on the making of Under a Killing Moon and the convergence of Hollywood and Silicon Valley

Bagnall, Brian (2007) *On the Edge: The Spectacular Rise and Fall of Commodore*. Winnipeg, Manitoba, Canada: Variant Press
History of the CDTV

Bechtold, Alan R. (1991) 'Not just Nintendo'. *Compute!*, Issue 134, October 1991, p12
Launch of the Magnavox CD-i

Bellatti, Andy (1999) 'Roberta Williams, Sierra On-Line'. *Adventure Classic Gaming*, 25 October 1999. [Online] www.adventure-classicgaming.com/index.php/site/interviews/127 [Last accessed: 13 March 2010]
Interview with Roberta Williams

Bielby, Matt (1991) 'The force: The Lucasfilm interview'. *Amiga Power*, Issue 7, November 1991, pp70-74
Interview with Lucasfilm Games' Doug Glen

Bormann, Adam (no date) 'Interviews with past Cinemaware employees'. *Just Adventure*. [Online] www.justadventure.com/Interviews/Cinemaware/Cinemaware_Interview.shtm [Last accessed: 13 March 2010]
Interviews about the CD version of It Came from the Desert and Cinemaware's demise

Busch, Kurt (1991) 'Multi-media comes home'. *InterAction*, No. 1, Fall 1991, pp40-41
Report on the rise of multimedia PCs

Campbell, Colin (1991) 'Just what is so special about CDTV?'. *Amiga Power*, Issue 3, July 1991, pp52-54
Preview of the CDTV

Campbell, Stuart (1994) 'Microcosm'. *Amiga Power*, Issue 36, April 1994, pp34-35
Review of Psygnosis' Microcosm

'Cosmic thing' (1992) *Amiga Power*. Issue 12, April 1992, pp66-67
Background on the development of Microcosm

Daly, Steve (1994) 'The land of Myst opportunity'. *Entertainment Weekly*, Issue 243, 7 October 1994. [Online] www.ew.com/ew/article/0,,303937,00.html [Last accessed: 13 March 2010]
Review of Myst

DeMaria, Rusel and Wilson, Johnny L. (2002) *High Score! The Illustrated History of Electronic Games*. Berkeley, California: McGraw-Hill/Osborne
History of Cinemaware, Sierra On-Line and Lucasfilm Games

Ferrell, Keith (1990) 'The Japan Factor'. *Compute!*, Issue 123, November 1990, p22
Report on Japanese console scene with references to the adoption of CD via the PC Engine

Fox, Barry (1991) 'Technology: Newcomer ahead in multimedia race'. *New Scientist*, Issue 1753, 26 January 1991. [Online] www.newscientist.com/article/mg12917533.400-technology-newcomer-ahead-in-multimedia-race.html [Last accessed: 13 March 2010]
Report on the CDTV

Free, John (1989) 'Interactive CDs'. *Popular Science*, Volume 235, No. 5, November 1989, pp92-94
An early look at the potential of CD for video games

Hall, Lee (editor) (2003) *PC Gamer Presents The Ultimate Guide to PC Games*. Bath, UK: Future Publishing
Reviews of video-based CD-ROM games

Hetherington, Ian (1992) 'Psygno Analysis'. *Amiga Power*. Issue 12, April 1992, pp62-64
Profile of Psygnosis

Information Processing Society of Japan (no date) 'Historical computers in Japan'. [Online] http://museum.ipsj.or.jp/en/computer/other/index.html [Last accessed: 13 March 2010]
Information about the FM Towns

Jacob, Bob (1991) 'If I'd known then…'. *Amiga Power*, Issue 7, November 1991, pp76-77
Bob Jacob looks back on Cinemaware and its closure

Kent, Steven L. (2001) *The Ultimate History of Video Games*. New York, New York: Three Rivers Press
The US launch of the PC Engine

Miller, Laura (1997) 'Riven rapt'. *Salon*, 6 November 1997. [Online] www.salon.com/21st/feature/1997/11/cov_06riven.html
[Last accessed: 13 March 2010]
Reaction to Myst and its sequel Riven

Nelson, Robin (1990) 'Video Games aim at reality'. *Popular Science*, Vol. 237, No. 6, December 1990, pp90-93
Feature looking at how CD will change the nature of video games

Noonan, Damien (1990) 'Gimmie CDTV'. *Commodore Format*, Issue 1, October 1990, pp16-17
Preview of the CDTV

'Out of the mysts' (2005) *Edge*, Issue 154, October 2005, pp66-71
Interview with Rand Miller

'Popcorn and joysticks' (1998) *The Economist*, 14 May 1998 [Online] www.economist.com/displaystory.cfm?story_id=E1_
TGRQSP [Last accessed: 11 March 2010]
The convergence of Hollywood and Silicon Valley

Randall, Ronnie (1992) 'Mindscape – 'not successful just by accident''. *Amiga Power*, Issue 19, November 1992, pp82-87
Profile of Mindscape including a look at Wing Commander

Rosenthal, Marshal M. (1990) 'Ant Attack!!'. *The Games Machine*. Issue 28, March 1990, pp80-81
Information on the CD version of It Came from the Desert

Shannon, Lorelei (1991) 'Lights! Camera! Interaction!'. *InterAction*, No. 1, Fall 1991. pp46-47
Report on Sierra On-Line's investment in CD games

'Sierra's Multimedia Upgrade Kit' (1991) [Advert] *InterAction*, No. 1, Fall 1991, p43
Costs of multimedia PC hardware

Smith, Rob (2008) *Rogue Leaders: The Story of LucasArts*. London, UK: Titan Publishing Group
Background on Star Wars: Rebel Assault and George Lucas' letter congratulating the developers

'Super-tape kills CD?' (1987) *Advanced Computer Entertainment*, Issue 4, Christmas 1987/January 1988, p8
Digital Audio Tape touted as alternative to CD-ROM

Thackray, Rachelle (1998) 'Welcome to their worlds'. *The Independent*, 3 March 1998. [Online] www.independent.co.uk/life-style/
welcome-to-their-worlds-1147967.html [Last accessed: 13 March 2010]
Interview with the creators of Myst

'The making of Bandersnatch' (2002) *Edge*, Issue 118, Christmas 2002, pp106-109
Psygnosis' links to Imagine

'The making of It Came from the Desert' (2007) *Retro Gamer Collection Volume 1*. Bournemouth, UK: Imagine Publishing.
pp18-22
Includes information about the CD version of It Came from the Desert

'The making of The Gabriel Knight Trilogy' (2009) *Edge*, Issue 198, February 2009, pp104-107
Interview with Jane Jensen

Tucker, Tim (1993) 'Hey, Mr Songwriter'. *Amiga Power*, Issue 30, October 1993, pp58-60
The impact of CD on game audio

Williams, Ken (1989) 'President's corner: Multimedia – An advance look'. *Sierra News Magazine*. Volume 2, No. 2, Autumn
1989. pp4 & 30
Opinion piece on the Multimedia PC standard

Williams, Ken (1990) 'President's corner'. *Sierra News Magazine*. Volume 3, No. 3, Fall 1990. pp4-5
The Sierra On-Line founder discusses the potential of CD-ROM for video games

20 | THE ULTIMATE DISPLAY

Antoniades, Alexander (1994) 'Monsters from the Id: The making of Doom'. *Game Developer*, Issue 1, January 1994. [Online] www.gamasutra.com/php-bin/news_index.php?story=21405 [Last accessed: 13 March 2010]
The making of Doom

Bellamy, Ron (1989) 'City in search of its identity'. *The Eugene Register-Guard*, 13 December 1989, p1B
Report on Shreveport around the time John Romero joined Softdisk

Brooks Jr., Frederick P. (1999) 'What's real about virtual reality?'. *IEEE Computer Graphics and Applications*. November/December 1999, pp16-27
Examines the legacy of the virtual reality boom of the early 1990s

Carlson, Wayne (no date) 'A critical history of computer graphics and animation'. [Online] http://design.osu.edu/carlson/history/ID797.html [Last accessed: 13 March 2010]
Historical information on the development of 3D graphics

Chaplin, Heather and Ruby, Aaron (2005) *Smartbomb: The Quest for Art, Entertainment and Big Bucks in the Videogame Revolution*. 2006 paperback edition. Chapel Hill, North Carolina: Algonquin Books
Profiles of John Carmack and John Romero

DeMaria, Rusel and Wilson, Johnny L. (2002) *High Score! The Illustrated History of Electronic Games*. Berkeley, California: McGraw-Hill/Osborne
The history of Id Software

Digital Eclipse (2004) *Midway Arcade Treasures 2* [PlayStation 2] Midway Games
Information about Hard Drivin'

Edwards, Benji (2009) 'From the past to the future: Tim Sweeney talks'. *Gamasutra*, 25 May 2009. [Online] www.gamasutra.com/view/feature/4035/from_the_past_to_the_future_tim_.php [Last accessed: 13 March 2010]
Interview with Tim Sweeney of Epic Games

Hunt, Stuart (2009) 'The making of Castle Wolfenstein 3D'. *Retro Gamer*, Issue 65, June 2009, pp76-82
The making of Wolfenstein 3D

'Hurt me plenty' (2006) *Retro Volume 2: The Ultimate Retro Companion from Games*™. Bournemouth, UK: Imagine Publishing, pp220-225
A look back at John Romero-era Id Software

Kent, Steven L. (1997) 'Cyberplay: Why do so many games have violence and devil imagery?'. *CNN*, 30 May 1997 [Online] www.cnn.com/SHOWBIZ/9705/29/cyber.lat [Last accessed: 13 March 2010]
Doom and its violent content

Kent, Steven L. (2001) *The Ultimate History of Video Games*. New York, New York: Three Rivers Press
Background on Wisdom Tree

King, Brad and Borland, John (2003) *Dungeons and Dreamers: The Rise of Computer Game Culture from Geek to Chic*. New York, New York: McGraw-Hill/Osborne
The history of Id Software

Kushner, David (2003) *Masters of Doom: How Two Guys Created an Empire and Transformed Pop Culture*. 2004 edition. London, UK: Judy Piatkus Publishing
A detailed history of Id Software

Kutner, Lawrence and Olson, Cheryl K. (2008) *Grand Theft Childhood: The Surprising Truth About Violent Video Games*. New York, New York: Simon & Schuster
Claims of a link between Doom and the US high-school shootings of the late 1990s

Looking Glass Studios (2000) 'To all the fans and supporters of Looking Glass'. 24 April 2000. [Online] http://web.archive.org/web/20000619044721/http://www.lglass.com [Last accessed: 13 March 2010]
Looking Glass Studios' parting message

'Louisana's chief learns to adapt' (1989) *The Milwaukee Journal*, 28 February 1989, p2A

Recession strikes in Louisana and Shreveport

McCall, Kevin (1988) 'Uneasy calm in Shreveport'. *The Gettysburg Times*, 23 September 1988, p7A
Riots in Shreveport

McCartney, Scott (1989) 'Louisiana: Oil bust puts once-prosperous state on bankruptcy's doorstep'. *Anchorage Daily News*, 5 March 1989, pF1 & F8
The impact of the oil and gas industry decline on Louisana

Neurath, Paul (2000) 'The story of Ultima Underworld'. [Online] www.ttlg.com/articles/uw1.asp [Last accessed: 13 March 2010]
The development of Ultima Underworlds and information on Looking Glass Studios

O'Toole, Mary Ellen (2000) *The School Shooter: A Threat Assessment Perspective*. Quantico, Virginia: National Centre for the Analysis of Violent Crime, FBI Academy.
FBI report dismissing the link between video games and high-school shootings

Shannon, L.R. (1987) 'Peripherals: New look of magazines'. *The New York Times*, 27 October 1987, pC11
The popularity of disk magazines in the 1980s

Rheingold, Howard (1991) *Virtual Reality*. 1992 edition. London: Mandarin Paperbacks.
The history and hype surrounding virtual reality

Romero, John (no date) 'Games'. *Planet Romero* [Online] http://planetromero.com/games [Last accessed: 13 March 2010]
List of John Romero's games

'Scattered violence threatens fragile peace in Shreveport' (1988) *The Argus Press*. Owosso, Michigan. 22 September 1988, p22
News report on riots in Shreveport

Sevo, David (no date) 'History of computer graphics'. [Online] www.danielsevo.com [Last accessed: 13 March 2010]
Background on the development of 3D graphics

Siegler, Joe (2000) 'A look back at Commander Keen'. *3D Realms* [Online] www.3drealms.com/keenhistory [Last accessed: 13 March 2010]
History of Id Software's Commander Keen

Smith, Steve (1994) *PC Pilot: The Complete Guide to Computer Aviation*. New York, New York: Avon Books
Information on flight simulators

Sutherland, Ivan E. (1965) 'The Ultimate Display'. *Proceedings of IFIP Congress 1965*, pp506-508
Ivan Sutherland's talk that kicked off research into virtual reality

'The pixel revolution' (2009) *Edge*, Issue 205, September 2009, pp74-81
Background on the adoption of GPUs

'Through the looking glass' (2004) *Edge*, Issue 138, July 2004, pp90-95
Feature on the legacy of Looking Glass Studios

'Uneasy calm in Shreveport' (1988) *The Argus Press*. Owosso, Michigan. 23 September 1988, p7
News report on riots in Shreveport

21 I WE TAKE PRIDE IN RIPPING THEM TO SHREDS

'A twist in the tail' (2003) *Retro: The Collector's Series*, Edge Specials, Issue 10. Bath, UK: Future Publishing, pp92-103
Retrospective look at the Sega Saturn

'An interview with Shigeru Miyamoto' (1996) *Next Generation*, Issue 2, August 1996, p58
Shigeru Miyamoto discusses the development of Super Mario 64

Asakura, Reiji (2000) *Revolutionaries at Sony*. New York, New York: McGraw-Hill
History of the original PlayStation

'Back from the dead' (2004) *Edge*, Issue 135, April 2004, pp52-58

Preview of Resident Evil 4

Boctok Inc and Fukuda, Miki (editors) (2000) *Bit Generation 2000*. Kobe, Japan: Kobe Fashion Museum
The development of the PlayStation and reaction in Japan

'Brand Nu' (1999) *Edge*, Issue 72, June 1999, pp56-69
Wipeout's design and marketing

Brown, Damon (2008) *Porn & Pong: How Grand Theft Auto, Tomb Raider and Other Sexy Games Changed Our Culture*. Port Townsend, Washington: Feral House
Covers the development of Tomb Raider and Lara Croft as a video game icon and sex symbol

Brown, Janelle (1997) 'All-girl Quake clans shake up boys' world'. *Wired*, 5 February 1997 [Online] www.wired.com/culture/lifestyle/news/1997/02/1885 [Last accessed: 13 March 2010]
The rise of all-female online game clans

Brown, Janelle (1997) 'GameGirlz turns industry on to female gamers'. *Wired*, 11 November 1997. [Online] www.wired.com/culture/lifestyle/news/1997/11/8434 [Last accessed: 13 March 2010]
The growing interest in making games for female players and coverage of the game development conference with several sessions on girl games

Bryan, Kevin (2005) *PSX: The Guide to the Sony Playstation*. 2007 edition. Digital Press. [Online] www.digitpress.com/products/psxbook.htm [Last accessed: 10 March 2010] pp6-15 & 79
History of the PlayStation, sales data and the European nightclub marketing campaign

Burns, Axel (2002) 'Resource centre sites: The new gatekeepers of the web?'. [PhD paper] University of Queensland
Interview with the founder of the Grrlgamer website and its links to the riot grrrl movement

Burr, Ty (1995) '1995 The Best & Worst: Multimedia'. *Entertainment Weekly*, Issue 307-308, 29 December 1995 [Online] www.ew.com/ew/article/0,,300154,00.html [Last accessed: 11 March 2010]
Chop Suey named the best multimedia CD of the year

Cassell, Justine and Jenkins, Henry (editors) (1998) *From Barbie to Mortal Kombat: Gender and Computer Games*. Cambridge, Massachusetts: The MIT Press
Covers the girl game movement of the mid- to late-1990s

Cousins, Mark (2004) *The Story of Film*. 2008 paperback edition. London, UK: Pavilion Books
Background on horror films, including Japanese horror movies

Crandall, Robert W. & Sidak, J. Gregory (2006) *Video Games: Serious Business for America's Economy*. Washington, DC: Entertainment Software Association
Paper highlighting the economic importance of video games, produced by the US's national video game publisher trade association

Debord, Matthew (1998) 'From girls to glamour'. *Salon*, 24 September 1998. [Online] www.salonmagazine.com/21st/feature/1998/09/24feature2.html [Last accessed: 11 March 2010]
Interview with Chop Suey co-creator Theresa Duncan

'Gently does it' (1999) *The Economist*, 6 May 1999. [Online] www.economist.com/culture/displaystory.cfm?story_id=E1_GDSVTQ [Last accessed: 11 March 2010]
Report on the emergence of stealth video games

Heianna, Sumiyo (2005) 'Interview with Roy Lee, matchmaker of the macabre'. *Kateigaho International Edition*, Winter 2005. [Online] http://int.kateigaho.com/win05/horror-lee.html [Last accessed: 13 March 2010]
Background on Japanese horror films

Hirabayashi, Yoshiaki (2005) 'The graphical styling of Resident Evil 4'. *Game Developer*, Volume 12, Number 9, October 2005, pp26-33
The Capcom game developer looks back at the visual approach of Resident Evil 4

Jones, Sandra (1996) 'Nintendo could beat itself at its own game'. *New Straits Times*, 27 May 1996, p36
Nintendo of America's Howard Lincoln compares the PlayStation to a Chevrolet and the Nintendo 64 to a Cadillac

Karlyn, Kathleen Rowe (2003) 'Scream, popular culture and feminism's third wave: 'I'm not my mother'. *Genders Journal*, Issue 38 [Online] www.genders.org/g38/g38_rowe_karlyn.html [Last accessed: 11 March 2010]

The links between the riot grrrl and game grrrl movements

Kennedy, Helen W. (2006) 'Illegitimate, monstrous and out there: Female *Quake* players and inappropriate pleasures'. In: Hollows, Joanne & Moseley, Rachel (2006) (Editors) *Feminism in Popular Culture*. New York, New York: Berg
Details about the games for girls movement

Kent, Steven L. (2001) *The Ultimate History of Video Games*. New York, New York: Three Rivers Press
The success of the PlayStation in the USA

Kohler, Chris (2005) *Power-Up: How Japanese Video Games Gave the World an Extra Life*. Indianapolis, Indiana: BradyGames
Historical overview of the Final Fantasy series

McRoy, Jay (editor) (2005) *Japanese Horror Cinema*. Honolulu, Hawaii: University of Hawai'i Press
Background on Japanese horror films

'Nintendo, Sega & Sony under one roof' (1996) *Next Generation*, Issue 2, August 1996, pp6-12
Report on the PlayStation, Nintendo 64, Saturn console war and Psygnosis' contribution to the PlayStation

'Nintendo's Yamauchi speaks out' (1996) *Next Generation*, Issue 2, August 1996, p30
The Nintendo chairman on delaying the Nintendo 64 so Shigeru Miyamoto could perfect Super Mario 64

Nutt, Christian (2007) 'Infiltrating Kojima Productions: Ryan Payton talks Metal Gear Solid'. *Gamasutra*, 15 October 2007. [Online] www.gamasutra.com/view/feature/1954/infiltrating_kojima_productions_.php [Last accessed: 13 March 2010]
Interview about the Metal Gear Solid series

Pegoraro, Rob (2007) 'Faster forward: Theresa Duncan has signed off'. *Washington Post*, 1 August 2007. [Online] http://blog.washingtonpost.com/fasterforward/2007/08/post_6.html [Last accessed: 11 March 2010]
Theresa Duncan obituary

Poole, Steven (2000) *Trigger Happy: The Inner Life of Videogames*. London, UK: Fourth Estate
Examines of Lara Croft's popularity

'Reasons to be playful' (2005) *The Economist*, 27 January 2005. [Online] www.economist.com/business-finance/displaystory.cfm?story_id=E1_PGDDDGV [Last accessed: 11 March 2010]
Differences in adoption of game consoles in US, UK, France and Germany

'Scare tactics' (2005) *Edge*, Issue 147, March 2005, pp68-73
Feature on horror games

'Screen play' (2005) *Edge*, Issue 155, November 2005, pp70-81
Feature on games turned into movies including Tomb Raider

'Silent Hill 2' (2000) *Edge*, Issue 91, December 2000, pp44-48
Interview with the creators of Silent Hill 2

'Soft sell' (2005) *Games^TM*, Issue 29, February 2005, pp70-75
Lucozade's use of Lara Croft in its advertising

'Solid states' (2007) *Edge*, Issue 173, March 2007, pp54-61
Feature on the Metal Gear Solid series

'Special report: You've come a long way baby...' (1997) *BBC News Online*. 30 December 1997 [Online] http://news.bbc.co.uk/1/hi/entertainment/38786.stm [Last accessed: 11 March 2010]
Report on girl power

Szczepaniak, John (2005) 'Hardware legend: 3DO'. *Games^TM*. Issue 37, October 2005, pp140-143
History of the 3DO console

'Teenage murderer gets life term' (2004) *BBC News Online*, 3 September 2004 [Online] http://news.bbc.co.uk/1/hi/england/leicestershire/3624654.stm [Last accessed: 11 March 2010]
UK murder connected to the game Manhunt

'Teiyu Goto' (2001) *Edge*, Issue 105, Christmas 2001, pp44-45

Interview with the designer of the PlayStation controller

'The making of Alone in the Dark' (2005) *Edge*, Issue 150, June 2005, pp104-107
The creation of Alone in the Dark

'The making of Final Fantasy VII' (2003) *Edge*, Issue 123, May 2003, pp108-113
Interview with the creators of Final Fantasy VII

'The making of PlayStation' (2009) *Edge*, Issue 200, April 2009. pp126-133
History of the PlayStation's development

'Videogame violence returns to the agenda' (2004) *Edge*, Issue 133, February 2004, pp12-13
The controversy surrounding Manhunt

Weil, Elizabeth (1997) 'The girl-game jinx'. *Salon*, 10 December 1997. [Online] www.salonmagazine.com/21st/feature/1997/12/cov_10feature2.html [Last accessed: 11 March 2010]
The video game industry gets interested in female game players

'Wipeout 3' (1999) *Edge*, Issue 72, June 1999, pp52-55
Feature on the Wipeout series of games

22 | BEATMANIA

'All the way to 11' (2008) *Edge*, Issue 194, November 2008, pp72-77
Interview with Guitar Hero creators Harmonix

Ashcraft, Brian with Snow, Jean (2008) *Arcade Mania! The Turbo-Charged World of Japan's Game Centers*. Tokyo, Japan: Kodansha International
A guide to the Japan's arcade scene

'Dance dance revolution' (2003) *Edge*, Issue 124, June 2003, pp54-61
The legacy of Dance Dance Revolution and how it boosted Konami's earnings

'Daytona USA' (2000) *Edge*, Issue 91, December 2000, pp50-55
Interview with the creator of the coin-op racing game

GameCity Squared (2009) 'Masaya Matsuura to deliver BAFTA vision statement at GameCity Squared'. [Press release] 27 August 2009. [Online] www.lincolnbeasley.co.uk/pressreleases/2009/aug/27/masaya-matsuura-deliver-bafta-vision-statement-gam [Last accessed: 13 March 2010]
Biography of Masaya Matsuura

Hawkins, Matthew (2005) 'Interview: Rodney Greenblat, the mother of Sony's almost Mario'. *Gamasutra*, 5 July 2005. [Online] www.gamasutra.com/view/feature/2340/interview_rodney_greenblat_the_.php [Last accessed: 13 March 2010]
Interview with the PaRappa the Rapper artist

'Playing along' (2008) *The Economist*, 9 October 2008. [Online] www.economist.com/business-finance/displaystory.cfm?story_id=E1_TNPSGRRJ [Last accessed: 11 March 2010]
How music games boosted music sales

'Rhythm Faction' (2005) *Edge*, Issue 156, December 2005, pp62-69
Interview with Guitar Hero creators Harmonix

'State of the Industry Report 2008' (2009) *Play Meter*. Metairie, Louisiana: Skybird Publishing
Figures on weekly income of various times of coin-operated machines

'The book of arcade games' (2008) *Games^TM*, Issue 78, Christmas 2008, pp70-77
Overview of arcade gaming and its decline

'The book of rhythm action' (2007) *Games^TM*, Issue 53, February 2007, pp76-83
Guide to music-based games

'The home of disco' (1999) *Edge*, Issue 74, August 1999, p140
Dance games make the leap to home consoles

Weyhirch, Steven (2010) 'Apple II history'. [Online] http://apple2history.org/history/ah14.html [Last accessed: 13 March 2010]
Information about the Apple II demo Kaleidoscope

Williams, Dmitri (2006) 'A (brief) social history of video games'. In: Vorderer, Peter & Bryant, Jennings (editors) (2006) *Playing Computer Games: Motives, Responses and Consequences.* Mahwah, New Jersey: Lawrence Erlbaum. [Online] http://dmitriwilliams.com/research.html [Last accessed: 6 March 2010]
Figures on the size of the arcade video game business in the US

Wright, Chris (2008) 'A brief history of mobile games'. *Pocketgamer,* 22 December 2008. [Online] www.pocketgamer.co.uk/r/PG.Biz/A+Brief+History+of+Mobile+Games/feature.asp?c=10619 [16 January 2010]
The history of video games on mobile phones

23 | YOU HAVEN'T LIVED UNTIL YOU'VE DIED IN MUD

Anderson, Brooke P. (editor) (1997) *How to Fly and Fight in Air Warrior.* 21 March 1997 version. [Online] www.electraforge.com/brooke/flightsims/air_warrior/awtaman.txt [Last accessed: 13 March 2010]
Air Warrior etiquette

Bartle, Richard A. (1985) 'Introducing the wizards'. *Micro Adventurer,* Issue 16, February 1985. p23 & 25
MUD co-creator Richard Bartle on the players of the game

Bartle, Richard A. (1990) 'Interactive multi-user computer games'. [Research paper] December 1990 [Online] ftp://ftp.lambda.moo.mud.org/pub/MOO/papers/mudreport.txt [Last accessed: 13 March 2010]
Research into the challenges facing multiplayer online games

Bartle, Richard A. (2004) 'Designing virtual worlds'. Indianapolis, Indiana: New Riders Publishing
Background on the evolution of MUDs

Barton, Matt (2008) *Dungeons & Desktops.* Wellesley, Massachusetts: A K Peters
Background on AOL's Neverwinter Nights

Brice, Katherine (2009) 'Blizzard reveals full scale of World of Warcraft operation'. *Gamesindustry.biz,* 18 September 2009 [Online] www.gamesindustry.biz/articles/blizzard-reveals-full-scale-of-world-of-warcraft-operation [Last accessed: 13 March 2010]
Stats on the scale of the World of WarCraft operation

Bunten Berry, Dani (no date) *Game Design Memoir* [Online] www.anticlockwise.com/dani/personal/biz/memoir.htm [Last accessed: 13 March 2010]
Her reflections on Modem Wars

Castronova, Edward (2005) *Synthetic Worlds: The Business and Culture of Online Games.* 2006 paperback edition. Chicago, Illinois: University of Chicago Press
Analysis of online multiplayer game economics

Daleske, John (2008) 'PLATO: Also an excellent platform to design games'. [Online] www.daleske.com/plato/plato-games.php#content [Last accessed: 13 March 2010]
Playing games on the PLATO system

Daleske, John and Fritz, Gary (2008) 'How Empire came to be'. [Online] www.daleske.com/plato/empire.php [Last accessed: 13 March 2010]
The history of the PLATO multiplayer game Empire

'Eve Online' (2001) *Edge,* Issue 106, January 2002, pp38-43
Preview of Eve Online

Farmer, F. Randall and Morningstar, Chip (1990) 'The lessons from Lucasfilm's Habitat'. In: Salen, Katie and Zimmerman, Eric (editors) (2006) *The Game Design Reader: A Rules of Play Anthology.* Cambridge, Massachusetts: The MIT Press
The creators of Habitat reflect on its successes and failures

Ferrell, Keith (1987) 'The future of computer games'. *Compute!,* Issue 90, November 1987, p14
Discusses Habitat and online gaming's potential

Greely, Dave and Sawyer, Ben (1997) 'Has Origin created the first true online game world?'. *Gamasutra*, 19 August 1997 [Online] www.gamasutra.com/view/feature/3220/has_origin_created_the_first_true_.php [Last accessed: 13 March 2010]
Report on the creation of Ultima Online

Geraci, Vince (1991) 'The nation embraces TSN'. *InterAction*, Fall 1991, p58
News story on The Sierra Network

Guest, Tim (2007) *Second Lives: A Journey Through Virtual Worlds.* London, UK: Hutchinson
Second Life and its players

'Interview with Rainz, the man who killed Lord British' (2002) *Ultima Online Travelogues*, September 2002. [Online] www.aschulze.net/ultima/stories9/beta.htm [Last accessed: 10 May 2009]
Interview with Rainz about his actions in Ultima Online

King, Brad and Borland, John (2003) *Dungeons and Dreamers: The Rise of Computer Game Culture from Geek to Chic.* New York, New York: McGraw-Hill/Osborne
Background on online games, Richard Garriott and Ultima Online

'Lord British was assassinated in Ultima Online beta' (2009) *Birdinforst'blog* [Online] www.birdinforest.com/blog/?p=193 [Last accessed: 13 March 2010]
Details about Lord British's murder in Ultima Online

Maloy, Deirdre (1982) 'Long distance gaming: Games via The Source and Compuserve'. *Computer Gaming World*, Volume 2, No. 3, May/June 1982, pp14-15 & 22
Report on online gaming in the early 1980s

Moriarty, Brian (1998) 'Dani Bunten: Lifetime Achievement Award'. [Presentation] Computer Game Developers Association Awards, Long Beach, California, 7 May 1998. [Online] www.anticlockwise.com/dani [Last accessed: 12 March 2010]
Mentions Dani Bunten Berry's Modem Wars and her thinking on online gaming

'News of the realm' (1993) *Neverwinter Nights*, Issue 2, Week ending: 18 August 1993. [Online] www.bladekeep.com/nwn/newsrealm/1993-09.txt [Last accessed: 13 March 2010]
News item on a virtual wedding within Neverwinter Nights

'News of the realm' (1993) *Neverwinter Nights*, Issue 5, Week ending: 9 October 1993. [Online] www.bladekeep.com/nwn/newsrealm/1993-10.txt [Last accessed: 13 March 2010]
News item on a couple who met in Neverwinter Nights and got engaged in real life

'Nuits brûlantes dur Funitel le Minitel se Dévergonde' (1985) [In French] *Tilt*, Issue 19, March 1985, p6
News report on Minitel gaming

Rheingold, Howard (1993) *The Virtual Community.* Electronic version. Cambridge, Massachusetts: The MIT Press. [Online] www.rheingold.com/vc/book/intro.html [Last accessed: 13 March 2010]
Information on early online networks and Minitel

Rossignol, Jim (2005) 'Interview: Evolution and risk: CCP on the freedoms of EVE Online'. *Gamasutra*, 23 September 2005 [Online] www.gamasutra.com/features/20050923/rossignol_01.shtml [Last accessed: 13 March 2010]
Interview with Eve Online's creators CCP

Rubin, Michael (2006) *Droidmaker: George Lucas and the Digital Revolution.* Gainesville, Florida: Triad Publishing
Recounts the creation and beta test of Habitat

Smith, David (1980) 'Letter from America'. *Liverpool Software Gazette*, No. 3, March 1980, p42
Briefly discusses the earliest online network services

Smith, Rob (2008) *Rogue Leaders: The Story of LucasArts.* London, UK: Titan Publishing Group
The history of Habitat

'Sony Online Entertainment' (2001) *Edge*, Issue 102, October 2001, pp56-61
Interview with the publisher of EverQuest

Spear, Peter (1991) 'The online games people play'. *Compute!*, Issue 135, November 1991, p96
Looks at just-launched The Sierra Network and the potential for online gaming

Swain, Meg (2008) *Career Building Through Alternate Reality Gaming*. New York, New York: The Rosen Publishing Group
Overview of alternate reality games and The Beast

'Time extend: Phantasy Star Online' (2005) *Edge*, Issue 155, November 2005. pp114-119
A look back at Phantasy Star Online

The 3DO Company (1996) '3DO enters into agreement to acquire Archetype Interactive' [Press release] *Business Wire*, 13 May 1996. [Online] www.allbusiness.com/media-telecommunications/internet-www/7229149-1.html [Last accessed: 13 March 2010]
Background details on Archetype Interactive and Meridian 59

The 3DO Company (1996) '3DO's Meridian 59 now available worldwide initial response to product overwhelming'. [Press release] *Business Wire*, 7 October 1996. [Online] www.thefreelibrary.com/3DO%27S+Meridian+59+Now+Available+Worldw ide+Initial+Response+to+Product...-a018734646 [Last accessed: 13 March 2010]
Details about the early days of Meridian 59

'The book of massively multiplayer online role-playing games' (2008) *Games^TM*, Issue 67, February 2008, pp92-99
Overview of online multiplayer role-playing games

'The making of MUD' (2004) *Edge*, Issue 141, October 2004, p114
The story of how MUD was created and the failed attempts to turn it into a commercial product

'The pioneers: MUDs, MMORPGs and mayhem' (2003) *Gamespy*, 26 September 2003. [Online] http://archive.gamespy.com/ amdmmog [Last accessed: 13 March 2010]
The history of online role-playing games

'Ultima Online release' (2002) *Ultima Online Travelogues*, September 2002 [Online] www.aschulze.net/ultima/stories9/release1. htm [Last accessed: 10 May 2009]
Account of the murder of Lord British in Ultima Online

Williams, Ken (1991) 'A view from the inside: The interactive film industry is a virtual reality'. *InterAction*, Fall 1991, pp4, 6 & 10
The launch of The Sierra Network and Sierra On-Line co-founder Ken Williams' vision for the service

Yakal, Kathy (1986) 'Habitat: A look at the future of online games'. *Compute!*, Issue 77, October 1986, p32
Preview of Habitat

24 | SECOND LIVES

'50 illegal electronic games banned' (2005) *China View*, 26 January 2005. [Online] http://news.xinhuanet.com/eng-lish/2005-01/26/content_2511068.htm [Last accessed: 13 March 2010]
China bans video games, including FIFA 2005

Ahonen, Tomi T. and O'Reilly, Jim (2007) *Digital Korea*. London, UK: Futuretext
Overview of South Korea's embrace of the internet, including a look at its online gaming culture

'Average monthly wages of staff and workers 1978-2007' (2008) *China Statistical Yearbook*. [Online] www.stats.gov.cn [Last accessed: 13 March 2010]
The figures on the average earnings of Chinese workers

Bigpoint (2008) 'Bigpoint, Europe's leading multiplayer game publisher, expands into US market with rapid-fire launch sched-ule'. [Press release] 3 December 2008 [Online] www.marketwire.com/press-release/Bigpoint-Europes-Leading-Multiplayer-Game-Publisher-Expands-Into-US-Market-With-Rapid-926487.htm [Last accessed: 13 March 2010]
Background on German online games publisher Bigpoint

Bolande, M. Asher (1997) 'Patriotic computer games hope to win over China's youth'. *The Nation*, Bangkok, Thailand. 21 May 1997, pB4.
China's pre-internet game scene

Castronova, Edward (2002) 'Virtual worlds: A first-hand account of market and society on the Cyberian Frontier'. In: Salen, Katie and Zimmerman, Eric (editors) (2006) *The Game Design Reader: A Rules of Play Anthology*. Cambridge, Massachusetts: The MIT Press
Edward Castronova's groundbreaking study of EverQuest's economy

'China censors online video games' (2004) *BBC News Online*. 1 June 2004. [Online] http://news.bbc.co.uk/1/hi/technology/3766023.stm [Last accessed: 13 March 2010]
Chinese government censors online games

Heeks, Richard (2008) 'Current analysis and future research agenda on 'gold farming': Real-world production in developing countries for the virtual economies of online games'. *Development Informatics*, working paper 32, Institute for Development Policy and Management, University of Manchester. [Online] www.sed.manchester.ac.uk/idpm/research/publications/wp/di/di_wp32.htm [Last accessed: 13 March 2010]
Research into gold farming and its effects

Jansen, Marius B. (2000) *The Making of Modern Japan*. Cambridge, Massachusetts: Belknap Press, Harvard University Press
Historical background on Japan's relationship with Korea and China

Jenkins, David (2009) 'Chinese online games market grew 63% in 2008'. *Gamasutra*, 9 April 2009. [Online] www.gamasutra.com/php-bin/news_index.php?story=23133 [Last accessed: 13 March 2010]
News report on the growth of the Chinese video game business

Kim, Min-kyu and Park, Tae-soon (2006) *Between Censorship and Rating: State of Global Screening Systems*. Seoul, South Korea: Korea Game Development Institute
White paper examining efforts to censor and control the content of video games

Korea Game Development & Promotion Institute (2002) *The Rise of Korean Games*.
Report on South Korea's game industry

Korea Game Development & Promotion Institute (2004) *Trends in Game Immersion (Addiction) According to Changes in Game Environments and Their Significance*.
Research into levels of game playing in South Korea

Korea Game Development & Promotion Institute (2005) *The Rise of Korean Games: Guide to Korean Game Industry and Culture 2005*.
Report on South Korea's game industry

Korea Game Development & Promotion Institute (2006) *The Rise of Korean Games: Guide to Korean Game Industry and Culture 2006*.
Report on South Korea's game industry

'Lei Feng becomes online game hero' (2006) *China View*, 16 March 2006. [Online] http://news.xinhuanet.com/english/2006-03/16/content_4308138.htm [Last accessed: 13 March 2010]
News story on the Chinese game Learn from Lei Feng Online

Roberts, Dexter (1996) 'Let a hundred Chinese video games bloom'. *Business Week*, 23 December 1996. [Online] www.businessweek.com/1996/52/b35076.htm [Last accessed: 13 March 2010]
News report on Chinese government investment in video games

Sanders, Myke (2009) 'Analysis and control of gold farming transaction activities in the online gaming environment'. [Online] www.gamersrage.com/goldfarming.html [Last accessed: 6 March 2010]
Research into gold farming

'Swedish video game banned for harming China's sovereignty' (2004) *China Daily*, 29 May 2004. [Online] www.chinadaily.com.cn/english/doc/2004-05/29/content_334845.htm [Last accessed: 13 March 2010]
China bans Hearts of Iron

'The Professionals' (2003) *Games*™, Issue 5, May 2003, pp44-49
Feature on professional gaming that compares the US, European and South Korean scenes

25 | LITTLE COMPUTER PEOPLE

Alexander, Christopher et al (1977) *A Pattern Language*. New York, New York: Oxford University Press
The architectural theory book that inspired The Sims

'Big trouble in LittleBigPlanet' (2009) *Edge*, Issue 197, January 2009, pp8-11
Report on volume of LittleBigPlanet player creations that breach copyright law

Brown, Damon (2008) *Porn & Pong: How Grand Theft Auto, Tomb Raider and Other Sexy Games Changed Our Culture*. Port Townsend, Washington: Feral House
The Hot Coffee controversy

'Building a better world' (2008) *Games™*, Issue 68, March 2008, pp14-15
Interview with Thomas Vu, producer of Spore

Chaplin, Heather and Ruby, Aaron (2005) *Smartbomb: The Quest for Art, Entertainment and Big Bucks in the Videogame Revolution*. 2006 paperback edition. Chapel Hill, North Carolina: Algonquin Books
Profile of Will Wright

DeMaria, Rusel (2008) *Spore: The Evolution*. Roseville, California: Prima Games
Overview of the creation of Will Wright's Spore, including details of the influence of social networks on the game

Hill, Steve (2004) 'Hooray for Hollywood…The Movies'. *PC Zone*, Issue 141, May 2004, pp46-48
Preview of Peter Molyneux's machinima-inspired The Movies

Hölldobler, Bert and Wilson, Edward O. (1990) *The Ants*. London, UK: Springer-Verlag
The encyclopaedic guide to ant biology that inspired Will Wright's Sim Ant and indirectly influenced The Sims

Kutner, Lawrence and Olson, Cheryl K. (2008) *Grand Theft Childhood: The Surprising Truth About Violent Video Games*. New York, New York: Simon & Schuster
Covers the Hot Coffee controversy

'LittleBigPlanet community reaches one million creates' (2009) *PlayStation Blog*, 22 July 2009. [Online] http://blog.us.playstation.com/2009/07/littlebigplanet-community-reaches-one-million-creations [Last accessed: 7 March 2010]
LittleBigPlanet player creations pass the million mark

McLean-Foreman, John (2001) 'Interview with Minh Le'. *Gamasutra*, 30 May 2001. [Online] www.gamasutra.com/view/feature/3072/interview_with_minh_le.php [Last accessed: 7 March 2010]
Interview with the co-creator of Counter-Strike

Musgrove, Mike (2005) 'Game turns players into indie moviemakers'. *The Washington Post*, 1 December 2005. [Online] www.washingtonpost.com/wp-dyn/content/article/2005/11/30/AR2005113002117.html [Last accessed: 13 March 2010]
The Movies and the machinima film The French Democracy

Salen, Katie (2002) 'Telefragging monster movies'. In: King, Lucien (editor) (2002) *Game On: The History and Culture of Videogames*. London, UK: Laurence King Publishing
Overview of the machinima movement

Seabrook, John (2006) 'Game master: Will Wright changed the concept of video games with The Sims. Can he do it again with Spore?'. *The New Yorker*, 6 November 2006. [Online] www.newyorker.com/archive/2006/11/06/061106fa_fact [Last accessed: 13 March 2010]
Profile of Will Wright

Shoemaker, Richie (2003) 'Games that changed the world: Counter-Strike'. *PC Zone*, Issue 125, February 2003, pp134-137
Overview of Counter-Strike and its impact

'The making of Desktop Tower Defense' (2009) *Edge*, Issue 206, October 2009, pp104-107
Recounts the birth of the tower defense genre

'The truth behind the biggest PC game ever' (2004) *Edge*, Issue 143, December 2004, pp24-25
Feature on the creation of The Sims

26 | ALL-ACCESS GAMING

'9: Don't do the easy' (2007) *Hobo Nikkan Itoi Shinbun*, 12 September 2007. [Online] www.1101.com/iwata/2007-09-12.html [Last accessed: 28 March 2010]
Interview with Nintendo president Satoru Iwata

Alexander, Leigh (2009) 'Iwata: 'Essence of fun' can overcome gap between Japanese, Western culture'. *Gamasutra*, 6 February 2009. [Online] www.gamasutra.com/view/news/22192/Iwata_Essence_Of_Fun_Can_Overcome_Gap_Between_Japanese_Western_Culture.php [Last accessed: 13 March 2010]

Nintendo's president makes the case for simpler games

Annual Report 2000 (2000) Sega Corporation. [Online] www.segasammy.co.jp/english/ir/pdf/ir/kako/sega_AR_all_2000.pdf [Last accessed: 29 March 2010]
Sega losses in 1997/98. Converted from yen to US dollars using historical exchange rate data from Oanda.com

Asakura, Reiji (2000) *Revolutionaries at Sony*. New York, New York: McGraw-Hill
The history of the PlayStation

Boctok Inc and Fukuda, Miki (editors) (2000) *Bit Generation 2000*. Kobe, Japan: Kobe Fashion Museum
Background on Nintendo and Japanese video game culture

Caoili, Eric (2009) 'Iwata: 'Something is wrong' with dwindling Japanese console market'. *Gamasutra*, 5 February 2009. [Online] www.gamasutra.com/view/news/22187/Iwata_Something_Is_Wrong_With_Dwindling_Japanese_Console_Market.php [Last accessed: 13 March 2010]
Nintendo president on the decline of console retail game sales in Japan

Edge Presents Equip: The Insider's Guide to the Future of the Xbox (2003) *Edge* Specials Issue 9. Bath, UK: Future Publishing
Details about the Xbox and its hard drive

'Facing the camera' (2003) In: *Edge Presents Equip: The Insider's Guide to the Future of the PlayStation 2* (2003), *Edge* Specials Issue 6. Bath, UK: Future Publishing, pp12-19
The development of the EyeToy

'Inside…Nintendo Co Ltd' (2001) *Edge*, Issue 100, August 2001, pp60-67
Profile of Nintendo

Iwabuchi, Koichi (2002) *Recentering Globalization: Popular Culture and Japanese Transnationalism*. Durham, North Carolina: Duke University Press.
Examines Japan's cultural influence in the west and in Asia

Iwata, Satoru (2005) 'Message from the president'. *Nintendo Annual Report 2005*. Kyoto, Japan: Nintendo Co. Ltd.
Nintendo's president sets out the company's vision for games that appeal beyond existing players

Japan External Trade Organisation (2007) *Japanese Video Game Industry*. [Research report]
Data on the Japanese video game business

Juul, Jesper (2010) *A Casual Revolution: Reinventing Video Games and Their Players*. eBook edition. Cambridge, Massachusetts: The MIT Press.
An overview of the audience and development of 'casual' video games

Kalata, Kurt (no date) 'Segagaga – Dreamcast (2001)'. *Hardcore Gaming 101*. [Online] http://hg101.kontek.net/segagaga/sega-gaga.htm [Last accessed: 7 March 2010]
A retrospective look at Segagaga

Kelts, Roland (2006) *Japanamerica: How Japanese Pop Culture has Invaded the U.S.*. 2007 paperback edition. New York, New York: Palgrave Macmillan
The influence of Japan on American culture

Kent, Steven L. (2001) *The Ultimate History of Video Games*. New York, New York: Three Rivers Press
Mentions Nintendo's early exploration of motion control during the development of the Nintendo 64

Keveney, Bill (1998) 'Japan's latest export to U.S.: Pokemon'. *Milwaukee Journal Sentinel*, 18 February 1998, p8B
Pokémon reaches the USA

Kim, Tom (2008) 'Eye to eye – The history of the EyeToy'. *Gamasutra*, 6 November 2008. [Online] www.gamasutra.com/php-bin/news_index.php?story=20975 [last accessed: 7 March 2010]
The development and impact of the EyeToy

Masuyama (2002) 'Pokémon as Japanese culture'. In: King, Lucien (editor) (2002) *Game On: The History and Culture of Videogames*. London, UK: Laurence King Publishing
Pokémon's significance and Japan's video game culture

Kohler, Chris (2005) *Power-Up: How Japanese Video Games Gave the World an Extra Life*. Indianapolis, Indiana: BradyGames
Overview of Japanese video games, including Pokémon

Larimer, Tim (1999) 'The ultimate game freak'. *Time Asia*, Volume 154, No. 20, 22 November 1999. [Online] www.time.com/
time/asia/magazine/99/1122/pokemon6.fullinterview1.html [Last accessed: 14 March 2010]
Interview with Satoshi Tajiri, the creator of Pokémon

'Making Waves' (2005) *Edge*, Issue 156, December 2005, pp82-89
An early pre-release look at the Wii, when it was still being called the Revolution

Molokh (2009) 'Interview d'Okano Tatsu'. [In French] *Dream-Storming Mag*, Issue 9, pp7-8. [Online] http://sd-1.archive-host.
com/membres/up/107817311824861617/DSMagn9-SGGG.pdf [Last accessed: 28 March 2010]
Information about Segagaga

Nintendo of America (2008) 'Biography of Satoru Iwata'. [Press release] July 2008.
Biography of the Nintendo president

'Nintendo stands firm with Revolution' (2005) *Games^TM*, Issue 34, July 2005, p14
Early reactions to the Wii, then still being called the Revolution, ahead of its launch

'Note perfect' (2009) *Games^TM*, Issue 82, April 2009, pp78-83
Interview with music game developers Harmonix and Neversoft

Orecklin, Michele et al (1999) 'Pokemon: The cutest obsession'. *Time*, 10 May 1999. [Online] www.time.com/time/magazine/
article/0,9171,990959,00.html [Last accessed: 14 March 2010]
Report on the success of Pokémon

Palmer, Edwina (editor) (2005) *Asian Futures, Asian Traditions*. Folkestone, Kent: Global Oriental
Japan's cultural relations with Asia

Pattison, Louis (2003) 'Peripheral vision'. *PSNext*, Issue 1, June/July 2003, pp44-47
Report on the EyeToy

'Pokémania v globophobia' (1999) *The Economist*, 18 November 1999 [Online] www.economist.com/world/united-states/dis-
playstory.cfm?story_id=E1_NGDDJT [Last accessed: 11 March 2010]
Report on the international success of Pokémon and Japan's rising cultural influence

'Pokemon zaps US cinemas' (1999) *BBC News Online*, 15 November 1999. [Online] http://news.bbc.co.uk/1/hi/entertain-
ment/520900.stm [Last accessed: 14 March 2010]
News story on the success of Pokémon: The First Movie

Porter, Michael E. et al (2009) *The Video Games Cluster in Japan*. [Online] www.isc.hbs.edu/pdf/Student_Projects/Japan_Video_
Games_2009.pdf [Last accessed: 14 March 2010]
Figures on the Japanese game industry including the popularity of role-playing games

'Remote composer' (2008) *Edge*, Issue 196, Christmas 2008, pp58-62
Interview with Shigeru Miyamoto

Rheingold, Howard (1991) *Virtual Reality*. 1992 edition. London: Mandarin Paperbacks.
The development of the Power Glove

Sage, Adam (2009) 'Console yourself if you can't afford a DS: a pencil and paper will get your brain working'. *The Times*, 26
January 2009. London, UK. p4
Research questions the benefits of the Nintendo DS game Dr Kawashima's Brain Training

'Saudi Arabia bans Pokemon' (2001) *BBC News Online*. 26 March 2001. [Online] http://news.bbc.co.uk/1/hi/world/middle_
east/1243307.stm [Last accessed: 28 March 2010]
Fatwa against Pokémon

'Sega scraps the Dreamcast' (2001) *BBC News Online*, 31 January 2001. [Online] http://news.bbc.co.uk/1/hi/business/1145936.
stm [Last accessed: 7 March 2010]
Sega stops production of the Dreamcast and abandons the console manufacturing business

'The final countdown' (2009) *Edge*, Issue 201, May 2009, pp64-71
Feature on Final Fantasy XIII and a look back at the Japanese role-playing games

'This is how you make successful games' (2008) *Edge*, Issue 191, August 2008, p68-73
Feature on Segagaga

Tobin, Joseph (editor) (2004) *Pikachu's Global Adventure: The Rise and Fall of Pokémon*. Durham, North Carolina: Duke University Press
Extensive examination of Pokémon's global success, meaning and influence

'Viva la revolution' (2005) *Games*[TM], Issue 30, March 2005, pp10-11
Pre-launch report on the Wii while it was still being called the Revolution

'What caused Japan's recession?' (2002) *BBC News*, 14 August 2002. [Online] http://news.bbc.co.uk/1/hi/business/2193853.stm [Last accessed: 28 March 2010]
Background on Japan's 1990s recession

'Who dares wins' (2007) *Edge*, Issue 177, July 2007, pp62-71
Feature on Nintendo's Wii and how it broke with video game tradition

'Wiitness the fitness' (2008) *Edge*, Issue 185, February 2008, pp12-13
Wii Fit preview

27 | THE GROOVIEST ERA OF CRIME

Barson, Michael and Heller, Steven (2001) *Red Scared! The Commie Menace in Propaganda and Popular Culture*. San Francisco, California: Chronicle Books
Overview of Cold War US propaganda

Barton, Matt (2008) *Dungeons & Desktops*. Wellesley, Massachusetts: A K Peters
Background information on The Elder Scrolls series of role-playing games

Birdwell, Ken (1999) 'The Cabal: Valve's design process for creating Half-Life'. In: Salen, Katie and Zimmerman, Eric (editors) (2006) *The Game Design Reader: A Rules of Play Anthology*. Cambridge, Massachusetts: The MIT Press
Valve founder Ken Birdwell looks back on the development of Half-Life

'Computer games: classification' (1998) House of Lords debates, *Hansard*, 13 January 1998, Vol. 584, cc931-3. [Online] http://hansard.millbanksystems.com/lords/1998/jan/13/computer-games-classification#S5LV0584P0_19980113_HOL_7 [Last accessed: 13 March 2010]
The UK Parliament debates Grand Theft Auto

Cousins, Mark (2004) *The Story of Film*. 2008 paperback edition. London, UK: Pavilion Books
Hollywood's move towards blockbuster movies in the late 1970s and 1980s

Dailly, Mike (2008) 'Mike's Homepage: About Me' [Personal website] [Online] www.javalemmings.com/miked/aboutme.htm [Last accessed: 13 March 2010]
Background on the development of the original Grand Theft Auto

'Dark Knight' (2009) *Edge*, Issue 201, May 2009, pp78-83
Interview with Chris Avellone, the writer of Planescape Torment

Den Uyl, Douglas J. and Rasmussen, Douglas B. (1984) 'Ayn Rand on rights and capitalism'. In: Boaz, David (editor) (1997) *The Libertarian Reader: Classic & Contemporary Writings from Lao-Tzu to Milton Friedman*. 1998 paperback edition. New York, New York: The Free Press. pp169-180
Background on Ayn Rand and her beliefs

Jones, Darran (2008) 'The history of Elite'. *Retro Gamer*, Issue 47, January 2008, pp24-31
Mentions Elite's influence on Grand Theft Auto

MacDonald, Ryan (1998) 'GTA: Grand Theft Auto: European version review'. *GameSpot*, 6 May 1998. [Online] http://uk.gamespot.com/ps/adventure/grandtheftauto/review.html [Last accessed: 7 March 2010]
Review of the original Grand Theft Auto that highlights the level of violence

'On the road: Adventures in the Capital Wasteland' (2009) *Games*™, Issue 82, April 2009, pp84-89
Feature on the world of Fallout 3

Pavitt, Jane (2008) 'The bomb in the brain'. In: Crowley, David and Pavitt, Jane (editors) (2008) *Cold War Modern: Design 1945-1970*. London, UK: V&A Publishing. pp100-121
Attitudes to atomic power and nuclear war as well as Cold War propaganda, plus information on real life plans to build a network of nuclear bunkers in the USA

Rand, Ayn (1957) *Atlas Shrugged*. 2007 edition. London, UK: Penguin Classics
The novel that inspired BioShock

Rockstar North (2002) *The Degenatron Page* [Promotional website] [Online] www.degenatron.com [Last accessed: 13 March 2010]
Mock website for the Degenatron video game console advertised in Grand Theft Auto: Vice City

'Sex violence and videogames' (2001) *Edge*, Issue 94, February 2001, pp62-69
Feature examining controversial video games with references to the first two Grand Theft Auto games

Smith, Rob (2008) *Rogue Leaders: The Story of LucasArts*. London, UK: Titan Publishing Group
Details about the making of Grim Fandango

'The making of Grand Theft Auto' (2008) *Edge*, Issue 187, April 2008, pp60-75
The history of the Grand Theft Auto series. Source for Sam Houser quotes, used with the kind permission of Future Publishing

'The vision of Mass Effect' (2007) [DVD] In: BioWare (2007) *Mass Effect: Limited Edition* [Xbox 360] Electronic Arts
Mass Effect's development team discusses the influences on the game

Toffler, Alvin (1964) 'The *Playboy* interview with Ayn Rand'. In: Boaz, David (editor) (1997) *The Libertarian Reader: Classic & Contemporary Writings from Lao-Tzu to Milton Friedman*. 1998 paperback edition. New York, New York: The Free Press. pp161-168
Background on Ayn Rand and her beliefs

Walker, Sophie (1997) 'Computer Nerds discover sex, drugs and rock 'n' roll'. *The Independent on Sunday*. 6 July 1997. [Online] www.independent.co.uk/news/computer-nerds-discover-sex-drugs-and-rock-n-roll-1249248.html [Last accessed: 7 March 2010]
Report on Sensible Software's Sex, Drugs and Rock 'n' Roll

'Who's the daddy now?' (2009) *Games*™, Issue 82, April 2009, pp62-71
Feature on BioShock

28 I MAGIC SHOOTING OUT OF PEOPLE'S FINGERS

'An audience with Jenova Chen' (2009) *Edge*, Issue 206, October 2009, pp62-67
Interview with the co-founder of Thatgamecompany

Barton, Laura (2005) 'The question: Have the Artic Monkeys changed the music business?'. *The Guardian*, London, UK. 25 October 2005. [Online] www.guardian.co.uk/music/2005/oct/25/popandrock.arcticmonkeys [Last accessed: 7 March 2010]
The Arctic Monkeys challenge the music business

Cousins, Mark (2004) *The Story of Film*. 2008 paperback edition. London, UK: Pavilion Books
Hollywood's move towards summer blockbuster movies in the 1980s

Donovan, Tristan (2003) 'Top 20 Publishers'. *Game Developer*, Volume 10, Number 9, September 2003, pp30-35
Figures on the world's largest video game publishers measured by annual turnover

Experimental Gameplay Workshop presentations (2002) [Online] www.experimental-gameplay.org/2002/index.html [Last accessed: 7 March 2010]
Summary of the first Experimental Gameplay Workshop

Experimental Gameplay Workshop presentations (2005) [Online] www.experimental-gameplay.org/2005/index.html [Last accessed: 20 March 2010]
Summary of the 2005 Experimental Gameplay Workshop that featured Rag Doll Kung Fu, Braid and the Experimental Gameplay Project Kyle Gabler helped start

PopCap Games (2006) 'Survey: Casual computer games as TV replacement'. [Press release] 13 September 2006
Survey shows demographics of PopCap's audience is outside the video game norm

Purchese, Robert (2010) 'Bejeweled sales reach 50m worldwide'. *Eurogamer*, 10 February 2010. [Online] www.eurogamer.net/articles/bejeweled-sales-reach-50m-worldwide [Last accessed: 7 March 2010]
Bejeweled hits 50 million sales landmark after 10 years

Reynolds, Christopher (2009) 'Braid has been "very profitable" for creator'. *Now Gamer*, 18 March 2009. [Online] www.nowgamer.com/news/307/braid-has-been-very-profitable-for-creator [Last accessed: 7 March 2010]
Interview with Jonathan Blow on the success of Braid

ACKNOWLEDGEMENTS

This book would not have been possible without the support and help of many other people. My partner Jay Priest tops the thank you list for his design work on the book, support in getting the whole project off the ground, sound advice, help with the research and business aspects, and – above all – for his patience and unfailing support over the year-and-a-half it took me to write it.

Keith Grimes and David McCullough's insights and suggestions were incredibly helpful, as was their willingness to help me sift through dozens and dozens of good (and many not so good) games when putting together the book's game guide.

Tom Homewood gets a big thank you for his work on the cover design and his advice on typography as does my crack proofreading team of Ruth Smith, Cathy Wallace and Rachael Wood, who were willing to make their way through a book on a subject they knew nothing about in meticulous detail. Cathy Wallace gets an extra mention for her tweets.

My Japanese translator Klara Ellefson went beyond the call of duty by providing me with numerous useful insights in Japanese culture and through her willingness to help track down several of the Japanese developers I interviewed.

Thanks also go to: Cathy Campos of Panache PR for her patience with the delays to the book's completion and her PR work, Jess McAree for the legal pointers, and Jon Savage for inspiration (although he doesn't know it).

Lastly, and by no means least, I'll like to thank the hundreds of game industry veterans, PRs and agents who I contacted while researching this book for the time they took in answering my many questions, helping me track down the information I needed and for digging out photos or images for the book. A special thanks goes to Ralph Baer, Dona Bailey, Dave Nutting, Cynthia Franco (of the DeGolyer Library at Southern Methodist University, Dallas), Richard Garriott, Eugene Jarvis, Ed Logg and Philip Oliver who all went the extra mile.

INDEX

CPSIA information can be obtained
at www.ICGtesting.com
Printed in the USA
LVHW112033080123
736729LV00001B/29